BOILER OPERATOR'S WORKBOOK

Third Edition

AMERICAN TECHNICAL PUBLISHERS, INC.
HOMEWOOD, ILLINOIS 60430-4600

R. Dean Wilson

American Technical Publishers, Inc., Editorial Staff

Editor in Chief:
Jonathan F. Gosse
Production Manager:
Peter A. Zurlis
Technical Editor:
Eric F. Borreson
Copy Editor:
Richard S. Stein
Cover Design:
Sarah E. Kaducak
Illustration/Layout:
Sarah E. Kaducak
Thomas E. Zabinski
William J. Sinclair
Jennifer M. Hines
Gianna C. Butterfield

3 4 5 6 7 8 9 – 05 – 9 8 7 6 5 4 3 2 1

Printed in the United States of America

ISBN 0-8269-4495-7

Acknowledgments

Technical information and assistance was provided by the following companies, organizations, and individuals.

Aurora Pump
Babcock & Wilcox Co.
Bell & Gosset
Clark-Reliance Co.
Cleaver-Brooks
Crane Valves
Crosby Valve, Inc.
Dresser Industries, Inc.
Endress+Hauser
Gateway Community College
Hayes Republic Corp
Honeywell, Inc.
ITT Bell & Gosset
Jenkins Bros.
Johnston Boiler Co.
Keeler Boiler, Inc.
Kewaunee Boiler Co.
Lunkenheimer Co.
McDonnell & Miller
Nebraska Boiler
Omron Electronics
Riley Stoker
Rockwell Automation, Allen-Bradley Company, Inc.
Spirax Sarco, Inc.
Superior Boiler Works, Inc.
The Foxboro Company
Trerice, H.O., Co.
Watts Regulator Co.
Weil-McLain
Wheelabrator Air Pollution Control

Carroll Hooper, P.E., CEM
President
Steam Solutions, Inc.
American Society of Power Engineers, National Board of Directors

Byron D. Nichols
Power Plant Operations Superintendent
Virginia Polytechnic Institute and State University
American Society of Power Engineers, Region 4 Board of Directors

Contents

- *Using This CD-ROM*
- *Quick Quizzes™*
- *Illustrated Glossary*
- *Sample Licensing Exams*
- *Master Math™ Problems*
- *Media Clips*
- *Reference Material*

Introduction

Boiler Operator's Workbook 3rd Edition, provides information on virtually all facets of steam boiler operation, maintenance, and troubleshooting. Common boiler auxiliaries and operating techniques are covered in detail, and safety and efficiency of operation are stressed. Information is included on modern boiler water treatment techniques, construction and repair methods, waste heat recovery, controls, fuels, and draft. This new edition of *Boiler Operator's Workbook* includes the latest combustion control technology, instrumentation, operations and maintenance practices, and an interactive CD-ROM.

Boiler Operator's Workbook is arranged in a convenient question-and-answer format specifically designed for use in preparation for obtaining a boiler operator's license. This format makes the book easy to use as a refresher for boiler licensing examinations and for industrial and vocational boiler classes. The expanded answers, many with illustrations, explain the theory and practice of steam plant operations in a clear, concise manner. Key terms are italicized and defined in the text for additional clarity.

Throughout *Boiler Operator's Workbook,* math calculations are used to obtain information. On many licensing examinations, the ability to set up math problems, assign values, and solve the problems is assumed. Questions on licensing examinations may also test the ability to transpose and use basic math equations to obtain the answers. Examples of these math equations are included in the chapter where the equation is introduced. Additional explanation and sample problems are given on the CD-ROM included with the text.

Boiler Operator's Workbook contains 11 chapters. Chapters 1 through 10 contain information on the operation of boilers and related equipment. At the end of each chapter, a Trade Test presents students with sample questions to reinforce the information presented in the chapter. Tests include true-false, multiple choice, completion, matching, and essay questions. Chapter 11, "Licensing," contains three Sample Licensing Examinations as well as information to help students develop good test-taking skills. The sample examinations in Chapter 11 contain a mixture of objective and essay questions. The answers to the Trade Test and Sample Licensing Examination questions are in a separate Answer Key.

An extensive Glossary and comprehensive Appendix provide useful information in an easy-to-find format. Of particular importance are the boiler formulas with examples. A comprehensive Index simplifies navigation and makes it easy to find information of interest.

The *Boiler Operator's Workbook* CD-ROM is located in the back of the book. This CD-ROM is designed as a self-study aid to enhance information presented in the book, and includes Quick Quizzes™, Master Math™, an Illustrated Glossary, Sample Licensing Exams, Media Clips, and Reference Material. The Quick Quizzes™ provide an interactive review of topics in a chapter. The Master Math™ problems provide a review of typical math problems encountered by boiler operators. The Illustrated Glossary provides a helpful reference to terms commonly used in industry. The Sample Licensing Exams button accesses four 25-question exams that provide practice in preparing for a boiler operator's license examination. The Media Clips button accesses a collection of video clips and animated graphics. The Reference Material is accessed through Internet links to manufacturer, association, and American Tech resources. Clicking on the American Tech web site button (www.go2atp.com) or the American Tech logo accesses information on related boiler operation training products.

Features

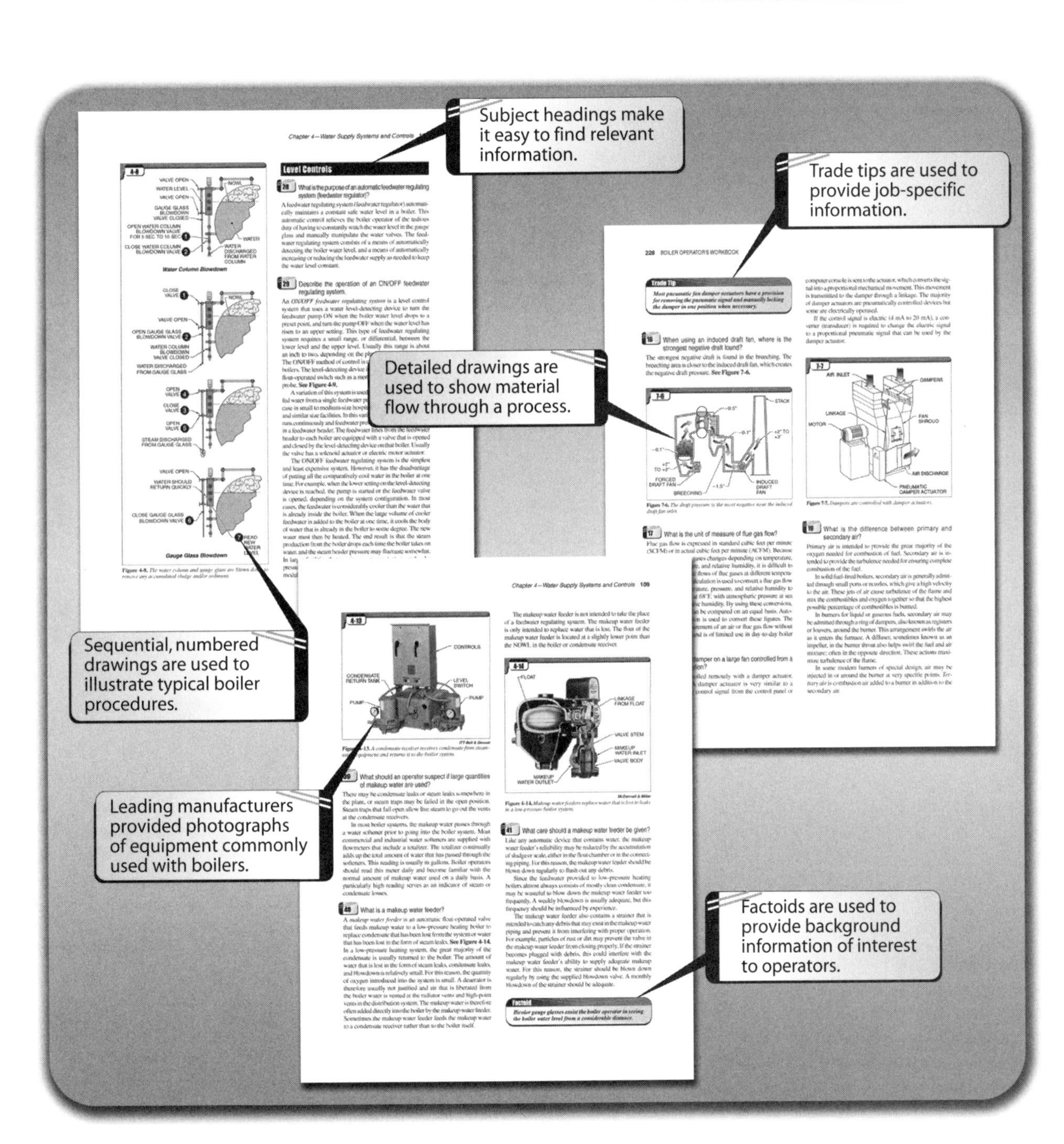
Subject headings make it easy to find relevant information.
Trade tips are used to provide job-specific information.
Detailed drawings are used to show material flow through a process.
Sequential, numbered drawings are used to illustrate typical boiler procedures.
Leading manufacturers provided photographs of equipment commonly used with boilers.
Factoids are used to provide background information of interest to operators.

Chapter 1

Boiler Theory and Principles

Steam boilers have been called the workhorses of America. Steam is used in one way or another in over 95% of all manufactured products. Steam is also vital to comfort heating in commercial, institutional, and industrial facilities. In addition, the great majority of the electric generators producing electricity today are driven by steam turbines using steam produced by high-pressure boilers.

There are several primary types of boilers and many individual design configurations, but most of the terminology and operating principles associated with boilers apply to all of them. The basic functional subsystems needed for boiler operation are consistent among the various designs as well. The safe generation and use of steam requires that boiler operators have a clear understanding of these principles and subsystems. In addition, rising fuel costs dictate that boiler operators be more knowledgeable today than ever in operating their boiler systems efficiently.

Principles of Boilers

What is a steam boiler?

A *steam boiler* is a closed vessel in which water is transformed into steam under pressure through the application of heat.

What is steam?

Steam is the gaseous form of water. It is odorless, colorless, and tasteless. Technically, steam is an invisible gas. For example, the visible whitish cloud that blows from a teapot is not actually steam, it is water vapor. As pure steam blows out of the teapot, it is invisible for a distance of about ¼″. The steam quickly gives up heat and cools in the surrounding air. The resulting cloud of vapor is actually made up of countless tiny droplets of water, which are what makes the vapor visible. *Live steam* is steam in its pure, invisible form. **See Figure 1-1.**

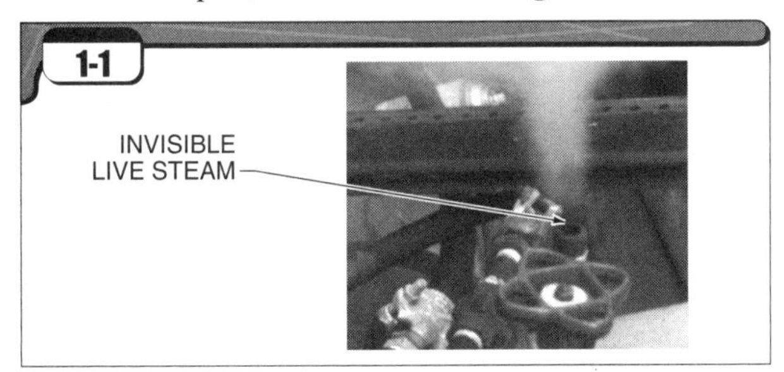

Figure 1-1. *Steam is the invisible, gaseous form of water.*

How is steam from boilers used?

- Heating buildings
- Papermaking
- Driving steam turbines (which in turn drive fans, compressors, pumps, generators, etc.)
- Generating hot water for laundries and institutional uses
- Cooking and processing in steam-jacketed kettles
- Evaporating refrigerant in absorption-type chillers
- Humidification for controlled-atmosphere environments
- Sterilization in hospitals (autoclaves)
- Pasteurizing food products in dairies, breweries, etc.
- Heat-treating manufacturing processes such as forming

Why is steam the most common medium for transporting heat in institutions and industry?

Steam is clean and odorless. It transports heat very well. It is easily distributed and controlled. It gives up its heat at a constant temperature. In its raw form (water), it is inexpensive and plentiful. Leaks are not an environmental hazard or fire hazard.

What laws apply to the operation of boilers?

Thermodynamics is the science of thermal energy (heat) and its conversion to and from other forms of energy. *The first law of thermodynamics* states that energy cannot be created or destroyed, but can be changed from one form to another. This is also known as the law of conservation of energy. There is thus no loss or gain of energy, there is only conversion of energy. A steam boiler provides a means by which the Earth's raw energy resources may be converted into useful forms. **See Figure 1-2.** For example, in a power plant, boilers convert chemical energy in the form of fuels into heat energy in the form of steam. The heat energy in the steam is converted into mechanical energy in a steam turbine. The steam turbine drives a generator, which converts the mechanical energy into electrical energy. The electrical energy may then be converted into heat energy again in an oven or water heater, and so on.

The *second law of thermodynamics* states that heat always flows from a body having a higher temperature toward a body having a lower temperature. For example, in a boiler, the heat from the burning fuel is transferred into the water in the boiler.

6 How is heat transferred from one body to another?

Radiation is the transfer of heat from a hotter body to a colder body without the aid of physical contact or a conveying medium. For example, the sun transfers heat to the Earth by radiation. Convection is the transfer of heat by a conveying medium such as air. For example, a forced air furnace transfers heat by convection. Conduction is the transfer of heat by actual physical contact from molecule to molecule. For example, boiler metal transfers heat by conduction. **See Figure 1-3.**

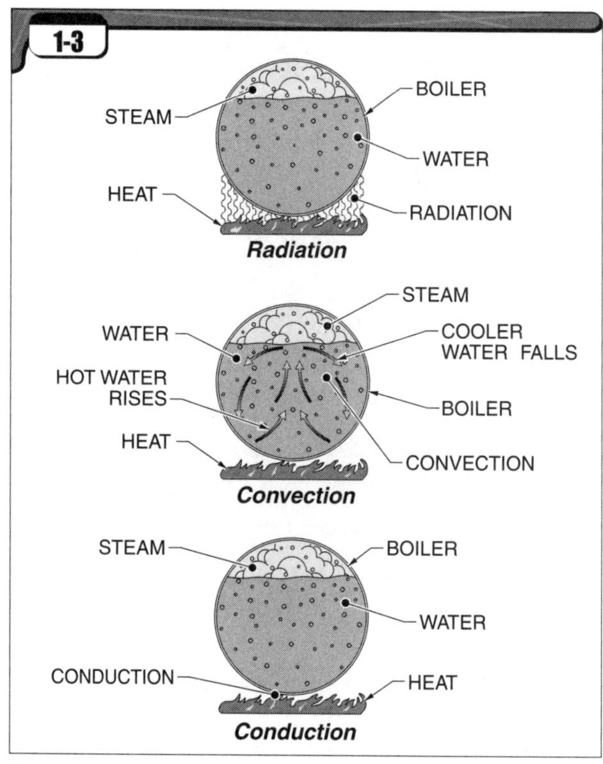

Figure 1-3. *Heat transfer occurs through radiation, convection, and conduction.*

7 What is the heating surface of a boiler?

A *boiler heating surface* is any part of the boiler metal that has hot gases of combustion on one side and water on the other. Thus, the heat energy from the combustion of fuel is transferred into the water in the boiler through the heating surface.

8 How is the amount of heating surface of a boiler expressed?

The heating surface area is expressed as square feet. A square foot is 144 sq in. For example, a piece of steel measuring 12″ × 12″ equals 1 sq ft (12″ × 12″ = 144 sq in.) as does a piece of steel measuring 24″ × 6″ (24″ × 6″ = 144 sq in.).

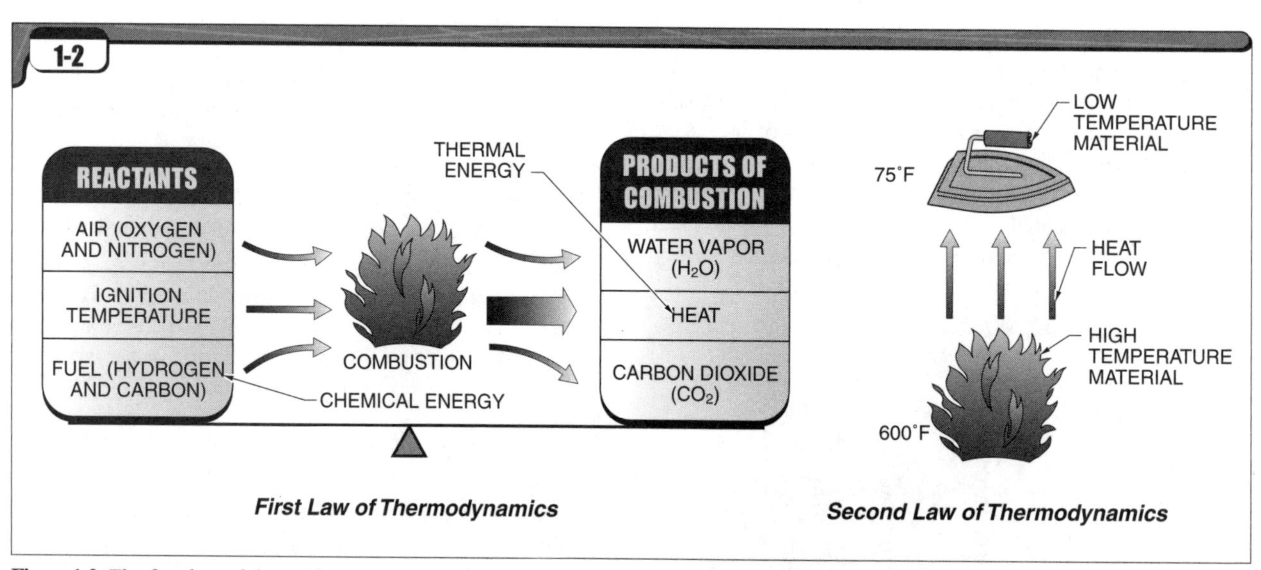

Figure 1-2. *The first law of thermodynamics states that energy can be changed from one form to another. The second law of thermodynamics states that heat always flows from a material at a high temperature to a material at a low temperature.*

What is the advantage of using tubes and flues in a boiler?

Tubes and flues increase the heating surface area, thus increasing efficiency of heat transfer to the water in a boiler. **See Figure 1-4.**

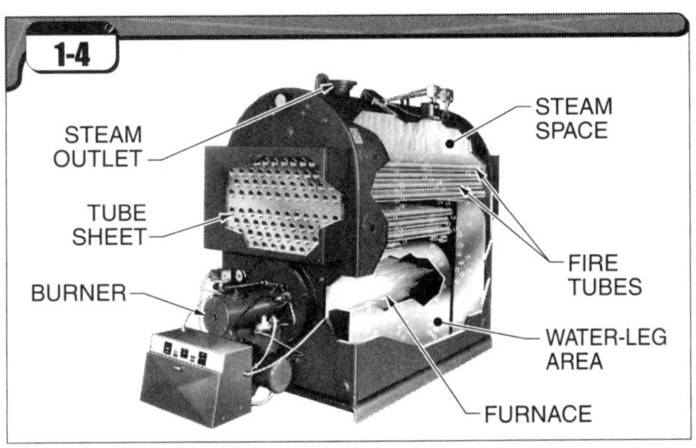

Kewanee Boiler Co.

Figure 1-4. *Tubes and flues increase the heating surface area, thus increasing heat transfer efficiency to the water in a boiler.*

What is the difference between tubes and flues?

Tubes are specified by outside diameter (OD). They are less than 4″ OD. Flues are specified by inside diameter (ID). They are larger than 4″ ID. Dimensions for tubes and flues are given in inches. For example, a tube may have a 3″ OD and a flue may have a 24″ ID.

How does the heating surface relate to efficiency?

The more heating surface a boiler has for its size, the more efficient it is in transferring heat from the combustion process into the boiler water.

Describe the three main types of boilers.

In a firetube boiler, the hot gases of combustion pass through the tubes. The boiler water contacts the outside surfaces of the tubes. **See Figure 1-5.** In a watertube boiler, the boiler water passes through the tubes. Hot gases of combustion pass over the outside surfaces of the tubes. A cast iron boiler is usually fabricated of hollow sections that contain the boiler water. The gases of combustion pass through openings in the hollow sections, thus transferring heat into the water. The individual sections are interconnected to form one unit.

What four systems apply to any steam boiler?

The *feedwater system* consists of all the equipment, controls, and piping that prepare and treat the water for use in the boiler, put the water into the boiler, and maintain a normal, safe amount of water in the boiler. The *fuel system* consists of all the equipment, controls, and piping that deliver the fuel to the boiler's combustion equipment and control the combustion process. The *draft system* consists of the equipment, controls, and ductwork that deliver air to the boiler's furnace area for combustion of the fuel, and then conduct the spent combustion gases to the atmosphere. The *steam system* consists of the equipment, controls, and piping that carry the steam generated by the boiler to its points of use.

Describe the major parts of a basic steam distribution and condensate return system.

The boiler provides a place for combustion of the fuel and then transfers the heat into the water. The boiler also provides a location for steam to accumulate under pressure until it passes out of the boiler through the main steam outlet line. **See Figure 1-6.** The *main steam outlet lines* consist of the piping and valves that direct the steam from the boilers to the steam header.

The *steam header* is a manifold that receives steam from two or more boilers and provides a single location from which steam may be routed through the steam mains and branch lines to various points of use. The steam header also provides a means by which steam may be directed to any steam-using equipment in a facility by any of the boilers connected to the steam header. This allows for redundancy, or backup, in the operation. For example, if one boiler must be shut down to make a repair, another boiler may be used to provide steam. If only one boiler is used, a steam header and the main steam line are the same.

A *steam main* is the piping that carries steam to a section of a building or plant. Steam mains are similar to the main branches of a tree. A *branch line* is the piping that takes steam from the steam mains to individual pieces of steam-using equipment. A *heat exchanger* is any piece of equipment that transfers the heat of the steam into some other material. For example, radiators in a steam heating system are heat exchangers. Paper machines in a paper mill or sterilizers in a hospital are also examples of heat exchangers. Heat exchangers are also known as heat transfer equipment.

A *steam trap* is a mechanical device used for removing condensate and/or air from steam piping and heat exchange equipment. As steam gives up its heat to the material being heated, the steam changes back into water. *Condensation* is the process of steam being changed back into water. *Condensate* is the water formed when steam condenses to water. Condensate must be removed from the steam distribution system and heat exchangers or it will interfere with proper operation of the system. Air must be removed from the system because it interferes with efficient heat transfer in the heat exchangers. Air in a steam system also contributes to corrosion. Steam traps automatically discharge air out of the steam system and into the condensate return lines.

A *condensate return line* is a pipe that carries the condensate and air discharged by the steam traps. The air is routed to a point where it can be discharged to the atmosphere. The condensate is routed to a point where it can be recovered and reused. A condensate receiver is a tank or other collection point where condensate is accumulated and saved for reuse. The condensate return lines usually carry the condensate and air to a condensate receiver. Condensate receivers normally contain a vent line that allows air to be vented to the atmosphere.

A *makeup water line* is a city water pipe or well water pipe through which makeup water is added to a boiler. *Makeup water* is water used to replace condensate that is not returned to the boiler. For example, water in the form of condensate or steam may be lost due to leaks. A constant volume of water and steam must be maintained in a boiler and steam system at all times, therefore water that is lost from the system must be replaced. *Feedwater* is a mixture of condensate and makeup water that is provided to the boiler to make steam. A *feedwater pump* is a pump that sends the returned condensate and any makeup water into the boiler. A *feedwater line* is the piping that carries the feedwater from the feedwater pump to the boilers.

Factoid

In most cases, only the heat in the steam is transferred into the material being heated. The steam itself is usually kept separate from the heated material.

15 List common fuels used in steam production.

- Natural gas
- Fuel oil
- Coal

Natural gas, fuel oil, and coal are fossil fuels. A *fossil fuel* is any fuel derived from the remains of plants and other organic matter that have been stored below the Earth's surface.

16 What other materials are used as fuels?

Wood may be used as boiler fuel in the form of sawdust, bark, or scrap. Municipal refuse is sometimes incinerated in large boilers as a means of reducing waste volume and generating steam at the same time. Many industries such as refineries, steel mills, paper mills, and other manufacturing industries produce wastes and byproducts that may be used as inexpensive boiler fuels. Some materials are produced in agricultural processing that may be used as boiler fuels. These include bagasse (sugar cane waste), seeds, and pits. Electricity is used in some boilers

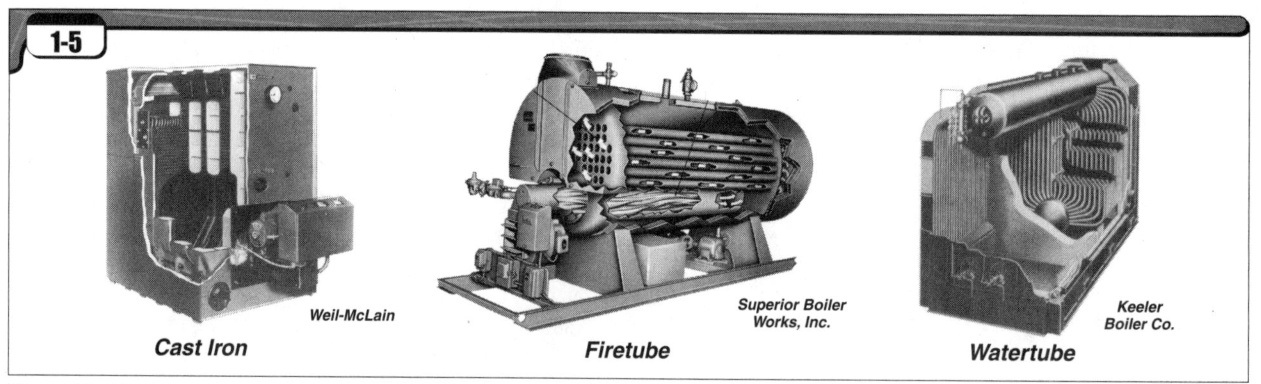

Figure 1-5. *The three main types of boilers are cast iron, fire tube, and water tube.*

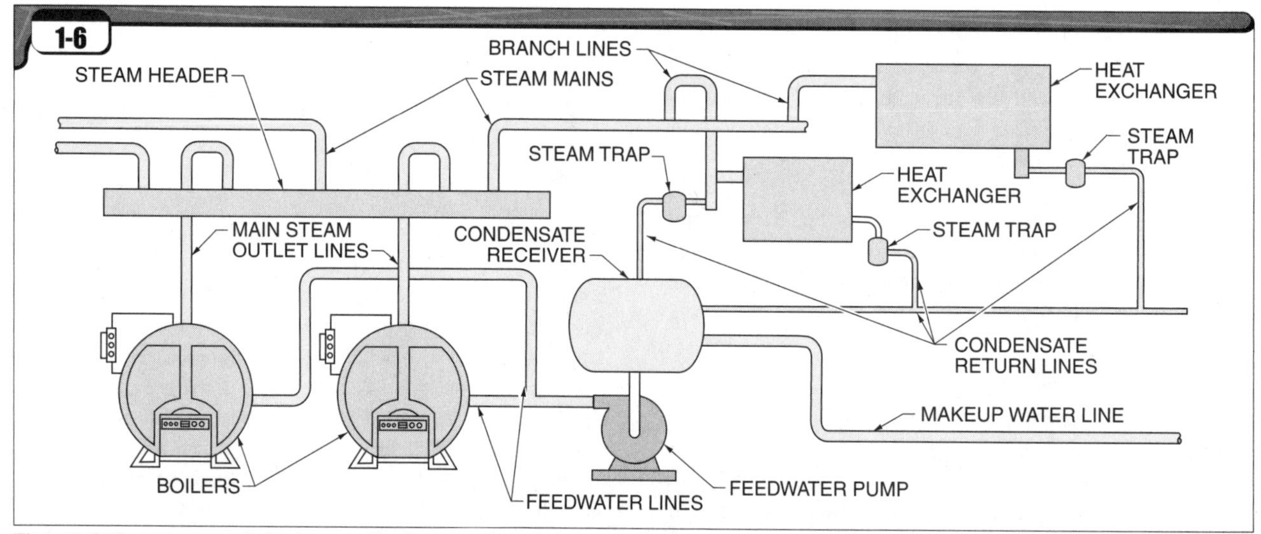

Figure 1-6. *The main parts of a basic steam distribution and condensate return system are the boilers, main steam outlet lines, steam header, steam mains, branch lines, heat exchangers, steam traps, condensate return lines, condensate receiver, makeup water line, feedwater pump, and feedwater lines.*

as a heat source to generate steam. Electricity is generally confined to smaller boilers due to its cost. Electric boilers use the primary voltage supplied to the plant, up to about 16,000 V (16 kV).

17 What is a combustion chamber?

A *combustion chamber* is the area of a boiler where the burning of fuel occurs. The combustion process is completed before the mixture comes into contact with the heating surfaces of the boiler. Soot and inefficiency will result if the combustion process is not complete.

18 What is breeching in a boiler installation?

Breeching is the ductwork that carries cooled flue gases from the exit of the boiler to the stack.

Principles of Pressure

19 What is a fluid?

A *fluid* is any material that can flow from one point to another. Fluids can be liquids or gases. For example, air is a fluid and so are water, natural gas, fuel oil, and steam.

20 Which way does a fluid flow?

A fluid will flow from a point of higher pressure to a point of lower pressure. In addition, fluids will take the path of least resistance.

21 What are the measurement units of pressure?

Pressure is often measured in psi. Psi is an acronym for the unit of pressure, pounds per square inch, and is the number of pounds of force exerted on an object divided by the surface area of the object exposed to that force. For example, a 5 pound weight resting on 1 square inch exerts 5 psi.

pressure = *force* ÷ *area*
pressure = 5 pounds ÷ 1 sq in.
pressure = **5 psi**

22 How does pressure relate to total force?

Total force is the pressure (in psi) being exerted on the surface multiplied by the area of the surface. For example, a pipe is pressurized to 100 psi and a valve in the pipe is closed. There is 100 psi on one side of the valve and 0 psi on the other side. If the surface area of the valve plug (the part holding the pressure back) is 12.5 sq in., then the total force is 1250 lb (100 psi × 12.5 sq in. = 1250 lb). This is the equivalent of a 1250 lb weight pushing against the valve. Total force is always expressed in pounds, not psi.

Factoid

Every pressure-containing portion of a steam and condensate system must be capable of withstanding the internal pressure without bursting. Therefore, pressure-containing components such as valves, steam traps, piping, and pipe fittings are also rated for MAWP. The MAWP ratings for such components are found in various industry standards published by organizations such as the American National Standards Institute (ANSI).

23 What is the total force on a 14 sq in. valve surface with a pressure of 125 psi acting against it?

total force = *pressure* × *area*
total force = 125 psi × 14 sq in.
total force = **1750 lb**

24 What is MAWP of a boiler?

MAWP (maximum allowable working pressure) is the highest pressure at which a boiler or pressure vessel may be safely operated. Every boiler and pressure vessel is rated for MAWP. The MAWP of a pressure vessel is listed on the nameplate and is stamped on the pressure vessel itself.

25 What is working pressure of a boiler?

Working pressure is a shortened term for maximum allowable working pressure, but it may also be used to mean the pressure at which the boiler is normally operated. For example, the normal operating pressure of a boiler is typically about 10% to 25% below the MAWP. *Note:* When using this term, always clarify the way in which it is being used.

26 What industry rule determines whether a boiler is a low-pressure or high-pressure boiler?

A low-pressure boiler has an MAWP of 15 psi or less. A high-pressure boiler has an MAWP above 15 psi.

27 What is differential pressure?

Differential pressure is the difference between two pressures at different points. For example, a valve may have 250 psi on one side and 100 psi on the other side. The differential pressure across the valve is 150 psi (250 psi – 100 psi = 150 psi). *Note:* The term differential is also used to express the difference between temperatures or other variables at two points.

The Greek letter delta (Δ) is used to represent differential. Thus, the differential pressure between two points may be referred to in the plant as the "delta *P*" (Δ*P*). Many operators abbreviate even further, using only the letter "d" for differential. For example, an operator may refer to the differential pressure between two points as the "dp."

28 What is atmospheric pressure?

Atmospheric pressure is the force exerted by the weight of the atmosphere bearing on the Earth's surface. The atmospheric pressure at sea level is 14.696 psi, usually rounded to 14.7 psi. **See Figure 1-7.** Atmospheric pressure decreases at elevations above sea level.

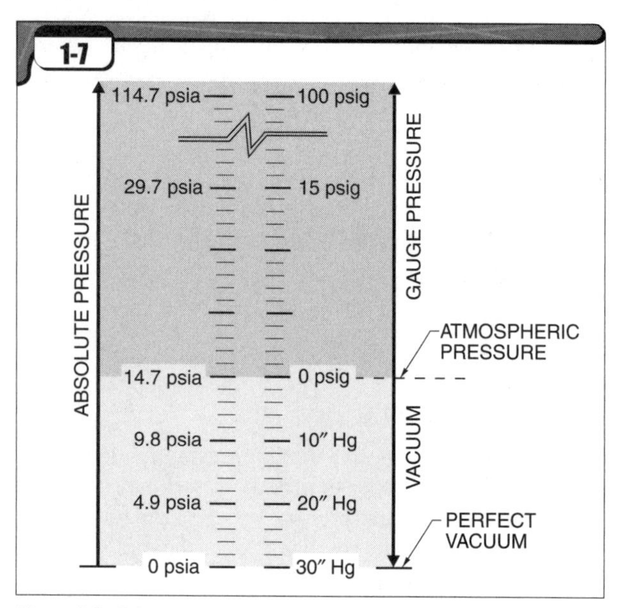

Figure 1-7. *Gauge pressure is pressure above atmospheric pressure. Absolute pressure is pressure above a perfect vacuum. Vacuum is pressure below atmospheric pressure.*

29 What is gauge pressure?

Gauge pressure is the pressure above atmospheric pressure. It is expressed as psig (pounds per square inch gauge). In the operation of boilers and steam systems, the boiler operator is primarily concerned with pressures above atmospheric pressure. Gauge pressure uses atmospheric pressure as the zero of the measurement scale. Pressure given in psi is normally understood to mean psig.

30 What is absolute pressure?

Absolute pressure is the pressure above a perfect vacuum. It is expressed as psia (pounds per square inch absolute). For example, the atmospheric pressure at sea level is 14.7 psia (or 14.7 psi above a perfect vacuum). Absolute pressure is often used in design engineering so that pressures both above and below atmospheric pressure may be expressed on the same measurement scale in the same units.

31 What is vacuum?

Vacuum is a pressure lower than atmospheric pressure. A perfect vacuum is a complete absence of any pressure. Vacuum is usually expressed as inches of mercury, abbreviated as in. Hg or ″ Hg. Hg is the chemical symbol for mercury.

The term, "inches of mercury" is used in two contexts. In one context, it refers to a vacuum measurement with 0 psig as the reference point from which the measurement is made. When a differential pressure measurement combines units of psi and in. Hg, the vacuum measurement must be converted to psi so that the differential pressure may be expressed in psi. "Inches of mercury" also refers to barometric pressure measurements with a perfect vacuum as the reference point. In this context, it is used by meteorologists in reporting weather.

32 How is a vacuum reading converted to psi when it will be used to calculate a differential pressure?

A vacuum reading of 7″ Hg is converted to psi as follows:

$P = in.\ Hg \times 0.491$

$P = 7 \times 0.491$

$P = \mathbf{3.4\ psi}$

Note: This is actually –3.4 psig because it is below zero.

33 What is the differential pressure across a steam trap when the inlet pressure is 5.0 psi and the outlet pressure is 7″ Hg (vacuum)?

The differential pressure across a steam trap is the difference between the inlet and outlet pressure measurements. **See Figure 1-8.**

$P = P_{in} - P_{out}$

$P = 5.0 - (-3.4)$

$P = \mathbf{8.4\ psi}$

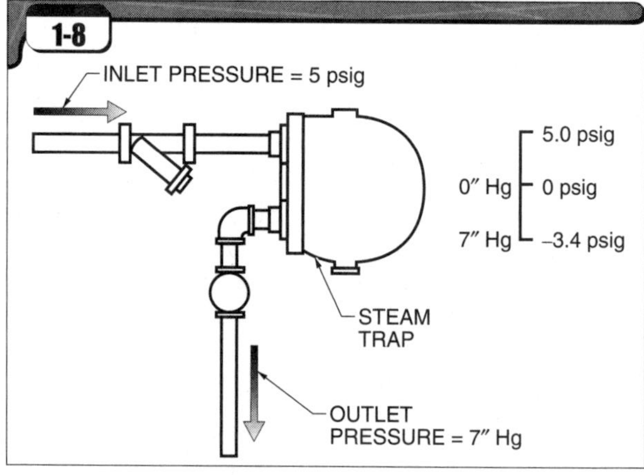

Figure 1-8. *The differential pressure across a steam trap is the difference between the inlet pressure and outlet pressure measurements.*

34 Convert the atmospheric pressure at sea level, 14.696 psia, to in. Hg.

$in.\ Hg = P \div 0.49116$

$in.\ Hg = 14.696 \div 0.49116$

$in.\ Hg = \mathbf{29.921\ in.\ Hg}$

This is usually rounded to 29.92 in. Hg.

How is pressure created by the elevation of water?

Because of the density and weight of water, pressure is exerted as water is pulled downward by gravity. For example, scuba divers encounter greater and greater pressures as they dive deeper in a body of water. The same is true when water is elevated above the surface of the Earth. For example, water in a vertical city water pipe inside a high-rise building will have more pressure at the bottom than at the top of the pipe. When referring to the pressure exerted at the bottom (or at some specified elevation) of a body of water in this way, the phrase "column of water" is often used. A *column of water* is water of some specified depth or height. The pressure due to a column of water depends only on the elevation, not the shape of the container. For example, the pressure in psi at the bottom of a 24″ pipe that is 100′ tall and full of water will be the same as the pressure in psi at the bottom of a 2″ pipe that is 100′ tall and full of water.

Factoid

Gauge pressure is psi above atmospheric pressure, while absolute pressure is psi above a perfect vacuum. Pressure given in psi is normally understood to mean psig.

36 What pressure is exerted at the bottom of a column of water 1″ high?

The pressure due to the weight of the water is 0.433 psi. This pressure is additive. That is, it increases by 0.433 psi for every additional foot of water. For example, the force exerted at the bottom of a column of water 2′ high is 0.866 psi (0.433 psi/ft × 2′ = 0.866 psi). Municipal water towers use this principle to produce water pressure without requiring pumps to run continuously. **See Figure 1-9.**

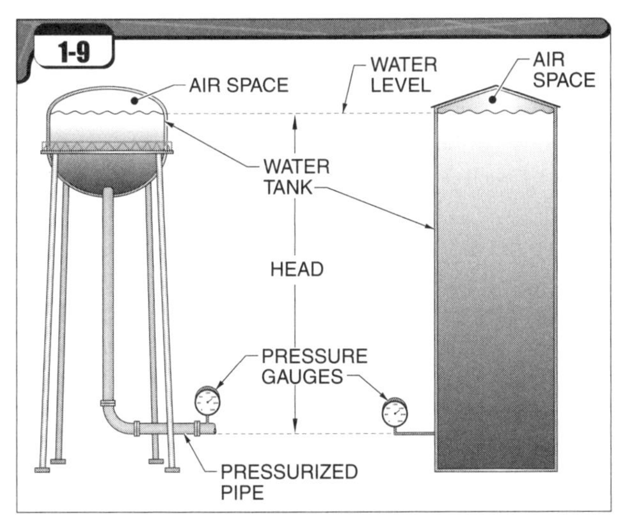

Figure 1-9. *A column of water exerts a pressure of 0.433 psi for each foot of elevation. The shape of the column has no effect on the pressure.*

How high is a column of water that exerts 1 psi at the bottom?

$water\ column = P \div 0.433$

$water\ column = 1\ \text{psi} \div 0.433\ \text{psi/ft}$

$water\ column = \mathbf{2.309\ ft}$

How much pressure is found at the bottom of a column of water 125′-0″ high?

$P = water\ column \times 0.433$

$P = 125\ \text{ft} \times 0.433\ \text{psi/ft}$

$P = \mathbf{54.125\ psi}$

Trade Tip

When installing a steam pressure gauge below the steam line to which it is connected, one must consider the pressure that will be created as the intervening pipe fills with condensate. For example, if a pressure gauge is connected to and located 25 ft below a steam line that contains 100 psig, the gauge will read the steam pressure plus the pressure created by the column of condensate [100 + (25 × 0.433)], or 110.8 psig. In some cases, gauges are calibrated to compensate for this head pressure.

Principles of Heat

What are the common measurement units used to express the intensity of heat?

The Fahrenheit scale is commonly used with the English system of measurement in use in the United States. It is designed with 32°F as the freezing point of water and 212°F as the boiling point at normal atmospheric pressure (14.7 psi).

The Celsius scale is commonly used with the metric system of measurement and is used almost universally in scientific measurements. It is designed with 0°C as the freezing point of water and 100°C as the boiling point at normal atmospheric pressure (14.7 psi). This scale was formerly known as the Centigrade scale. **See Figure 1-10.**

Convert 78°F to Celsius.

$°C = (°F - 32) \div 1.8$

$°C = (78 - 32) \div 1.8$

$°C = 46 \div 1.8$

$°C = \mathbf{25.6°C}$

Convert 40°C to Fahrenheit.

$°F = (1.8 \times °C) + 32$

$°F = (1.8 \times 40) + 32$

$°F = 72 + 32$

$°F = \mathbf{104°F}$

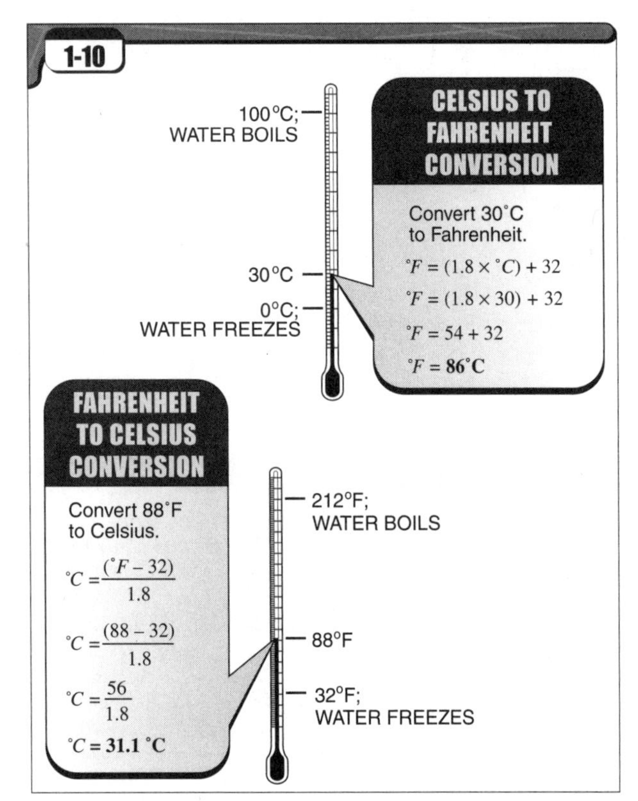

Figure 1-10. *The intensity of heat is expressed in degrees Fahrenheit (°F) or degrees Celsius (°C).*

42 How many pounds of water are in a gallon?

There are 8.33 lb of water in a gallon.

43 What is the weight of 5 gal. of water?

8.33 lb/gal. × 5 gal. = **41.65 lb**

44 What is differential temperature?

Differential temperature is the difference in temperature between two different points. For example, what is the differential temperature of water that enters a heat exchanger at 60°F and leaves at 200°F?

$T = T_1 - T_2$

$\Delta T = 200 - 140$

$\Delta T =$ **140°F**

45 The flue gas temperature entering a boiler is 1900°F and is 750°F leaving the boiler. What is the differential temperature (T) across the boiler?

$T = T_1 - T_2$

$T = 1900°F - 750°F$

$T =$ **1150°F**

46 What is the measurement unit used to express the quantity of heat?

The quantity of heat is measured in Btu. A *British thermal unit (Btu)* is the amount of heat necessary to raise the temperature of 1 lb of water by 1°F. *Note:* 1 lb of water = 16 oz.

Factoid

In the English system of measurement, there is a very slight difference between an ounce by weight and an ounce by volume (a fluid ounce) of water. In order to avoid confusion, the metric system of measurement is generally used in laboratory settings and other applications where measurements must be precise.

47 How much heat is added to 100 lb of water to raise the temperature from 60°F to 70°F?

The heat required to change the temperature of an object is equal to the temperature change times the weight times the heat capacity. The heat capacity of water has a value of 1 Btu/lb per °F. Since the heat capacity of water equals 1, it is sometimes ignored in calculations. Therefore, the heat required equals the temperature change times the weight.

Btu added = (70°F – 60°F) × *weight* × *heat capacity*

Btu added = 10 × 100 × 1.0

Btu added = **1000 Btu**

48 How much heat is given up when the temperature of 26,525 lb of water is lowered from 212°F to 190°F?

Btu loss = (212°F – 190°F) × *weight* × *heat capacity*

Btu loss = 22 × 26,525 × 1.0

Btu loss = **583,550 Btu**

49 How is heat expressed in the metric system?

A *calorie* is the amount of heat necessary to raise the temperature of 1 g of water by 1°C. The metric system of measurement is generally used in laboratory applications. For example, a *calorimeter* is a laboratory instrument used to measure the heat content of a substance, such as a sample of a fuel to be used for a boiler.

50 What is the temperature-pressure relationship?

For every pressure that can be applied to water, there is a corresponding temperature at which water boils and begins to make steam. For example, at atmospheric pressure, water boils at 212°F. When a pressure of 100 psi is applied, water boils at approximately 338°F. For steam at 5 psig the corresponding temperature is 227°F, and for steam at 175 psig, the temperature is 377°F. Boilers are therefore often selected for the steam temperature needed for a process, not for the pressure needed. Steam tables are used to look up the steam temperature-pressure relationship.

51 Why is it important to maintain a constant pressure in a boiler?

Constant pressure is important because most heating systems and manufacturing processes for which steam is produced require a constant temperature. Since the temperature corresponds to the pressure, the pressure must be kept constant to keep the temperature constant. Thus, control of pressure in most steam systems is really only an indirect way to control the temperature of the steam.

52 How many pounds of steam can be produced from 1 lb of water?

One pound of water makes 1 lb of steam. A pound is a unit of mass (weight); therefore, 1 lb of steam has the same mass as 1 lb of water.

53 In what unit of measure is the amount of steam produced by a boiler expressed?

Pounds of steam per hour is the unit of measure in which the amount of steam produced by a boiler is expressed. This is abbreviated as lb/hr. Even small boilers produce thousands of lb/hr. The letter K can be used to signify 1000 in the abbreviation. For example, the steam production of a boiler producing 24,000 lb/hr can be stated 24 Klb/hr.

54 What is sensible heat?

Sensible heat is the Btu content of a substance that represents the heat absorbed or given up as it changes temperature. Sensible heat can be measured with a thermometer. For water, sensible heat is normally expressed as Btu added or given up relative to water at 32°F. For example, 180 Btu of sensible heat must be added to 1 lb of water to raise the temperature of the water from 32°F to 212°F (32°F + 180°F = 212°F). Therefore, we can say that water at 212°F has a heat content of 180 Btu.

55 What is latent heat?

Latent heat is the Btu content of a substance that represents the heat absorbed or given up as it changes state between solid, liquid, and gas. Latent heat cannot be measured with a thermometer. At normal atmospheric pressure, 970.3 Btu is added to 1 lb of water at 212°F to change it into steam at 212°F. It takes 144 Btu of heat to melt 1 lb of ice at 32°F to water at the same temperature. Since we say that water at 32°F has a heat content of 0 Btu, we can say that ice at 32°F has a heat content of –144 Btu. This heat is given as a negative number since the amount of heat in ice is less than the amount of heat in water.

56 What is total heat?

Total heat is the sum of sensible heat and latent heat. It is expressed as the total number of Btus contained in each pound of steam. Steam tables provide a tabulation of the total heat with 32°F as a reference. For example, if 180 Btu of sensible heat is added to 970.3 Btu of latent heat, 1 lb of steam at atmospheric pressure contains 1150.3 Btu of total heat (970.3 Btu + 180 Btu = 1150.3 Btu). **See Figure 1-11.**

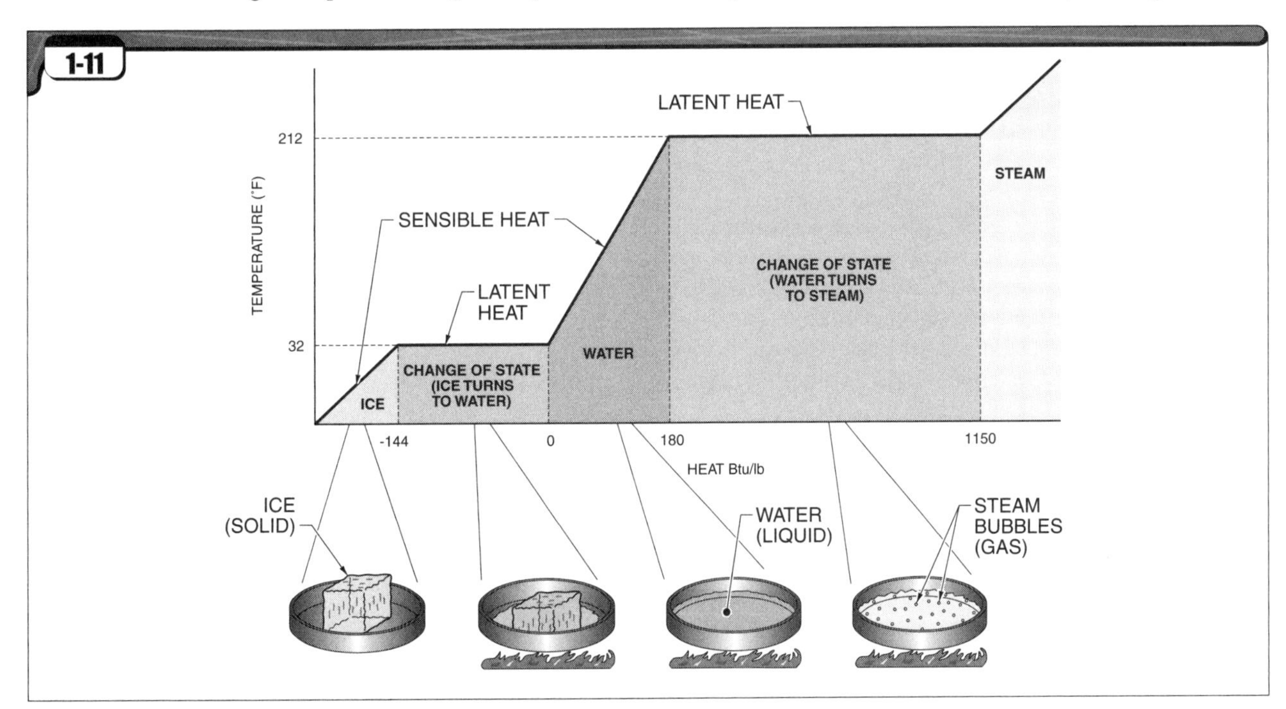

Figure 1-11. *Total heat is the sum of sensible heat and latent heat.*

57 Why do almost all steam heating systems use low-pressure steam?

Heating a building does not require high-temperature steam. For example, steam at 5 psi is 227°F, which is adequate for building heat. In addition, less energy is required to generate low-pressure steam than high-pressure steam.

58 What is specific volume?

Specific volume is the space occupied by a fluid or gas of a particular unit of weight under specified conditions of pressure and temperature. The specific volume of steam, for example, is expressed in terms of the volume occupied by a pound of steam at a given pressure and temperature.

Steam under pressure is a compressed gas. The volume of compressed gases is reduced as the pressure is increased. For example, a volume of air the size of an average bedroom may be reduced to the point that it will fit into a scuba tank when the air is compressed under very high pressure. One pound of water occupies 0.00167 cu ft of space while that same pound of water converted to steam at atmospheric pressure occupies 26.8 cu ft of space. However, when the same steam is under a pressure of 300 psi, it only occupies 1.47 cubic feet. Conversely, when a compressed gas is released from a higher pressure to a lower pressure, the volume of the gas increases accordingly. **See Figure 1-12.**

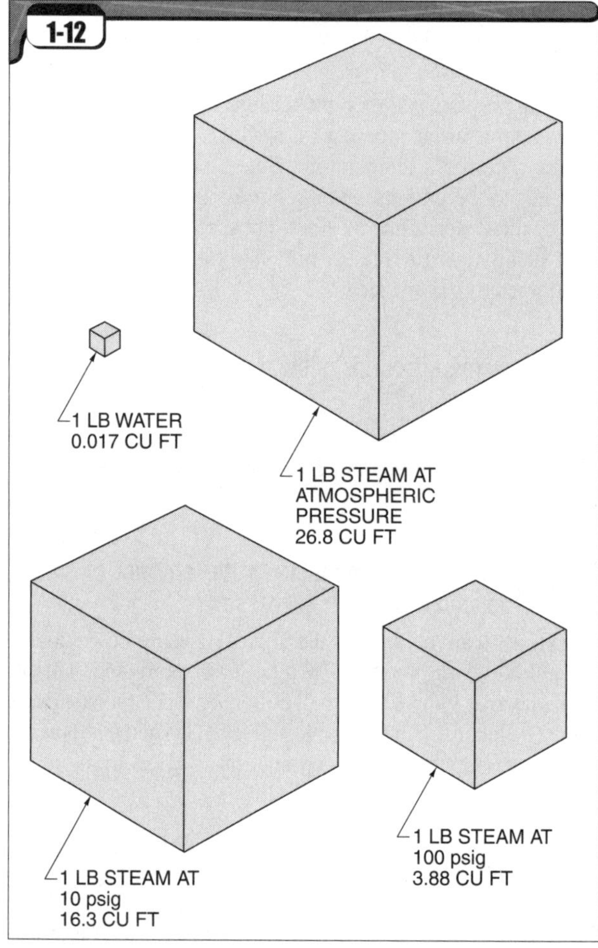

Figure 1-12. *The specific volume of steam is reduced as the pressure is increased.*

59 What is boiler horsepower?

Boiler horsepower (BHP) is the energy required for the evaporation of 34.5 lb of water at 212°F into steam at atmospheric pressure and at 212°F in 1 hr. This is the "from and at" rule: from 212°F water into steam at 212°F.

The "from and at" rule refers to steam made at atmospheric pressure and at 212°F. Boilers in actual practice, however, do not operate under these conditions. They operate under a higher pressure and temperature.

Because boilers operate under higher pressure and temperature, the factor of evaporation is used. The *factor of evaporation (FE)* is the heat added to the water in an actual boiler in Btu/lb and divided by 970.3. This formula allows the performance of all boilers to be broken down into equivalent terms for comparison. It is used to determine how much steam a boiler would make if it were operating from and at 212°F.

In modern boiler system design, boiler horsepower is generally only used for smaller boilers (up to about 1000 HP) as a rough estimate of how much steam a boiler will produce. For example, a 100 HP boiler will produce approximately 3450 lb/hr (100 HP × 34.5 lb of steam per horsepower). In using horsepower as a rough estimate, the factor of evaporation is often ignored. When designing larger boiler installations, designers state the maximum continuous steam production in lb/hr at a given pressure and temperature, and at a specified feedwater temperature.

60 What is the boiler horsepower of a boiler generating 20,000 lb of steam per hour at 150 psi? The factor of evaporation is 1.078.

$BHP = (lb/hr \times FE) \div 34.5$

$BHP = (20{,}000 \times 1.078) \div 34.5$

$BHP = 21{,}560 \div 34.5$

$BHP = \mathbf{624.9\ BHP}$

61 Water enters a boiler at 225°F and 100 psig. The heat of the steam (total heat) at these conditions is 1190.6 Btu. Assuming that there is no moisture in the steam, what is the factor of evaporation?

$FE = (TH - SH) \div 970.3$

$FE = [1190.6 - (225 - 32)] \div 970.3$

$FE = (1190.6 - 193) \div 970.3$

$FE = 997.6 \div 970.3$

$FE = \mathbf{1.028}$

Factoid

Absolute zero is the temperature at which molecular movement is at a minimum. Absolute zero measures –273.15° on the Celesius scale and –459.67° on the Fahrenheit scale. Temperature scales with absolute zero as their zero point, such as the Kelvin and Rankine scales, are called absolute temperature scales.

62 How much steam will a 70 HP boiler make in 2 hrs?

$Steam = HP \times 34.5 \times time$
$Steam = 70 \times 34.5 \times 2$
$Steam =$ **4830 lb**

Note that time in hours is used in this calculation because of the way BHP is defined.

63 What determines how much steam must be generated by a boiler?

The boiler must generate steam at the same rate at which the steam is consumed in the heat transfer equipment. As steam gives up its heat in the heat transfer equipment, it condenses. Each pound of steam that is consumed (condensed) in the heat transfer equipment must be replaced by an additional pound of steam in order to maintain a constant pressure within the equipment. If the boiler does not replace the steam at the same rate at which it is consumed, the pressure in the overall steam system will drop. If the boiler puts steam into the steam distribution system at a faster rate than the rate at which the steam is consumed in the heat transfer equipment, the pressure in the overall steam system will rise. If multiple boilers are used to supply steam, the sum of the steam production from the boilers must equal the total steam consumption in the heat transfer equipment.

In addition, the boiler(s) must continually replace steam that is consumed due to losses and leaks. For example, steam condenses as heat is lost through uninsulated pipes. Steam may be lost through leaks and through broken steam traps as well.

64 What is the term for the amount of steam produced by a boiler?

The *boiler load* is the amount of steam being produced by a boiler. The load on a boiler is equal to the demand for steam from the steam-using equipment, plus the total of the steam consumed due to losses and leaks.

65 What is saturated steam?

Saturated steam is steam that contains no liquid water and is at the temperature of the boiling water that formed the steam. Saturated steam is saturated with heat energy, not with moisture.

In the sciences, *saturated* means that a substance has absorbed as much of another substance as it can absorb. For example, water will dissolve table salt until the resulting solution contains approximately 26% salt. At this point, the saltwater solution is saturated and cannot dissolve any more salt.

Inside a boiler, saturated steam cannot absorb any more heat while it is still in contact with the water from which it was generated. The steam temperature is the same as the temperature of the boiling water.

As steam is removed from the boiler, saturated steam is at the same temperature and pressure that it was at in the boiler. At this point, the removal of any heat energy causes some of the steam to condense back to water. The addition of any heat energy causes the steam to become superheated.

66 What is superheated steam?

Superheated steam is steam that has been heated above the saturation temperature (the temperature that corresponds to the steam pressure). Once the steam has left the boiler, it can be passed through a superheater to produce superheated steam. A *superheater* is a bank of tubes through which only steam passes, not water. **See Figure 1-13.** While the superheater is confined within the same enclosure as the boiler, it is considered a separate and distinct component and not part of the boiler itself. Superheaters, if used, are almost always used with watertube boilers. Superheaters are rarely used with firetube boilers.

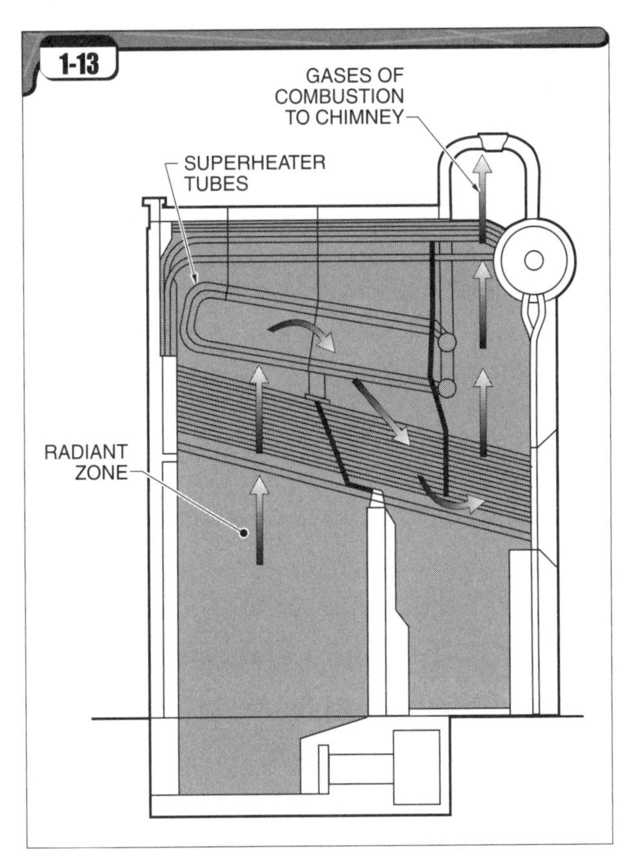

Figure 1-13. *A superheater raises the temperature of steam above the saturation point.*

Boiler Basics

List several design criteria for boilers.

- Ability to expand and contract without undue strain
- Adequate circulation of water
- Quality of materials and construction
- Ability to respond to changes in steam demand smoothly
- Simplicity of construction, inspection, and repair
- Furnace volume
- Heating surface
- Selection of fuel-burning equipment

What is a package boiler?

A *package boiler* is a boiler that is supplied from the manufacturer complete with controls, burner(s), and appliances attached. **See Figure 1-14.** Firetube, watertube, and cast iron boilers are commercially available as package boilers. Most new boilers are package units. It is less expensive to fabricate a boiler in the manufacturer's shops, due to the availability of automated assembly jigs and fixtures. Package watertube boilers may be quite large in capacity, up to approximately 350,000 lb/hr.

Cleaver-Brooks

Figure 1-14. *Package boilers are supplied complete with controls, combustion equipment, and other appliances attached.*

What is a field-assembled boiler?

A *field-assembled boiler* is a boiler of large size that cannot be shipped as a completed unit by the manufacturer to the site where it will be placed in service. Assembly of a very large boiler on site involves considerable expense due to additional labor, crane and equipment rental, delays due to inclement weather, etc. In order to reduce such costs, the boiler is usually prefabricated in the manufacturer's shops in large component modules to the greatest degree feasible. This minimizes the amount of field work necessary.

Factoid

The majority of new boilers are package boilers. A package boiler is a boiler supplied complete with controls, combustion equipment, and other appliances already attached.

What is a miniature boiler?

A *miniature boiler* is a small boiler that meets several criteria for dimensions and capacity. The design criteria are as follows: (a) The inside diameter of the shell does not exceed 16″; (b) it has not more than 20 sq ft of heating surface; (c) it does not exceed 5 cu ft in gross volume (including internal flue gas passages); and (d) the MAWP does not exceed 100 psig. Small boilers find many uses in dry cleaning shops, tailoring shops, restaurants, health spas, and other applications.

What are passes of the gases of combustion?

Passes are the number of times the gases of combustion flow the length of the boiler as they transfer heat to the water. **See Figure 1-15.** For example, in a four-pass firetube boiler, one pass of the gases is made over the length of the boiler in the furnace. The gases then pass through various groups of fire tubes, traveling the length of the pressure vessel three more times before leaving the boiler.

What causes the water to circulate naturally?

The water circulates naturally because of the difference in density between water at different temperatures. Hot water is lower in density than cool water. Hotter water in a boiler will rise toward the surface of the water while the cooler water will descend toward the bottom of the boiler. **See Figure 1-16.** This principle is observed in boiler design, and careful consideration is applied to make sure the circulation in the boiler is adequate to control the temperature of the heating surfaces.

What is forced circulation?

Forced circulation is a variation in watertube boiler design in which boiler water circulation through the tubes is enhanced by a pump. This design is used to avoid steam blanketing and subsequent tube damage in some very high pressure utility units that have extremely high heat transfer rates. Forced circulation boilers are often of one-pass design. A one-pass boiler is a watertube boiler in which the water passes through the tubes only once and then emerges as steam.

Factoid

Superheated steam is usually used for driving large steam turbines because large turbines require steam that contains no moisture droplets. The rotative speed at the outer edges of the rotor in a large turbine may exceed 1500 mph. Moisture droplets in the steam would erode the turbine rotor.

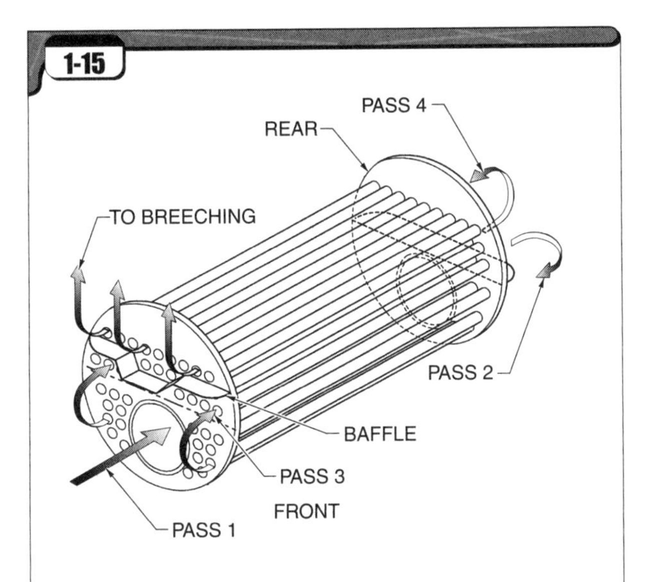

PASS 1 - FROM BURNER THROUGH TUBULAR FURNACE TO REAR

PASS 2 - FROM TUBULAR FURNACE THROUGH LOWER FIRE TUBES TO FRONT

PASS 3 - FROM FRONT THROUGH INTERMEDIATE FIRE TUBES TO REAR

PASS 4 - FROM REAR THROUGH UPPER FIRE TUBES TO FRONT AND BREECHING

Scotch Marine Package Boiler

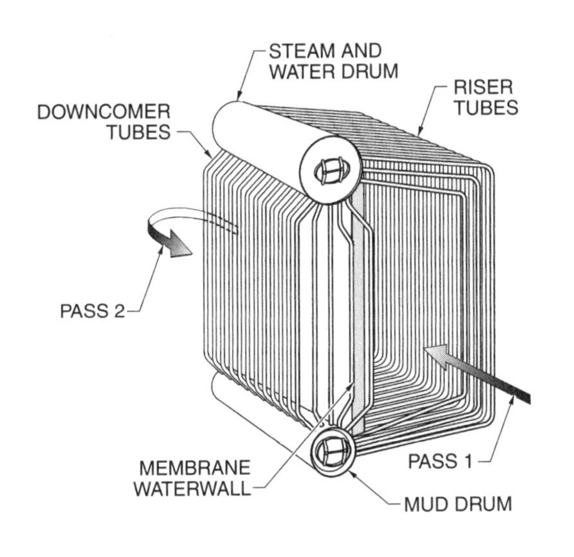

PASS 1 - FROM BURNER THROUGH OPEN PORTION WITH RISER TUBES TO REAR

PASS 2 - FROM REAR THROUGH DOWNCOMER TUBES TO FRONT

"D" Style Package Boiler

Figure 1-15. *The passes of gases of combustion in boilers are designed for efficiency.*

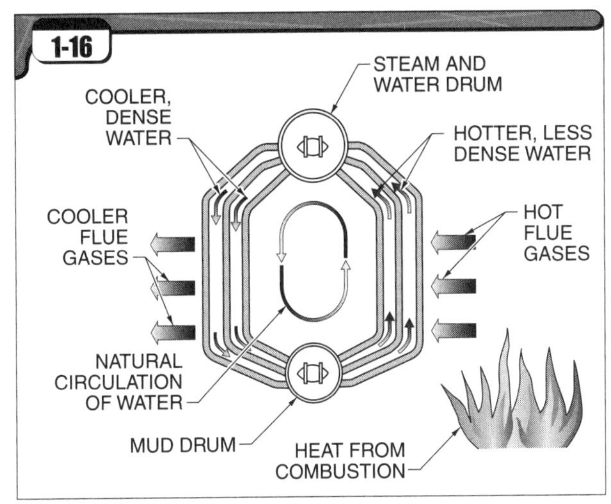

Figure 1-16. *Hot water rises and cool water descends.*

What are the relative advantages and disadvantages of firetube and watertube boilers?

The main advantages of firetube boilers are that they are simpler in design and generally lower in initial cost compared to watertube boilers of equal capacity. The main disadvantages are that they are limited in pressure and capacity. For example, firetube boilers are generally limited to 350 psig and capacities of 70,000 lb/hr. In addition, scotch marine firetube boilers are limited to gaseous and liquid fuels and are slower to respond to demands for steam compared to watertube boilers. Also, stresses are greater in firetube boilers because of their rigid design and subsequent inability to expand and contract easily.

The main advantages of watertube boilers are that they can withstand much higher operating pressures and steam temperatures than firetube boilers and have much higher capacities. For example, watertube boilers can be designed to withstand pressures of up to 3850 psig, and can generate over 6,000,000 lb/hr. Watertube boilers are rapid steamers, respond quickly to changes in demand for steam, and can be designed to burn almost any available fuel. Watertube boilers expand and contract more easily than firetube boilers and therefore generally have a longer service life. The main disadvantages are that they are initially more expensive due to more sophisticated construction techniques required. They also require more complicated furnaces and repair techniques.

Firetube boilers

What is an internally fired firetube boiler?

An *internally fired firetube boiler* is a boiler with a furnace area surrounded by the pressure vessel. A scotch marine boiler is an example of an internally fired firetube boiler.

76 What is an externally fired firetube boiler?

An *externally fired firetube boiler* is a boiler with a separate furnace area that is usually built of refractory brick. A horizontal return tubular boiler is an example of an externally fired firetube boiler.

77 What is a scotch marine boiler?

A *scotch marine boiler* is a firetube boiler that has a flue furnace and horizontal shell. **See Figure 1-17.** A scotch marine boiler is commonly a package design. The flue furnace is composed of a large tube or pipe, often corrugated, that passes the length of the shell. The fuel burner is located at the front end of the flue so that the combustion flame extends most of the length of the flue. Gases of combustion make up to three passes through the smaller fire tubes before going out to the stack (chimney). The fire tubes are attached to tube sheets at both ends. The original scotch marine boilers were used on ships. A scotch marine boiler uses natural gas or fuel oil to maximize the convenience of automatic operation.

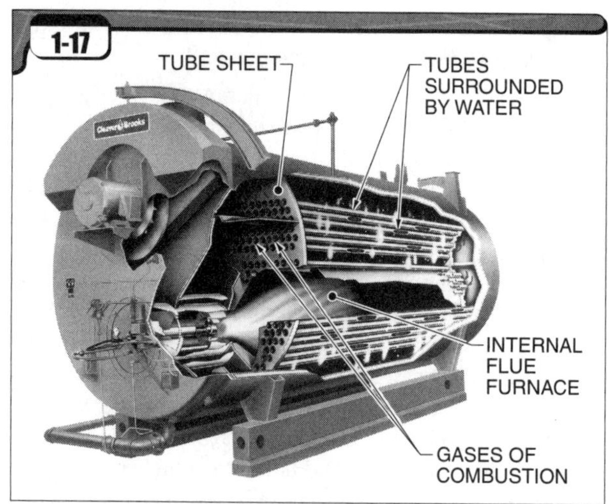

Figure 1-17. *A scotch marine boiler is a firetube boiler that has a flue furnace and a horizontal shell.*

78 What is a baffle?

A *baffle* is a metal or refractory-covered panel that directs the flow of gases of combustion for maximum boiler heating surface contact. Baffles are found in firetube, watertube, and cast iron boilers.

79 What is a dryback scotch marine boiler?

A *dryback scotch marine boiler* is a firetube boiler with a refractory-lined chamber at the rear of the boiler that is used to direct the combustion gases from the flue furnace to the first pass of tubes. This chamber, usually in the form of a large door, may also reverse the flue gas flow from one section of tubes into another. In this case, the chamber is divided inside by one or more baffles.

80 What is a wetback scotch marine boiler?

A *wetback scotch marine boiler* is a firetube boiler with a water-cooled reversing chamber used to direct the combustion gases from the flue furnace to the first pass of tubes. The wetback design utilizes less refractory (firebrick) at the rear than the dryback design and thus necessitates less refractory maintenance. However, this design has limited access to perform maintenance on both the water side and the fire side of the reversing chamber, so repairs tend to be more difficult and costly. Some manufacturers use the term waterback scotch marine boiler for this type of design.

81 What is a horizontal return tubular boiler?

A *horizontal return tubular (HRT) boiler* is a firetube boiler consisting of a horizontal shell set above a refractory brick-lined furnace. **See Figure 1-18.** This design is externally fired. The pressure vessel contains many fire tubes and has a suspended baffle at the rear of the unit that directs the gases of combustion into the tubes. Early HRT boilers burned coal. Modern HRT boilers and those that have been converted for automatic operation burn natural gas or fuel oil. Most modern HRT boilers are package designs that are small in size and used for commercial applications such as laundries.

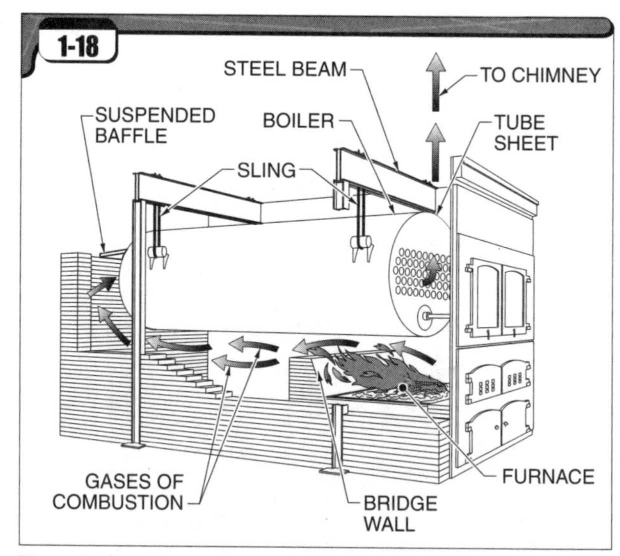

Figure 1-18. *A horizontal return tubular (HRT) boiler is heated by gases of combustion directed by baffles for maximum efficiency.*

82 What is a firebox boiler?

A *firebox boiler* is a firetube boiler in which an arch-shaped furnace is surrounded on the sides by a water leg area. The water space is extended downward so that the furnace walls are surrounded by water. **See Figure 1-19.** These flat, side water leg areas are supported by staybolts to prevent them from bulging. Most firebox boilers are low-pressure boilers used for comfort heating. These boilers usually burn natural gas or fuel oil. Early firebox boilers burned coal. Many of these have been converted for convenience. Other fuels (for example, wood waste and paper waste) may also be burned, depending upon the design of the furnace.

83 What is the difference between an exposed-tube vertical boiler and a submerged-tube vertical boiler?

An *exposed-tube (dry-top) vertical boiler* is a firetube boiler with fire tubes extending several inches through the steam space at the top before ending at the tube sheet. **See Figure 1-20.** This exposes steam in the top of the boiler to extra heat, which helps to dry it. However, it also increases the possibility that the tubes may become overheated. A *submerged-tube (wet-top) vertical boiler* is a firetube boiler with fire tubes completely covered with water all the way to the upper tube sheet. The area between the conical smoke box and the outside shell at the top of the boiler composes the steam space.

Vertical firetube boilers are an attractive option when floor space is at a premium. They are generally used for commercial applications such as in dry cleaning shops, cafeterias, fitness centers, and manufacturing facilities with limited steam applications.

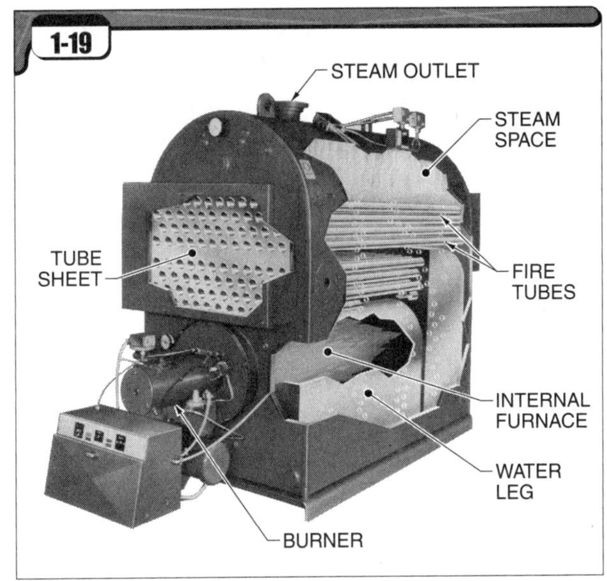

Kewanee Boiler Co.

Figure 1-19. *A firebox boiler is a firetube boiler that uses an arch-shaped furnace surrounded by a water leg area.*

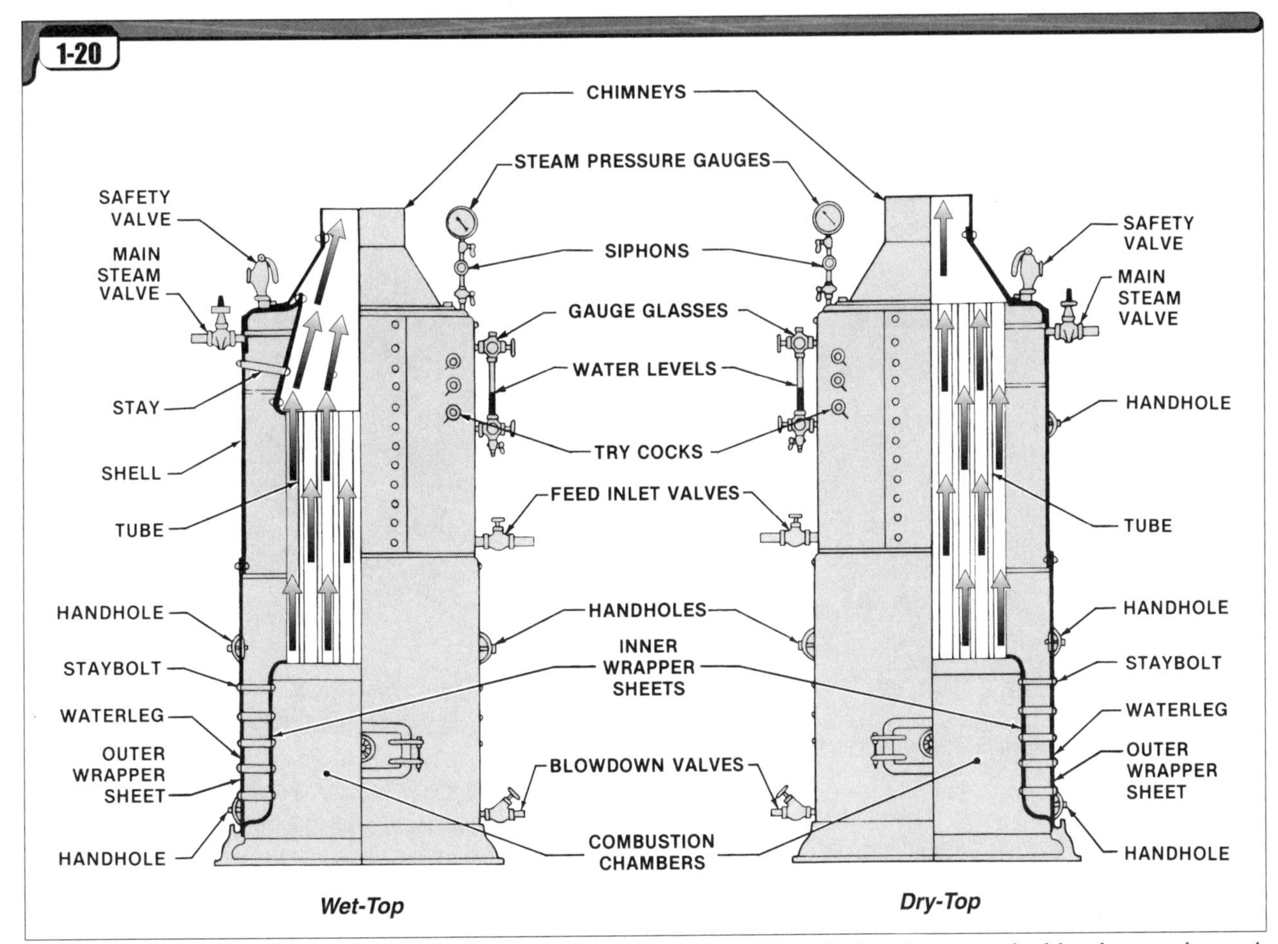

Figure 1-20. *Vertical firetube boilers are used where floor space is limited. In wet-top vertical boilers, the upper ends of the tubes are submerged. In dry-top vertical boilers, the upper ends of the tubes are exposed.*

Watertube boilers

Why do watertube boilers generate steam more quickly from a cold start and respond more quickly to load changes than firetube boilers?

Watertube boilers generate steam more quickly than firetube boilers because the heating surface is arranged to provide more intimate contact of the water and gases of combustion with the heating surface. This allows for faster conduction of heat into the water. In addition, the curvature of the tubes in a watertube boiler allows for greater flexibility during expansion; therefore, watertube boilers generally may be brought up to normal operating temperature more quickly than firetube boilers.

A watertube boiler can handle load changes more quickly than a firetube boiler because the water circulation is better in a watertube boiler. This allows faster heat transfer from the hot exhaust gases to the water, which results in more rapid changes in the steam generation rate.

85 What is a straight-tube watertube boiler?

A *straight-tube watertube boiler* is a boiler design in which the steam-generating tubes are straight rather than bent or curved. The tubes in straight-tube watertube boilers are inclined to aid in creating more positive circulation of the water and the steam bubbles. **See Figure 1-21.** Straight-tube boilers often have only one drum. In such cases, headers are used to supply water from the drum to the lower ends of the tubes, and to collect the heated water and steam from the upper ends of the tubes. Baffles are used to enhance contact of the combustion gases with the tubes.

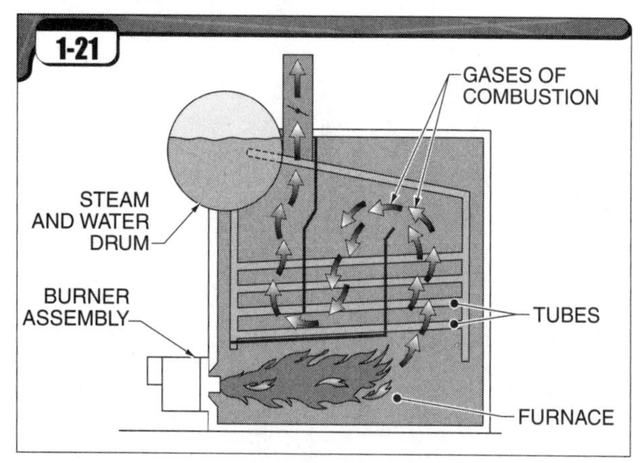

Figure 1-21. *Straight-tube watertube boilers have straight, inclined tubes to aid in water circulation.*

What is a bent-tube watertube boiler?

A *bent-tube watertube boiler* is a boiler design in which the tubes are bent (curved) to some degree. In bent-tube boilers, there are almost always two or more drums. The tubes usually penetrate the drums individually, and much of the bending of the tubes is done so that the tubes may enter the drum at a perpendicular angle. Most industrial and large commercial watertube boilers are of bent-tube design, but differ somewhat in configuration.

What is a flexible-tube watertube boiler?

A *flexible-tube watertube boiler* is a boiler design in which the tubes exposed to the combustion gases are sharply bent to provide the maximum possible flexibility. The tubes often have multiple bends in a zigzag pattern. This design is intended to maximize the boiler's resistance to the thermal strains that would otherwise occur as the steel tubes expand during fast startups. These boilers may be low-pressure or high-pressure boilers. They are most commonly used for steam heating and for commercial applications.

What is a membrane watertube boiler?

A *membrane watertube boiler* is a boiler design in which rows of tubes are formed into solid panels through the use of welded steel strips that fill the spaces between the tubes. **See Figure 1-22.** In this configuration, the boiler is highly compact. The welded tube panels also serve as baffles. Due to their compactness, membrane watertube boilers are often sold as replacement boilers where other designs would be too difficult to move into existing mechanical rooms. These boilers are used for heating and small process applications.

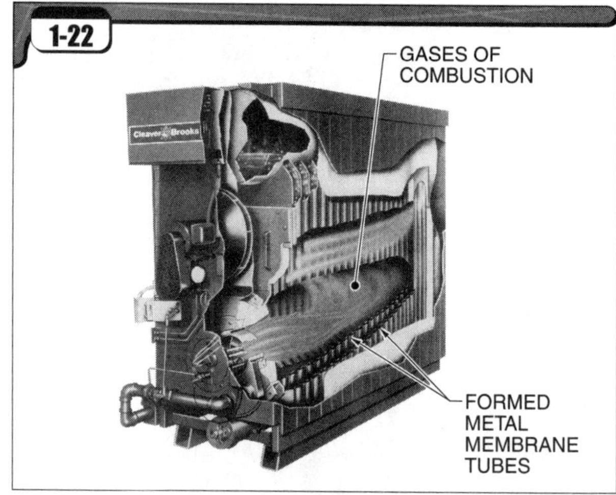

Cleaver-Brooks

Figure 1-22. *The tubes in a membrane watertube boiler are formed into solid panels by welding a steel strip between each tube.*

What is a coil watertube boiler?

A *coil watertube boiler* is a boiler design in which the tubes are formed into a continuous coil, with the combustion gases passing through the interior of the coil. These boilers use an external drum in which the steam and water separate, and the steam leaves the boiler from this drum. The drum also serves as the reservoir from which water is directed into the inlet end of the coils. In some coil watertube boilers the coil is oriented horizontally, while in other designs it is oriented vertically. The boiler water is forced through the coiled tubes by a circulating pump. This positive circulation allows these boilers to be started up very rapidly, from a cold start to full steam-generating conditions, within a few minutes.

90 What is an "O" style watertube boiler?

An *"O" style watertube boiler* is a watertube boiler design with a top steam and water drum and a bottom mud drum that are interconnected by banks of symmetrical tubes in an "O" shape. **See Figure 1-23.** The burner is in the center of the "O." This unit is almost always a package type. An "O" style boiler commonly burns natural gas or fuel oil. Larger "O" style boilers may be designed to generate as much as 200,000 lb/hr, at pressures of up to 1000 psig.

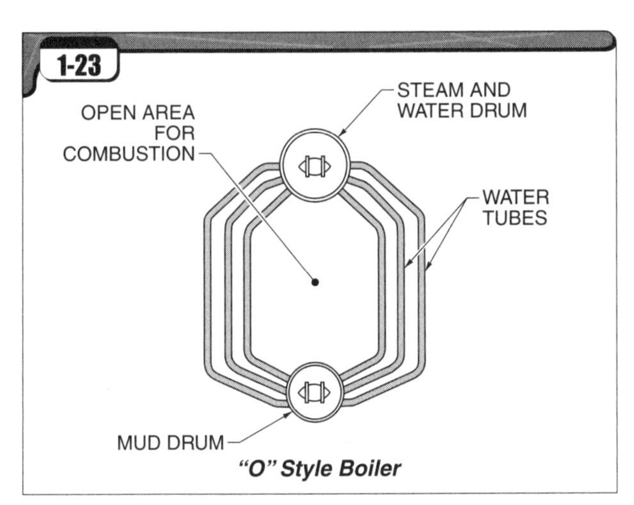

Figure 1-23. *An "O" style boiler is a watertube boiler with a top steam and water drum and a bottom mud drum interconnected by banks of symmetrical tubes.*

91 What is an "A" style watertube boiler?

An *"A" style watertube boiler* is a watertube boiler design with a top steam and water drum and two smaller bottom mud drums. **See Figure 1-24.** These are interconnected by banks of symmetrical tubes in an "A" shape. This is a popular configuration for package types. An "A" style boiler commonly burns natural gas or fuel oil. "A" style boilers have capacities similar to "O" style boilers.

92 What is a "D" style watertube boiler?

A *"D" style watertube boiler* is a watertube boiler similar to the "O" style, except that the steam-generating tubes on one side are extended to leave an open area close to the center. **See Figure 1-25.** This area, in a "D" shape, is for combustion of the fuel. The two sides are separated by a baffle so that the gases pass to the rear on the combustion side, and then turn back toward the front for the convection side. There is either a top or side outlet for the gases to leave the unit. A "D" style boiler commonly burns natural gas or fuel oil. "D" style boilers have capacities similar to "A" style and "O" style boilers.

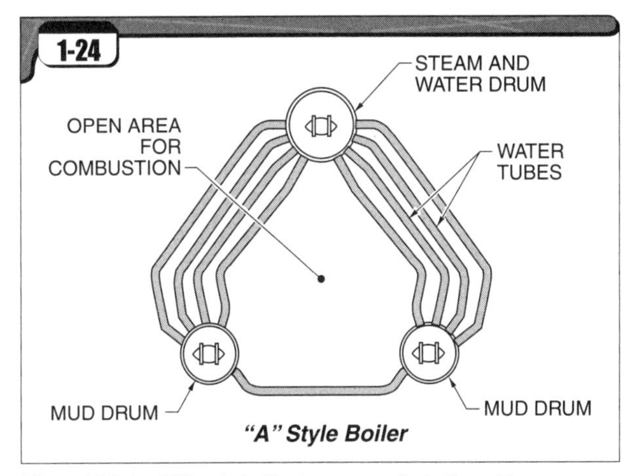

Figure 1-24. *An "A" style boiler is a watertube boiler with a top steam and water drum and two bottom mud drums.*

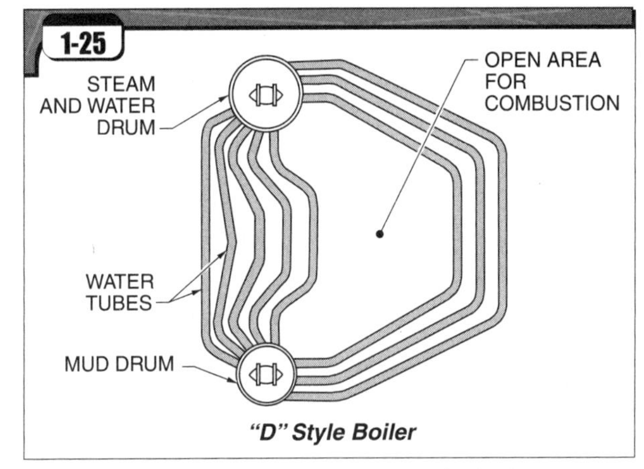

Figure 1-25. *A "D" style boiler is a watertube boiler that has a top steam and water drum and bottom mud drum connected by tubes to form a "D" shape. "D" style boilers are available as left or right hand units.*

93 How are large field-assembled industrial watertube boilers configured?

Field-assembled industrial watertube boilers vary widely in design because field-assembled boilers are designed to meet the particular needs of a specific plant or facility. Most are two-drum designs, with a top steam and water drum and a bottom mud drum. These two drums are normally configured with symmetrical banks of tubes in an "O" design. In contrast to a package "O" style watertube boiler, however, the combustion chamber area is outside the "O." The shape and size of the combustion chamber varies widely, depending on the type of fuel. Depending on the nature of the steam-using processes, these large boilers may have superheaters or other special auxiliary equipment.

94 What is a Stirling boiler?

A *Stirling boiler* is a watertube boiler design with three steam and water drums on the top and a mud drum beneath, interconnected by a large number of water tubes. **See Figure 1-26.** Water enters the top rear drum and passes downward through the rear bank of tubes to the mud drum. It then goes upward to the top front drum and top center drum. As the steam separates from the water, equalizing tubes connecting the steam spaces of all three drums allow the steam pressure to equalize. The steam then flows either to the rear or center drum (depending on the design) to leave the boiler.

Circulating tubes connect the water spaces of the top front and top center drums, allowing the water levels in these drums to balance out. Usually there are very few or no circulating tubes between the top center and the top rear drums. Circulating tubes in this location allow water to short-circuit to the other drums without first flowing through the water tubes.

There are several variations of the Babcock and Wilcox Stirling boiler. The original four-drum Stirling boiler is an older design, though many remain in use. Depending on the furnace design, a Stirling boiler can burn any of the common fuels.

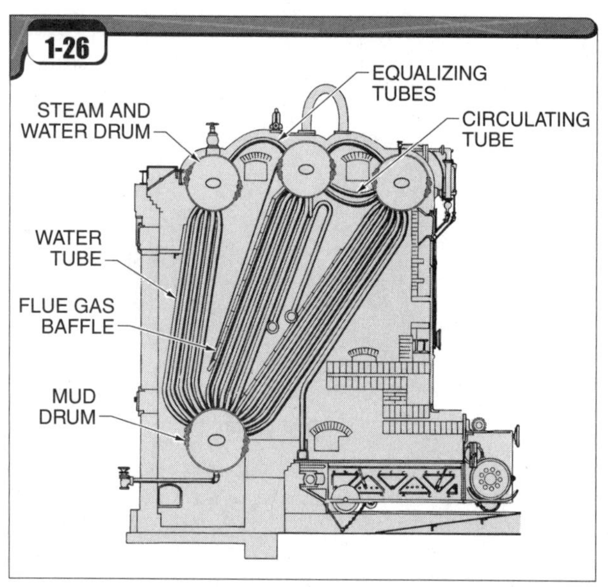

Figure 1-26. *A Stirling boiler is a watertube boiler with three steam and water drums on the top and a mud drum beneath, interconnected by water tubes.*

95 What is a utility watertube boiler?

A *utility watertube boiler* is an extremely large watertube boiler that generates steam at a very high pressure and temperature. Utility watertube boilers almost always generate steam for the purpose of driving large steam turbines. These steam turbines in turn drive electric generators or very large process equipment. Steam turbines for such applications require superheated steam at up to 1050°F and up to 3800 psig or more. Utility watertube boilers are used by electric power companies as well as a variety of heavy industries such as paper mills, refineries, and very large chemical plants.

Cast Iron Boilers

96 What are cast iron boilers used for?

Cast iron boilers are used for closed, low-pressure steam systems only. A *closed system* is a steam system in which the condensate is recovered and returned to the boiler.

97 What is a push-nipple cast iron sectional boiler?

A *push-nipple cast iron sectional boiler* contains hollow cast iron sections joined with tapered nipples and pulled tightly together with tie rods or bolts. **See Figure 1-27.** The push nipples are similar in function to the copper ferrules that are used on faucets and other plumbing applications. They are pressed between each section, forming a watertight fit. If a push-nipple cast iron sectional boiler develops a leak, the boiler usually must be dismantled to some degree to replace the defective section.

The sectional design allows the manufacturer to keep the cost of the boiler low by using the same cast iron section for a number of boilers in the manufacturer's line. For example, a smaller boiler may use nine sections while a larger boiler may use 13 sections. The size of the burner and other support equipment is changed to match the boiler design.

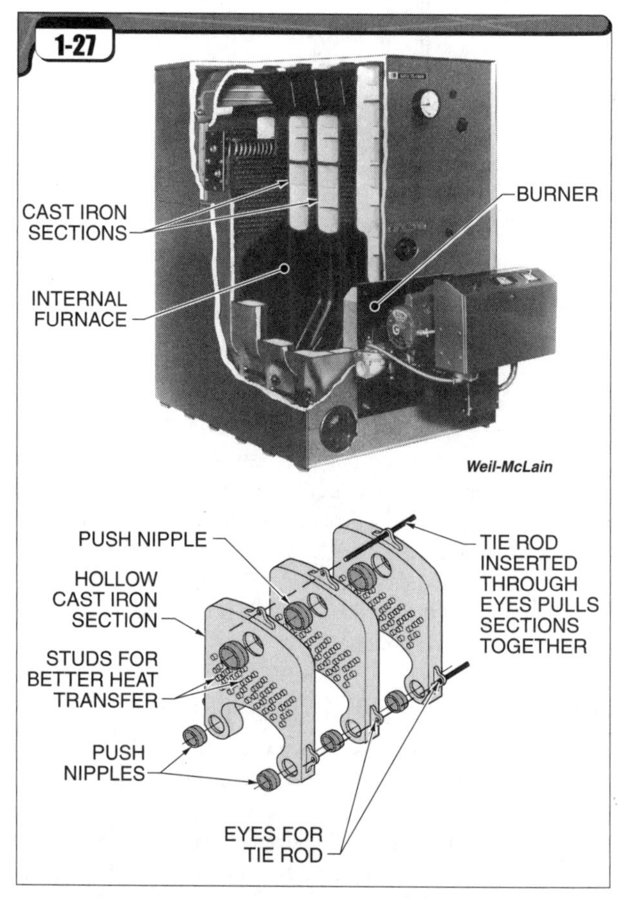

Figure 1-27. *A push-nipple cast iron sectional boiler contains hollow sections joined with tie rods or bolts.*

98 What is an external header cast iron sectional boiler?

An *external header cast iron sectional boiler* contains cast iron sections individually connected to external manifolds (headers) with screwed (threaded) nipples. These units usually have a lower header on each side and a central top header. The lower headers supply water to the sections, and the top header collects the steam from the sections. External header cast iron sectional boilers are sometimes referred to as pork chop boilers, because the sections resemble the shape of a pork chop. These boilers have an advantage over the push-nipple style, because the sections are configured in a left-hand and right-hand pattern, and each is individually connected to the headers by a threaded fitting. If a section develops a leak, that individual section can be replaced without disturbing the other sections.

Other Boilers

99 What is a hot water boiler?

A *hot water boiler* is a boiler that generates hot water, not steam. The heated water produced by a hot water boiler is usually between approximately 170°F and 190°F. The water is pumped through the boiler by a circulating pump while the burner provides the heat to bring the water up to the specified temperature. **See Figure 1-28.** Hot water boilers are considered low-pressure heating boilers if the MAWP does not exceed 160 psig and the maximum operating temperature does not exceed 250°F.

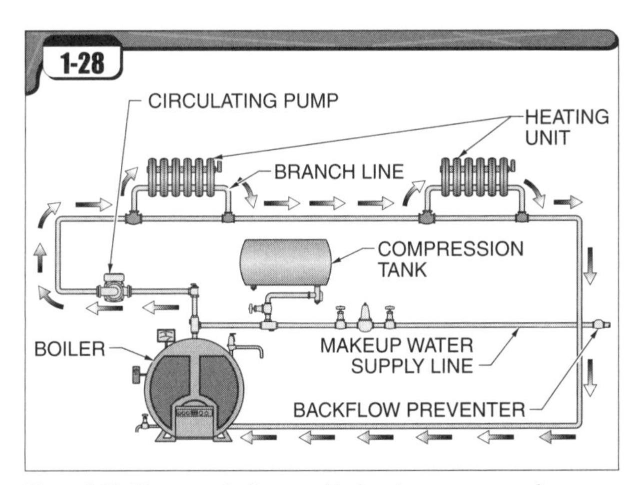

Figure 1-28. *Hot water boilers used in heating systems use hot water to transport heat to a building space.*

100 What is a high temperature water (HTW) boiler?

A *high temperature water boiler* is a boiler in which the maximum operating temperature of the water may reach temperatures in excess of 250°F and the operating pressure may exceed 160 psig. These boilers are used for very large central heating systems. A single HTW boiler may provide hot water to many buildings on a university campus or military base.

101 What is a waste heat recovery boiler? Name three cases where the use of this boiler can be an advantage.

A *waste heat recovery boiler* is a fire tube or watertube boiler in which heat that would otherwise be discarded is used to make steam. The heat may be in the form of hot gases from a separate process, or may be from the combustion of a waste fuel in the boiler furnace. These boilers operate at low cost and conserve natural resources such as natural gas and coal. Waste heat recovery boilers are also referred to in some cases as heat recovery steam generators (HRSGs).

Waste heat recovery boilers are efficient in plants that produce industrial waste by-products such as wood pulp, by-product gases, or flammable chemicals that can serve as sources of heat. Municipal refuse is a common fuel for steam plants, as well as methane gas from landfills and wastewater treatment plants.

102 What is a combined-cycle boiler system?

A *combined-cycle boiler system* is an electric power generating system that uses both a gas turbine-driven generator and a steam turbine-driven generator. A gas turbine is similar to a jet engine, but is fixed in place and coupled to a generator. The hot exhaust gases from the gas-driven turbine are typically 900°F to 1200°F. These exhaust gases serve as the primary heat source for a waste heat recovery boiler. Sometimes the waste heat is supplemented by burning additional gas or oil. The steam produced from the waste heat recovery boiler drives a second generator.

103 What is a thermal liquid boiler?

A *thermal liquid boiler* is a firetube or watertube boiler that uses a liquid chemical solution instead of water. Thermal liquids are manufactured products that resemble motor oil in appearance. Thermal liquids have certain thermal properties that allow them to be heated to very high temperatures at low pressures. For example, a typical thermal liquid may be able to be heated to 400°F while the system operating pressure remains as low as 35 psig. This offers an advantage in safety over using steam for the same purpose, since saturated steam would have to be about 240 psig to provide the same temperature. The disadvantage of thermal liquid systems is that if a leak occurs, a fire or environmental incident may result. While thermal liquid boilers are not steam boilers, they may be used in conjunction with steam boilers.

104 How may steam boilers be classified by their applications?

Boilers used for steam heating may be firetube, watertube, or cast iron units, and are most commonly package boilers. Low-pressure boilers are the most prevalent when the steam is used for heat only within a single building. For multiple buildings, high-pressure steam may be used for distribution and then converted to low-pressure steam for use in the actual heating equipment.

Examples of commercial processes include small laundries, dry cleaning shops, light manufacturing, and restaurant kitchen operations. These boilers provide steam at medium pressures, up to about 125 psig. Package firetube and small package watertube boilers are the most common for these applications. Most electric boilers are commercial process boilers as well.

Most industrial and institutional boilers are either large package firetube or large package watertube units. Some are large field-assembled watertube boilers. Boilers in these facilities provide steam at pressures up to about 600 psig. Industrial steam applications include almost all manufacturing. Institutional boiler applications include hospitals, universities, and reformatories. In almost all plants that utilize these boilers, part of the steam is used for comfort heating as well. Many industrial boilers are waste heat recovery boilers.

Utility boilers are the largest watertube boilers and operate at very high pressures. Utility boilers are used for power plants and heavy industries.

Boiler Safety Issues

 List at least six categories of responsibilities of a boiler operator.

- Human safety
- Safety of the plant's equipment
- Maintaining compliance with Environmental Protection Agency (EPA) regulations on stack emissions
- Keeping the plant on-line
- Operating efficiently
- Maintaining complete and accurate records

 What keeps boiler metal from being destroyed by the intense furnace heat?

Circulating water in the boiler carries heat away from the heating surfaces as fast as it is conducted through the metal. This is the only factor that prevents destruction of the boiler metal by furnace heat.

 What is the most common cause of failure of steam and hot water boilers?

Tubes, flues, and other boiler components may fail from overheating as a result of a low water level in the boiler. If the heating surfaces in a boiler are not kept covered with water, the metal surfaces soften and melt. In many cases, the pressure in the boiler can then rupture these overheated portions of the boiler and an explosion can result.

Factoid

Utility watertube boilers in major power plants may be well over 200′ tall and may burn over six thousand tons of coal per day.

 What is steam blanketing of boiler heating surfaces?

Steam blanketing is a condition that occurs when steam bubbles are generated so quickly from a boiler heating surface that a layer of steam is formed between the water and the heating surface. Steam blanketing is also known as film boiling. In this situation, the natural circulation of the water is not brisk enough to wash the steam from the heating surface and maintain cooling of the surface.

Steam blanketing is usually the result of a design error or an attempt to generate an excessive amount of steam from a boiler. Steam blanketing is likely to cause the affected heating surface to overheat and become damaged.

The current trend, especially in package boiler design, is toward very high heat transfer rates from the heating surfaces in order to maximize steam production while minimizing the size of the boiler. In some jurisdictions, boilers above a certain threshold size require licensed boiler operators on duty constantly. In order to avoid these requirements, boiler designs are sometimes taxed to unrealistic limits. Because of these high heat transfer rates, boiler operators must observe the manufacturer's operating instructions for the boiler and adhere to sound boiler water treatment procedures in order to prevent damage.

 Why do scale deposits form in a boiler?

Water contains impurities in the form of natural salts and minerals. Some of these impurities settle out (precipitate) as the water is heated to high temperatures. In addition, these materials do not evaporate as the water evaporates. When the water is evaporated in the form of steam bubbles, these salts and minerals crystallize as a hard crust on boiler heating surfaces. **See Figure 1-29.**

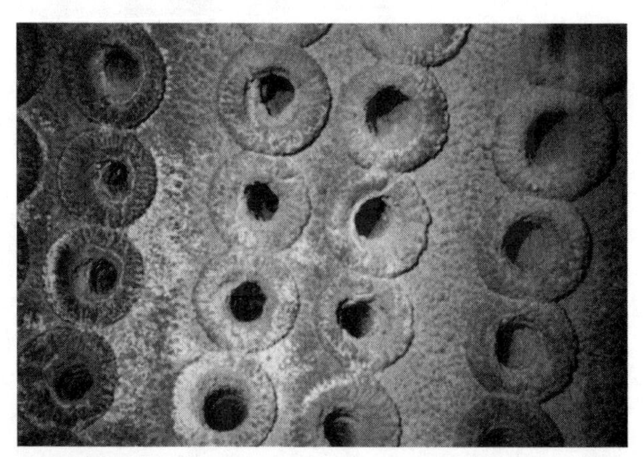

Figure 1-29. *Scale insulates boiler heating surfaces. This can cause the surfaces to overheat and fail.*

110 How does scale contribute to overheating?

The natural impurities that form scale (mostly calcium and magnesium compounds) are excellent insulators. As scale accumulates and increases in thickness, this insulating effect interferes with heat transfer through the heating surfaces. As a result, the heating surfaces overheat and can fail.

111 What is thermal shock?

Thermal shock is the stress imposed on boiler metal by a sudden and drastic change in temperature. For example, boiler surfaces may fail if they are overheated due to low water and then relatively cold water is added to the boiler. This could cause an explosion.

112 How may boiler water cause direct deterioration of the boiler metal?

Corrosion and pitting result from poor water treatment. Corrosion and pitting cause thin spots to develop in the tubes, flues, and pressure vessel walls. **See Figure 1-30.** This may cause failure as a result of internal pressure in water tubes and drums, and external pressure in flues and fire tubes.

Figure 1-30. *Corrosion and pitting cause thin spots to develop in tubes, flues, pressure vessel walls, and piping.*

113 How is a steam boiler protected from excessive pressure?

Safety valves are used on steam boilers to prevent the boiler pressure from exceeding the boiler's MAWP. Though boiler explosions due to excessive pressure were frequent in the 1800s and early 1900s, advances in safety valve design and standards of construction have made them rare today.

114 What is the primary hazard associated with the fuel-burning equipment in a boiler?

An explosion or fire can result if the fuel delivery system allows a significant volume of air-fuel mixture to accumulate in the combustion chamber without being ignited. If this flammable mixture is exposed to an ignition source such as a flame, spark, or hot refractory, it can ignite all at once and cause an explosion. This potential exists in boilers fired with gaseous fuels, liquid fuels, or pulverized coal. It may be the result of leaking fuel delivery valves or control failure.

115 What is flash steam?

Flash steam is steam that is instantly produced when very hot water is released to a lower pressure and, thus, a lower boiling temperature. Water under a given pressure will not boil and change to steam until it reaches the temperature that corresponds to that pressure, as given in the Properties of Saturated Steam tables.

If this very hot water is suddenly released to a lower pressure area, a portion of the water will instantly flash into steam because the water has a lower boiling point at the lower pressure. **See Figure 1-31.** The water expands many times in volume when it changes to steam. When water changes to steam at atmospheric pressure, it expands approximately 1600 times in volume.

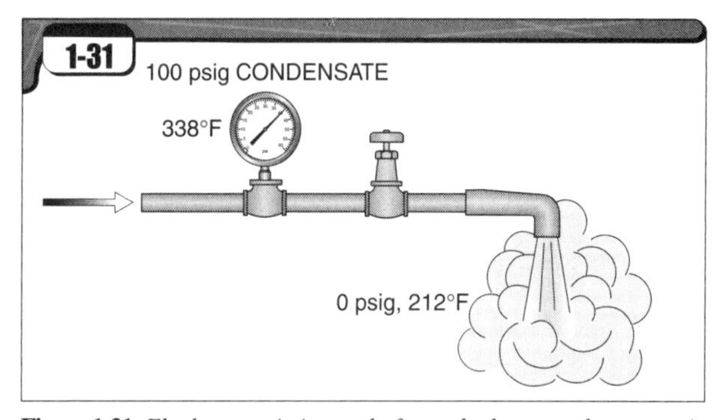

Figure 1-31. *Flash steam is instantly formed when very hot water is released to a lower pressure.*

Factoid

Overheating due to low water is by far the most common cause of damage to steam and hot water boilers.

116 Why is an explosion generally less disastrous in a watertube boiler than in a firetube boiler?

In a watertube boiler, the large volume of water is distributed into many small tubes, and the volume of water in the drum is comparatively small. If one water tube bursts, the pressure in the entire pressure vessel is bled off slowly, and the tendency of the large body of water to flash into steam is minimized. Usually only one water tube bursts. This may steam-cut a small number of nearby water tubes.

In a firetube boiler, a sudden crack in the shell or a flue pulled loose at the end causes the entire body of water to be subjected to a sudden and substantial drop in pressure. The huge volume of flash steam that results can cause an explosion of tremendous force.

1 Chapter

Boiler Theory and Principles

Name: Date:

_______________ 1. ___ is the gaseous form of water.

_______________ 2. Water boils at ___°F under normal atmospheric pressure.
A. 100 B. 144
C. 212 D. 430

3. How does sensible heat differ from latent heat?

T F 4. A boiler is a closed vessel in which water is transformed into steam under pressure by the application of heat.

T F 5. Steam from a boiler may be used to drive steam turbines.

_______________ 6. Any part of the boiler metal that has hot gases of combustion on one side and water on the other side is the heating ___.

_______________ 7. Water boils at ___°C under normal atmospheric pressure.
A. 100 B. 144
C. 212 D. 430

_______________ 8. The three most common fuels used in steam production are ___.
A. natural gas, fuel oil, and kerosene B. natural gas, kerosene, and wood
C. natural gas, fuel oil, and coal D. natural gas, wood, and coal

T F 9. Water passes through the tubes of a firetube boiler.

T F 10. Flues are specified by their inside diameter.

_______________ 11. A temperature of 71°F equals ___°C.
A. 7.44 B. 21.67
C. 39.44 D. 159.8

T F 12. There are 8.33 lb of water in 1 cu ft of water.

_______________ 13. A temperature of 35°C equals ___°F.
A. 95 B. 100
C. 105 D. 110

_______________ 14. The ___ of heat is expressed in Btu.

_______________ **15.** When 24,666 lb of water is lowered from 212°F to 186°F, ___ Btu are given up.
A. 641,316 B. 4,587,876
C. 5,229,192 D. 9,817,068

_______________ **16.** When 23,330 lb of water is lowered from 212°F to 196°F, ___ Btu are given up.
A. 373,280 B. 4,572,680
C. 4,945,960 D. 9,518,640

_______________ **17.** The formula for finding pressure is ___.
A. pressure = force × area B. pressure = (force × area) ÷ 2
C. pressure = force ÷ area D. pressure = force ÷ (area × 2)

_______________ **18.** A force of 220 lb exerted on 8 sq in. equals ___ psi.

_______________ **19.** A 12 lb weight resting on 4 sq in. exerts ___ psi.
A. 3 B. 8
C. 16 D. 48

_______________ **20.** The flue gas temperature entering a boiler is 1845°F. The temperature of the same flue gases as they leave the boiler is 715°F. The differential temperature across the boiler is ___°F.

T F 21. One Btu is the amount of heat necessary to raise the temperature of 1 lb of water 1°F.

T F 22. The total force on a 16 sq in. surface with a pressure of 140 psi against it is 2240 lb.

_______________ **23.** A valve has a surface area of 12 sq in. exposed to a pressure of 120 psi. The total force on the surface is ___ lb.

_______________ **24.** A force of 518.4 lb exerted on 1 sq ft exerts ___ psi.

T F 25. Condensate is water that results from steam losing its heat.

_______________ **26.** 28.2 in. Hg is the equivalent of ___ psi.

_______________ **27.** A pressure of 14.1 psi is equal to ___ in. Hg.

_______________ **28.** The atmospheric pressure at sea level is ___ psia.

T F 29. Gauge pressure is the pressure below atmospheric pressure.

_______________ **30.** The formula for finding boiler horsepower (BHP) is ___.
A. BHP = lb/hr ÷ Factor of evaporation × 34.5
B. BHP = lb/hr × Factor of evaporation × 34.5
C. BHP = lb/hr × Factor of evaporation ÷ 34.5
D. none of the above

_______________ **31.** The boiler horsepower of a boiler generating 18,000 lb of steam per hour at 140 psi is ___. The factor of evaporation is 1.08.
A. 483.09 B. 563.48
C. 670,680 D. none of the above

_______________ **32.** A 75 HP boiler will make ___ lb of steam in 3 hr.

_______________ **33.** If the gauge pressure on a boiler is 120 psig, the absolute pressure is ___ psia.

T F **34.** Thermal shock is the stress imposed on the boiler metal by a sudden and drastic change in pressure.

T F **35.** An HRT boiler is a watertube boiler.

______________ **36.** A scotch marine boiler has a(n) ___ furnace and a horizontal shell.

______________ **37.** In a watertube boiler, the ___ pass(es) through the tubes and the ___ pass(es) across the outside surface of the tubes.

A. hot gases of combustion; water
B. water; hot gases of combustion
C. hot gases of combustion; hot gases of combustion
D. water; water

______________ **38.** MAWP stands for ___.

A. minimum achievable working pressure
B. maximum achievable working pressure
C. minimum allowable working pressure
D. maximum allowable working pressure

______________ **39.** The maximum voltage for electric boilers is ___V.

A. 480
B. 4160
C. 13,800
D. 16,000

______________ **40.** Converting a pound of water at 212°F to a pound of steam at 212°F at atmospheric pressure requires ___ Btu.

T F **41.** Flash steam is formed when water at a high temperature and pressure is released to a lower pressure.

______________ **42.** Superheaters are almost always found in ___ boilers.

A. watertube
B. firetube
C. cast iron
D. hot water

T F **43.** An "A" style boiler is a firetube boiler with a top drum and two bottom mud drums.

______________ **44.** Cast iron boilers are used for ___ heating systems only.

A. open, high-pressure
B. closed, high-pressure
C. open, low-pressure
D. closed, low-pressure

______________ **45.** A pressure gauge reads 150 psig. The absolute pressure is ___ psia.

______________ **46.** A high-pressure boiler has an MAWP above ___ psi.

______________ **47.** ___ heat is the Btu content of water between the freezing and boiling points.

______________ **48.** A boiler produces 24 Klb/hr of steam. The boiler is producing ___ lb of steam per hour.

A. 24
B. 240
C. 2400
D. 24,000

______________ **49.** A pressure of 27 in. Hg is equal to ___ psi.

______________ **50.** A ___ watertube boiler is the largest in both size and steam-generating capacity.

A. membrane
B. package
C. flexible tube
D. utility

T F **51.** Watertube boilers generate steam from a cold start faster than firetube boilers.

T F **52.** A "D" style boiler is a watertube boiler.

T F **53.** Straight nipples are pressed between each section of a push-nipple cast iron sectional boiler with tie rods or bolts.

______________ **54.** A column of water that exerts 3.5 psi at the bottom is ___′ high.

______________ **55.** A piece of steel plate that is 8″ long and 6″ wide has an area of ___ sq in.

T F **56.** Tubes are specified by their inside diameter.

T F **57.** The Fahrenheit scale is commonly used with the metric system of measurement.

______________ **58.** ___ in the boiler keeps the boiler metal from being destroyed by the intense furnace heat.
- A. Forced air
- B. Circulating water
- C. Circulating steam
- D. all of the above

T F **59.** An explosion is generally less disastrous in a watertube boiler than in a firetube boiler.

______________ **60.** A(n) ___ fuel is any fuel formed below the Earth's surface from plant and animal remains.

______________ **61.** The total force on a 3.75 sq in. surface with 125 psi acting against it is ___ lb.

T F **62.** A low-pressure boiler has an MAWP of 15 psi or less.

T F **63.** A firebox boiler is a watertube boiler in which the furnace is surrounded on the sides by a water-leg area.

______________ **64.** The original Stirling boiler is a watertube boiler with ___ steam and water drum(s) on the top and a mud drum beneath.
- A. one
- B. two
- C. three
- D. four

T F **65.** A combined-cycle boiler system includes a waste heat recovery boiler that uses hot exhaust gases as its heat source.

______________ **66.** A temperature of 21°C equals ___°F.
- A. 37.8
- B. 53
- C. 58.8
- D. 69.8

______________ **67.** The total force on a 12 sq in. valve with 116 psi acting against it is ___ lb.
- A. 104
- B. 128
- C. 1392
- D. none of the above

______________ **68.** A piece of 48″ × 96″ steel plate has an area of ___ sq in.
- A. 48
- B. 96
- C. 144
- D. 4608

______________ **69.** A temperature of 72°F equals ___°C.
- A. 22.22
- B. 40
- C. 72
- D. 104

T F 70. Steam is odorless, colorless, and tasteless.

_______________ 71. The statement ___ is not true.
A. burning fuel creates energy
B. energy can be moved
C. there are several forms of energy
D. energy cannot be created or destroyed

T F 72. Heat always travels from a body at a high temperature toward a body of low temperature.

_______________ 73. The purpose of a steam trap is to ___.
A. automatically vent steam from the steam system
B. prevent condensate from escaping from the steam system
C. control the steam pressure in the steam system
D. automatically drain condensate and vent air from the steam system

_______________ 74. Makeup water is used in a steam boiler system to ___.
A. replace condensate that is lost from the system due to leaks or drainage
B. replace water lost from the system in the form of steam leaks
C. replace water in the form of steam that is directly injected into a product or for humidification
D. all of the above

_______________ 75. A column of water 160 ft high exerts a pressure at the bottom of ___ psig.
A. 69.28
B. 78.56
C. 160
D. 369.5

_______________ 76. A 55-gal. drum contains ___ lb of water when full.
A. 127
B. 458
C. 1544
D. 1897

Matching

_______________ 77. Fluid
_______________ 78. Vacuum
_______________ 79. Steam
_______________ 80. Square foot
_______________ 81. Radiation
_______________ 82. Centigrade
_______________ 83. In. Hg
_______________ 84. Convection
_______________ 85. BHP
_______________ 86. Conduction

A. Boiler horsepower
B. Transfer of heat from a hot body to a cold body without physical contact or a conveying medium
C. Pressure less than atmospheric pressure
D. Inches of mercury
E. Gaseous form of water
F. Transfer of heat by actual physical contact
G. 144 sq in.
H. Transfer of heat by conveying medium
I. Any material that is able to flow from one point to another
J. Obsolete term for Celsius temperature

Chapter 2

Boiler Construction and Design

The inherent hazards associated with pressure-containing vessels demands that the vessels be fabricated with superior quality materials and workmanship. The assurance of satisfactory results in welded construction is especially important. The ASME Boiler and Pressure Vessel Code defines the metallurgical requirements, procedural standards, and standards for qualification of welding personnel that ensure these objectives are met. Those who design or specify boilers for particular applications endeavor to strike a balance among initial cost of the equipment, cost of installation, life expectancy, suitability of the boilers for their intended use, and many other factors.

An understanding of the construction features of boilers helps the boiler operator monitor the integrity of the equipment and spot potentially unsafe conditions as they develop. Boiler operators should be able to interact knowledgeably with boiler inspectors, plant engineers, maintenance and repair personnel, and department supervision in order to maintain a safe, efficient operation. Quality operating personnel are also able to perform the mathematical calculations and conversions that are required to surmount the occasional challenges faced in planning and problem solving in a boiler plant.

Regulatory Organizations

1 What is a badge plate?

A *badge plate (nameplate)* is a data plate attached to a boiler. The information on the badge plate usually includes the MAWP, heating surface area, manufacturer, date built, ASME symbol stamp, and National Board of Boiler and Pressure Vessel Inspectors' registration number. **See Figure 2-1.** The badge plate is commonly attached to the front of the boiler. On large field-assembled boilers, it may be attached to a steel column or other structure.

2 What is the function of the ASME?

ASME International is the new name for the American Society of Mechanical Engineers. In common usage, the name is often shortened to ASME. *ASME* is an educational association dedicated to the advancement of the art and science of mechanical engineering and related sciences. Among other objectives, the various ASME committees establish codes and standard practices and distribute them in periodicals, technical papers, and other publications. ASME has been responsible for establishing standards for boilers and pressure vessels since 1911. The ASME Boiler and Pressure Vessel Code Committee and its many subcommittees consist of designers, manufacturers, inspection authorities, metallurgical and testing specialists, and others. These groups convene regularly and make appropriate updates to the Code so that the Code continually reflects sound engineering practices.

The governmental agencies that are concerned with boiler safety vary substantially from one community to another. For example, some municipalities have stringent boiler inspection standards while the state in which the municipality is located does not. Not all states or other jurisdictions recognize the ASME Boiler and Pressure Vessel Code. Some recognize certain portions of the ASME Code and not other portions.

Factoid

Regulatory organizations ensure safe operation of boilers by requiring tested boiler designs and by ensuring that only qualified persons operate boilers.

3 What is the function of the National Board of Boiler and Pressure Vessel Inspectors?

The National Board of Boiler and Pressure Vessel Inspectors promotes uniform recognition and enforcement of the ASME Boiler and Pressure Vessel Code. They administer uniform qualifications testing for inspectors who assure that the construction, maintenance, and repair of boilers and other pressure vessels are consistent with ASME Code requirements. The National Board also administers the application of the various ASME approval stamps on boilers, pressure vessels, safety valves, and other parts that comply with the ASME Code. In addition, new boilers are registered with the National Board and receive a registration number. This ensures that a permanent record of the boiler specifications is kept on file and available for future reference by inspectors, repair personnel, owner-users, and jurisdictional officials. This organization is often referred to as the National Board and is abbreviated NB on boiler and pressure vessel data plates.

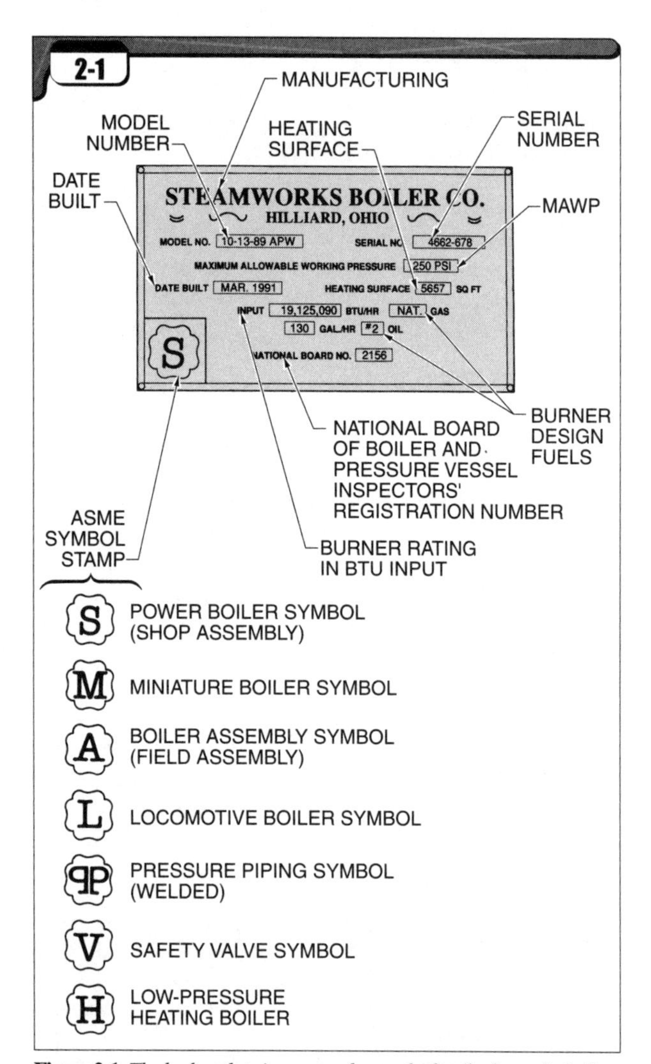

Figure 2-1. *The badge plate is commonly attached to the front of a boiler.*

Trade Tip

Boiler operators should develop a good working relationship with insurance companies and jurisdictional inspectors. These inspectors are usually an excellent source of knowledge regarding steam system safety, efficiency, and troubleshooting tips.

How does the insurance industry advance the safety of boilers and pressure vessels?

Individual insurance companies provide boiler and machinery insurance coverage to companies that use boilers and other machinery that involve inherent risk. Because of the potential for catastrophic loss, insurance companies employ loss prevention inspectors to inspect the insured equipment on a scheduled basis. Usually, these inspectors have received qualification through training offered by the National Board or other training providers.

Another layer of insurance coverage is provided by reinsurance organizations. *Reinsurance* is insurance purchased by insurance companies for large-loss contingencies. In consideration of additional insurance protections, such organizations often mandate certain additional loss-prevention strategies. For example, they may require fire sprinkler systems in the boiler room or a specific configuration of fuel control valves in a piping system providing natural gas to a boiler.

Metallurgical Issues

How is the quality assured for steel components used for boilers?

ASME publishes standards for materials of construction for boiler tubes, steel plate, and other components to be used for boilers and pressure vessels. ASME also provides standard material certification forms that the manufacturer completes and provides with such components, showing the type and grade of material used. The companies that use these materials in fabrication or repair of pressure vessels provide the material certification forms to the inspection agency (insurance company or governmental jurisdiction) as documentation that appropriate quality materials were used in completing the work. Upon completion of the new vessel fabrication or the field repair, ASME standard data sheets or repair/alteration forms are submitted to the National Board and/or the governmental agency having jurisdiction to provide a permanent record of the work.

What is ductility?

Ductility is the plasticity exhibited by a material under tension loading. It is measured by the amount the material can be permanently elongated, or stretched, before breaking. **See Figure 2-2.** For example, boiler steel should be able to adjust itself slightly to compensate for local stress concentrations.

What is malleability?

Malleability is the ability of a material to deform permanently under compression without rupture. For example, the steel used to make boiler tubes must be able to withstand the beading or flaring process on tube ends without cracking. Tubes in watertube boilers must be able to be bent to the proper curvature.

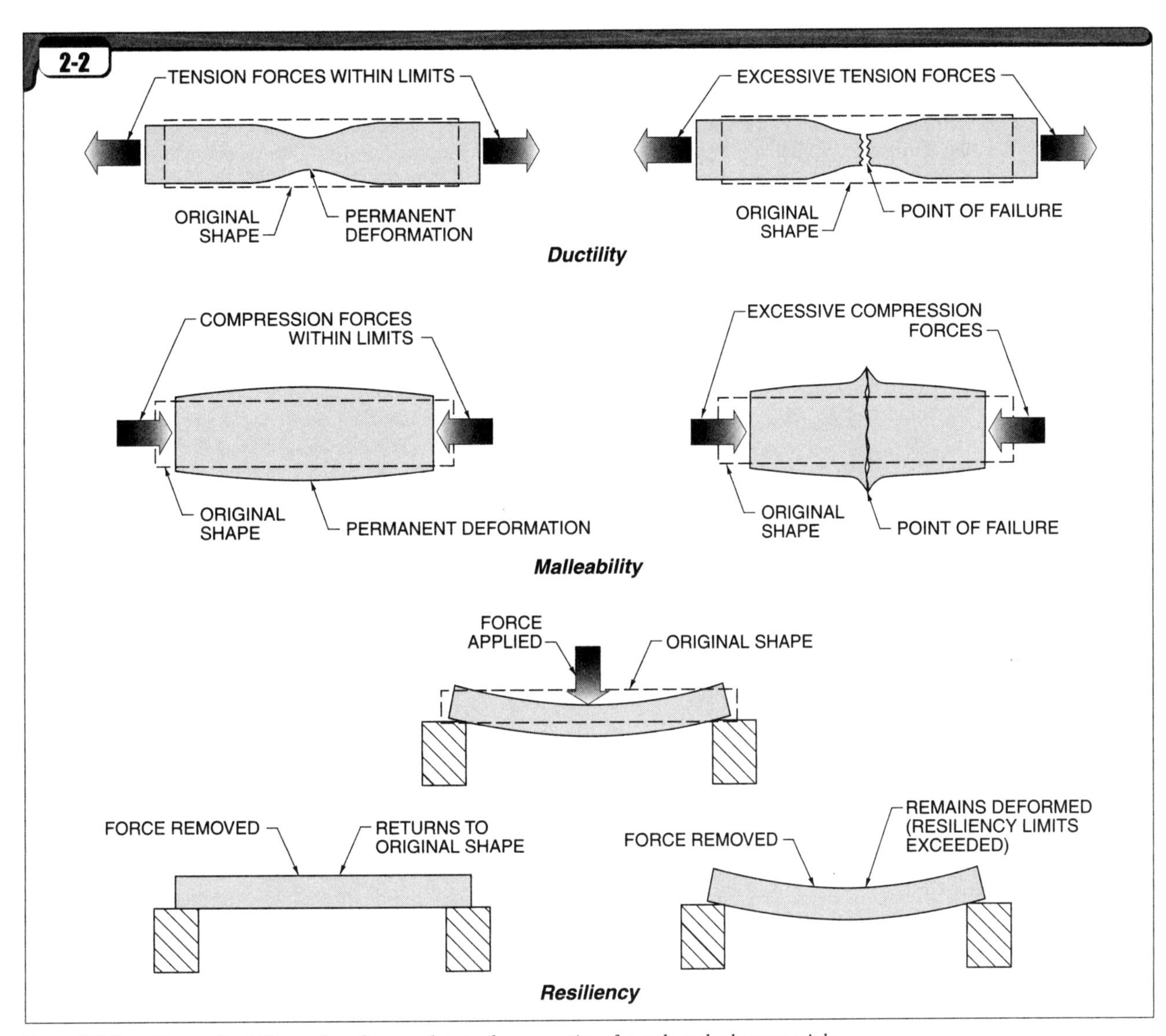

Figure 2-2. *Ductility, malleability, and resiliency relate to the properties of metals and other materials.*

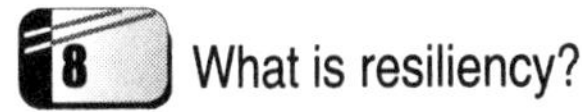

8 What is resiliency?

Resiliency is the ability of a material to return to its original shape after being deformed. For example, bolts, studs, and staybolts must be able to return to their original dimensions when stresses on them are removed.

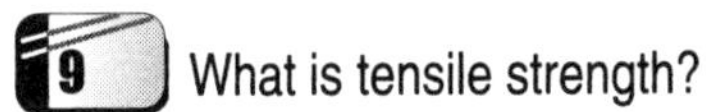

9 What is tensile strength?

Tensile strength is the amount of force required to pull an object apart. It is the limit of force that the object can withstand. Tensile strength refers to the force in a testing machine used to stretch a uniform sample to the point of failure. Tensile strength is found by applying the formula $S = F \div A$ as follows:

where

S = tensile strength (in psi)

F = force (in lb)

A = area (in sq in.)

For example, what is the tensile strength if a hydraulic tester is used to apply a force of 55,000 lb to fail a test sample with an original cross-sectional area of 0.75 sq in.?

$S = F \div A$

$S = 55{,}000 \div 0.75$

$S = \mathbf{73{,}333\ psi}$

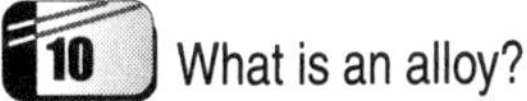

10 What is an alloy?

An alloy is a blend of two or more different metals. The metals are fused together to produce alloys with certain characteristics. For example, boiler tubes in utility boilers must be capable of withstanding very high temperatures. Stainless steels are alloys engineered for corrosion resistance. Cupronickel is an alloy of copper and nickel that benefits from much of the corrosion resistance of nickel and the heat transfer qualities of copper. Alloys are also engineered for other properties such as high strength and abrasion resistance.

11 What is compression?

Compression is the exertion of equal forces from opposite sides of an object that push toward the middle. **See Figure 2-3.** For example, most fire tubes are under compression when they try to expand faster than the boiler shell does because the shell limits their expansion.

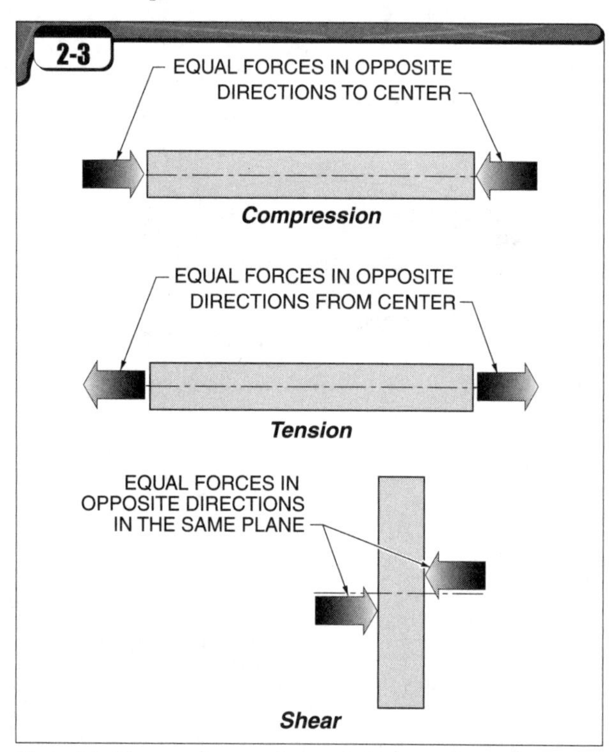

Figure 2-3. *Compression, tension, and shear are forces that act on boiler metal.*

12 What is tension?

Tension is the exertion of equal forces pulling in opposite directions that can stretch an object. For example, a boiler drum or shell is under tension because the pressure inside exerts a force as if to stretch the steel.

13 What is shear?

Shear is the exertion of equal forces in opposite directions in the same plane. It is a scissor-like force that can cut objects. For example, the rivets joining two plates may be cut by a shearing force.

14 Why is steel preferred over cast iron for boiler fittings?

Steel is preferred over cast iron for boiler fittings because it can withstand thermal stresses from expansion and contraction better than cast iron. Additionally, cast iron fittings are more brittle than steel, and may have unseen flaws such as air bubbles or impurities below the surface of the fitting. Cast iron is generally limited to applications of 250 psi and less.

15 Does temperature or pressure create the greater stress on a boiler shell?

Temperature creates the greater stress on a boiler shell because of the uneven expansion and contraction caused by heat. The pressure is evenly distributed and does not cause uneven stresses. The pressure is the same throughout a boiler shell, except for the slight additional pressure at lower points in the boiler caused by the force exerted by the weight of the water.

16 How does the diameter of a boiler drum or shell affect its safe working pressure?

The pressure that the boiler drum can withstand is inversely proportional to the diameter of the boiler drum. The larger the diameter of the boiler drum, the lower the pressure it can withstand. This relationship assumes that all other factors, such as thickness of the metal, are the same. For example, if the diameter of a boiler drum were doubled, the pressure would be halved.

17 What are the two main types of joints (seams) in relation to their orientation on a boiler shell?

A longitudinal joint runs along the length of a boiler drum or shell. **See Figure 2-4.** A circumferential (girth) joint runs around the circumference of a boiler drum or shell. A circumferential joint is also known as a girth joint.

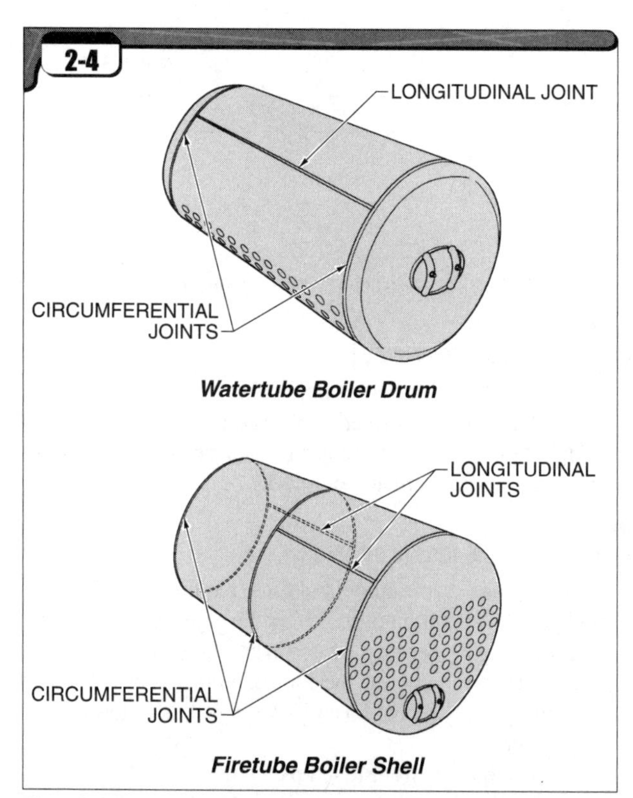

Figure 2-4. *Longitudinal joints (seams) run along the length of a boiler shell. Circumferential joints (seams) run around a boiler shell.*

What is the efficiency of the longitudinal joint?

Efficiency of the longitudinal joint is the comparison of the strength of the longitudinal joint to the strength of the rest of the plate used to make the shell of a boiler. It is expressed as a percentage. For example, if the longitudinal joint is 95% as strong as the rest of the shell, the efficiency of the longitudinal joint is 95%. *Note:* Whether riveted or welded, a joint may also be called a seam.

Why must the longitudinal joint of a boiler shell be stronger than the circumferential joints?

In a cylindrical vessel, the force pushing the longitudinal joint apart is twice the force that pushes the circumferential joints.

What are nozzles in boiler design?

A *nozzle* is one of the short stubs of piping that are connected to the boiler during construction. These provide attachment points for piping and appliances to be connected to the boiler, including the feedwater piping, safety valves, etc.

Welded Construction

What types of welding are commonly used in boiler construction and repair?

Submerged arc welding (SAW) is a welding process in which an electric arc is submerged or hidden beneath a granular material (flux). **See Figure 2-5.** SAW is the standard method for construction of boiler drums. The electric arc provides the heat necessary to melt and fuse the metal being welded. The flux completely surrounds the electric arc, shielding the arc and the metal from the atmosphere. A metallic wire is fed into the welding zone underneath the flux. The wire is the positive electrode and the vessel is the negative electrode. The wire and the flux must meet stringent ASME standards.

Shielded metal arc welding (SMAW) is a welding process that uses an electric arc to heat the metal in the weld area; metal from the electrode is added to the weld pool. SMAW is the standard method for making boiler repairs. Temperatures for the SMAW process range from 4000°F to 10,000°F.

Gas tungsten arc welding (GTAW) is a welding process in which a virtually nonconsumable tungsten electrode is used to provide the arc for welding. GTAW is also known as tungsten inert gas welding (TIG). GTAW is often used to make the first pass on higher pressure vessels (above 400 psi). This is because this welding process leaves almost no protrusion of weld material into the interior of boiler tubes or other passages. During the welding cycle, a shield of inert gas expels the air from the welding area and prevents oxidation of the electrode, weld pool, and surrounding heat-affected zone. In GTAW, the electrode is used only to create the arc. The electrode is not consumed in the weld. For joints where additional weld metal is needed, a filler rod is fed into the weld pool.

Oxyacetylene welding (OAW) is a welding process that uses acetylene (C_2H_2), which is combined with oxygen (O_2) to produce a flame with a temperature over 6000°F. OAW is used commonly as a fusion welding process. OAW filler metal is usually added as the weld area melts and fuses together. While oxyacetylene welding is an accepted method for use in the construction and repair of boilers, shielded metal arc welding (SMAW) is now the standard method for boiler repair.

Welding on new boilers or pressure vessels is done within ASME guidelines by a welder certified for that particular welding process. For example, the welder may be tested for welding a circumferential weld on an SA 192 tube with a wall thickness of 0.175″ using 3/32″ #7010 electrodes. The welder would have to perform the weld and samples would have to pass a destructive test. By passing the welding test, the welder is certified for this particular type of weld.

Factoid

Welding on pressure vessels is to be performed only by certified personnel.

List three reasons why welding is preferred over riveting for a longitudinal joint.

Welded longitudinal joints are stronger than riveted longitudinal joints. In fact, in some cases welded joints may actually be stronger than the original solid plate. Construction of high-pressure boilers by riveting requires the drum walls to be extremely thick. This creates practical limitations on the pressure that a boiler can withstand, because riveting cannot produce an adequately tight joint.

Riveted joints can contain a "shelf" created by overlapping steel plates where sludge can accumulate.

A leak in a riveted joint can concentrate chemicals and impurities between the layers of metal that compose the joint. These chemicals can deteriorate the interior of the riveted joint in such a way that it is difficult to visually detect.

List three issues with of welded joints.

Code requirements, practices, and procedures are strict and complicated. A boiler repair company must perform many of these repairs to cover the expense of maintaining the approvals. These complications limit the number of companies available to perform repairs on welded joints.

The welder must be highly trained and experienced. An individual who welds on pressure vessels must pass a qualification test for every type of weld to be made, and these qualifications must be continually kept updated. The costs of these tests are ultimately passed on to the companies that hire the welding companies.

Stresses often develop in welded areas of the boiler. The concentration of intense heat by welding creates stresses in the steel that can result in cracking of the area years later. The welds must be properly stress relieved.

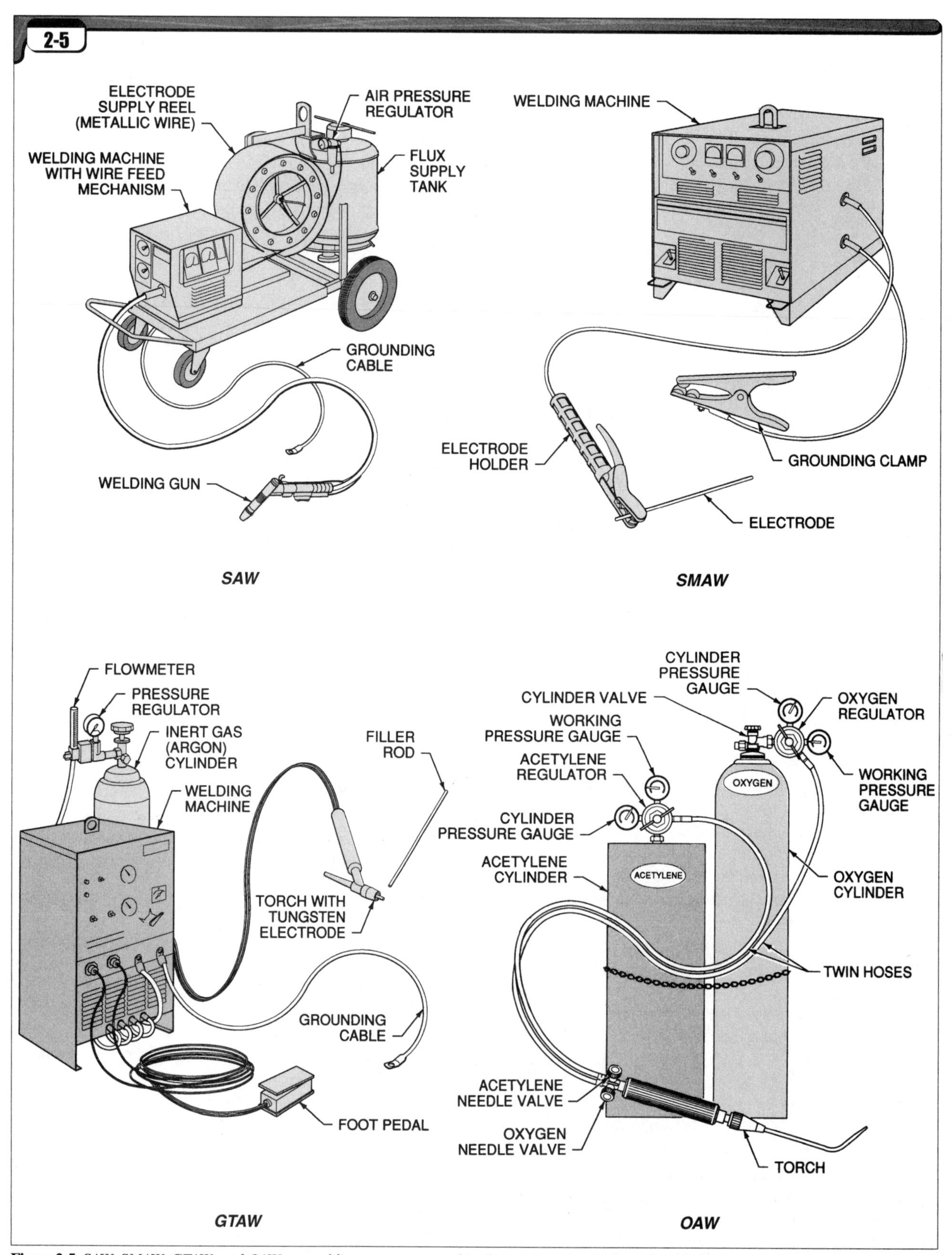

Figure 2-5. *SAW, SMAW, GTAW, and OAW are welding processes used in the construction and repair of boilers.*

24 How is a welded boiler stress relieved?

A boiler is stress relieved after construction by heating the entire vessel to a high temperature in a large stress relief oven. **See Figure 2-6.** This allows the concentrated stresses created by welding to dissipate. For example, a typical firetube boiler shell may be heated to 1100°F or 1200°F for a specified time, depending on the metal thickness and other factors. After heating, the oven is shut off and the boiler is allowed to cool slowly in a still atmosphere. After a field weld, a circumferential band around the boiler shell may be heated to relieve stresses created by the welding process.

Note that in some welding procedures, preheating of the steel to specific temperature conditions may also be required in order to ensure that the welding procedure produces satisfactory results.

Johnston Boiler Co.

Figure 2-6. *Boiler drums and shells are heated in stress relief ovens to dissipate stresses created by welding.*

25 How are welded joints on boilers and pressure vessels inspected for quality?

Depending on the ASME International standard procedure for the type of weld involved, specific welds may undergo a visual inspection or nondestructive testing. Visual inspection is performed by an authorized inspector who visually inspects the welded joints for defects such as surface porosity or slag inclusions. Surface porosity is entrapped air bubbles in the weld. Slag inclusions are welding slag that has become trapped in the weld metal.

Nondestructive testing (NDT) may be mandatory for certain types of welds. Nondestructive testing of the welding work requires testing equipment. The most common form of NDT in pressure vessel work is to take an X ray of the welds. **See Figure 2-7.** An X ray is required for longitudinal and circumferential joints in watertube boiler drums and firetube boiler shells. X rays are made by a qualified technician and are considerably more effective in exposing weld deficiencies than visual inspection. X rays reveal defects that may be internal to the weld or on the back side of the weld where it is impossible to visually inspect. In addition to porosity and slag inclusions, these defects may include incomplete weld penetration, interior cracking, concavity of the underside of the weld, and other problems.

Johnston Boiler Co.

Figure 2-7. *Welds of longitudinal or circumferential joints must be X-rayed to detect any imperfections.*

The X-ray technician tapes small metal markers on the back side of the weld to be tested. These markers show up in the final X-ray image and thus prove that the X ray penetrates the metal completely. Small lead numbers are also taped to the X-ray film. These provide a true-scale location reference for any defects discovered. For example, if a slag inclusion is detected that is 3¼″ from reference number 6 on the film, the technician can mark the exact location of the defect on the boiler. A welding technician can then grind out the defect and repair it.

Riveted Construction

26 What is a lap joint?

A *lap joint* is a riveted joint with two overlapping plates that are drilled through and riveted together at the edges. **See Figure 2-8.**

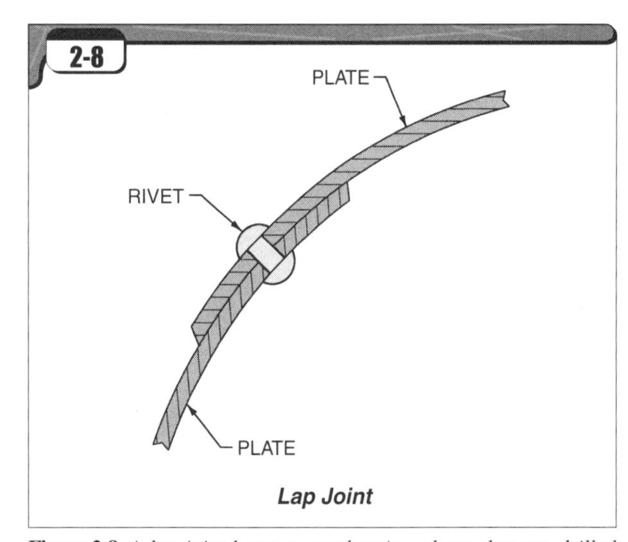

Figure 2-8. *A lap joint has two overlapping plates that are drilled through and riveted together at the edges.*

27 What is a double strap and butt joint?

A *double strap and butt joint* is a riveted joint made by rolling the steel plate into a cylinder to make the shell of the boiler (or drum) and butting the two edges of the plate together to form the seam of the cylinder. **See Figure 2-9.** Then, a strap of steel is placed lengthwise over the joint on the inside, and another strap is placed over the joint on the outside. Holes are drilled for rivets. Rivets are placed through the two straps and the shell, forming a reinforced seam.

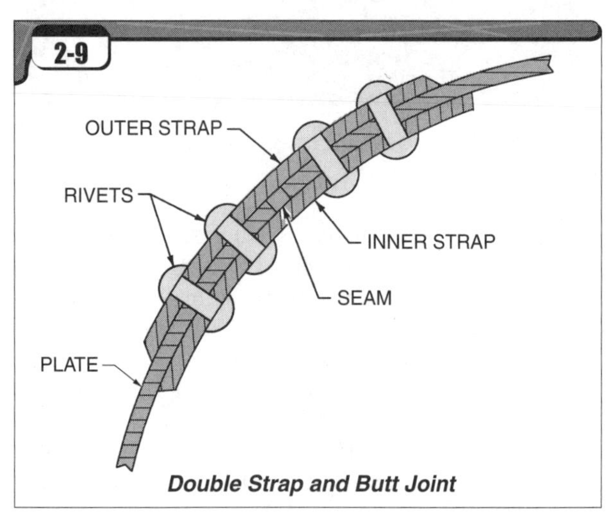

Figure 2-9. *A double strap and butt joint has inner and outer straps.*

28 What is rivet pitch?

Rivet pitch is the distance from the center of one rivet to the center of the next rivet in the same row. **See Figure 2-10.**

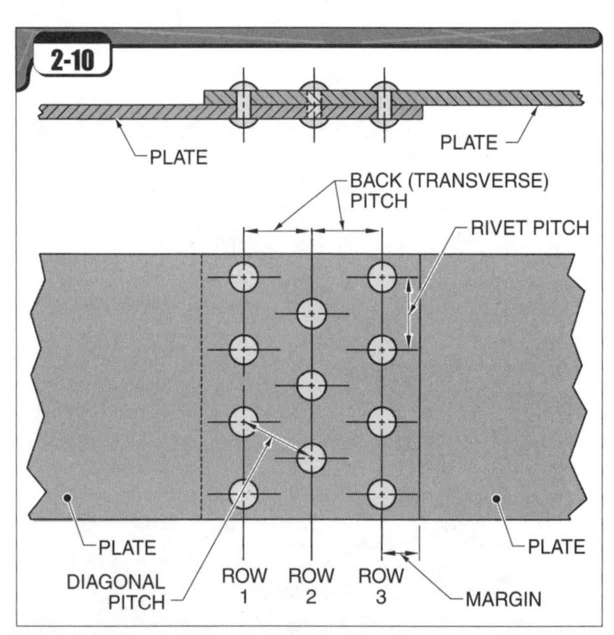

Figure 2-10. *Rivet pitch is the distance from the center of one rivet to the center of the next rivet in the same row.*

29 What is a firecrack?

A *firecrack* is a crack in a riveted joint that runs from the rivet to the edge of the plate or from rivet to rivet. In riveted boilers, some rivet-to-edge firecracks are acceptable. Rivet-to-rivet firecracks are unacceptable.

Boiler Configuration

30 How are tube ends secured in a firetube boiler?

Tube ends in a firetube boiler are usually rolled into place with a mandrel and beaded over with a beading tool. **See Figure 2-11.** Alternatively, the tube ends may be welded. A mandrel is a device that consists of a tapered steel pin that passes through a steel cage containing rollers. The rollers are not perfectly parallel with the tapered pin, but are cocked a few degrees from the axis of the pin. Thus, when the pin is turned, the rollers roll "uphill" on the pin and the taper of the pin forces the rollers outward. This expands the end of the fire tube and presses it tightly against the inside surface of the hole in the tube sheet.

When the tube is rolled into place, approximately ¼″ of the tube end is left protruding from the tube sheet. A beading tool is then used to roll this protruding metal over toward the outside of the tube, thus creating a rounded lip or band that prevents the tube from pulling out of the hole. Beading the end of the tube also causes the beaded lip to be pressed into contact with the tube sheet. This ensures that the beaded lip is cooled by the water on the opposite side of the tube sheet.

31 How are tube ends secured in a watertube boiler?

Water tubes are typically rolled into place in the holes of the drums with a mandrel. In contrast to a firetube boiler, the ends of the tubes in a watertube boiler are flared, rather than beaded, to keep them from pulling out. A flared end of a tube is spread outward.

32 Why are the tube ends normally beaded in a firetube boiler and flared in a watertube boiler?

The protruding end of a fire tube is susceptible to damage from hot gases of combustion passing across it. Beading the end of the fire tube so it contacts the tube sheet allows the water on the other side of the tube sheet to carry away the majority of the heat from the tube end. Also, the protruding ends are beaded to provide additional resistance against being pulled loose from the tube sheet by internal pressure.

The protruding end of a water tube, located inside the drum, is covered with water and not susceptible to damage from the hot gases. The protruding ends are flared to provide additional resistance against being pulled loose from the drum wall by internal pressure.

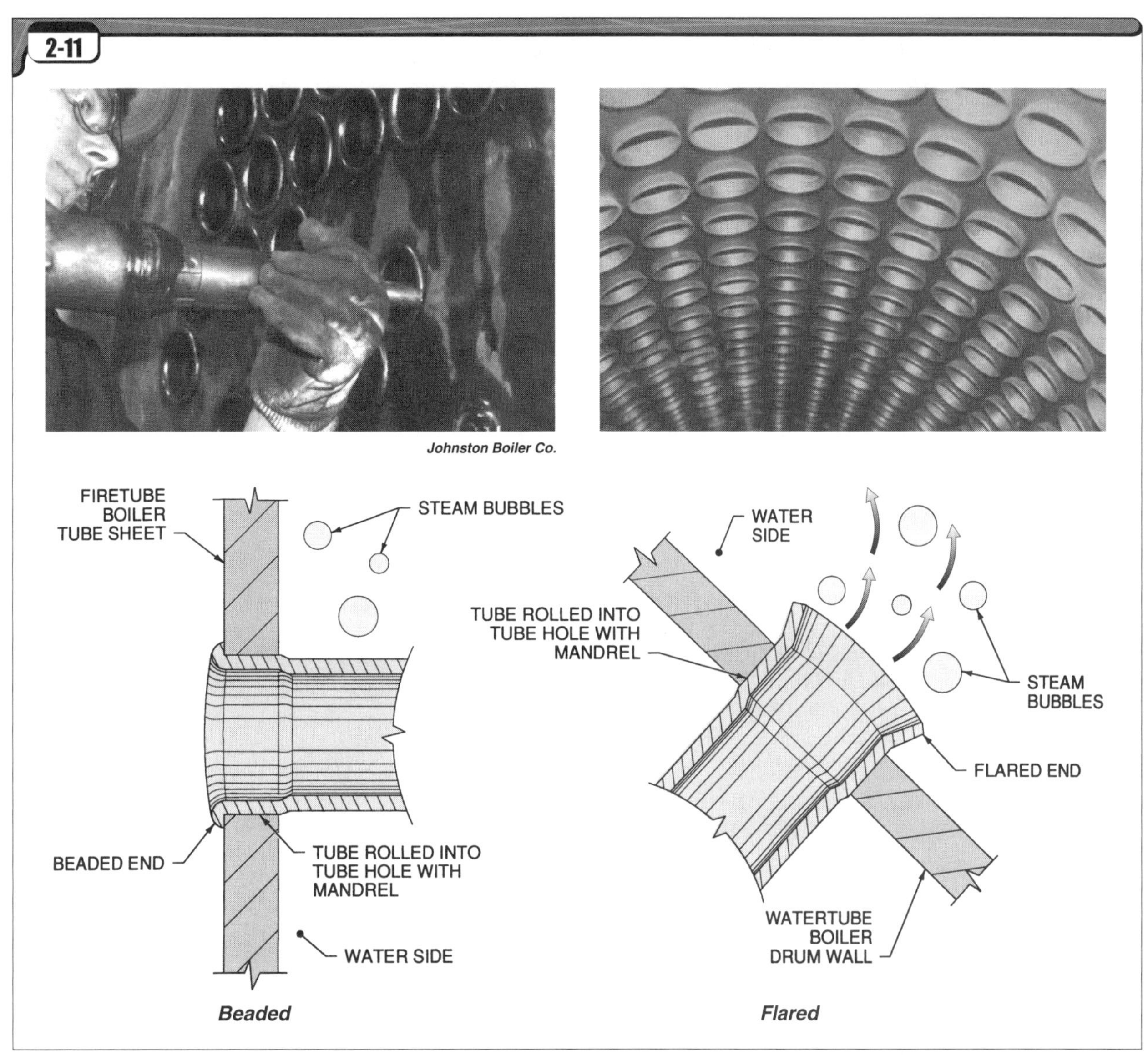

Figure 2-11. *Tube ends are beaded in a firetube boiler and flared in a watertube boiler.*

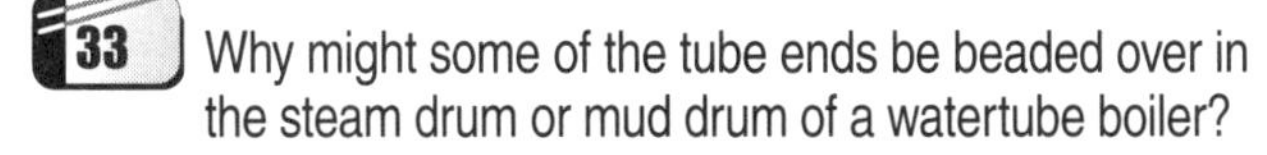

33 Why might some of the tube ends be beaded over in the steam drum or mud drum of a watertube boiler?

When the boiler is filled, flared tube ends could extend down far enough into the mud drum to prevent air from being expelled from the top of the drum. Also, when draining the boiler, flared tube ends in the belly of the steam drum may not allow all the water to drain out. Beaded tube ends, which do not extend as far as flared tube ends, overcome these problems.

34 Why do nearly all tube leaks in a firetube boiler occur around the tube ends?

Tubes expand and contract to a greater degree and more quickly than the boiler shell because they are exposed to hot gases. Because firetube boilers have straight tubes, a push-pull effect is created as the boiler alternately heats and cools and the tubes expand and contract. The alternating forces cause the ends of the tubes to come loose. The area most affected by the push-pull effect is the lower rear of the boiler, which is exposed to the hottest gases.

35 Why might a row of tubes be left out of a firetube boiler?

A row of tubes may be left out of a firetube boiler to allow better circulation of the water in areas of very rapid heat transfer, and to provide for inspection of critical heating surfaces. It is very common to leave an open space above the flue furnace in a scotch marine firetube boiler, for example, for these reasons.

What is the dry sheet of a firetube boiler?

The *dry sheet* is the metal that is an extension of the cylindrical shell of a firetube boiler. **See Figure 2-12.** The dry sheet forms the smoke box area between the tube sheet and the fire door(s) on some firetube boilers. The *smoke box* is the area at the end of a firetube boiler where the flue gases are allowed to reverse direction for a subsequent pass. This method of construction was common at the front of horizontal return tubular boilers and is frequently used with scotch marine boilers. A firetube boiler may have more than one smoke box.

What are diagonal stays and horizontal through stays?

Diagonal stays and *horizontal through stays* are braces that are installed in a firetube boiler to keep the upper portions of the tube sheets above the tubes from bulging outward due to the internal pressure. **See Figure 2-13.** Diagonal stays are installed from the shell to the upper portions of the tube sheet. Horizontal through stays are installed from one tube sheet to the other, and may be installed above or below the tubes.

Diagonal stays are welded in modern boilers, but may be riveted in older units. Horizontal through stays are welded in modern boilers, but were often bolted (threaded all the way through the tube sheet) on older boilers. Both diagonal stays and horizontal through stays are installed only in firetube boilers. They are not used in watertube boilers.

What helps to stay the tube sheets of a firetube boiler other than diagonal stays and horizontal through stays?

The beaded (or welded) ends of the fire tubes also help keep the tube sheets from bulging.

What is the purpose of a hammer test? How is it conducted?

A *hammer test* is a test performed on the stays inside the boiler to check the integrity of each, relative to the others. Each of the stays is struck with a hammer or mallet and produces a metallic tone. A noticeably different tone from the others warns the inspector that the stay could be defective. A stay can be defective due to a crack or a broken weld.

Figure 2-13. *Diagonal stays prevent the upper portion of the tube sheet from bulging.*

What is a staybolt? In what types of boilers are staybolts used?

A *staybolt* is a short bolt brace that passes through the water leg of a boiler. **See Figure 2-14.** Staybolts are used in vertical firetube boilers, firebox boilers, and wetback (waterback) scotch marine boilers. Staybolts support the furnace area to keep it from collapsing. Staybolts are made as solid through stays, solid staybolts, and hollow staybolts. Modern staybolts are welded. Older staybolts were screwed or riveted.

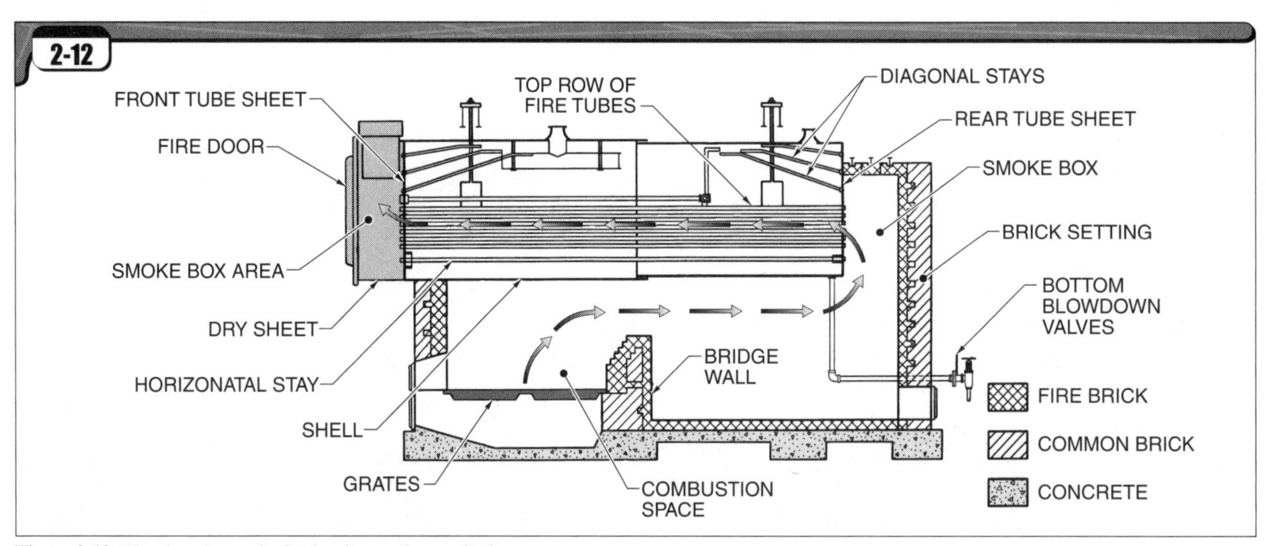

Figure 2-12. *The dry sheet of a boiler forms the smoke box area.*

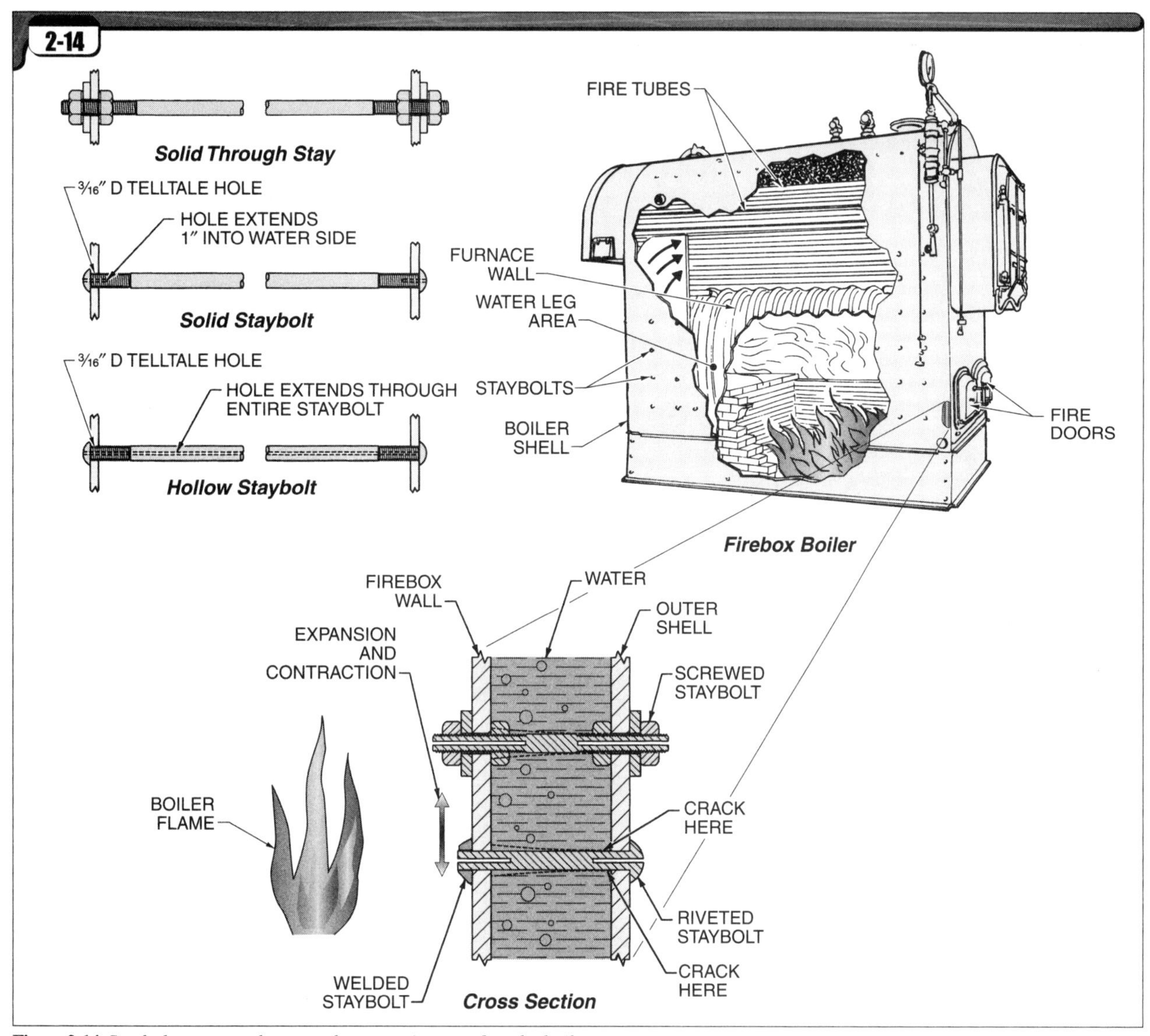

Figure 2-14. *Staybolts support the water leg areas in some firetube boilers.*

What is the purpose of a small hole drilled lengthwise through a staybolt?

The two surfaces that the staybolts pass through (the firebox wall and the outer shell) expand and contract at uneven rates because the firebox wall is exposed directly to the radiant heat of the fire. The expansion and contraction causes the staybolts to be worked up and down on the firebox end as the boiler heats and cools. Over a long period of time, the staybolts may crack and break.

A small hole, usually 3/16″ in diameter, is drilled from the outside end of the staybolt, extending about 1″ inward through the center. The hole is also known as a telltale hole. If the expansion and contraction are enough to cause a crack in the staybolt, water and flash steam from the water leg area comes out of the hole, which warns the operator of a defective staybolt.

Describe two methods of supporting horizontal return tubular (HRT) boilers.

The *lug and roller method* is a support method where steel lugs (support plates projecting horizontally from the boiler shell) are welded or riveted to the front and rear of the boiler shell. **See Figure 2-15.** The front lugs rest on the refractory brick walls. The rear lugs rest on a pad of cylindrical rollers on the refractory brick walls to allow expansion and contraction of the boiler shell.

The *overhead suspension method* is a support method where the boiler is suspended from an overhead steel beam structure by sling members. These sling members are attached to brackets that are welded or riveted to the boiler shell. This method has the advantage of removing the weight of the boiler from the brickwork.

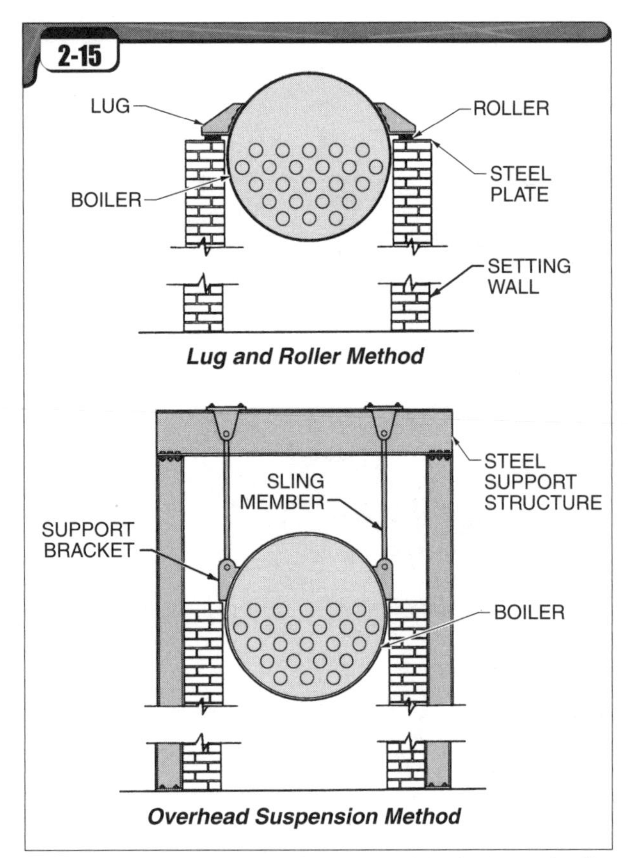

Figure 2-15. *HRT boilers may be supported by the lug and roller method or the overhead suspension method.*

43 Is an HRT boiler installed level or inclined?

An HRT boiler is installed inclined, or sloped, so that the rear of the boiler is 1″ lower than the front for every 10′-0″ of boiler length. This allows sludge and other impurities to settle to the rear, where they can be drained off.

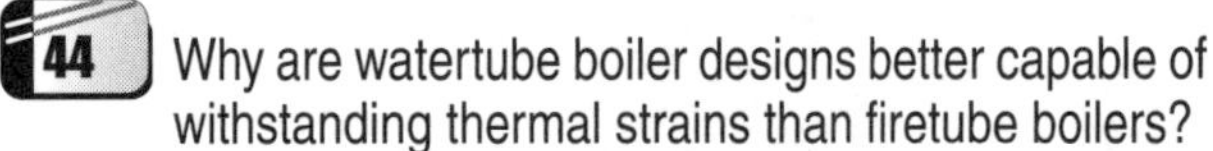

44 Why are watertube boiler designs better capable of withstanding thermal strains than firetube boilers?

Most watertube boilers have tubes with curvature. These curved tubes can absorb expansion and contraction better than straight tubes. Additionally, the heads of the drums of a watertube boiler are curved outward (convex), since the internal pressure would push them into that shape. Stays are not needed in a watertube boiler.

Factoid

Because the tubes in a firetube boiler are straight and rigidly attached at both ends, the strains that result from expansion and contraction of the tubes can be substantial. The bends and curves in water tubes provide relief for these strains.

Factoid

The great majority of modern watertube boilers contain waterwalls.

45 What are riser tubes and downcomer tubes?

Riser tubes are tubes exposed to the highest temperatures in the furnace area and contain rising water. *Downcomer tubes* are tubes that contain the cooler, descending water. Downcomer tubes continually replenish the water supply to the riser tubes. In many watertube boiler designs, downcomer tubes are located outside the boiler's steel casing so as to place them in the coolest environment possible.

These terms apply to watertube boilers. As the water in the boiler is heated, its density decreases and the hotter water rises toward the top of the boiler. Cooler water has a higher density and is heavier than hot water. The cooler water descends to the lowest points in the boiler. Continuous heating of water in the boiler creates a natural circulation.

46 What are waterwalls?

Waterwalls are many tubes placed side by side to create a large, flat surface against the furnace walls in a watertube boiler. **See Figure 2-16.** They increase the heating surface and steam-generating capacity of the boiler. In fact, for each square foot of waterwall heating surface, an average of about 12 lb/hr to 16 lb/hr of steam is generated. By absorbing heat in the furnace, waterwalls also help prevent refractory damage to the furnace walls because of overheating. Circulating boiler water enters the waterwall tubes through a bottom header and the heated water that leaves the top of the waterwall flows into the steam drum.

Riley Stoker Co.

Figure 2-16. *Waterwalls increase the heating surface and steam-generating capacity of a boiler.*

47 When is it advisable to use waterwalls?

It is advisable to use waterwalls if a great heat release rate in the furnace would, without the waterwalls, damage the refractory brick, or if a large steam capacity is needed from a boiler with minimum space.

What are the functions of studded waterwalls?

The steel studs hold the refractory on the waterwall tubes for thermal protection and erosion protection. **See Figure 2-17.** The area immediately encircling a large burner (the burner throat) is an area of extreme heat. The waterwall tubes in these throat areas are often protected from the excessive heat by a coating of refractory brick applied over the studs. In cases where the fuel may also be abrasive, such as pulverized coal, the refractory coating also protects the tubes from abrasion. The refractory is applied as a castable (trowelled) or sprayed-on coating.

The steel studs on the waterwall tubes help to dissipate heat from the castable or sprayed-on refractory cement coating and conduct this heat to the water inside the tubes.

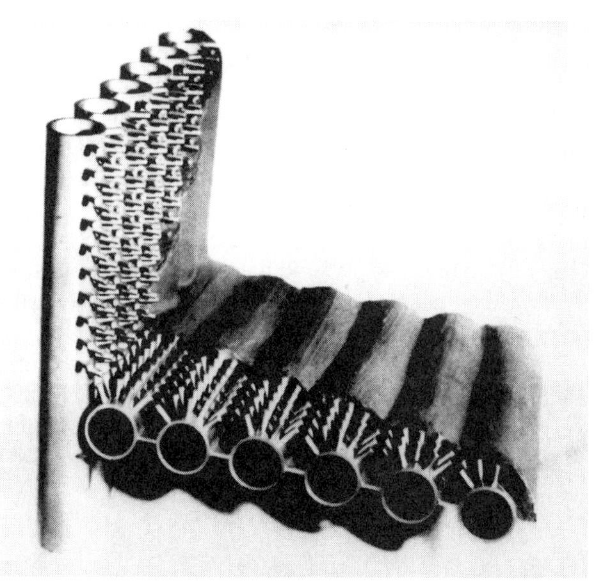

Babcock & Wilcox Co.

Figure 2-17. *Studded waterwalls hold refractory that protects the tubes from extreme heat and abrasion.*

What are membrane waterwalls?

Membrane waterwalls are waterwalls that are formed into a solid, airtight panel by welding a strip of steel between each of the waterwall tubes. **See Figure 2-18.** Most watertube boilers operate with a pressure in the furnace area that is very slightly less than the atmospheric pressure outside the furnace. This ensures that any leakage is air leaking into the furnace and not hot, noxious flue gases leaking out into the operating areas. However, it is more difficult to eliminate small leaks of air into the furnace of a larger boiler. Uncontrolled air leaks into the furnace area diminish the degree of control that the operators have over the combustion conditions. Excessive air leakage into the furnace cools the furnace area and causes heat to be lost up the chimney.

Air leaking into a furnace may also cause localized corrosion as the cooler air causes condensation of the moisture in the flue gases. The use of membrane waterwalls has the advantage of providing airtight furnace walls so that most air leakage is prevented. In addition, the use of membrane waterwalls greatly reduces the amount of refractory brick needed. Modern package watertube boilers usually use membrane waterwalls, as do some larger units.

Figure 2-18. *Membrane waterwalls minimize leakage of flue gases and air through the furnace walls and reduce the amount of refractory needed.*

What is a buckstay?

A *buckstay* is a metal brace used to attach a wall to a steel framework that supports the wall. **See Figure 2-19.** Buckstays were used on early HRT boilers to attach the refractory brick walls to the steel I beam framework around the boiler to keep the walls from bowing or sagging. Modern buckstays are steel beams that support large waterwall areas to keep them from bowing.

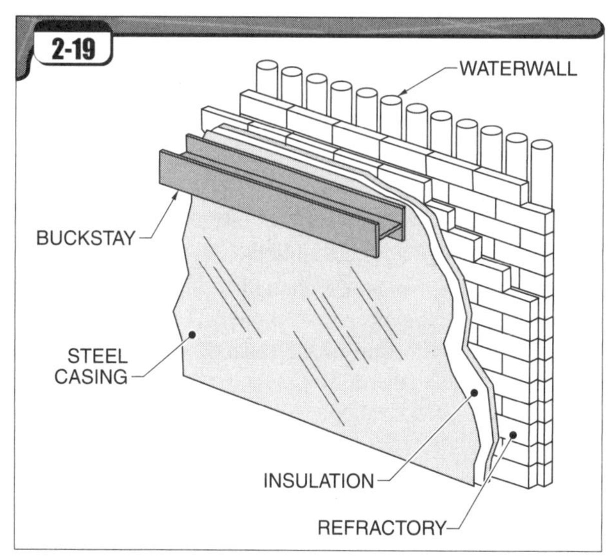

Figure 2-19. *A buckstay is a metal brace used to support a brick or block wall or waterwall.*

How are shells of firetube boilers and drums of watertube boilers formed?

The steel plate that is to become the firetube boiler shell or watertube boiler drum is cut to the proper size. Then a three-roll rolling machine is used to bend the steel plate to the proper curvature. **See Figure 2-20.** The spacing between the rolls and the angle at which the plate is guided into the rolls are calculated to produce a cylinder of the proper diameter.

Drum fabrication for high-pressure utility boilers may be performed by hydraulically pressing flat steel plate into half-cylinders. This creates the need to have two longitudinal joints, but in cases where the steel is very thick, rolling would be unfeasible.

Several rolled or pressed sections may need to be placed end to end in order to fabricate a drum of the required length.

The flat ends of a cylindrical firetube boiler form the tube sheets. Holes are drilled in the tube sheets to receive the tubes and, in most cases, the furnace flue. The dished (convex) heads that form the ends of watertube boiler drums are formed by pressing steel sheet in a die of the desired shape. Sometimes the plate is spun as the head is pressed in the die. The tube sheets or dished heads are welded to the cylindrical vessel to form a firetube boiler shell or watertube boiler drum.

Johnston Boiler Co.

Figure 2-20. *Steel plate is rolled to the proper curvature to form a boiler drum or shell.*

How are the drums of a watertube boiler prepared for having the tubes attached?

Engineering drawings are created to show the planned configuration of the tubes in the boiler. Then the tube holes are drilled into the wall of the finished drum. In modern boiler manufacture, this drilling process is often performed by computer-controlled equipment.

Since the drilling of the tube holes weakens the drum to some degree, the drum wall where the tubes penetrate is used in calculating the pressure-containing strength of the drum. A *ligament* is the portion of the drum wall between the tube holes. **See Figure 2-21.** Sometimes the ligament area is built up in thickness to add strength.

Riley Stoker Co.

Figure 2-21. *Watertube boiler drums are drilled to receive the tubes.*

What is the "fit-up" stage of assembly of a watertube boiler?

Fit-up is the process of fitting and rolling the tubes into the drums of a new watertube boiler. **See Figure 2-22.** Once the drums have been permanently placed on the frame of the boiler, the fit-up process begins. Each tube is placed into position so that it is symmetrical with the other tubes. An assembly technician rolls the tube into place. Refractory firebrick, insulation, steel cladding, and burner installation follow, as well as installation of valves and controls.

Nebraska Boiler Co.

Figure 2-22. *In the fit-up process, the tubes are fitted and rolled into place in the drums of a watertube boiler.*

What method is used to facilitate field assembly of the large number of tubes in a large utility boiler?

In order to expedite the field assembly of a large utility boiler and minimize the number of penetrations through the drum walls, large utility boilers are often assembled using a mini-header configuration. Groups of riser tubes are routed to small headers, which are then routed to the drum. These tube-and-header sections are made in the manufacturer's shops and delivered to the site intact. **See Figure 2-23.** This facilitates assembly of the boiler in a modular fashion, reducing labor costs and erection time.

Riley Stoker Co.

Figure 2-23. *Large utility boilers are prefabricated in sections and assembled in a modular fashion.*

55 Why are manholes elliptical in shape?

Manholes are elliptical in shape because the manhole cover (hatch) is placed against the manhole opening from the inside, so that the internal pressure holds it tightly in place. Therefore, the hatch must be slightly larger than the hole. If the hole were circular, the hatch could not be removed from the boiler.

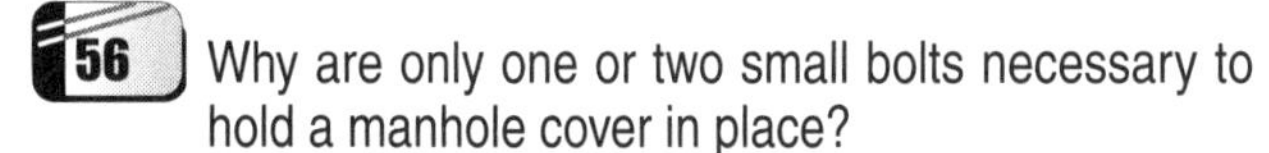

56 Why are only one or two small bolts necessary to hold a manhole cover in place?

Only one or two small bolts are necessary to hold a manhole cover in place because the pressure, after it is brought up on the boiler, holds the manhole cover in place, not the bolts.

57 What is a handhole?

A *handhole* is a small access hole used for looking and reaching into the boiler shell during inspections. A handhole is very similar to a manhole, but measures only about 4″ × 5″.

58 How is a boiler tested for leaks after construction is complete or after repairs?

The boiler is tested for leaks by application of a hydrostatic test. A *hydrostatic test* is a test in which the boiler is filled with approximately 70°F water and then pressurized to 1½ times its MAWP. Any leaks are exposed by observing water drips. Because water is not compressible, the use of water to apply a pressure test rather than a compressed gas reduces the danger of explosion.

59 What limits the amount of steam a given boiler will produce?

The ability of the boiler to absorb the heat released in the furnace limits the amount of steam that a boiler can produce. In the same way that a 1 HP electric motor cannot be expected to do the work of a 2 HP motor, 10,000 lb of steam per hour cannot be produced in a boiler designed for 5000 lb per hour without damage from overheating. Other limiting factors are the ability of the fuel system and draft equipment to produce the furnace heat, and the ability of the feedwater supply equipment to keep a safe water level in the boiler. Clean firing surfaces are essential because soot on the fire side of the tubes and scale on the water side act as insulation and limit heat transfer to the water.

60 What is meant by the maximum capacity of a boiler?

The *maximum capacity* of a boiler is the maximum rating in pounds of steam that a boiler is designed to produce in 1 hr at a given pressure and temperature. For example, a given watertube boiler may have a maximum design capacity of 40 Klb/hr at 150 psi.

61 Describe the differences between a steam boiler and a hot water boiler.

A steam boiler contains space at the top for steam. In a steam boiler system, steam transports heat. A hot water boiler is completely full of water. It includes an overhead compression tank to allow the water to change in volume slightly with changes in temperature. A hot water boiler requires that water be kept moving through the boiler when the burner is firing because the water could become overheated. In a hot water boiler system, hot water transports heat. **See Figure 2-24.**

Calculations

62 What is a circle?

A circle is a plane (flat) figure generated by drawing a line at equal distances about a centerpoint. **See Figure 2-25.** All circles contain 360°. Concentric circles are a series of two or more circles that have the same centerpoint but are different sizes. Eccentric circles are circles that have different centerpoints.

63 What are the diameter and radius of a circle?

Diameter is the distance from a point on the outside edge to another point on the outside edge of the circle through the centerpoint. The symbol for diameter is D. The abbreviation for inside diameter is ID. The abbreviation for outside diameter is OD.

Radius is one-half the diameter of a circle. It is the distance from the centerpoint of a circle to the circumference. The symbol for radius is r.

For example, what is the radius of a 16′-6″ circle?

$D = 16'\text{-}6''$

$D = 16' + \frac{6}{12}'$

$D = 16' + 0.5'$

$D = \mathbf{16.5'}$

$r = D \div 2$

$r = 16.5 \div 2$

$r = \mathbf{8'\text{-}3''}$

	STEAM BOILERS	HOT WATER BOILERS
Heating Medium	**Gaseous**	**Liquid**
Temperature of heating medium	215°F up to 1050°F	Up to 250°F in low-pressure heating boilers, up to 450°F in high temperature hot water (HTW) boilers
Water level	Has steam space in top and water level that must be monitored	Is completely full of water and has compression tank to allow expansion and contraction
Method of control	Control of combustion equipment is based on steam pressure	Control of combustion equipment is based on water temperature
Distribution system	Compressed gas that moves itself through system based on differential pressure; may require pumping system to return condensate	Water must be pumped through distribution system continually when boiler is on
Heating capacity (per pound)	Higher due to latent heat content	Lower due to only sensible heat content
Physical size vs. capacity	Smaller due to higher heat content (latent heat) per pound	Larger due to lower heat content (sensible heat) per pound

Figure 2-24. *A steam boiler and a hot water boiler have significant differences.*

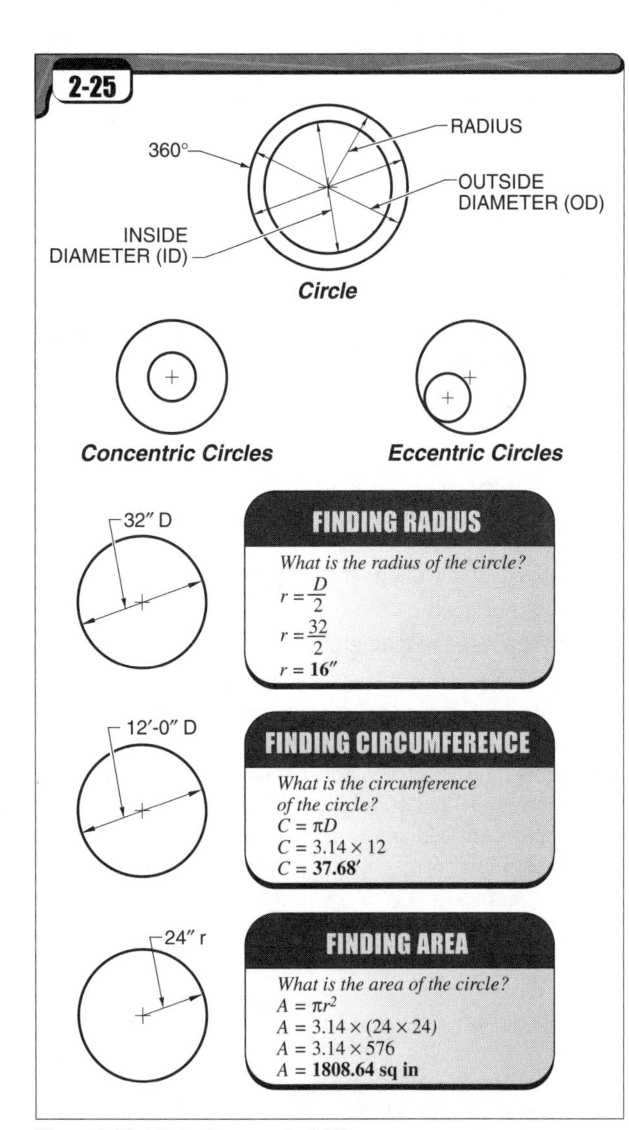

Figure 2-25. *All circles contain 360°.*

64 What is the circumference of a circle?

The *circumference* is the outside boundary of a circle. It is the distance around the circle.

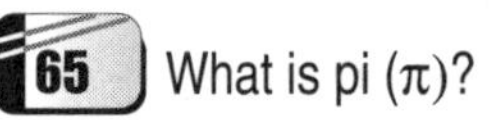

65 What is pi (π)?

Pi (π) is the ratio of the circumference of a circle to its diameter. The circumference of a circle is always equal to the diameter multiplied by pi. The Greek letter π represents this ratio. Pi is approximately equal to 3.14159265. This value is usually rounded off to 3.14 for solving common equations.

66 How is the circumference of a circle found?

The circumference of a circle is found by applying the formula $C = \pi D$

where

C = circumference

π = 3.14

D = diameter

For example, what is the circumference of a 12′-0″ OD tank?

$C = \pi D$

$C = 3.14 \times 12$

$C = \mathbf{37.68'}$

67 Determine the circumference of a boiler drum that is 5.25′ in diameter.

$C = \pi D$

$C = 3.14 \times 5.25$

$C = \mathbf{16.485'}$

68 What is area?

Area is the number of unit squares equal to the surface of an object. It is expressed in square feet, square inches, and other square units of measure. **See Figure 2-26.**

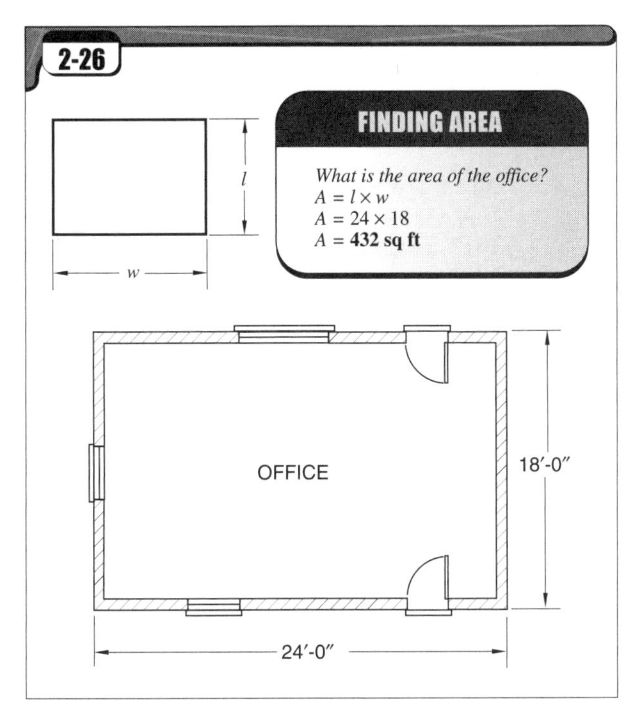

Figure 2-26. *Area is found by multiplying length times width.*

69 How is the area of a rectangular surface found?

The area of a square or rectangular surface is found by applying the formula $A = l \times w$

where

A = area

l = length

w = width

Consistent measurement units must be used. For example, inches × inches = square inches and feet × feet = square feet. If the measuremnt contains mixed feet and inches, the numbers must be converted to consistent units. The feet can be converted to inches or the inches can be converted to decimal feet. For example, what is the area of a room that is 22′-0″ × 16′- 0″?

$A = l \times w$

$A = 22 \times 16$

$A =$ **352 sq ft**

70 What is the area of a warehouse that is 120′-0″ long and 44′-0″ wide?

$A = l \times w$

$A = 120 \times 44$

$A =$ **5280 sq ft**

71 What is the area of a piece of steel that is 16″ long and 8″ wide?

$A = l \times w$

$A = 16 \times 8$

$A =$ **128 sq in.**

Factoid

To reduce the cost of site assembly, large boilers are prefabricated to the greatest degree feasible. This allows them to be assembled at the site in building block fashion.

72 How many square inches are in a square foot?

There are 144 square inches in a square foot. To convert square inches to square feet, divide square inches by 144. To convert square feet to square inches, multiply square feet by 144. For example, how many square feet are in 504 sq in.?

504 sq in. ÷ 144 = **3.5 sq ft**

73 How many square feet are in 5184 sq in.?

5184 sq in. ÷ 144 = **36 sq ft**

74 How many square inches are in 6.5 sq ft?

6.5 sq ft × 144 = **936 sq in.**

75 What is the total area of the storage area?

See Figure 2-27.

$A = l \times w$

$A = 36.5 \times 12$

$A = 438$ sq ft

$A = l \times w$

$A = 16 \times 12.5$

$A = 200$ sq ft

Total area = 438 + 200

Total area = **638 sq ft**

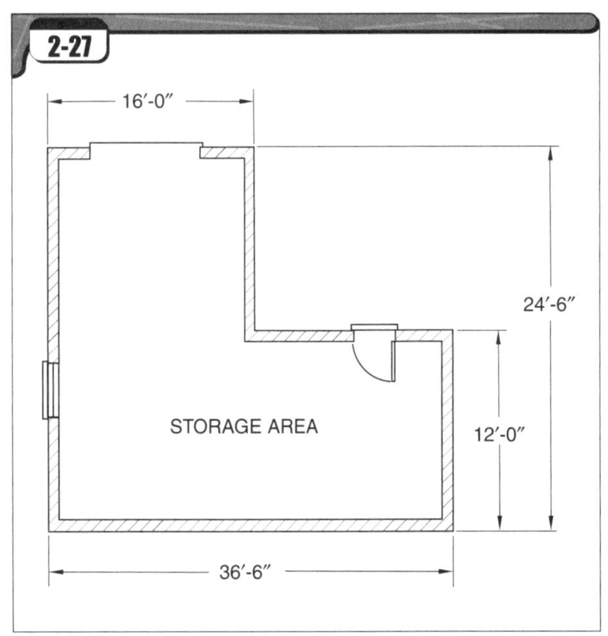

Figure 2-27. *To find the area of an L-shaped object, divide the object into two rectangular shapes.*

76 How is the area of a circle found?

The area of a circle is found by applying the formula $A = \pi r^2$

where

A = area

π = 3.14

r^2 = radius squared

For example, what is the area of a circle with a 24″ radius?

$A = \pi r^2$

$A = 3.14 \times (24 \times 24)$

$A = 3.14 \times 576$

$A =$ **1809 sq in.**

Trade Tip

The term Authorized Inspector, or A.I., is an industry term that refers to an inspector who is qualified to perform boiler inspections. Usually this person has been extensively trained and has received a commission from the National Board to perform pressure vessel inspections. An Authorized Inspector can help a boiler operator maintain a safe and productive boiler operation.

77 Find the area of the bottom of a 39″ OD vertical water tank.

The radius is half the diameter.

$A = \pi r^2$

$A = 3.14 \times (19.5 \times 19.5)$

$A = 3.14 \times 380.25$

$A =$ **1194 sq in.**

78 What is the area of a circle with a 16″ diameter?

The radius is half the diameter.

$A = \pi r^2$

$A = 3.14 \times (8 \times 8)$

$A = 3.14 \times 64$

$A =$ **201 sq in.**

79 How many square feet are in a circle with a 32″ OD?

The radius is half the diameter.

$A = \pi r^2$

$A = 3.14 \times (16 \times 16)$

$A = 3.14 \times 256$

$A = 804$ sq in.

$A = 804 \div 144$

$A =$ **5.58 sq ft**

Trade Tip

Any measurement in inches can be converted to feet by dividing by 12.

80 What is the area of the annular ring in sq ft?

The radius is half the diameter. **See Figure 2-28.**

36″ D Circle

$A = \pi r^2$

$A = 3.14 \times (18 \times 18)$

$A = 3.14 \times 324$

$A = 1017$ sq in.

$A = 1017 \div 144$

$A = 7.07$ sq ft

24″ D Circle

$A = \pi r^2$

$A = 3.14 \times (12 \times 12)$

$A = 3.14 \times 144$

$A = 452$ sq in.

$A = 452.16 \div 144$

$A = 3.14$ sq ft

Annular Ring

Area of X = 7.07 – 3.14

Area of X = **3.93 sq ft**

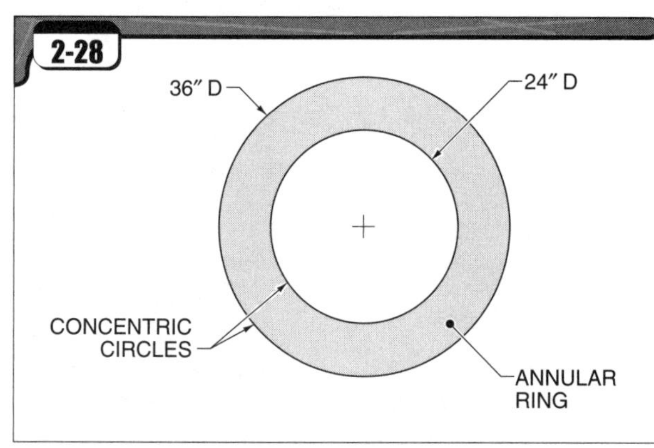

Figure 2-28. *To find the area of the annular ring, subtract the area of the smaller circle from the area of the larger circle.*

81 How is the area of a tubular object found?

The area of a tubular object is found by applying the formula $A = \pi D \times l$

where

A = area

π = 3.14

D = diameter

l = length

For example, what is the surface area in square feet of a boiler tube that is 2″ in diameter (0.167′) and 10′-0″ long?

Convert the inches to feet and apply the formula.

$A = \pi D \times l$

$A = 3.14 \times 2 \div 12 \times 10$

$A = 3.14 \times 0.167 \times 10$

$A =$ **5.24 sq ft**

82 What is the heating surface in square feet of a scotch marine firetube boiler that contains fifty-four 3″ tubes and a 24″ ID flue? The boiler is 12′-6″ long. The heating surface of the tube sheets may be disregarded.

Note: The heating surface of the tube sheets in a firetube boiler is negligible after the many holes for the tube ends are drilled. Also note that all dimensions must be consistent in feet or in inches.

Tubes

$A = \pi D \times l \times \textit{number of tubes}$

$A = 3.14 \times 0.25 \times 12.5 \times 54$

$A = 530$ sq ft (in tubes)

Flue

$A = \pi D \times l$

$A = 3.14 \times 2 \times 12.5$

$A = 79$ sq ft (in flue)

Heating surface = 530 + 79

Heating surface = **609 sq ft**

83 What is the heating surface of a horizontal return tubular firetube boiler that is 14′ long and 6′ in diameter, and contains eighty-six 2″ tubes? Half of the shell is exposed to the furnace. Disregard the heating surface of the tube sheets.

Note: In an HRT firetube boiler, the portion of the shell that is exposed to the heat of the furnace is heating surface.

Tubes

$A = \pi D \times l \times \textit{number of tubes}$

$A = 3.14 \times 0.167 \times 14 \times 86$

$A = 631$ sq ft (in tubes)

Shell

$A = \pi D \times l \div 2$

$A = (3.14 \times 6 \times 14) \div 2$

$A = 132$ sq ft (in shell)

Heating surface = 631 + 132

Heating surface = **763 sq ft**

84 How is the volume of a rectangular solid found?

The volume of a rectangular solid is found by applying the formula $V = l \times w \times h$

where

V = volume

l = length

w = width

h = height

For example, what is the volume of a container that is 6″ long, 4″ deep, and 2″ high? **See Figure 2-29.**

$V = l \times w \times h$

$V = 6 \times 4 \times 2$

$V =$ **48 cu in.**

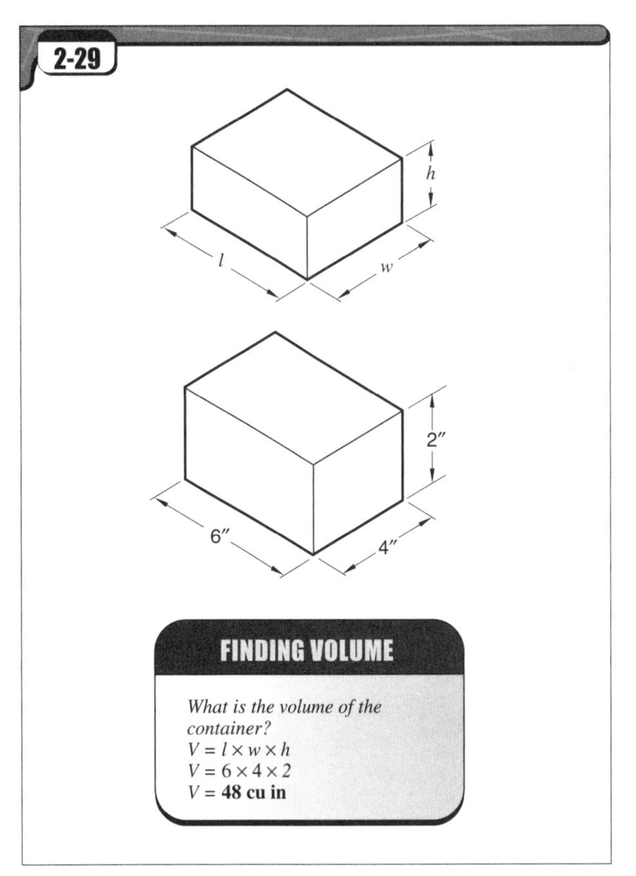

Figure 2-29. *The volume of a rectangular object is found by multiplying length times width times height.*

85 What is the volume, in cubic feet, of a pool that is 12′-0″ long, 8′-0″ wide, and 3′-0″ deep?

$V = l \times w \times h$

$V = 12 \times 8 \times 3$

$V =$ **288 cu ft**

Trade Tip

The answer to a calculation is slightly influenced by rounding of the individual factors in the calculation and by rounding of the intermediate steps. When performing calculations, such as during a boiler operator licensing examination in multiple choice format, rounding may account for a slight difference between the answer obtained by the person taking the examination and the correct answer selection shown on the examination. To illustrate this point, if the diameter of the tubes in question 83 is rounded up to 0.17 ft, the calculated area of the tubes is 642.7 sq ft. If the diameter of the tubes is rounded to 0.167 ft, the calculated area of the tubes is 631.4 sq ft. This value is rounded to 631 in the answer. As a rule of thumb, only round off the final answer to a question and not the intermediate calculations.

86 What is the volume, in cubic inches, of a piece of steel 6.25″ long, 3.0″ deep, and 4.5″ high?

$V = l \times w \times h$

$V = 6.25 \times 3.0 \times 4.5$

$V =$ **84 cu in.**

87 A 200′-0″ trench measures 24″ wide and 12″ deep. What is the volume, in cubic feet, of the trench?

$V = l \times w \times h$

$V = 200 \times 2 \times 1$

$V =$ **400 cu ft**

88 How is the volume of a cylindrical object found?

The volume of a cylindrical object is found by applying the formula $V = \pi r^2 \times l$

where

V = volume

π = 3.14

r^2 = radius squared

l = length

Where the cylindrical object is vertical, height (h) may be substituted for length. For example, what is the volume of a tank that is 6′-0″ in diameter and 11′-0″ long? **See Figure 2-30.**

$V = \pi r^2 \times l$

$V = 3.14 \times (3 \times 3) \times 11$

$V = 3.14 \times 9 \times 11$

$V =$ **311 cu ft**

89 What is the volume of a tank that is 4′-0″ in diameter and 12′-0″ high?

$V = \pi r^2 \times h$

$V = 3.14 \times (2 \times 2) \times 12$

$V = 3.14 \times 4 \times 12$

$V =$ **151 cu ft**

90 What is the volume of a tank that is 18″ in diameter and 48″ long?

$V = \pi r^2 \times l$

$V = 3.14 \times (9 \times 9) \times 48$

$V = 3.14 \times 81 \times 48$

$V =$ **12,208 cu in.**

91 How many cubic inches (cu in.) are in a cubic foot (cu ft)?

There are 1728 cu in. in a cubic foot. This conversion factor is found by multiplying the number of square inches (144) in 1 sq ft times the number of inches (12) in 1 foot (144 × 12 = 1728 cu in.). To convert cubic inches to cubic feet, divide cubic inches by 1728. To convert cubic feet to cubic inches, multiply cubic feet by 1728. For example, how many cubic feet are in 7776 cu in.?

7776 cu in. ÷ 1728 = **4.5 cu ft**

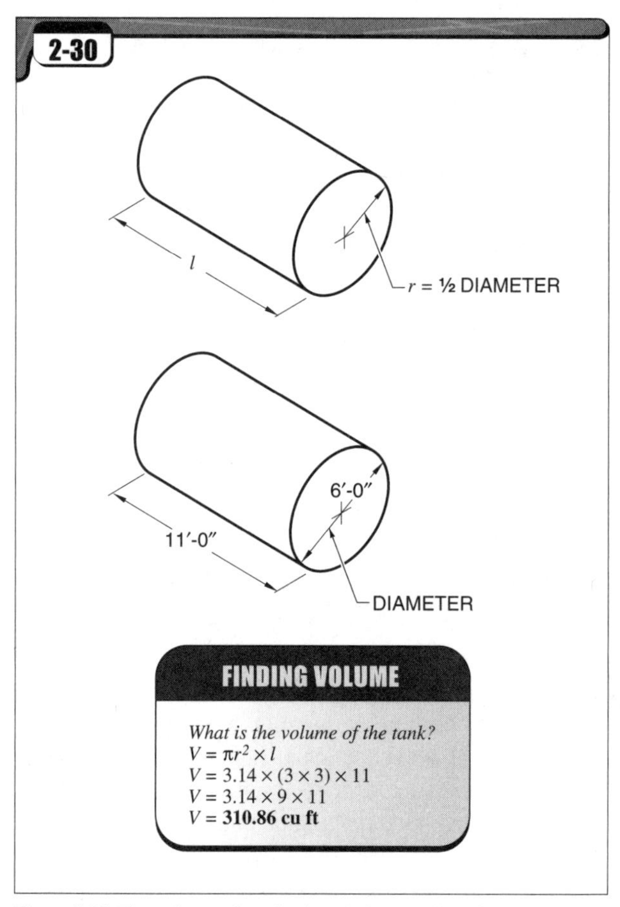

Figure 2-30. *The volume of a cylindrical object is found by multiplying π times radius squared times length.*

92 What is the volume, in cubic feet, of a pit that is 32″ in diameter and 6′-4″ deep?

$V = \pi r^2 \times l$

$V = 3.14 \times (16 \times 16) \times 76$

$V = 3.14 \times 256 \times 76$

$V = 61{,}091.84$ cu in.

$V = 61{,}091.84 \div 1728$

$V =$ **35.4 cu ft**

93 What is the volume, in cubic feet, of a 6″ ID pipe that is 8′-0″ long?

$V = \pi r^2 \times l$

$V = 3.14 \times (0.25 \times 0.25) \times 8$

$V = 3.14 \times 0.0625 \times 8$

$V =$ **1.57 cu ft**

94 What is the volume of a 42″ ID tank that is 6′-6″ tall?

$V = \pi r^2 \times h$

$V = 3.14 \times (21 \times 21) \times 78$

$V = 3.14 \times 441 \times 78$

$V = 108{,}010$ cu in.

$V = 108{,}010 \div 1728$

$V =$ **62.5 cu ft**

95 What is the net volume, in cubic feet, of the firetube boiler shell? Calculate the volume of the entire shell and subtract the volume taken up by the fire tubes.

Boiler shell **See Figure 2-31.**

$V = \pi r^2 \times l$

$V = 3.14 \times (40 \times 40) \times 216$

$V = 3.14 \times 1600 \times 216$

$V = 1{,}085{,}184$ cu in.

$V = 1{,}085{,}184 \div 1728$

$V = 628$ cu ft

Fire tubes

$V = \pi r^2 \times l \times \textit{number of tubes}$

$V = 3.14 \times (2 \times 2) \times 216 \times 84$

$V = 3.14 \times 4 \times 216 \times 84$

$V = 227{,}889$ cu in.

$V = 227{,}889 \div 1728$

$V = 132$ cu ft

$V = 628 - 132$

$V =$ **496 cu ft**

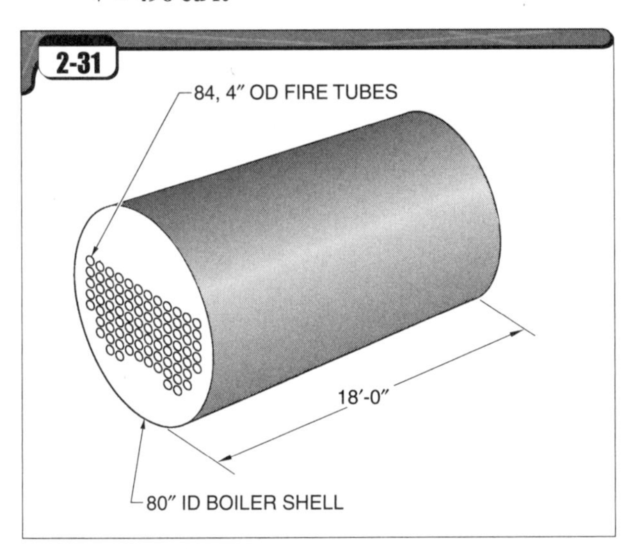

Figure 2-31. *To find cubic feet of boiler shell, subtract the volume of the fire tubes from the volume of the boiler shell.*

96 How many cubic inches are in a gallon?

There are 231 cu in. of water in a gallon. To convert cubic inches to gallons, divide cubic inches by 231. To convert gallons to cubic inches, multiply gallons by 231. For example, how many gallons fit into a tank with a volume of 14,750 cu in.?

$V = \textit{cu in.} \div 231$

$V = 14{,}750 \div 231$

$V =$ **63.9 gal.**

97 How many cubic inches of water are in 55 gal.?

$V = \textit{gal.} \times 231$

$V = 55 \times 231$

$V =$ **12,705 cu in.**

98 How many gallons are in a cubic foot?

There are 7.48 gal. in a cubic foot. To convert cubic feet to gallons, multiply by 7.48. To convert gallons to cubic feet, divide by 7.48. For example, how many gallons are contained in a tank with a volume of 68 cu ft?

$V = \textit{cu ft} \times 7.48$

$V = 68 \times 7.48$

$V =$ **509 gal.**

99 What is the volume, in cubic feet, of a 40-gal. water heater?

$V = \textit{gal.} \div 7.48$

$V = 40 \div 7.48$

$V =$ **5.35 cu ft**

100 How many gallons will the storage tank hold?

$V = \pi r^2 \times l \times 7.48$

$V = 3.14 \times 6 \times 6 \times 18 \times 7.48$

$V =$ **15,220 gal.**

See Figure 2-32.

101 The boilers in a boiler installation are consuming a total of 780 gal. of fuel oil per hour. The fuel oil storage tank is a vertical cylindrical tank that is 18′ in diameter. The tank currently contains 14′-6″ of oil. At the current fuel oil consumption rate, how many hours will the fuel oil supply in the tank last?

$\textit{Time} = (\pi r^2 \times l \times 7.48) \div 780$

$\textit{Time} = (3.14 \times 9 \times 9 \times 14.5 \times 7.48) \div 780$

$\textit{Time} = 27{,}586 \div 780$

$\textit{Time} =$ **35.4 hr**

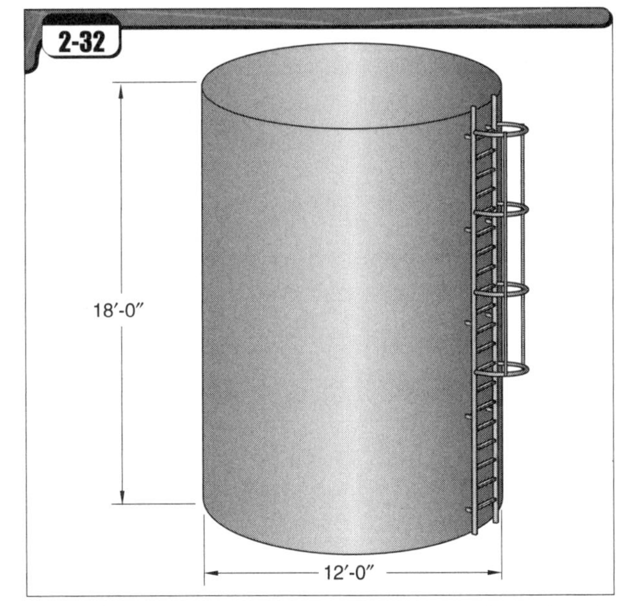

Figure 2-32. *Volume in cubic feet may be converted to gallons by multiplying by 7.48.*

2 Chapter

Boiler Construction and Design

Trade Test

Name: ______________________ Date: ______________________

______________ 1. The badge plate is commonly attached to the ___ of the boiler.

T F 2. Submerged arc welding is the standard method of construction of boiler drums.

______________ 3. ___ administers the application of various approval stamps on boilers, pressure vessels, and safety valves.

A. ASME International
B. The NB
C. The Boiler Inspector
D. none of the above

T F 4. GTAW is a welding process in which the electric arc is submerged or hidden beneath flux.

______________ 5. ___ is often used to make the first pass on higher pressure vessels.

A. SAW
B. SMAW
C. GTAW
D. OXY

______________ 6. A(n) ___ joint runs along the length of a boiler drum or shell.

T F 7. The electrode is consumed in GTAW.

T F 8. Welded longitudinal joints are stronger than riveted longitudinal joints.

______________ 9. A boiler is ___ after construction by heating it at 1100°F to 1200°F for a specified time and then allowing it to cool slowly in a still atmosphere.

______________ 10. A(n) ___ is a blend of two or more different metals.

______________ 11. ___ tubes of a watertube boiler are exposed to the highest temperatures in the furnace area.

T F 12. Cooler water descends to the lowest points in a watertube boiler.

______________ 13. A hydraulic tester applies a force of 52,000 lb that causes a test sample with an original cross-sectional area of 0.90 sq in. to fail. The tensile strength of the sample is ___ psi.

A. 46,800
B. 51,999.1
C. 52,000.9
D. 57,778

T F 14. Some rivet-to-edge firecracks are acceptable in riveted boilers.

______________ 15. The ___ forms the smoke box area between the tube sheet and fire door(s) of a firetube boiler.

T F 16. Rivet-to-rivet firecracks are acceptable in riveted boilers.

______________ 17. A ___ joint has two overlapping plates that are drilled through and riveted together at the edges.

A. butt
B. seam
C. lap
D. pitch

_______________ **18.** ___ pitch is the distance from the center of a rivet to the center of the next rivet in the same row.

A. Lap
B. Rivet
C. Margin
D. Center

_______________ **19.** ___ is the most common form of NDT (nondestructive testing) in pressure vessel work.

A. X raying
B. Magnetic particle testing
C. Dye penetrant testing
D. Eddy current testing

T F **20.** Temperature creates a greater stress than pressure on a boiler shell.

_______________ **21.** A(n) ___ is a short bolt brace that passes through the water leg of a boiler.

T F **22.** The beaded ends of the fire tubes in a firetube boiler help keep the tube sheets from bulging.

_______________ **23.** The telltale hole of a staybolt is usually ___″ in diameter.

A. 3⁄64
B. 3⁄32
C. 3⁄16
D. 3⁄8

_______________ **24.** Manholes are ___ in shape.

A. circular
B. elliptical
C. rectangular
D. all of the above

T F **25.** In a hydrostatic test, the boiler is filled with approximately 70°F water.

_______________ **26.** Alloys are engineered for the purpose of ___.

A. corrosion resistance
B. abrasion resistance
C. heat-transfer properties
D. all of the above

_______________ **27.** Handholes are generally about ___ in size.

A. 2″ × 3″
B. 4″ × 5″
C. 6″ × 8″
D. 10″ × 12″

_______________ **28.** The heads of the drums of a watertube boiler are ___.

A. curved inward
B. curved outward
C. flat
D. all of the above

_______________ **29.** The maximum capacity of a boiler is the ___.

A. area of the floor it occupies
B. volume of water its shell will hold
C. pounds of steam it will produce in 1 hr at a given pressure and temperature
D. gallons of water required to produce a given amount of steam at a given pressure and temperature

T F **30.** Water can be compressed to 75% of its original volume.

_______________ **31.** The formula for finding the radius of a circle is ___.

A. $r = D \times 2$
B. $r = D \div 2$
C. $r = D + 2$
D. none of the above

_______________ **32.** The radius of a 19″ diameter circle is ___″.

_______________ **33.** The radius of a 6′-10″ diameter circle is ___.

_______________ **34.** The formula for finding the circumference of a circle is ___.
A. $C = \pi \times D$ B. $C = \pi - D$
C. $C = \pi + D$ D. none of the above

_______________ **35.** The circumference of a circle that is 15″ in diameter is ___″.

_______________ **36.** The circumference of a boiler drum that is 5.5′ in diameter is ___′.

_______________ **37.** The formula for finding the area of a square or rectangular surface is___.
A. $A = l \times w$ B. $A = l \div w$
C. $A = l - w$ D. $A = l + w$

_______________ **38.** The area of a warehouse floor that measures 40′-0″ × 68′-0″ is ___ sq ft.

_______________ **39.** A piece of steel 7½″ long and 4¼″ wide contains ___ sq in.

T F **40.** One sq ft contains 144 sq in.

_______________ **41.** The formula for finding the area of a circle is ___.
A. $A = \pi r^2$ B. $A = \pi D^2$
C. $A = \pi^2 r$ D. $A = \pi^2 D$

_______________ **42.** A 22″ diameter circle has an area of ___ sq in.

_______________ **43.** A 31″ diameter circle has an area of ___ sq ft.

_______________ **44.** The formula for finding the area of a tubular object is ___.
A. $A = \pi D + l$ B. $A = \pi D - l$
C. $A = \pi D \div l$ D. $A = \pi D \times l$

_______________ **45.** The area of a boiler tube that is 3″ in diameter and 12′-0″ long is ___ sq ft.

_______________ **46.** The formula for finding the volume of a rectangular solid is ___.
A. $V = l \times w \times h$ B. $V = l + w + h$
C. $V = (l \times w) \div h$ D. $V = (l \times h) \div w$

_______________ **47.** The volume of a container that is 8″ long, 4″ deep, and 3″ high is ___ cu in.

_______________ **48.** The formula for finding the volume of a cylindrical object is ___.
A. $V = \pi r^2 + l$ B. $V = \pi r^2 - l$
C. $V = \pi r^2 \times l$ D. $V = \pi r^2 \div l$

_______________ **49.** The volume of a tank that is 2′-0″ in diameter and 6′-0″ long is ___ cu ft.

T F **50.** One cu ft contains 144 cu in.

_______________ **51.** Two methods of supporting HRT boilers are the ___ method and the ___ method.

_______________ **52.** ___ is the standard welding method for boiler repair.
A. GTAW B. SAW
C. SMAW D. none of the above

_______________ **53.** A circumferential joint is also known as a(n) ___ joint.

T F **54.** Waterwalls decrease heating surface of a boiler, helping to prevent refractory damage to the furnace walls because of overheating.

______ **55.** ___ is a scissor-like force that can cut objects.

______ **56.** ___ is the ability of a material to return to its original shape after being deformed.
A. Ductility
B. Resiliency
C. Malleability
D. none of the above

______ **57.** The downcomer tubes in a watertube boiler contain the ___ water.
A. hotter, rising
B. hotter, descending
C. cooler, rising
D. cooler, descending

______ **58.** Tube ends are normally ___ in a firetube boiler and ___ in a watertube boiler.
A. beaded; beaded
B. beaded; flared
C. flared; flared
D. flared; beaded

T F 59. Tension is the exertion of equal forces pulling in opposite directions that can stretch an object.

______ **60.** An HRT firetube boiler is 5′-6″ in diameter and 14′ in length. It contains sixty 3″ tubes. Half of the shell is exposed to the furnace area. What is the heating surface of the boiler? The tube sheets may be disregarded. *Note:* Assume that the tubes and the shell are the same length.
A. 207 sq ft
B. 659 sq ft
C. 780 sq ft
D. 902 sq ft

T F 61. Diagonal stays may be riveted or welded.

______ **62.** The radius of a 13.5′ diameter circle is ___′.
A. 1.35
B. 6.75
C. 13.5
D. 27.0

______ **63.** A drum with a 24″ OD has a circumference of ___″.
A. 12
B. 27.14
C. 75.36
D. 576

______ **64.** Six cu ft contains ___ cu in.
A. 12
B. 144
C. 864
D. none of the above

______ **65.** A bar that is 26″ long, 4″ deep, and 3″ high contains ___ cu in.
A. 33
B. 34.67
C. 312
D. none of the above

T F 66. The MAWP is given on the nameplate of a boiler.

T F 67. The flame in oxyacetylene welding exceeds 10,000°F.

______ **68.** The area of a 3′-0″ OD circle is ___ sq ft.
A. 3.00
B. 6.14
C. 7.065
D. none of the above

______ **69.** ___ is the plasticity exhibited by a material under tension loading.

T F 70. Steel is preferred over cast iron for boiler fittings.

T F 71. The majority of leaks in a firetube boiler occur near the tube ends.

_______________ **72.** ___ are the short stubs of piping connected to a boiler during construction for the purpose of attaching piping, safety valves, etc.

A. Downcomers
B. Membranes
C. Mandrels
D. Nozzles

T F **73.** The smaller the diameter of a boiler drum, the lower the pressure it can withstand, with all other factors being equal.

_______________ **74.** A scotch marine firetube boiler is 16′-0″ in length and has one hundred four 2″ tubes and a 2′-0″ diameter flue. What is the heating surface of the boiler in sq ft? *Note:* Assume that the tubes and flue are the same length.

A. 888
B. 971
C. 2575
D. 5275

Matching

_______________ **75.** Circle

_______________ **76.** Radius

_______________ **77.** Diameter

_______________ **78.** Circumference

_______________ **79.** π

A. Distance from centerpoint to outside boundary

B. 3.14

C. Distance from outside boundary to outside boundary through the centerpoint

D. Plane figure generated about a centerpoint

E. Outside boundary

Chapter 3

Steam Systems and Controls

One of the boiler operator's primary responsibilities is to ensure that high-quality steam is always available at the correct pressure and temperature for the equipment that uses the steam. This ensures smooth and efficient operation of the heating and production equipment. A clear understanding of the pressure control devices on boilers helps the boiler operator optimize steam-generating performance.

The fuel costs associated with modern steam generation dictate that boiler operators have an appreciation for steam system efficiency. The piping systems that transport the steam to the steam-using equipment should be monitored to ensure that the heat in the steam reaches the equipment. Expensive losses can occur due to uninsulated piping, steam leaks, air contamination in the steam, defective steam traps, and losses of hot condensate from the piping system.

Boiler operators also play an important role in ensuring the safety of the plant personnel whose work exposes them to the hazards associated with steam systems. Good boiler operators observe safe operating techniques to ensure high integrity of steam and condensate piping, valves, and other components. Safe practices help protect personnel from dangers such as water hammer and the extreme temperatures of superheated steam.

Pressure Measurement

How is steam pressure created in a boiler?

Once the water in a boiler reaches 212°F, any additional heat absorbed by the water is latent heat that causes the boiling water to change to steam. The volume of the steam is much greater than the volume of the water. Soon the fixed internal volume of the boiler is filled with steam at atmospheric pressure, and continued generation of steam results in compression in order to make room for additional steam.

How is a constant pressure maintained in a steam boiler?

If the flow rate of steam leaving the boiler is equal to the rate of steam generation in the boiler, then the number of pounds of steam occupying space in the boiler remains constant. In this case, the pressure in the boiler also remains constant. If the steam leaves the boiler faster than the rate at which new steam is being generated, the pressure in the boiler drops. If steam is generated in the boiler faster than the rate at which it is leaving the boiler, the pressure rises. The rate at which the steam leaves the boiler is determined by the rate at which the steam is being condensed or consumed in the steam-using equipment, as well as radiant heat losses through piping and steam losses through leaks.

Automatic pressure controls adjust the fuel-burning rate of the combustion equipment in a constant effort to match the steam generation rate to the steam consumption rate, and thus maintain a constant steam pressure. Many years ago, maintaining a constant steam system pressure was the job of the boiler operator. One of the boiler operator's primary responsibilities was to watch the main steam pressure gauge on the boiler and shovel coal or wood at a rate that would maintain a constant steam pressure.

Maintaining a constant steam system pressure ensures that the steam temperature and the steam volume remain constant. The steam temperature needs to remain constant so that the steam-using processes work correctly. The steam volume should remain constant because the piping system is designed for steam of a particular volume (cu ft/lb). For example, if the steam pressure is allowed to drop, and the lower pressure steam is larger in volume than higher pressure steam, the piping may not be large enough to pass the needed quantity of steam (lb/hr).

What is the name of the apparatus that makes a mechanical pressure gauge work?

A *Bourdon tube* is the tube inside a mechanical pressure gauge. It is a bronze or stainless steel tube bent into a question mark shape and flattened into an elliptical shape. **See Figure 3-1.** As pressure enters the gauge, it causes the tube to tend to straighten out. The amount of movement of the tube is proportional to the amount of pressure entering the tube. The tip of the tube is connected to a series of small gears and springs (adjustable linkage) that transfers the movement of the tube to movement of the pointer on the face of the pressure gauge. Most pressure gauges are calibrated to read in psi.

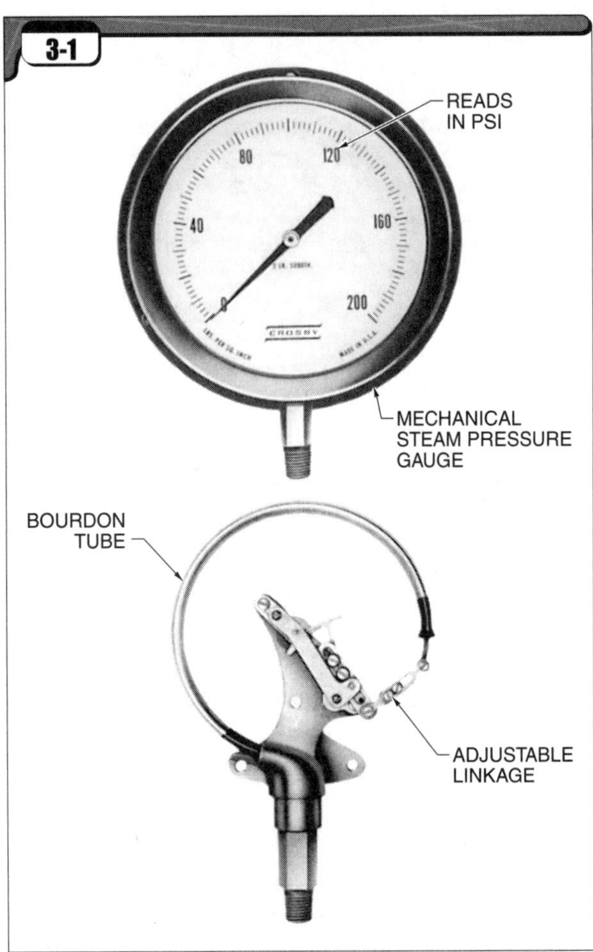

Crosby Valve Inc.

Figure 3-1. A Bourdon tube and an adjustable linkage are the internal parts of a mechanical steam pressure gauge.

What protection is needed in a pressure gauge to prevent steam from damaging the Bourdon tube?

A siphon loop or tube is placed between the steam pressure gauge connection and the steam pressure gauge on the boiler or on steam piping. The loop in a siphon forms a pocket where condensate gathers, similar to the drain trap under a kitchen sink. This condensate allows the pressure of the steam to be transmitted to the pressure gauge, but protects the delicate Bourdon tube from warping due to the high temperature of the steam. The pressures on both sides of the siphon loop equalize, so that the condensate in the loop remains stationary.

The tube is a small piece of pipe that is curled into a loop. This loop is usually an "O" shape (pigtail siphon) but may be a "U" shape (U-tube siphon). One end is connected to the steam pressure gauge and the other end is connected to the pressure gauge connection on the boiler. **See Figure 3-2.**

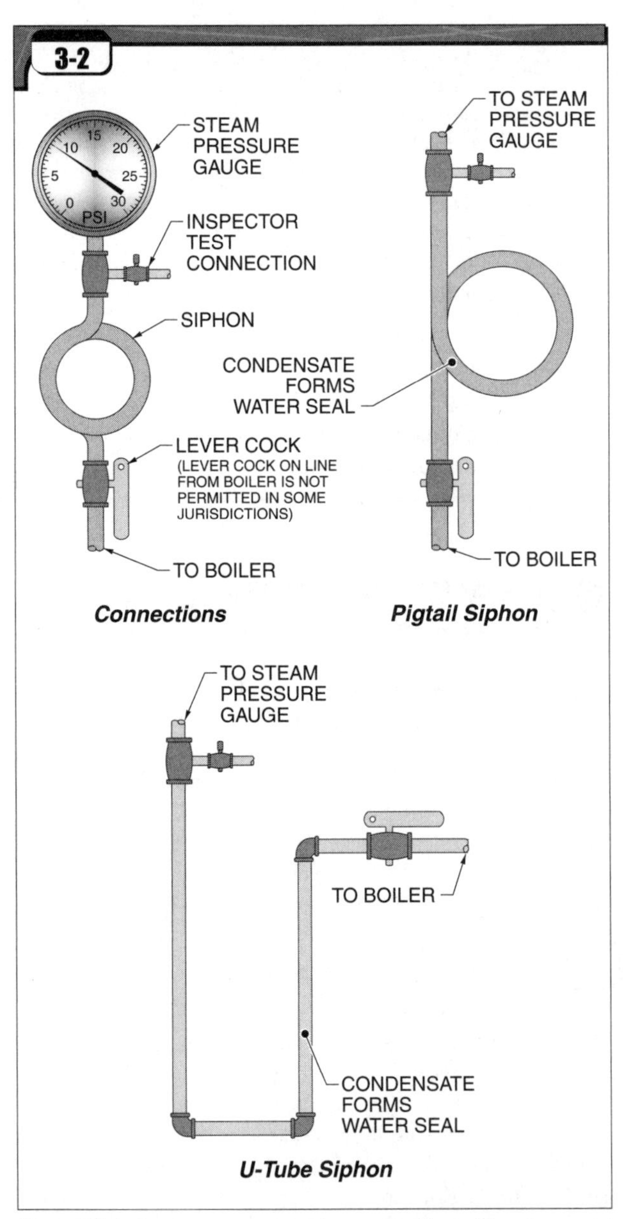

Figure 3-2. A steam pressure gauge is protected by a siphon to prevent steam from damaging the Bourdon tube.

What should be the range of a pressure gauge installed on a steam boiler?

The pressure gauge should be able to read at least 1½ times the MAWP of the boiler. Two times the MAWP is preferable for two reasons. First, a Bourdon tube pressure gauge is the most accurate in the middle third of its range. For example, a pressure gauge that reads 0 psi to 60 psi is the most accurate between approximately 20 psi and 40 psi. A pressure gauge that reads 0 psi to 300 psi is the most accurate between approximately 100 psi and 200 psi.

Second, a hydrostatic test may be applied to the boiler periodically. A hydrostatic test uses water pressure to check for leaks. This test is applied at 1½ times the MAWP and the pressure gauge should be able to read at least this much pressure.

What is the required accuracy of mechanical pressure gauges compared to the working pressure?

The gauge should be accurate to within 2% of the working pressure. If a steam pressure gauge is not accurate to within 2% of the working pressure, it should be recalibrated or replaced.

Factoid

Constant steam pressure is necessary to ensure that the steam temperature and volume remain constant. Constant temperature and volume are needed for many processes.

Describe two methods by which a steam pressure gauge can be tested for accuracy.

A pressure gauge can be compared to a test gauge. A test gauge is a very rugged, accurate, and expensive gauge used for calibrating other gauges. The ASME Code requires that a connection be provided in the piping to the steam pressure gauge for installing the boiler inspector's test gauge.

A pressure gauge can also be calibrated with a deadweight tester. **See Figure 3-3.** A deadweight tester is filled with distilled water or very light oil. Many designs and ranges of deadweight testers are available. Precisely calibrated weights are placed on a weight platform or balance beam, and this exerts exact hydraulic pressure on the distilled water or oil, which is in turn transmitted to the steam pressure gauge. For example, if 100 lb of weight is placed on the platform and exerted on a piston face area of 1 sq in, the reading on the gauge should accurately reflect 100 psig. If not, an instrument technician can recalibrate the gauge.

Lightweight equipment is also available that combines the features of the test gauge and the deadweight tester into a convenient calibration system that is easily transported around the plant. These devices utilize a manual, handheld hydraulic pump with which the user can pump up high pressures using distilled water or other clean liquid. A test gauge is attached, and the actual pressure gauge being tested is compared to the test gauge.

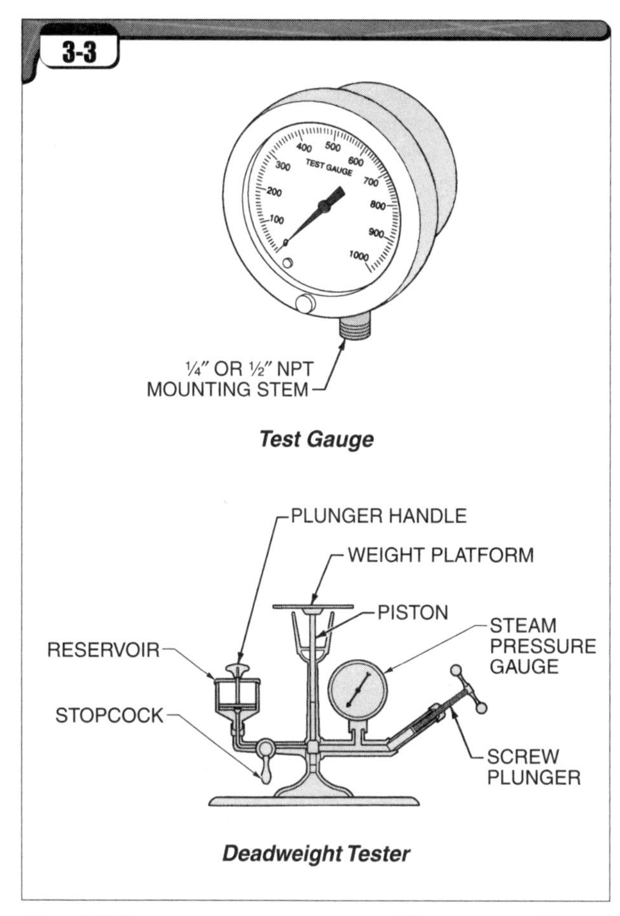

Figure 3-3. *Steam pressure gauges are tested for accuracy with a test gauge or a deadweight tester.*

What is a vacuum gauge?

A vacuum gauge is used to measure pressures below atmospheric pressure. It also uses a Bourdon tube, but vacuum causes the Bourdon tube to curl tighter, rather than extending. A vacuum gauge is calibrated to read in inches of mercury (in. Hg).

Trade Tip

It is a good idea to use compound gauges on low-pressure steam piping and heat exchangers so the operator can tell if a vacuum condition has occurred inside.

How may a vacuum be formed inside steam piping and heat exchange equipment?

When the steam supply to a section of piping or a piece of heat exchange equipment is shut off, the steam that is captured within the piping or equipment continues to transfer heat through the pipe wall or heat exchange surfaces. As this steam gives up its latent heat and condenses, its specific volume is greatly reduced. If there is nothing available to flow in and fill the space previously occupied by the steam (i.e., air or more steam), a vacuum is formed.

10 What is a compound gauge?

A compound gauge is a combination pressure gauge and vacuum gauge. It contains a single Bourdon tube. The compound gauge has a scale that shows both pressure measurement and vacuum measurement. **See Figure 3-4.** The vacuum portion of the scale occupies part of the left side of the scale, and the pressure portion occupies the remainder of the scale.

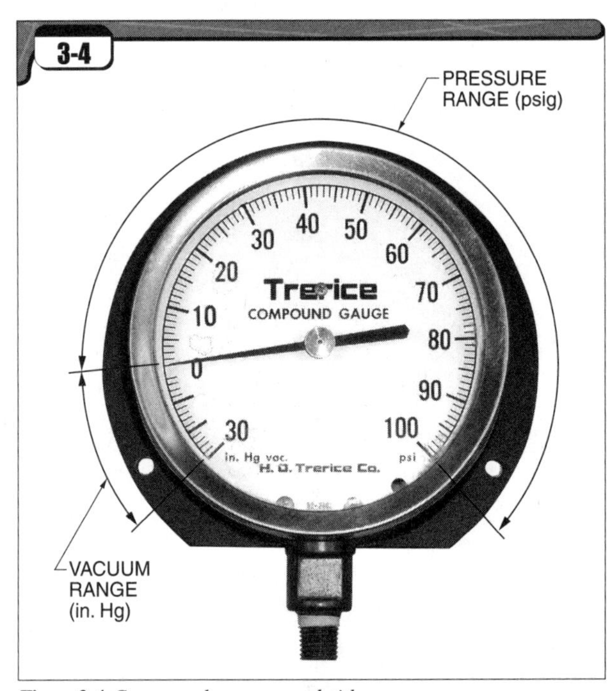

Figure 3-4. *Compound gauges read either pressure or vacuum.*

Pressure Control

11 What is the function of pressure controls?

Pressure controls on steam boilers relieve the operator of the tedious duty of regulating the combustion equipment to maintain a constant pressure in the steam header. The pressure controls automatically measure the steam pressure and then cause an appropriate change in the fuel firing rate.

12 What are the basic approaches to control of the steam pressure in boiler systems?

ON/OFF control is an approach to control in which a burner is either ON or OFF. **See Figure 3-5.** ON/OFF control is primarily used for smaller package boilers serving heating systems, small manufacturing facilities, and other smaller operations such as dry cleaning operations and soft drink bottlers. This simple type of control is similar to that used in the operation of a residential furnace. A pressure-sensing device (pressurestat) senses the pressure in the boiler. It starts or stops the burner depending on the pressure inside the boiler.

OFF/low/high control is an approach to control where a burner is either OFF or operating with a low flame or with a high flame. OFF/low/high control is often used with slightly larger boilers, i.e., up to about 20 HP. OFF/low/high control is similar in concept to being able to shift gears with a two-speed bicycle as opposed to a single-speed bicycle. Since burners using this type of control are somewhat larger than those used with ON/OFF control, it would create thermal strains on the boiler steel or cast iron to have the large burner come on all at once. This control approach allows about half of the burner capacity to come on during startup (low fire). If the steam pressure continues to drop after the first half of the burner comes on, the second half of the burner is started (high fire). As the steam pressure rises to meet the steam demand, half of the burner shuts off. If the pressure drops, the second half of the burner comes on again, and so on. If the steam pressure continues to rise with only half of the burner in service, the burner shuts off completely and waits for the steam pressure to fall again.

Larger burners using the OFF/low/high control scheme often include a combustion air blower or fan. The process of burning almost anything requires oxygen, and the combustion of a large amount of fuel requires a large amount of oxygen. Simply drawing the combustion air from the room (as in the case of a residential furnace) may not be feasible, so a fan or blower may be included with the burner equipment to provide adequate oxygen.

ON/OFF with modulation control is an approach to control in which the amount of flame is changed to a degree that is proportional to the need. For example, if the water from a residential shower is slightly too warm, the person showering reduces the flow of hot water to the shower a slight amount. If the shower is much too hot, a larger change is made. In this example, the temperature of the water can be adjusted more or less infinitely within the range from very cool to very hot.

Most commercial and smaller industrial boilers use ON/OFF with modulation control. These systems use a combustion air fan or blower. This method uses a pressure-sensing device (pressurestat) to start the burner(s) when the steam pressure reaches a minimum acceptable point. Once the burner is started, another control device is used to automatically turn the burner up or down as needed in a continuous effort to maintain a constant steam pressure. This control system does not actually measure or meter the flow of fuel or air to the burner. The fuel valve and air damper are operated through mechanical linkages from the control system. Once the linkages are adjusted, the assumption is made that moving the linkages in turn moves the fuel valve and air damper the correct amount. If the steam pressure rises slightly, the burner is turned down slightly. If the steam pressure drops slightly, the burner is turned up slightly. If the steam pressure rises by a larger amount, the burner is turned down more, and so on. If the steam pressure continues to rise with the burner turned down as far as it will go, the burner is shut off and the controls wait for the pressure to fall to the starting point again.

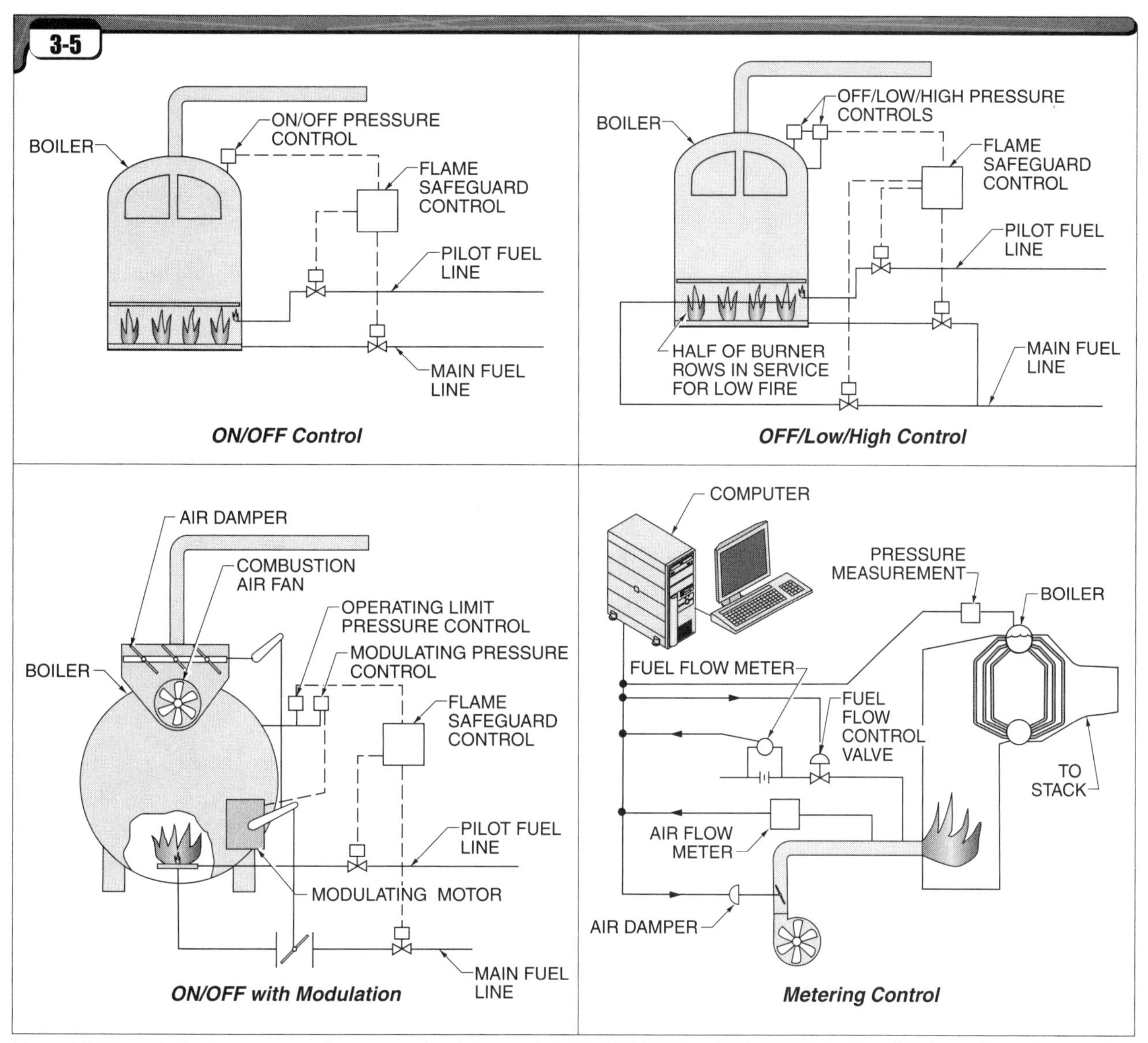

Figure 3-5. *The methods of pressure control on most modern boilers include ON/OFF, OFF/Low/High, ON/OFF with modulation, and metering control.*

Positioning control is an approach to control in which a master device, or controller, senses the pressure in the steam header and uses compressed air to modulate power units, or actuators, which in turn position control linkages. Most positioning control systems are pneumatic control systems and most are used in larger industrial plants. For example, if the steam header pressure drops slightly, the master controller applies air pressure to move one actuator to increase the fuel feed and another to open the combustion air fan damper more. The controller incorporates an automatic/manual selecting function that allows the boiler operator to take control of the system during system startup, shutdown, or upset conditions. Positioning control systems are subject to inaccuracy for a number of reasons and often require constant attention from the boiler operator to make manual adjustments. Conventional positioning control systems are generally considered obsolete and are becoming rare.

Positioning control systems are very similar to ON/OFF with modulation control systems with two basic exceptions. First, positioning control systems are pneumatic control systems while ON/OFF with modulation control systems operate electrically. Second, positioning control systems do not automatically turn the combustion equipment ON or OFF.

Metering control is an approach to control where the flows of fuel and air are precisely measured by flow-measuring devices and then adjusted by the control system so as to always be in the correct proportions. Metering control is generally used in larger industrial boilers where close control of the steam pressure and flow is required. Such plants include refineries, power plants, chemical process plants, and others. In modern boilers, metering control usually centers on programming software in a computerized control system. Based on the steam header pressure and the steam flow from the boiler, the software calculates the correct fuel flow rate. For any given fuel flow rate, the software also calculates the correct amount of air needed to make the fuel burn safely and efficiently. The actual changes to the flow rates are usually effected through pneumatically operated valves and dampers.

Metering control systems are expensive to install, but have the advantage of controlling the burner equipment so as to operate as efficiently as possible. This is very important in large boiler installations, because the cost of the fuel may run in the hundreds of thousands of dollars per month. Even a very small efficiency improvement in the utilization of that fuel may amount to significant savings.

13 How does a pressurestat control work?

Pressurestat controls are most commonly used on package boilers using ON/OFF, OFF/low/high, and ON/OFF with modulation control schemes. A pressurestat control starts and stops the burner(s) based on the steam pressure inside the boiler. Because of its function on a steam boiler, a pressurestat is very often referred to as the operating limit pressure control. That is, it operates the burner equipment to keep the steam pressure within acceptable limits. This control may contain a mercury switch or a single-pole switch activated by a lever. A *mercury switch* is a switch that uses the movement of mercury in a glass tube to start or stop electrical current flow in a circuit.

A pressurestat has a pressure connection on its base, which is connected to sense the pressure in the boiler. The pressure connection leads into a bellows assembly. A *bellows* is a flexible device that expands and contracts with changes in pressure. The bellows is cylindrical in shape and has convolutions around its circumference. These convolutions are similar to the convolutions in an accordion, and allow the bellows to expand and contract. Because the wall of the bellows must be thin in order to be flexible, the high temperature of the steam could warp the bellows and make the control inaccurate. Therefore, a siphon loop is used to connect the control to the boiler.

The top of the bellows is connected to a plunger, which is in turn connected to one end of a lever and fulcrum assembly, arranged like a seesaw on a playground. As the plunger pushes one end of the lever up, the other drops down. The movement of the lever is opposed by an adjustable spring. When the lever moves far enough, it actuates a switch that starts or stops the burner equipment.

If a mercury tube is used, the tube has electrical contacts molded into its end(s). The mercury tube is set on a pivot so that when the pressure in the boiler reaches a certain point, the tube tilts to let the mercury flow to the other end. For example, if the pressure drops, the mercury runs to the end with the contacts, which completes the electric control circuit and starts the burner. When the pressure rises to its upper setting, the mercury flows to the opposite end, which opens the electric control circuit and the burner shuts off.

Controls using mercury switches must be mounted in a level position so that the mercury switch opens or closes the electrical circuit at the proper time. For example, if the control is tilted off-center (out of plumb), the mercury switch may open or close at pressures that are inconsistent with the settings on the front of the control. For this reason, controls mounted with siphon loops should have the loop in the piping oriented front-to-back with the control, and not parallel with the face of the control. **See Figure 3-6.** If the loop is oriented in the latter fashion, expansion and contraction of the loop can cause the switch enclosure to tilt. This can make the switch inaccurate. Pressuretrol® is a Honeywell, Inc. trademark, but is widely used by operators in the industry to refer to all basic automatic pressurestat controls for boilers.

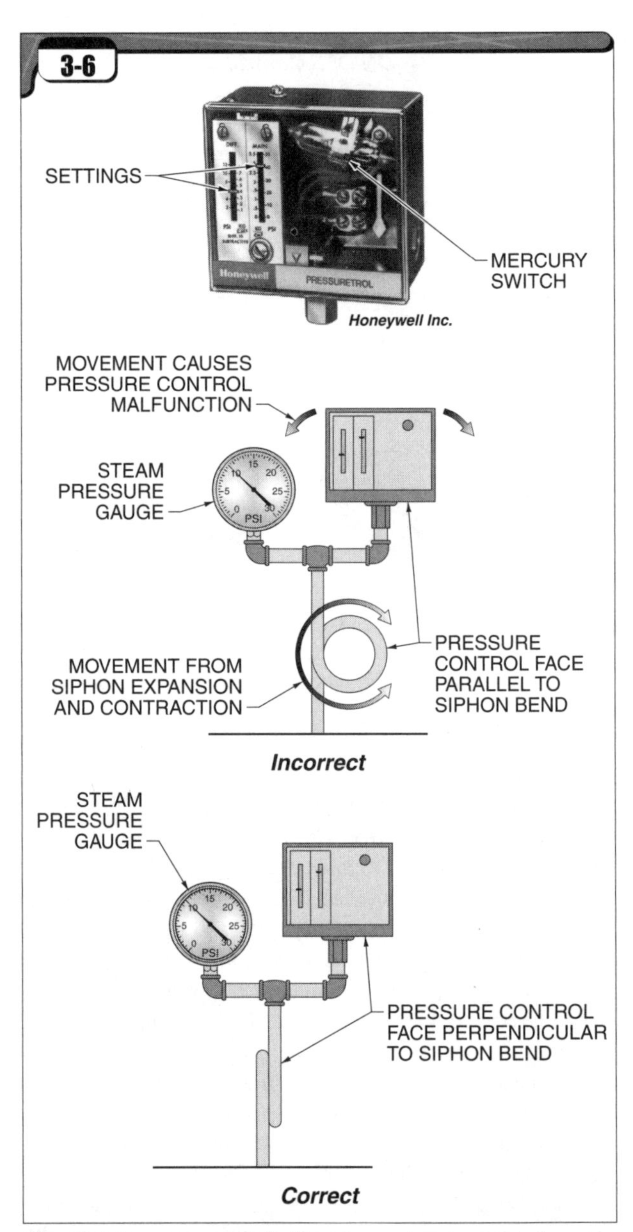

Figure 3-6. *A pressure control with an internal mercury switch must be properly mounted to compensate for movement of the siphon.*

What is the differential setting on automatic pressure controls?

The differential setting is the difference between the pressure at which the control turns the burner ON and the pressure at which the control turns the burner OFF. Differential settings may also be referred to as the cut-in and cut-out pressures. For example, a control may be adjusted so that the burner turns ON at 94 psi and turns off at 100 psi. The differential setting is 6 psi (100 – 94 = 6). Adjustment screws are provided to adjust both the ON/OFF settings and the differential setting as the particular system requires.

What is the differential setting of an automatic pressure control that turns the burner ON at 95 psi and OFF at 105 psi?

differential setting = *cut-out pressure* – *cut-in pressure*
differential setting = 105 – 95
differential setting = **10 psi**

How can safety be maintained if the pressurestat does not properly shut down the burner?

A back-up pressurestat, set for a slightly higher pressure than the operating pressurestat, is often provided. If the pressure rises beyond the setting of the operating limit control, this back-up pressurestat activates and shuts down the burner. The back-up pressurestat generally has a manual reset button. For example, an operator has to press a reset button or lever on the control before the burner can be fired again. The purpose of this design is to ensure that the operator is made aware of the problem.

What is a modulating pressure control?

A *modulating pressure control* is a control device that regulates the burner for a higher or lower fuel-burning rate, depending on the steam pressure in the boiler. **See Figure 3-7.** As the pressure rises above or falls below the pressure setting (setpoint) of the modulating pressure control, the control adjusts the burner firing rate accordingly to bring the pressure back to the setpoint. This control may be called a modulating pressure control, modulating pressurestat, or firing rate control.

A modulating pressure control adjusts the firing rate of the burner by operating a modulating motor. A *modulating motor* is a small electric motor and reduction gear assembly enclosed in a metal box. The reduction gear assembly converts the high speed of the small motor to a very low-speed movement of an output shaft. This low-speed shaft may rotate in either direction. The modulating motor contains electrical switches that limit the total arc of rotation of the output shaft to either 90° or 160°. The direction and duration of the shaft's travel is determined by the modulating pressure control in response to changes in the boiler pressure. An arm is attached to the modulating motor output shaft, and movement of the arm in turn moves the linkages to the fuel valve and fan damper.

The modulating pressure control contains an active winding. The winding is a coil of very fine wire tightly wound around a bolt. This active winding is a potentiometer winding that works like a dimmer switch. The bolt holds the potentiometer winding in its proper horizontal position.

The control has a connection on the bottom that is piped to sense the pressure in the boiler. The pressure connection leads into a bellows enclosure. The bellows expands and contracts with changes in pressure and moves a lever and fulcrum assembly as in the operating limit pressurestat. The movement of the lever is opposed by a spring. Instead of opening or closing a switch, the lever moves a conductive copper wiper arm to the left or right, depending on the boiler pressure.

A short wire soldered to the end of the wiper arm wipes across the coil. The pressure in the boiler determines the wiper's position on the coil. As the boiler pressure changes, the wiper moves across the potentiometer winding.

A red wire is connected from the terminal block to the wiper arm. As the wiper moves toward the end of the potentiometer where the blue wire is connected, the modulating motor, which moves the fuel and air linkages, adjusts the burner and fan damper to a higher burner firing rate.

As the wiper moves toward the end of the potentiometer where the white wire is connected, the modulating motor is driven in the other direction and reduces the firing rate. It reduces the combustion air and natural gas or oil flow rates by closing down on the damper and fuel valve.

Where are steam pressure controls connected to a package boiler?

Steam pressure controls on a package boiler are connected to the top of the boiler at the highest point of the steam space or to the top of the water column. This helps prevent debris from entering the controls.

How is the actuation of the pressure controls on a battery of package boilers usually arranged?

A *battery* is a group of boilers that feed steam into the same steam header. The differential settings on the pressure controls are staggered so that the boilers "lead and lag." One boiler is set to come on first. If it cannot satisfy the steam demand, another comes on, and the pattern follows through the battery.

There are unavoidable energy losses in the operation of boilers. For example, each time a boiler's burner lights, it must go through a purge period. The *purge period* is a short period of time (typically 30 sec to about 2 min) when air is blown through the furnace area to ensure that no volatile fuel vapors are present. Lighting the burner when such vapors are present could cause an explosion. During the purge period, heat is lost from the boiler water to the cool purge air and thus to the atmosphere. Therefore, the fewer the boilers that are cycling on and off to satisfy the steam demand, the lower the total losses. If desired, the lead/lag controls may be reset periodically to reconcile the hours of operation on the units.

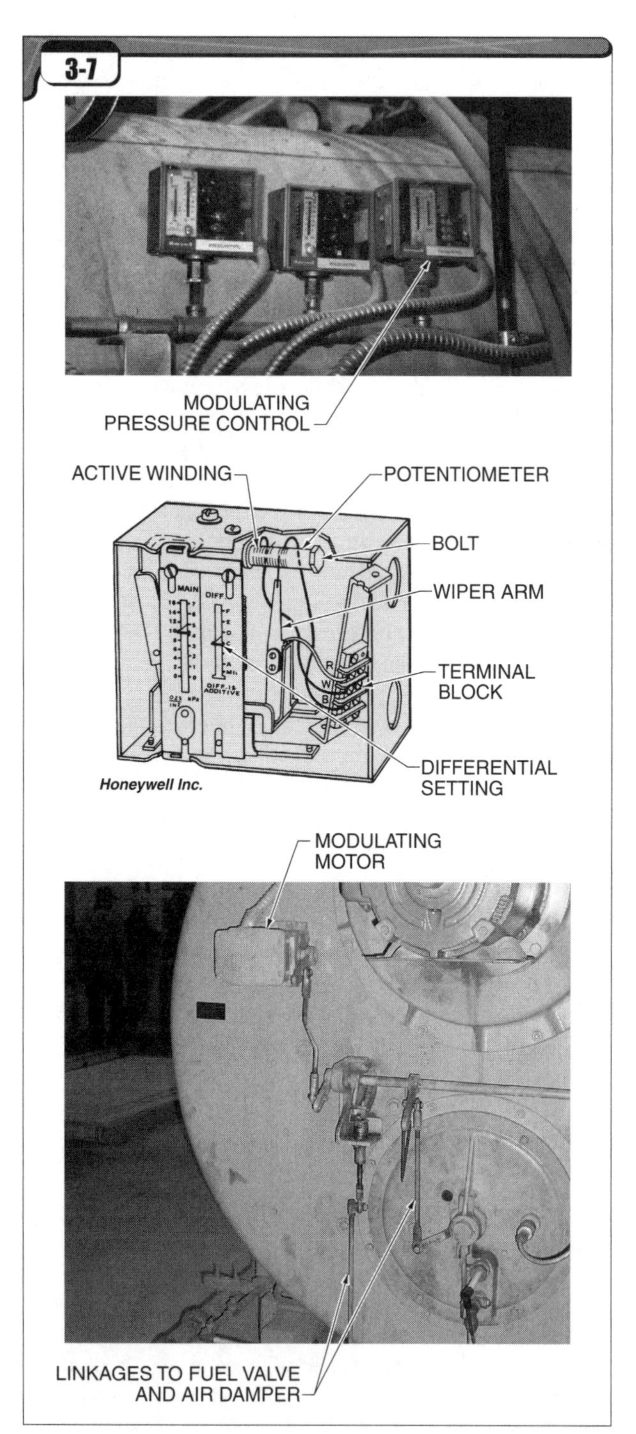

Figure 3-7. A modulating pressure control uses a modulating motor to regulate the burner for a higher or lower fuel-burning rate, depending on the pressure in the boiler.

Factoid

Automatic pressure controls adjust the fuel burning rate to maintain a constant boiler pressure and/or steam header pressure.

Factoid

Safety valves prevent the boiler pressure from exceeding the MAWP.

Safety Valves

20 What is the function of a safety valve on a boiler?

A safety valve keeps the steam pressure in the boiler from exceeding the boiler's MAWP in order to prevent an explosion. **See Figure 3-8.**

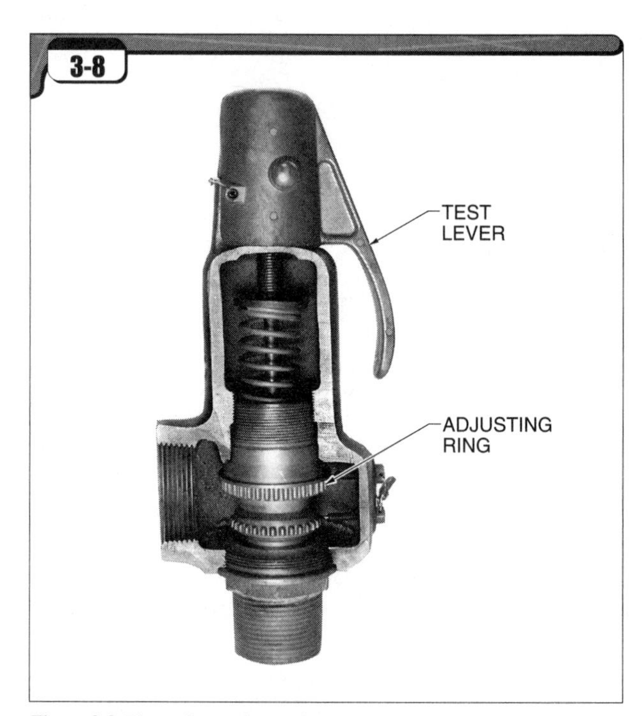

Figure 3-8. The safety valve is the most important valve on a boiler.

21 Why might some boilers have two safety valves?

Boilers with more than 500 sq ft of heating surface require at least two safety valves per the ASME Code. **See Figure 3-9.** Some boilers may have as many as seven safety valves. If a boiler has more than one safety valve, the valves are generally set to lift at different pressures. This helps minimize the upset to the boiler water level and the thrust on the adjacent piping and fittings that is caused by the popping of the valve. If the boiler has more than one safety valve, at least one must lift at not higher than the MAWP and the highest valve setting must be not more than 3% above the MAWP.

22 What type of safety valve is used on modern boilers?

Spring-loaded pop-off safety valves with manual lifting levers are used on modern boilers.

Figure 3-9. Boilers with more than 500 sq ft of heating surface require two or more safety valves under the ASME Code.

23 What is the difference between a safety valve and a relief valve?

A *safety valve* is a valve that opens fully and instantly and causes a definite, measured drop in pressure before closing. For example, a safety valve may pop open at 400 psig and reseat at 388 psig. Safety valves are used for compressed gases.

A *relief valve* is a valve that opens in proportion to the excess pressure, rather than popping open fully. Relief valves are generally used for liquid service. They simply prevent the pressure from getting too high. In addition, relief valves do not cause a reduction in pressure before they reseat.

A *safety relief valve* is a valve specially designed to serve as either a safety valve or relief valve, depending on the application for which it is used.

24 Why does a safety valve pop fully open instantly?

A safety valve pops open instantly because the valve disc that closes against the seat of a safety valve is made in a two-stage pattern. The area of the valve disc that is exposed to the pressure in the boiler or pressure vessel increases when the valve disc lifts from the seat. This increase in area is due to the tapered or staggered design of the valve disc and seat. **See Figure 3-10.** The valve disc is subjected to a higher total force when it opens than when it is closed. This sudden increase in area that the steam can push against when the valve opens causes the valve disc to pop open fully and quickly.

The total force is the pressure of the steam multiplied by the area exposed to the steam. Total force is found by applying the following formula:

$$TF = P \times A$$

where

TF = total force (in lb)

P = pressure (in psi)

A = area of valve disc exposed to steam (in sq in.)

For example, a boiler has a pressure of 280 psi and a 2.5″ diameter safety valve disc. What is the total force acting on the safety valve disc?

$$TF = P \times A$$

$$TF = 280 \times 3.14 \times (1.25 \times 1.25)$$

$$TF = 280 \times 3.14 \times 1.5625$$

$$TF = \mathbf{1374\ lb}$$

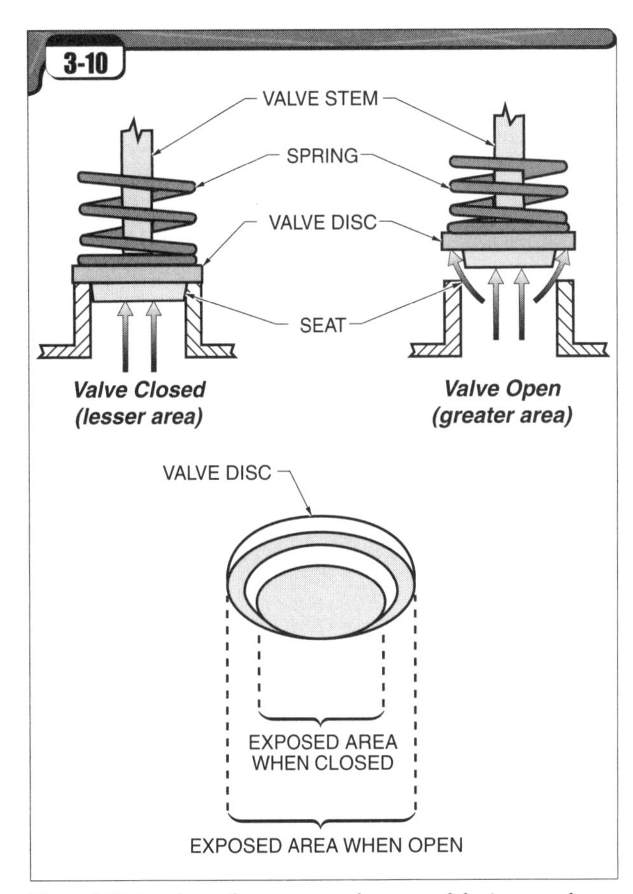

Figure 3-10. A safety valve pops open because of the increased area of the disc exposed to pressure.

Factoid

Many veteran boiler operators refer to a pressure gauge that reads higher than the actual pressure as a fast gauge, much the same as a wristwatch that reads a later time than the actual time. A pressure gauge that reads lower than the actual pressure is often called a slow gauge.

A boiler has a pressure of 265 psi and a 3″ diameter safety valve disc. What is the total force acting on the safety valve disc?

$TF = P \times A$
$TF = 265 \times 3.14 \times (1.5 \times 1.5)$
$TF = 265 \times 3.14 \times 2.25$
$TF = \mathbf{1872\ lb}$

What is blowdown of a safety valve?

Blowdown is the amount of pressure in a pressure vessel that must be reduced before a safety valve reseats. It is sometimes referred to as blowback. Blowdown is expressed as a percentage of the popping pressure. For example, if a safety valve is set for 100 psi and has a blowdown of 3%, it reseats when the pressure drops to 97 psi (100 psi – 3% = 97 psi). *Note:* As a guideline, blowdown should not exceed 4% of the popping pressure. For spring-loaded pop-type safety valves set for popping pressures of 100 psig to 300 psig, the ASME Code requires that the blowdown be not less than 2% of the popping pressure.

Why does a safety valve blow back?

A safety valve blows back when the pressure in the boiler drops sufficiently. The total force on the disc must drop below the force exerted by the spring on the valve. When the pressure drops to this point, the valve is throttled down until the steam can no longer exert force on the larger area of the seat, only on the smaller area. This change causes the valve to snap shut suddenly. The adjustment of the huddling ring determines the amount of the pressure drop that must take place before the valve closes.

What is the percentage of blowdown for a safety valve set to pop (lift) at 250 psi and reseat at 241 psi?

percent of blowdown = (popping pressure – reseat pressure) ÷ popping pressure × 100
percent of blowdown = (250 – 241) ÷ 250 × 100
percent of blowdown = (9 ÷ 250) × 100
percent of blowdown = 0.036 × 100 = **3.6%**

How can the blowdown be changed?

The blowdown can be changed by adjusting the huddling ring of the safety valve. A *huddling ring* is an adjustment in a safety valve that controls the degree to which the escaping steam is directed against the safety valve disc. This determines the percentage of blowdown of the valve. The huddling ring is also called the adjusting ring or blowback ring.

How may the safety valves used for utility boilers differ from those on most other steam boilers?

Very high-pressure safety valves used in utility plants must close quickly and positively and seat perfectly. Steam allowed to blow through or leak through the safety valves for any extended period could quickly erode the valve disc and seat. This causes the valve to leak. For this reason, safety valves used on such boilers are often of the nozzle-reaction type. A nozzle-reaction safety valve is constructed so that some of the escaping steam is directed through passageways in the valve in such a way as to build pressure above the disc and assist the spring in closing the valve. Nozzle-reaction type safety valves have greater discharge capacities but shorter blowdown periods.

What is the lift of a safety valve?

The lift of a safety valve is the distance the valve disc lifts from the valve seat when the valve is fully open. The lift is expressed in inches. For example, if the valve lift equals ⅜″, the lift is specified as 0.375″.

Where is important information about a safety valve found?

Each safety valve is supplied with a data plate that provides information about the valve. **See Figure 3-11.** The data plate is usually attached to the side of the safety valve. Information that may be found on the data plate includes the following:

- manufacturer's name or trademark
- manufacturer's design or model number
- size of valve, seat diameter (in in.)
- popping pressure setting
- blowdown (in psi)
- capacity (in lb/hr)
- lift of valve (in in.)
- year built or code mark
- ASME symbol stamp
- serial number

What is capacity of a safety valve?

Capacity is the amount of steam in pounds per hour that the safety valve is capable of venting at the rated pressure of the valve. The capacity, not just the popping pressure, must be considered when selecting a safety valve. The capacity of the valve(s) must be great enough to vent all the excess steam the boiler can produce without allowing the boiler pressure to build to over 6% above the MAWP.

Assuming that a low-pressure boiler and a high-pressure boiler have identical ratings in pounds of steam per hour, which boiler would require the larger safety valve?

The low-pressure boiler would require the larger safety valve (greater cross-sectional area of the inlet and outlet ports) because low-pressure steam has more volume per pound than higher-pressure steam. The low-pressure valve must have a larger passage to pass the same number of pounds of steam per hour through the safety valve.

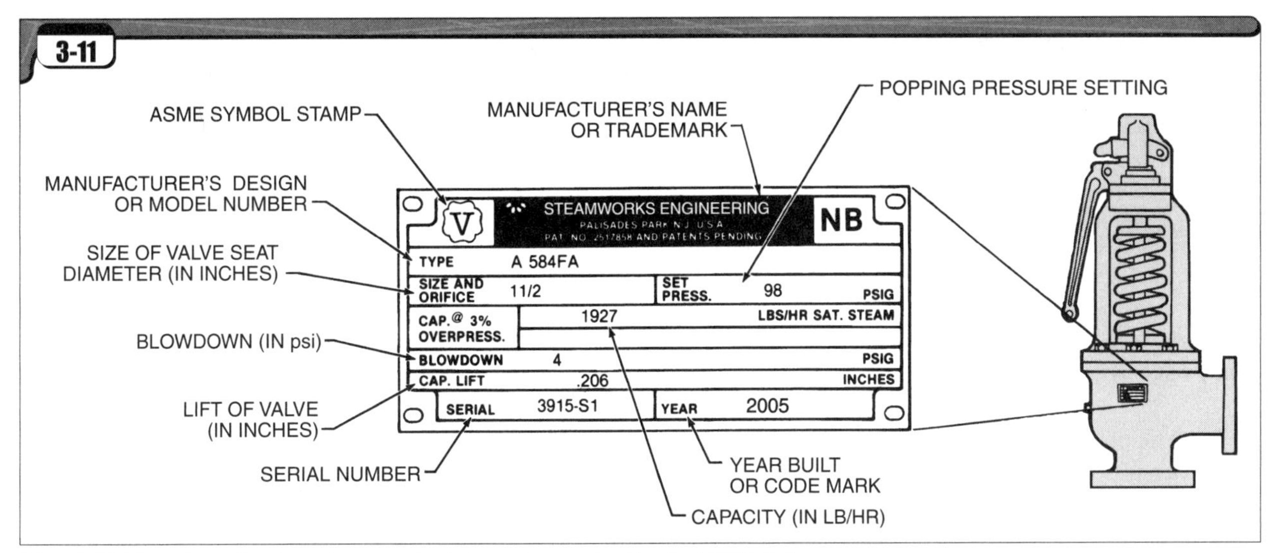

Figure 3-11. *The safety valve data plate contains information required by the ASME Code.*

35 What is the difference between the "V" and "UV" ASME symbol stamps on safety valves?

The "V" symbol stamp on a safety valve indicates that the valve conforms to the ASME Code requirements for use on fired vessels. A *fired vessel* is a pressure vessel that includes a burner or combustion equipment of some kind. Fired vessels include boilers, direct-fired hot water storage tanks, and similar vessels. The "UV" stamp indicates that a safety valve conforms to the ASME Code requirements for use on unfired vessels. An *unfired vessel* is a pressure vessel without combustion equipment, such as compressed air tanks, feedwater heating tanks, steam piping, steam-jacketed heat exchange equipment, and similar vessels.

A safety valve with the "V" symbol stamp will pass the flow of steam (capacity) stamped on its data plate when the pressure in the vessel reaches 3% over the safety valve's popping pressure. A safety valve with the "UV" symbol stamp will not reach its rated capacity until the pressure in the pressure vessel reaches 10% over the safety valve's popping pressure. For this reason, a safety valve with the "V" stamp may be used on unfired pressure vessels, but the safety valve with the "UV" stamp may not be used on fired pressure vessels.

36 How often should safety valves be tested manually and by pressure?

Recommended testing frequencies for safety valves vary. Some jurisdictions may set specific requirements for testing. Check with the local agency having jurisdiction and the plant's insurance company when outlining the plant's testing procedures. FM Global, formerly known as Factory Mutual, recommends that for boilers up to 400 psig, including low-pressure boilers, the safety valves should be manually tested by pulling the lever once per month, and tested by actually raising the pressure to the safety valve's popping point once per year. The pressure test may be performed on the boiler or on a test bench in an authorized valve repair shop. Testing frequencies for boilers operating at over 400 psig should be determined from actual operating experience.

Safety valves and other safety devices on boilers should be tested often enough to demonstrate reliability. The test frequency is affected by the quality of the water treatment in the boiler and the steam pressure and superheat. For example, a safety valve may be adversely affected by frequent testing if the pressure and temperature of the steam are high enough to damage the disc and seat.

It is good practice to conduct the pressure tests just before a scheduled boiler shutdown. If a safety valve does not perform correctly, it can be serviced while the boiler is shut down.

37 What is the pressure at which a high-pressure boiler's safety valve should be manually tested?

A high-pressure boiler's safety valve should be tested at about 75% of the normal popping pressure. The steam pressure should be high enough to blow out any debris that could prevent the valve from reseating properly, yet low enough that the valve immediately reseats when the testing lever is released.

38 What is the minimum pressure at which a low-pressure boiler's safety valve should be tested?

The minimum pressure should be 5 psig. This should provide enough steam pressure to blow out any debris that would keep the valve from reseating properly.

39 Why might a safety valve fail to operate as intended?

A safety valve may be stuck because of corrosion or scale, or it may have been tampered with or blocked closed.

Trade Tip

If a safety valve leaks after popping, it is often due to debris being caught in the valve. Manually popping the valve a few times often dislodges the debris.

Is it permissible to install shutoff valves between the boiler and the safety valve? Why or why not?

Shutoff valves must not be installed between the boiler and the safety valve because closing the shutoff valve renders the safety valve inoperative.

Why might a safety valve not reseat tightly after popping open?

A safety valve may not reseat tightly after popping if scale or other debris becomes stuck in the valve. Popping the valve manually with the lifting lever a few times usually blows out the debris. Having the reseating pressure set too close to the lifting pressure can also cause this problem.

Thermal expansion of a valve may also cause problems with the valve's operation. If there is insufficient provision for expansion, the expansion may warp the body of the valve. For example, the discharge piping may be attached too rigidly, allowing thermal expansion to bind the safety valve.

Factoid

The safety valve's discharge line must never create any external strain on the safety valve. The discharge piping must be solidly anchored to prevent any movement.

Describe the installation of a safety valve discharge line.

Typically, a safety valve discharge elbow is used. This elbow is piped to the discharge outlet of the valve. It has a round drip pan that catches condensate from the steam and an extension that enters the discharge piping to the roof. **See Figure 3-12.** This short extension is not to be attached to the discharge piping. The discharge piping telescopes over the short extension, and this separation allows the discharge piping to expand and contract independently of the safety valve.

A small hole in the bottom of the elbow allows condensate to drain out. This prevents water from corroding the valve and ice from accumulating in the discharge piping if the boiler is outdoors. The safety valve body also has a hole to allow any condensate in the valve itself to run out.

The inlet connection to the valve must never be restricted to any size smaller than the valve inlet itself, and the discharge piping must never be made smaller than the safety valve outlet size. In most cases, the discharge piping is made larger than the safety valve outlet size to ensure that there is no interference with the discharge of the valve. The discharge piping is rigidly anchored to keep the steam from moving it.

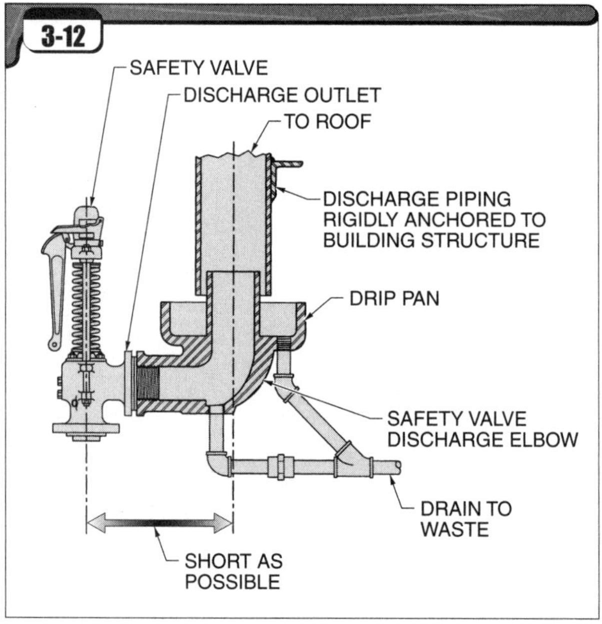

Figure 3-12. *The safety valve discharge elbow attaches to the discharge of the safety valve.*

What is the purpose of an accumulation test?

An *accumulation test* is a pressure test used to ensure that the safety valve has sufficient relieving capacity to vent all of the excess steam that the boiler can produce. This test is performed only on a new boiler or when a new safety valve is installed. As a rule, boilers equipped with superheaters should not be subjected to an accumulation test due to the potential for overheating the superheater.

Describe how to conduct an accumulation test.

The boiler is fired at its maximum firing rate and the steam outlet valve (nonreturn valve or main steam stop valve) is slowly closed. When the first safety valve lifts, the nonreturn valve or main steam stop valve is then fully closed, so the only path for the steam is through the safety valves. The safety valves are allowed to vent continuously, and the steam in the boiler is monitored. In high-pressure boilers, the pressure must not rise higher than 6% above the MAWP. For low-pressure boilers, the pressure must not rise to greater than 20 psig. If the pressure does continue to rise, the safety valves do not have enough capacity to vent all the excess pressure and must be replaced with valves having a greater venting capacity.

The water level in the boiler is monitored during the exercise. For each pound of steam that is vented through the safety valve, a pound of water must be replaced in the boiler. The operator must keep watch over the water supply to ensure that the boiler water is replaced. State or insurance company boiler inspectors should be notified and have the opportunity to be present to witness this test.

45 If a high-pressure boiler's MAWP is 250 psig, what is the maximum pressure that may be developed during an accumulation test?

The maximum pressure is 265 psig (250 × 1.06 = 265).

46 What are the acceptable alternatives to applying an accumulation test?

The maximum fuel-burning capacity of the boiler's combustion equipment directly reflects the steam-generating capacity of the boiler. Therefore, the maximum steam-generating capacity can be roughly determined from the heating value of the fuel (the number of Btu available in each unit of fuel) and the quantity of fuel that can be burned over a period of time. The boiler designers can determine how many Btu will be added to each pound of steam generated, and thus how many pounds per hour of steam will be generated.

Alternately, the maximum feedwater delivery to the boiler may also be used in determining the required safety valve capacity. Each pound of steam generated must be produced from a pound of water; therefore, the maximum quantity of water that may be delivered to the boiler directly reflects the maximum steam generation possible.

47 Where should safety valves be installed other than on steam boilers?

Safety valves are installed wherever appropriate to protect equipment from overpressure. For example, safety valves may be installed on heat exchange equipment and other pressure vessels or on piping where the pressure could become greater than the rated strength of the piping.

Boiler Vents

48 What is a boiler vent?

A *boiler vent* is a section of steel pipe about ½″ to 1″ ID coming off the top of the vessel with one or two valves in it. **See Figure 3-13.** The boiler vent is also known as a drum vent or air cock. The length of the pipe varies. Sometimes the boiler vent runs all the way above the roof. The boiler vent pipe should terminate in a safe location where its end can be seen. The boiler vent has several purposes.

As the boiler is filled with water, the vent allows air in the boiler to be displaced. Otherwise, the air would stay in the boiler and ultimately pass out of the boiler with the steam. Air interferes with proper heat transfer in the steam system, so it is desirable to remove the air.

When the water in the boiler is heated, the dissolved air starts to come out of the water. When the water reaches approximately 180°F, the air contained in the water begins to separate and bubble off relatively quickly. This is similar to the air bubbles that gather in the water from heating a pan on a stove. Air contains oxygen, which causes corrosion in the boiler. To prevent corrosion, the air must be allowed to vent off until only steam comes out. On high-pressure boilers, the boiler vent should be left open until there is approximately 25 psig in the boiler. On low-pressure boilers, the boiler vent should be left open until there is approximately 2 psig to 5 psig in the boiler. This ensures that all the air is vented.

When the boiler drain valves are opened, the boiler vent must also be opened. This allows air to be drawn into the boiler to prevent the formation of a vacuum as the water drains out.

The boiler is a pressure vessel and is constructed to withstand steam pressure on the inside, pushing out. If a vacuum forms inside the boiler, the atmospheric pressure outside the boiler pushes inward and can cause an implosion. An *implosion* is an inward collapse from external pressure. A vacuum can develop while the boiler drains, or when the cooling steam inside condenses after the boiler is shut down. While boiler implosions are rare, such an occurrence is possible. Formation of a strong vacuum can also cause leaks to develop or cause excessive water and treatment chemicals to be siphoned into the boiler by the vacuum. Venting allows air to enter the boiler and prevents a vacuum from forming in these cases. When shutting down a high-pressure boiler, the boiler vent should be opened when the pressure falls to about 25 psig. When shutting down a low-pressure boiler, the boiler vent should be opened when the boiler pressure falls to about 2 psig to 5 psig.

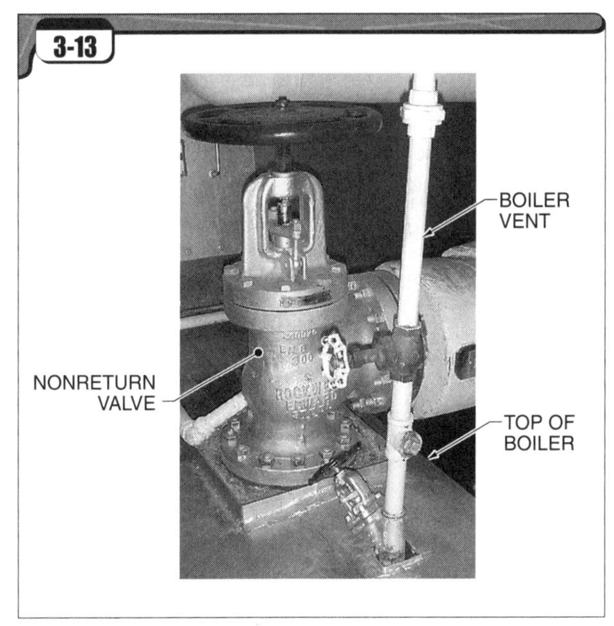

Figure 3-13. The boiler vent is used to remove air from the boiler during filling and startup, and to allow air into the boiler to prevent vacuum formation during shutdown.

49 Where is the boiler vent located?

The air in the boiler travels to the highest point in the steam space. Therefore, the boiler vent is connected to the highest point on the top of the shell or steam drum of the boiler.

Steam Quality

What is quality of steam?

Quality of steam is the dryness of the steam. Pure steam is a gas and contains no moisture. For example, if the quality of steam is 97%, it contains 3% moisture.

What is foaming in a steam boiler?

Foaming is the development of froth on the surface of the boiler water. Foaming typically occurs because some organic contaminant, such as oil, is in the boiler. For example, condensate returning from a leaking fuel oil heat exchanger could contain fuel oil. Foaming causes a film of impurities to develop on the surface of the water. The steam bubbles have difficulty breaking through this film. As a result, the water level in the boiler becomes violently erratic. The sight glass may be completely full at one moment and may be completely empty at the next moment.

Foaming may also be due to excessive concentrations of water treatment chemicals or other impurities in the water. To stop foaming, the source where the impurities entered the boiler water must be found and repaired. Foaming may be reduced by draining some of the contaminated water (also known as blowing down the boiler) and replacing it with uncontaminated, properly treated water.

What is carryover?

Carryover is the entrainment of small water droplets with steam leaving the boiler. *Entrainment* is the process where solid or liquid particles are carried along with steam flow. Carryover is most commonly due to sudden substantial demands for steam or to foaming. For example, if a sudden, large demand for steam is placed on the boiler, the pressure in the boiler drops and some of the water in the boiler immediately flashes into steam. The formation of many flash steam bubbles in the boiler causes the water level to rise, or swell, and this places the water level closer to the steam outlet nozzle. At the same time, the large rush of steam through the outlet nozzle creates very high steam velocity at this area, and boiler water can become entrained in the steam flow.

Overfiring the boiler and a high water level can also cause carryover. *Overfiring* a boiler is forcing the boiler beyond its designed steam-producing capacity.

What is priming?

Priming is a severe form of carryover in which large slugs of water leave the boiler with the steam. Priming is commonly a result of foaming where contaminants are present in the boiler water. Priming may also result from sudden and excessive surges of steam from the boiler. Priming is dangerous because it can produce water hammer.

What is water hammer?

Water hammer is the hydraulic shock in piping caused by the presence of liquids in the steam flow. **See Figure 3-14.** Water hammer is most commonly caused by condensate buildup in the steam piping. Steam is a gas, and though it travels through steam piping at up to about 135 miles per hour, it offers comparatively little resistance to change in direction. If condensate is allowed to accumulate in the steam piping, the friction created as the steam moves across the surface of the condensate creates a dragging effect and waves form on the surface of the condensate. If enough condensate accumulates so that these waves close off the entire area of the pipe, the condensate instantly forms a piston that is pushed through the piping at high velocity. The velocity of the water is accelerated as steam downstream of the piston condenses and the pressure drops. Water has considerable mass and resists directional changes. When the high-velocity slug of condensate encounters an elbow, tee, or other restriction in the pipe, the water tends to continue in a straight line and strikes the obstruction with great force. This is dangerous because the piping can burst, releasing steam and scalding condensate in the area.

Water hammer can also occur in piping because of carryover or priming from the boiler, an inadequate number of steam traps to remove condensate from the piping, failed traps, inadequate insulation, and other reasons.

Where does water hammer damage a steam line first?

Water hammer damages a steam line first at elbows, tees, or restrictions such as control valves in the piping.

Why should steam be admitted slowly to a cold heating system?

The high temperature of the steam causes a thermal strain (shock) on the cold piping, and the rapid condensation of the steam can result in water hammer.

Describe two methods of drying steam.

Cyclone separators are cylindrical in shape and are found in many watertube boilers. **See Figure 3-15.** The steam and water mixture coming into the drum from the riser tubes passes through a cyclone separator. The mixture is swirled rapidly by stationary vanes in the separator and the water is thrown out by centrifugal force. The water drains out the bottom of the cyclone separator and the steam passes out the top.

Steam scrubbers are a series of corrugated plates that the steam passes through when leaving the boiler drum. Steam scrubbers force the steam to follow a zigzag pattern as it passes through. The water resists these directional changes and is removed. Steam scrubbers are sometimes known as chevron scrubbers because of their shape.

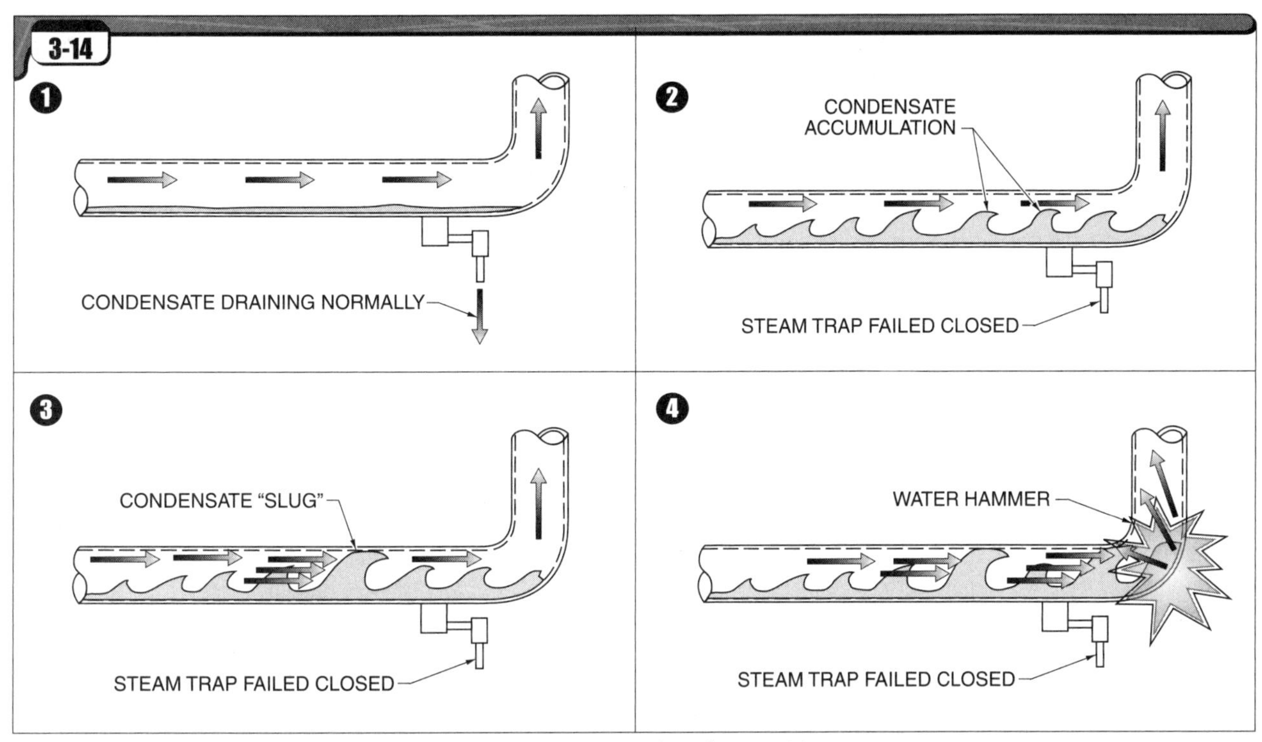

Figure 3-14. *Water hammer is dangerous because it can rupture steam and condensate piping.*

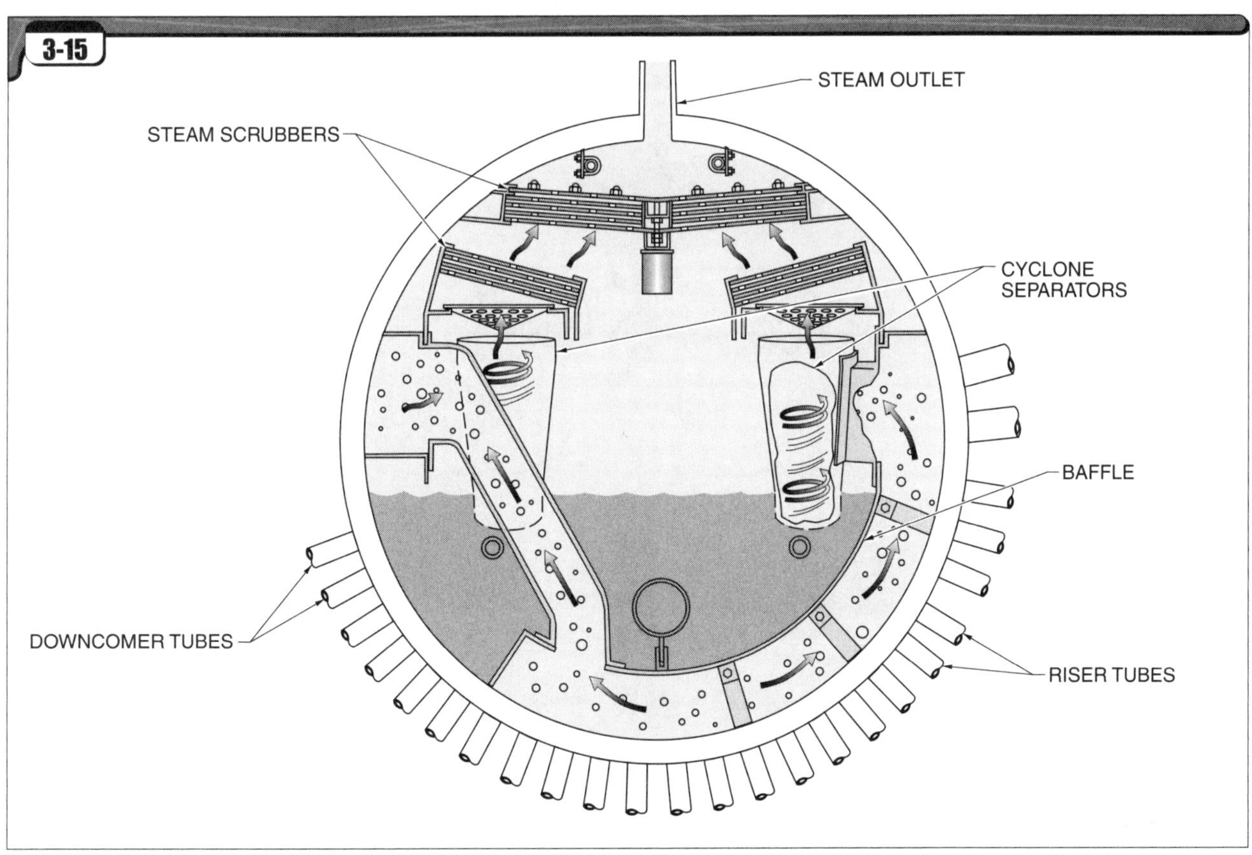

Figure 3-15. *Cyclone separators and steam scrubbers are usually found in large watertube boilers.*

Why are baffles used inside firetube boilers and in the drums of watertube boilers?

Baffles, like the various other types of separation equipment at the steam outlet, are used to create a rapid change in the direction of flow of the steam leaving the steam space of a boiler. This helps separate the water from the steam by centrifugal force, thus producing drier steam.

Baffles are also used to force the water and steam mixture to follow a desired path in the steam drum. For example, baffles may be used in the bottom of the drum to force the steam and water mixture to pass through the cyclone separators.

What is a dry pipe?

A *dry pipe* is an upside-down T-shaped pipe connected to the main steam outlet from either a firetube or watertube boiler. **See Figure 3-16.** The pipe contains many holes or slots cut into the top side and a water drain in the bottom. As the steam passes through the dry pipe, it is forced to change direction quickly several times before leaving the boiler. The water, being denser than the steam, continues in one direction and is thrown out of the steam by centrifugal force. The dry pipe is similar in this way to a baffle.

Factoid

Cyclones and dry pipes are commonly used to separate water droplets from steam.

What is a drum pressure control valve?

A *drum pressure control valve* is a control valve that is configured so as to maintain a constant pressure on the steam drum of a watertube boiler or the shell of a firetube boiler at all times. **See Figure 3-17.** The purpose of such an installation is to minimize carryover by minimizing the flashing of water into steam that occurs during periods when large amounts of steam are drawn from the boiler in a short period of time.

A drum pressure control valve is equipped with a sensing line that senses the pressure in the boiler. The valve itself is installed in the main steam outlet line between the boiler and the steam header. If the boiler pressure begins to drop, the valve starts to close and reduces the amount of steam leaving the boiler. If the boiler pressure rises, the valve starts to open and increases the amount of steam leaving the boiler. By maintaining a constant pressure in the boiler, flashing of the boiler water is greatly reduced and the water level remains fairly constant. This helps to significantly reduce carryover.

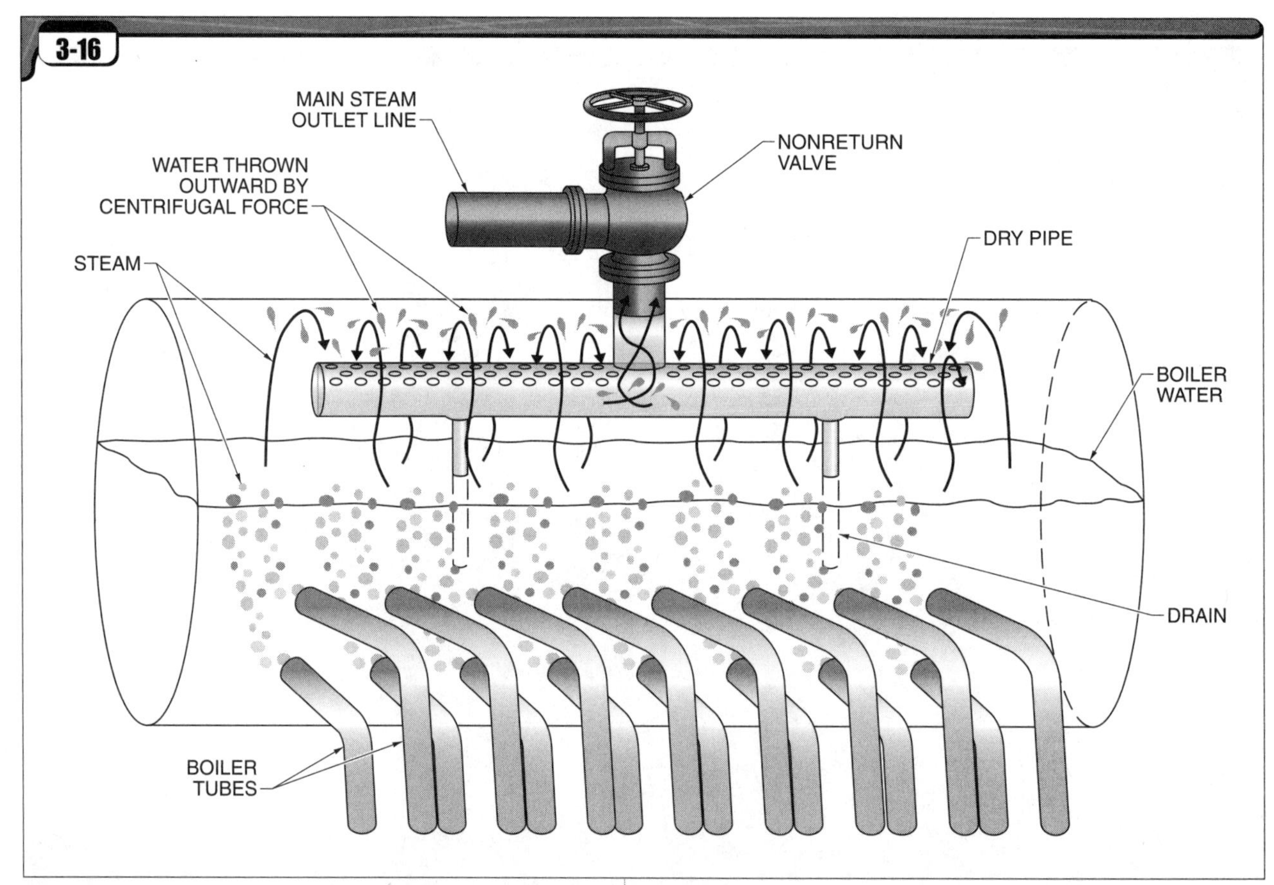

Figure 3-16. *Dry pipes force the steam to change direction quickly, throwing water out by centrifugal force.*

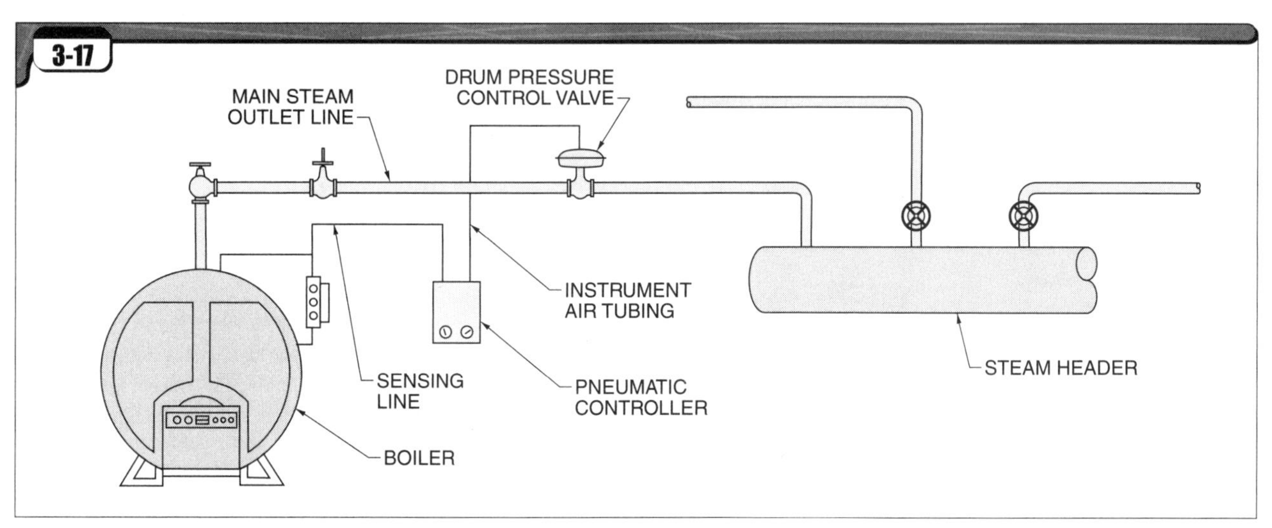

Figure 3-17. *Drum pressure control valves maintain a constant pressure on the boiler during operation. This prevents flashing boiler water from causing erratic water level fluctuations.*

Valves

61 What is a gate valve?

A *gate valve* is a valve used to stop or start flow. It has a wedge-like disc that is lowered into or raised out of the path of the fluid that flows through the valve. **See Figure 3-18.** The flow passes straight through the valve with no bends and minimal restrictions to the flow.

A gate valve should not be used to regulate flow because flow through a partially closed gate valve erodes the gate (wedge) and seats. This rounds the edges so that the valve does not seal tightly when closed. A gate valve also has poor control characteristics. Most of the flow control occurs in approximately the first 20% of the opening of the valve.

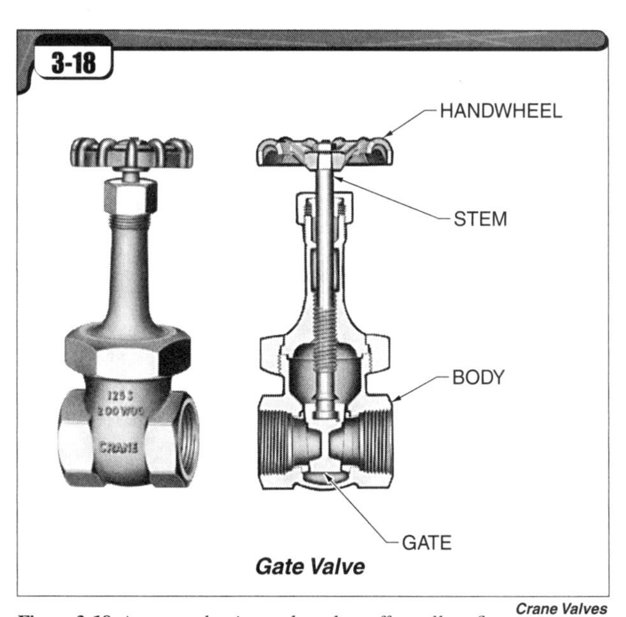

Figure 3-18. *A gate valve is used to shut off or allow flow.*

62 What is meant by throttling a valve?

Throttling is controlling the amount of flow that passes through a valve by partially closing it.

63 What is wire drawing?

Wire drawing is the erosion that occurs as steam or another high-velocity fluid flows through a small opening like a throttled valve. It is best avoided.

64 How is wire drawing avoided?

Wire drawing is avoided by throttling steam with a valve designed for that purpose instead of the gate valves in the steam lines. Control valves and pressure regulating valves are constructed of materials that resist wire drawing.

65 Why are gate valves used for steam service?

Gate valves are used for steam service because they offer little resistance to the flow. The wear on the valve and the pressure loss in the steam, which are due to friction in the valve, are small.

66 What do the letters os&y represent?

The letters os&y represent outside stem and yoke. An os&y valve is also known as an outside screw and yoke valve. In an os&y valve, the valve stem screws out from the center of the valve handwheel when the valve is opened. **See Figure 3-19.**

Factoid

In a cyclone, the heavier particles that are being carried along by the fluid are heavier than the fluid and have more momentum. More momentum means that the heavier particles are not able to make the turns and get thrown out of the flowing fluid stream.

Most high-pressure valves are of this design. The position of the stem allows the operator to visually determine whether the os&y valve is open or closed. This is convenient, especially when the os&y valve is installed where it is not easily reached.

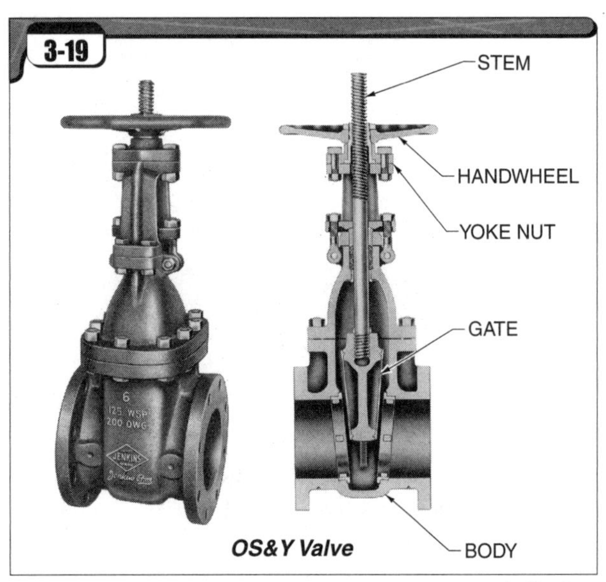

Figure 3-19. The stem of an os&y valve screws out from the center of the handwheel when the valve is opened.

67 What is a rising stem valve?

A *rising stem valve* is a valve that has a handwheel and stem that move outward from the body of the valve as the valve is opened. **See Figure 3-20.** This is the most common design for small and low-pressure valves. The terms rising stem and os&y are often used interchangeably; however, a valve with a rising stem is not necessarily an os&y valve.

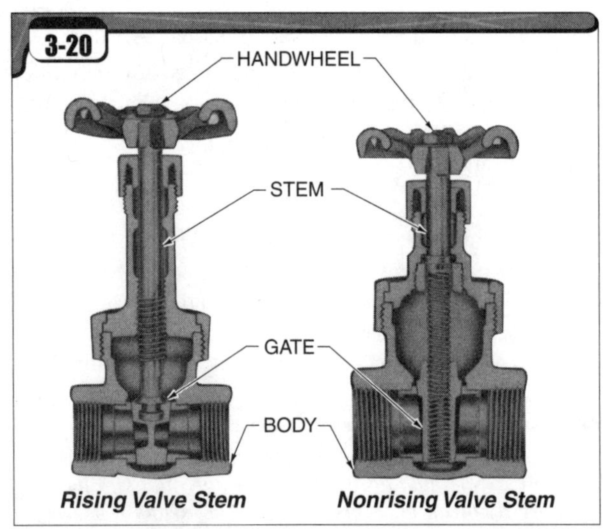

Figure 3-20. The position of a rising stem valve can be visually determined. The position of a nonrising stem valve cannot be visually determined.

68 What is a nonrising stem valve?

A *nonrising stem valve* is a valve that has a disc in the valve that threads up onto the stem as the stem is turned, and the stem does not back out of the valve. The position of a nonrising stem valve cannot be visually determined.

69 What is a ball valve?

A *ball valve* is a quick-acting, two-position shutoff valve. Ball valves contain a ball that is usually stainless steel or brass. The ball has a hole that the fluid passes through. Ball valves are opened and closed by moving a lever handle 90°. **See Figure 3-21.** When the ball valve is open, the hole through the ball is in line with the flow, and flow passes through. When the ball valve is turned 90°, the hole through the ball is perpendicular to the flow, and no flow can pass through.

Ball valves are inexpensive and simple in design. They offer little resistance to flow and are designed to be fully open or fully closed. Ball valves should not be used for throttling, which can erode the ball. Also, control is poor when throttling with a ball valve.

Ball valves are also made in three- and four-way patterns. These valves are used in order to minimize the number of valves required in a complicated piping system.

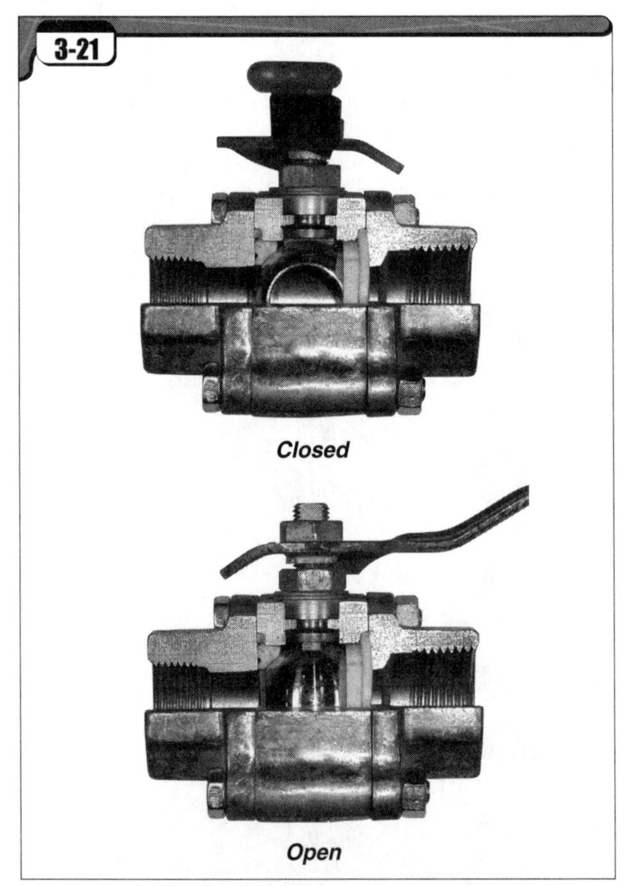

Figure 3-21. A ball valve is a quick-acting, two-position shutoff valve.

What is a plug valve?

A *plug valve* is a valve that is similar to a ball valve, except that a plug valve contains a semi-conical plug through which the flow passes. The passageway through the plug is a rectangular hole, rather than the round hole of a ball valve. The plug turns 90° to open or close, as does the ball in a ball valve. Plug valves are most commonly used for natural gas service, but are also often used for liquids. They are rarely used for steam.

What is a globe valve?

A *globe valve* is a valve that has a tapered, rounded, or flat disc held horizontally on the stem. Closing the valve pushes the disc down onto a matching seat that mates with the disc for a tight fit. **See Figure 3-22.** Globe valves are designed to control the flow of fluids such as water or compressed air. Standard globe valves are used occasionally for low-pressure steam service and they may be used to stop flow when a piece of equipment is taken out of service for maintenance.

Standard globe valves erode severely if used on high-pressure steam systems. Therefore, a globe valve to be used for steam service must have the interior parts (trim) constructed of special materials to resist erosion and wire drawing. For example, trim can be constructed of an alloy called Monel®, or the parts can be hardfaced with Stellite®. *Stellite* is an alloy of chromium, cobalt, and tungsten. In the process of applying Stellite hardfacing, the valve's interior components are coated with Stellite to make them very hard and erosion resistant.

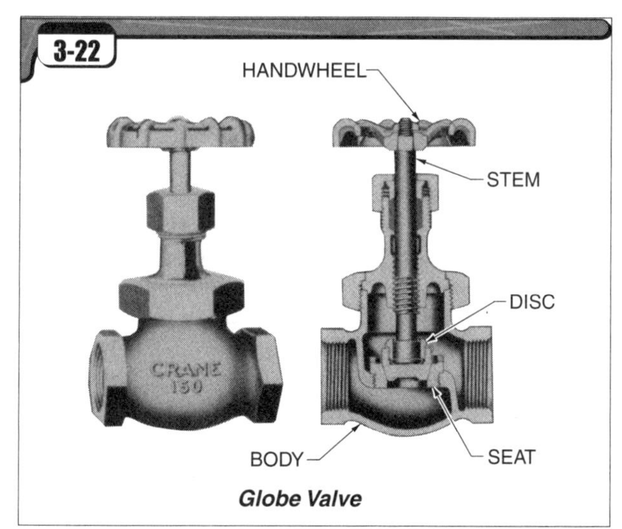

Figure 3-22. A globe valve has a tapered, rounded, or flat disc held horizontally on the stem.

Describe the flow through a globe valve.

The flow entering a globe valve turns upward and strikes the disc. The flow is then deflected 90° and passes out of the valve. A globe valve should always be installed so that the entering flow strikes the bottom of the disc. If the valve is installed backward, the valve may wear unevenly, be difficult to open, and chatter when in service.

Why is a globe valve suitable for throttling service?

A globe valve is suitable for throttling service because the flow through the valve is fairly proportional to the amount that the valve is open. Additionally, the flow is directed against the disc more evenly than with a gate valve and therefore the wear on the disc is more even.

What is a needle valve?

A *needle valve* is a valve that is very similar to a globe valve, except that the opening/closing mechanism on the end of the valve stem is usually a sharp tapered cone that seats in a matching cone-shaped seat. This design gives the needle valve a more precise degree of flow control than a globe valve. Needle valves are small in size and used in applications where only a small, precisely adjusted flow is needed. This includes small sampling flows, lubricant flows, and chemical flow control applications.

What is a piston valve?

A *piston valve* is a valve that contains a finely machined piston that moves up or down in the interior of a cylindrical steel cage. **See Figure 3-23.** At each end of the cage are sets of graphite packing rings, and the piston extends and retracts through the interior of these rings. As the valve is opened, the piston rises out of the lower set of graphite rings, allowing flow to pass through the valve. Once the piston reenters the lower set of graphite rings, flow is stopped. The valve does not need to be closed to the end of its travel. The graphite rings form the sealing surface against the piston. Piston valves work well for throttling service, provide close flow control, and are not as susceptible to seat damage as gate valves or globe valves.

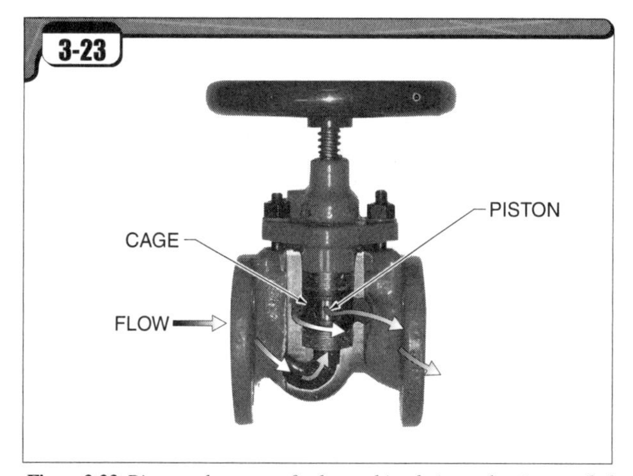

Figure 3-23. Piston valves use a finely machined piston that is extended or retracted inside of a cylindrical steel cage.

76 What is a butterfly valve?

A *butterfly valve* is a valve that consists of a circular disc that is rotated by the valve stem so that the disc is parallel to the flow through the valve, perpendicular to the flow, or somewhere in between. **See Figure 3-24.** Butterfly valves rotate from fully open to fully closed in 90° of travel. Butterfly valves are generally used for liquid service and for cooler gases, but rarely for steam.

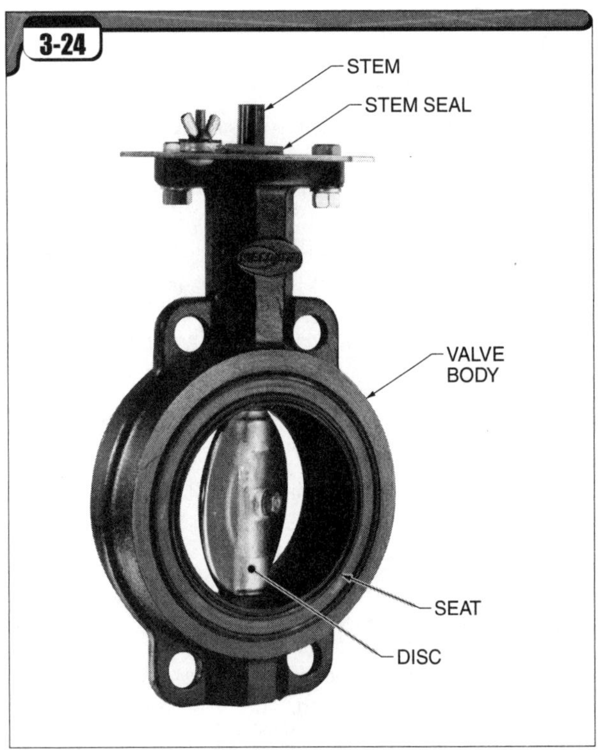

Watts Regulator Company

Figure 3-24. *Butterfly valves use a circular disc that is rotated by the valve stem.*

77 What is a diaphragm valve?

A *diaphragm valve* is a valve that uses a flexible diaphragm as the movable sealing surface. Diaphragm valves are mainly used for liquids such as water, oils, and liquid chemical solutions. When the valve is closed, the flexible diaphragm is pressed onto the top of a dividing barrier, called a weir, between the inlet and outlet ports. **See Figure 3-25.** When the valve is opened, the diaphragm is relaxed and withdrawn from the dividing barrier. The liquid can then flow from the inlet port, under the diaphragm, and into the outlet port.

Factoid

Steam valves should always be opened very slowly to allow piping to expand and to avoid water hammer.

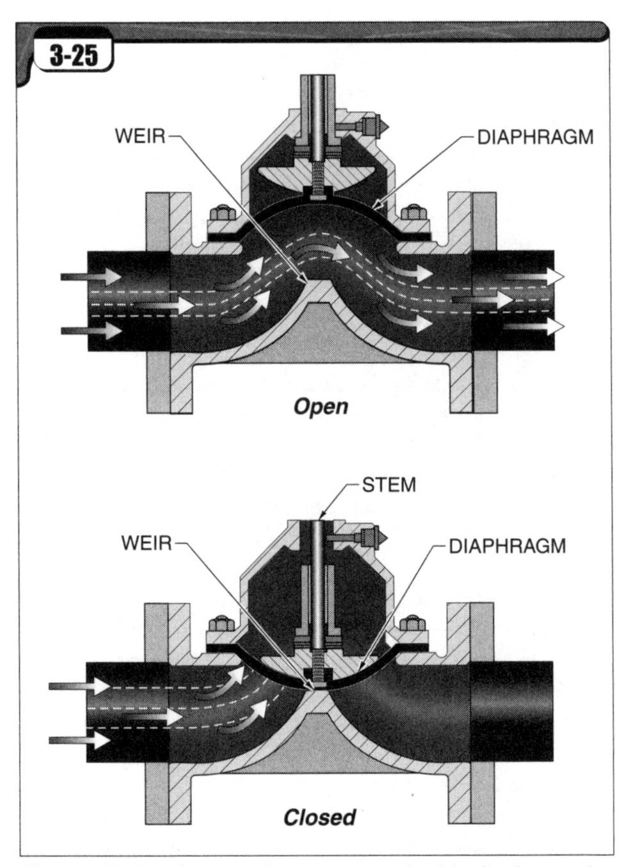

Figure 3-25. *Diaphragm valves contain a flexible diaphragm that closes against a dividing barrier, or weir. They are most commonly used for liquids.*

78 What is the purpose of the small bypass line attached to the body of a large manual valve?

The small bypass (equalizing) line equalizes the temperatures and/or pressures on both sides of a large valve before opening it. **See Figure 3-26.** Equalizing lines may be installed by the valve manufacturer or may be field installed.

If the pressure is much higher on one side of the valve than on the other side, the valve may be difficult to open. Also, a very slight opening of a large valve may cause a substantial flow, and a sudden rush of flow and pressure into the downstream piping can damage the piping.

If the temperature is much higher on one side of the valve than on the other side, opening the valve quickly can cause thermal strains on the downstream piping. If the material in the piping is steam in this case, the steam condenses rapidly as it contacts the cooler downstream piping. This rapid condensation may accumulate faster than the steam traps can remove it, and the piping begins to flood. The steam may then propel slugs of water through the piping. This makes the piping pound, hammer, and possibly rupture.

These conditions may be avoided by slowly equalizing the pressures and/or temperatures through the small bypass line. The large valve can then be opened more easily and safely.

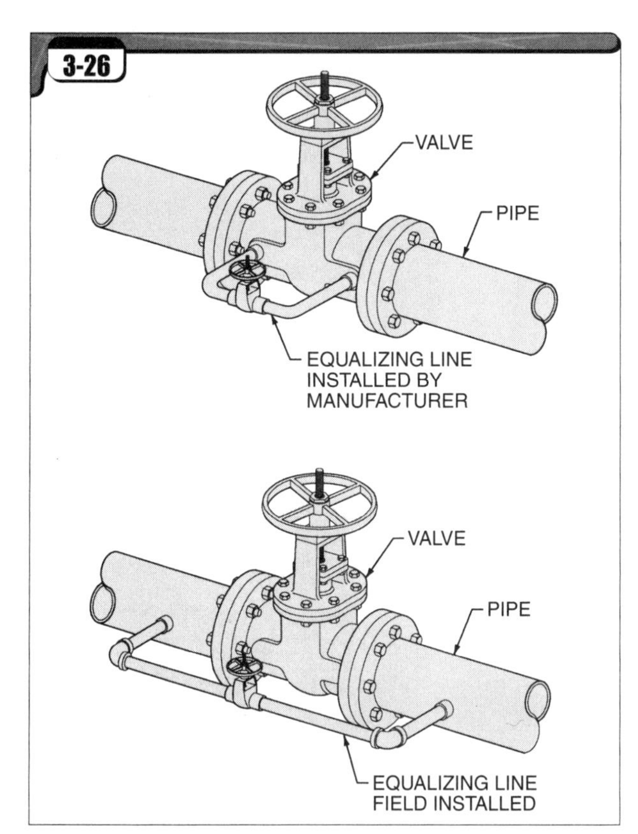

Figure 3-26. *The small bypass line equalizes the pressures and temperatures on both sides of a large valve before opening it.*

79 What is the main steam stop valve?

A *main steam stop valve* is a gate valve in the main steam line between the boiler and the steam header used for isolating the steam side of a boiler that is to be out of service.

80 What is a check valve?

A *check valve* is a one-way flow valve for fluids. It allows the flow to pass through in only one direction. If the direction of the flow reverses, the valve closes and prevents backflow. **See Figure 3-27.**

81 Describe two different classes of check valves.

A swing check valve has a flapper or disc that swings in one direction. If the flow reverses, the flapper falls onto its seat and stops the backflow.

A lift check valve has a piston, ball, or disc that is forced upward (sometimes against the compression of a spring) as the flow passes through. If the flow reverses, the piston, ball, or disc moves back onto its seat and stops the backflow.

82 What type of valves are used between the boiler outlet and the steam header on high-pressure boilers?

A nonreturn valve is generally installed directly on the boiler outlet nozzle flange. The main steam stop valve is installed between the nonreturn valve and the steam header.

83 What is a nonreturn valve?

A *nonreturn valve* is a combination shutoff valve and check valve that allows steam to pass out of the boiler. **See Figure 3-28.** However, if the pressure in a boiler drops below the header pressure, which happens in the event of a tube failure, then the backflow of steam from the header into that boiler causes the nonreturn valve to automatically close and isolate the damaged boiler.

The nonreturn valve also has a manual handwheel and stem that can be used to close the valve manually. The valve cannot be opened manually, although the stem can be backed out. This is because the valve disc, or piston, is not attached to the stem. Only pressure from inside the boiler opens the valve. When the boiler pressure rises slightly above the steam header pressure, the nonreturn valve opens and automatically places the boiler on-line with the other boilers.

The ASME Code requires at least two main steam stop valves or one main steam stop valve and one nonreturn valve on all boilers that are connected to a common steam header and have manhole openings. The gate valves should be the os&y type. There must be an adequate free-blowing drain valve provided between these two steam valves. The drain is used to remove condensate from the main steam line before placing the boiler on-line. It is also used to vent pressure from between the two steam stop valves when isolating the boiler for inspection.

This drain also serves to warn of unsafe conditions when the boiler is to be entered for inspection or maintenance. The discharge of this drain should be visible. If the main steam stop valve leaks steam backwards from the steam header toward the boiler, steam blows from the manual drain. This provides a visual indication that the main steam stop valve does not hold tightly.

84 What is a vacuum breaker?

A *vacuum breaker* is a check valve that prevents the formation of a vacuum in a tank, pressure vessel, or piping system. Vacuum breakers are installed where the piping or vessel could be damaged by vacuum conditions inside, where vacuum conditions interfere with proper drainage, or where a reversal of flow direction (backflow) could cause contamination of a piping system with foreign material.

For example, if a large tank were drained without opening the vent on top of the tank, the vessel could implode because of the vacuum inside. A vacuum breaker prevents a vacuum from developing by allowing air to be pulled into the tank to break the vacuum. If vacuum conditions form inside a steam heat exchanger, the vacuum prevents the condensate from being drained. Vacuum breakers allow air to enter the heat exchanger to break the vacuum. This in turn allows the condensate to be drained.

Vacuum breakers are usually spring-loaded check valves. A spring holds the check valve in the closed position unless a vacuum condition occurs. When the differential pressure between the atmosphere and the vacuum inside of the piping or vessel becomes great enough, the atmospheric pressure opens the valve against the compression of the spring. Air can then flow into the piping or vessel to break the vacuum.

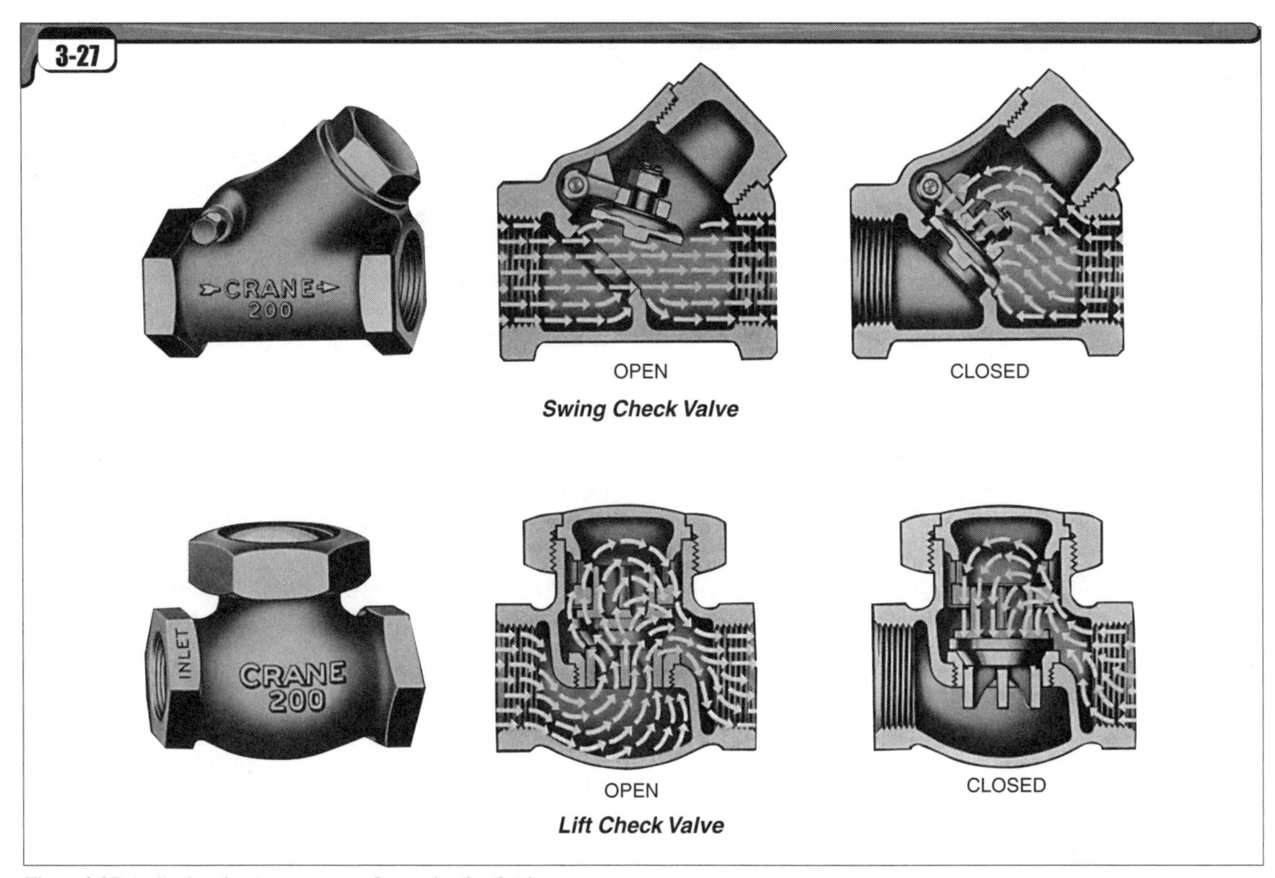

Figure 3-27. A check valve is a one-way flow valve for fluids. Crane Valves

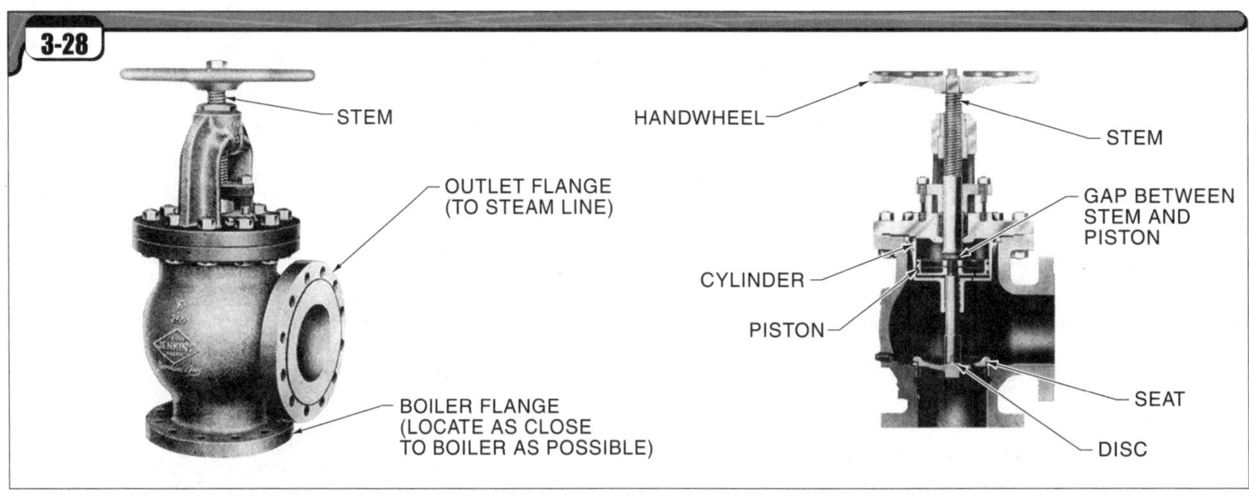

Figure 3-28. A nonreturn valve is a combination shutoff and check valve that allows steam to pass out of the boiler. Jenkins Bros.

Factoid

Some manufacturers use the term working steam pressure (WSP) instead of steam working pressure (SWP). The terms have the same meaning.

85 Why are check valves used on feedwater pumps?

Check valves are used in the boiler feedwater piping to keep the boiler water from draining out of the boiler when the feedwater pump stops. Check valves are also used when feedwater pumps are installed in parallel.

86 What do the letters SWP and WOG on the side of a valve represent?

Steam working pressure (SWP) is the maximum steam working pressure of the valve. *Water, oil, or gas (WOG)* is the maximum pressure under which the valve may be used with these fluids. For example, the side of a valve may read "150 SWP, 400 WOG." This indicates that the valve may be used for steam service at a maximum of 150 psig, and for water, oil, or gas service at up to 400 psig. The steam rating is lower because the high temperature of the steam weakens the valve.

87 What is a control valve?

A *control valve* is a valve used to modulate the flow of fluid. **See Figure 3-29.** A control valve is typically equipped with an actuator. An *actuator* is a device that receives a control signal from a controller and converts the signal into a proportional movement of the valve. For example, the control signal may consist of air pressure or an electric signal delivered over copper wires. Control valves are used for automatic control of a process, such as the liquid level in a tank or the steam pressure supplied to a heat exchanger. The majority are globe valve designs, but they may also be ball valves or other valve types.

Control valve actuators are operated by a local or remote controller or by an attached sensing device. Local control indicates that the device is installed at the location of the actual process. Remote control indicates that the device is installed some distance away from the actual process. The controller may be a pneumatic or electronic control mounted on a control panel, or it may consist of software in a computer.

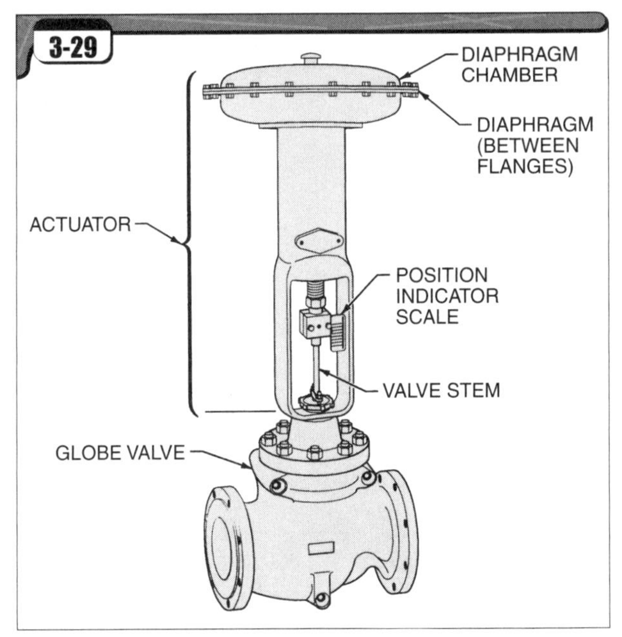

Figure 3-29. A control valve contains an actuator that receives a command signal from a controller and converts the signal into a proportional movement of the valve.

88 What is a solenoid valve?

A *solenoid valve* is a valve that is snapped open or closed by an electric actuator. **See Figure 3-30.** A *solenoid* is an electric actuator that consists of an iron plunger surrounded by an encased coil of wire. When the coil of wire is energized, a magnetic field is created that causes the iron plunger to be extended from or retracted into the solenoid housing. Whether the plunger is extended or retracted depends on the solenoid design. For example, if the plunger is to be extended when the solenoid is energized and retracted when the solenoid is de-energized, a spring is installed that returns the plunger into the retracted position when the solenoid is de-energized. The magnetic field is strong enough to compress the spring when the solenoid is energized.

A solenoid is attached to a valve that is designed to receive it. When the solenoid is energized, the plunger forces the valve to one position, and when the solenoid is de-energized, the spring returns the valve to the other position. The design of these valves varies. For example, the spring is often installed in the valve body rather than in the solenoid body. Whether the valve is open or closed when power is applied depends to a large degree on which position is safer. For example, a natural gas valve should close if there is a power outage. In this case, the valve is designed so that the spring forces the valve closed if the electrical power is lost.

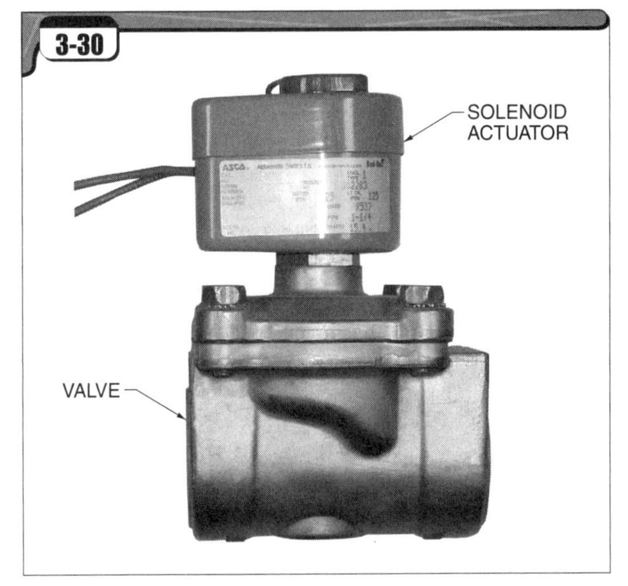

Figure 3-30. A solenoid valve is a valve that is snapped open or closed by an electrical solenoid actuator.

89 What is a pressure reducing valve?

A *pressure reducing valve (PRV)* is a valve that is designed to reduce the pressure of a fluid flowing through a pipe to a desired lower pressure, and constantly maintain this desired pressure downstream of the valve. **See Figure 3-31.** This can be done with a manual valve, but it is tedious and requires the presence of a person at the location of the valve. Thus, it is desirable to have this function performed automatically. A PRV is also known as a pressure regulating valve or a pressure regulator.

It may be necessary to reduce the pressure in particular parts of a steam system for any of several reasons. First, higher pressure steam is smaller in volume than lower pressure steam. Therefore, assuming equal flow rates in lb/hr, higher pressure steam may be passed through smaller diameter pipe than lower pressure steam. Steam is best distributed at higher pressures because the piping can be smaller. PRVs are then used to reduce the steam pressure just before the process that uses the steam.

Also, the temperature of steam corresponds to the pressure. Therefore, if a particular process requires a specific temperature, a specific pressure steam is desirable. For example, the steam may be used to cook a food product. If the steam is too hot, it may burn the product. A PRV allows the steam pressure for the process to be adjusted to the exact requirements of the process.

In addition, PRVs allow for the use of steam heat exchange equipment of lower pressure ratings than that of the boilers and steam distribution system. By using PRVs, the steam pressure is reduced so that heat exchange equipment of lower pressure ratings can be used. Thus, the use of pressure reducing valves allows the plant to use many different steam pressures and temperatures, even though the boilers provide steam at a constant pressure and temperature. In the event of failure of the PRV, such equipment is protected by safety valves.

PRVs can contribute to energy efficiency of the plant. Lower pressure steam has more latent heat per pound than higher pressure steam. This means that fewer pounds of steam are required to deliver the same amount of energy. Since less steam is required, the amount of fuel required to generate the steam is reduced. If the remaining sensible heat is being recovered from the condensate when using high-pressure steam, the efficiency gain from using low-pressure steam may be less. However, the use of low-pressure steam may still be necessary for the reasons mentioned in the previous paragraphs.

90 What is a temperature control valve?

A temperature control valve (TCV) is a valve designed to automatically maintain a constant temperature in a process by controlling the flow of the heating or cooling medium to the process. **See Figure 3-32.** For example, a temperature control valve may be used to control the flow of steam to maintain the water temperature in a tank. The TCVs used in most industrial applications are very similar to pressure reducing valves in that they control the flow of steam. A TCV uses a temperature-sensing device that is immersed in the process material.

The sensing device is connected to the valve actuator by a small copper or stainless steel tube called a capillary tube. The sensing device and the capillary tube are filled with a fluid that expands and contracts with changes in temperature, such as a refrigerant gas, mercury, or a special wax. The expansion and contraction of the fill material in the capillary tube determines the pressure applied to the actuator on the valve, and thus changes the valve position in response to temperature changes in the process material.

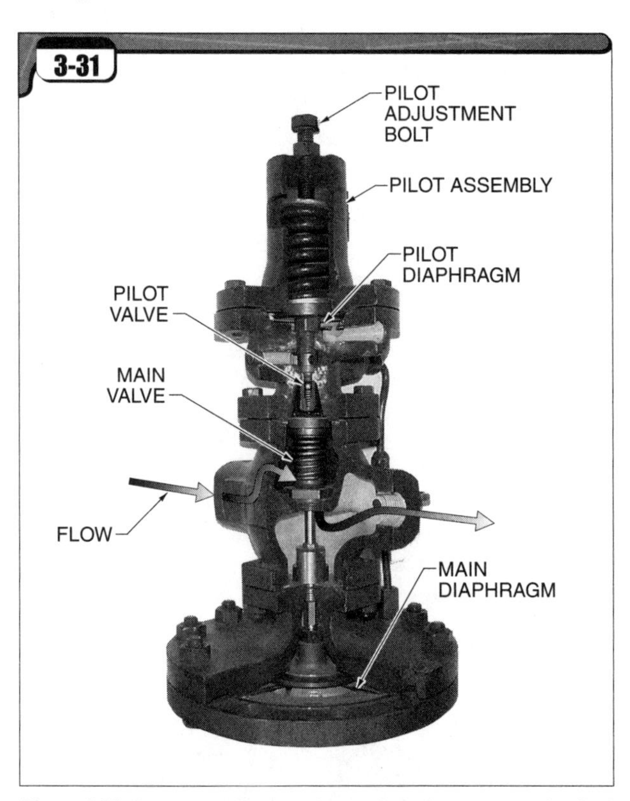

Figure 3-31. *Pressure reducing valves reduce the pressure of a fluid flowing through a pipe and maintain a constant downstream pressure.*

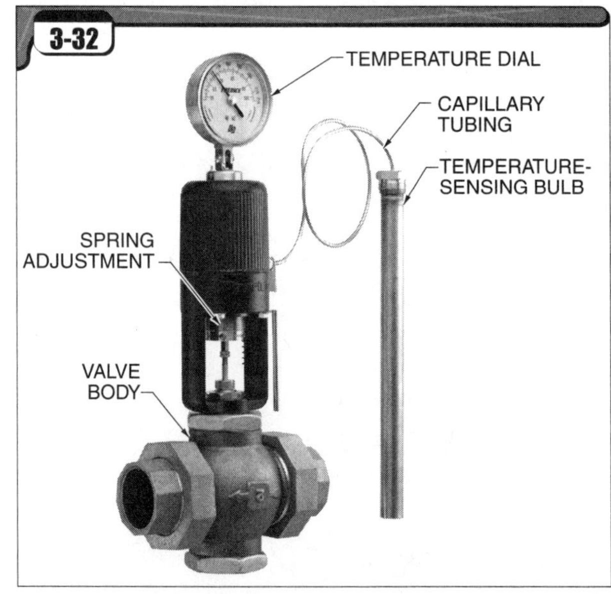

Trerice, H.O., Co.

Figure 3-32. *Temperature control valves automatically adjust the flow of a heating or cooling medium to maintain a constant temperature in a process.*

Piping

List four types of pipe fitting connections.

One type of connection is a threaded connection. **See Figure 3-33.** Male threads are cut on the ends of pipes, and female threads are cut into the pipe fitting. The pipes and fitting are then screwed together. Threaded piping connections are also commonly called screwed connections. Threaded pipe and fittings are generally limited to piping no bigger than about 4″ because it is too cumbersome to work with larger sizes.

Another type of connection is a flanged connection. Flanges are welded or screwed on the ends of pipes to be joined. Flanged fittings are also used. The flanges are then bolted together with a gasket between them.

Another type of connection is a socket welded connection. The socket ends of the pipe fitting slip over the ends of the pipes. Then welds are made around the circumference of the pipes where they enter the fitting.

Another type of connection is a butt welded connection. The ends of the pipe fitting are the same diameter (ID and OD) as the pipes. The ends of the pipes and fitting are beveled and butted together. Then the V-shaped grooves formed by the beveled ends are welded.

The type of pipe joining method used is dependent on several factors. The primary consideration is the pressure and temperature of the fluid involved. The American National Standards Institute (ANSI) and other industry bodies set standards for the strength of various pipe materials and fittings to ensure safe installations. Other factors include the necessity of opening the joints for maintenance purposes and the desire to minimize potential points of leakage.

Factoid

Solenoid-operated valves should not be used for steam service. The snap-open action may lead to water hammer.

List at least five materials from which pipe and tubing used in a steam plant are made.

Carbon steel is the most common material used for standard steel piping. For example, condensate, steam, feedwater, and city or well water are usually piped with carbon steel piping. Carbon steel pipe is commonly referred to as black pipe.

Stainless steel is used for corrosive fluids. For example, some concentrated boiler chemicals may corrode carbon steel piping.

Copper is commonly used for hot and cold water plumbing. For example, the piping to softeners or to the domestic water supply system may be plumbed in copper tubing. Copper and brass piping may be used for pressures and temperatures up to 250 psig or 406°F.

Polyvinyl chloride (PVC) is plastic piping suitable for many corrosive fluids. For example, caustic soda is often piped through PVC piping. PVC may also be used for water and other common fluids. The allowable pressure in PVC piping is greatly influenced by the temperature of the fluid, because the plastic becomes soft as it is heated. For this reason, PVC piping is intended for cooler fluids.

Fiberglass tubing is used for abrasive and/or corrosive fluids. For example, fiberglass is often used for salt slurries that are used in flue-gas cleaning equipment. Like PVC piping, fiberglass piping systems are intended for cooler fluids.

Describe the difference between a one-pipe heating system and a two-pipe heating system.

These terms are used with regard to low-pressure steam heating systems. In a one-pipe heating system, steam enters the radiator and the condensate from the steam returns from the radiator through the same pipe.

In a two-pipe heating system, the steam enters the radiator through one pipe and the condensate from the steam drains out of the radiator through a separate pipe.

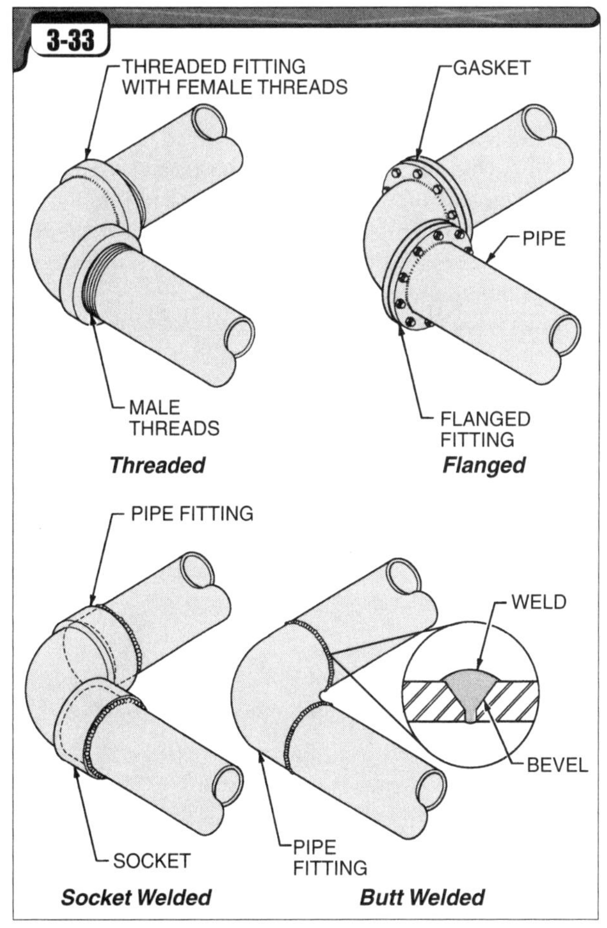

Figure 3-33. *Four types of pipe fitting connections are threaded, flanged, socket welded, and butt welded.*

94 List four methods of providing for expansion and contraction of piping.

Expansion loops in the piping often consist of a simple U-bend or circular loop in the piping. **See Figure 3-34.** For example, a "U" is commonly formed in the piping by welding four 90° elbows to pipes. This method provides flexibility that lets the piping expand inward toward the middle.

Flexible expansion joint fittings consist of a section of corrugated stainless steel tubing with short sections of pipe welded to each end. The pipe sections are either threaded or flanged on the ends. A sheath of braided stainless steel mesh is added for strength.

Sleeve-type expansion joints consist of a telescoping arrangement of one sleeve inside another sleeve, with packing between the two. This type of expansion joint is used where space is limited.

Corrugated expansion joints are specifically made to expand and contract slightly like a bellows or an accordion. This type of expansion joint is often installed at the suction and/or discharge side of a pump. This allows the strain induced by the expansion or contraction of the piping to be isolated from the pump.

The hangers used to suspend piping from steel supports also often have built-in springs, counterweights, or other means to provide for movement of the piping as it expands or contracts, while still providing adequate support of the weight. Lengthwise expansion and contraction of long runs of pipe may also be facilitated through the use of roller-type pipe hangers. In this case, the roller is suspended horizontally between two vertical steel rods, and the piping rests on the roller.

Many accidents have occurred due to inadequate provisions having been made for expansion or contraction of piping, boilers, or other components. The strains created by expansion or contraction can cause the piping to rupture, cast iron valves to be broken, or other serious damage. Large field-erected boilers and utility boilers are almost always suspended from a massive steel beam structure. This allows the entire boiler to expand downward as it is heated and prevents strains on the piping that is attached to the upper drum.

95 What is meant by coefficient of expansion?

Coefficient of expansion is the property of a given material that expresses how much a standard unit of length of the material expands or contracts under a specific change in temperature. For steel piping, the coefficient of expansion is the amount the specific type of steel expands per foot of original length per degree Fahrenheit change in temperature. The coefficient of expansion differs very slightly for different grades and types of steels. It also varies slightly within precise temperature ranges. For practical purposes, engineers use averages and safety factors when calculating the amount of expansion that occurs.

For carbon steels, the coefficient of expansion varies from 0.0000073 to 0.0000080 feet per foot of pipe per degree Fahrenheit temperature change. The linear (lengthwise) expansion is the important issue; the diameter or cross-sectional area of the pipe is of little concern.

Factoid

To avoid damage, provision must be made for steel piping to expand and contract with changes in temperature. An engineering thermal analysis must be performed to determine the proper supports and expansion.

96 How may the expansion and contraction of a length of carbon steel piping be calculated?

The linear expansion of carbon steel pipe may be predicted as follows:

$$E = (T_2 - T_1) \times L \times C.E. \times 12$$

where

E = expansion of the piping (in in.)

T_2 = the higher temperature of the steel (in °F)

T_1 = the lower temperature of the steel (in °F)

L = length of the pipe (in ft)

$C.E.$ = coefficient of expansion (in ft/ft-°F or in./in.-°F)

12 = inches in a foot

For example, what is the expansion that occurs when a 200 ft section of carbon steel piping is heated from 60°F to 350°F, assuming a coefficient of expansion of 0.00000734?

$$E = (T_2 - T_1) \times L \times C.E. \times 12$$

$$E = (350 - 60) \times 200 \times 0.00000734 \times 12$$

$$E = 290 \times 200 \times 0.00000734 \times 12$$

$$E = \mathbf{5.1''}$$

97 How does the use of insulation and lagging contribute to efficiency?

Insulation and lagging are installed on steam and hot liquid piping and vessels to prevent heat from being lost. Lagging generally refers to the aluminum jacketing or other covering that is wrapped around the insulation to protect it from the elements and from mechanical damage. **See Figure 3-35.** Fiberglass pipe insulation is very common. Calcium silicate, a hard, chalky material, is another widely used pipe insulation. This type of insulation is manufactured in two halves that fit around the pipe in a clamshell configuration. The two halves are held together with light wire or fiberglass-reinforced tape and then covered with lagging.

Cold water piping and vessels are also often insulated. Neoprene is usually used for this purpose. Neoprene is a synthetic rubber that resists oils and other liquids. Neoprene prevents the humidity in the air from condensing on the surface (sweating). This helps prevent corrosion of the exterior surfaces.

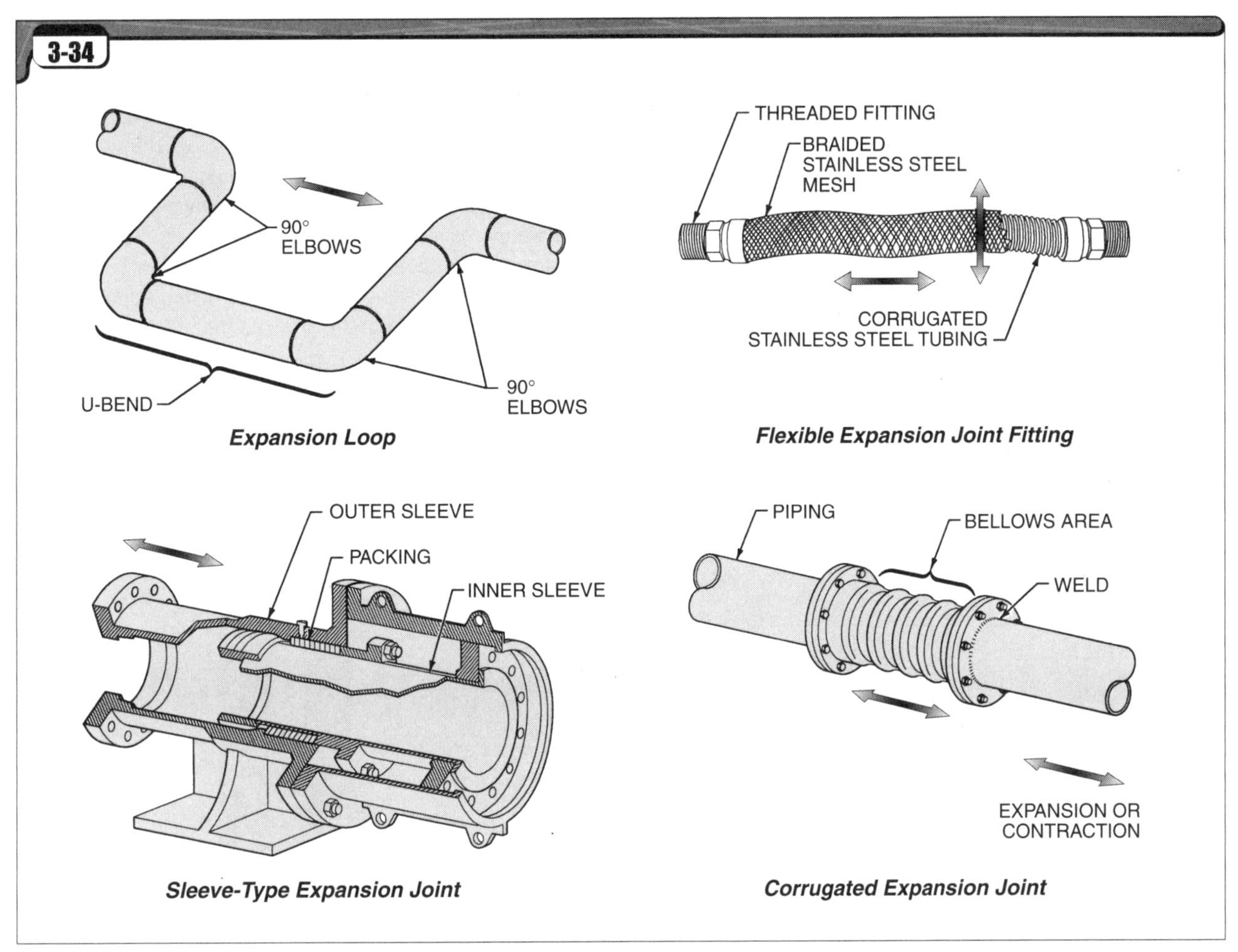

Figure 3-34. *Expansion loops, flexible expansion joint fittings, sleeve-type expansion joints, and corrugated expansion joints provide for expansion and contraction of piping.*

Figure 3-35. *Insulation and lagging prevent excessive heat loss.*

Describe the issues to be considered when piping a main steam line from the boiler to a major piece of steam-using equipment.

The steam piping should be anchored enough to prevent excessive movement, but a provision should be made for expansion in the form of expansion joints, loops, or pipe rollers. The steam piping should be large enough and strong enough to handle the steam flow and pressure required. It should include enough properly sized steam traps to remove condensate from the piping. Adequate valves should be provided and a shutoff (gate) valve should be supplied immediately before the steam-using equipment. A drain should be provided in the piping so that any remaining condensate may be drained and any remaining pressure may be relieved prior to maintenance work. The piping should not create a strain on the steam-using equipment because of expansion or weight. Insulation should be installed to prevent excessive heat loss.

Factoid

Sudden substantial demands for steam from a boiler should be avoided, as they can lead to carryover of boiler water droplets in the steam.

99 Why should branches from steam mains always be taken from the top of the main?

It is desirable to have the cleanest, driest steam possible for plant processes. If the branch line is connected to the side or bottom of a steam main, the steam may carry along condensate, rust and products of corrosion, and other contaminants. **See Figure 3-36.**

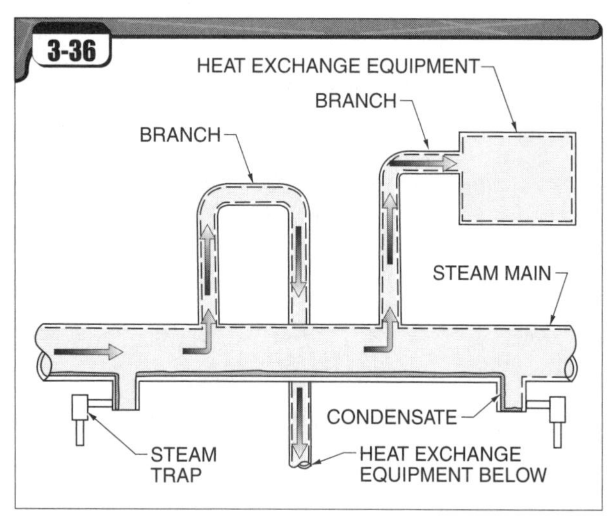

Figure 3-36. *Branch lines should be taken from the top of the steam main in order to obtain clean, dry steam.*

100 What is steam tracing?

Steam tracing is a small copper or steel tube which is supplied with steam and is usually run alongside a process pipe to keep the fluid within the pipe warm. The pipe and the tracing are wrapped together with insulation so that the steam tubing keeps the process pipe warm. **See Figure 3-37.** Steam tracing is used for freeze prevention or maintenance of process temperature in piping. Steam tracing is useful for moving viscous liquids like heavy fuel oils that might otherwise congeal in the piping. A *viscous liquid* is a liquid that is thick and resists flow.

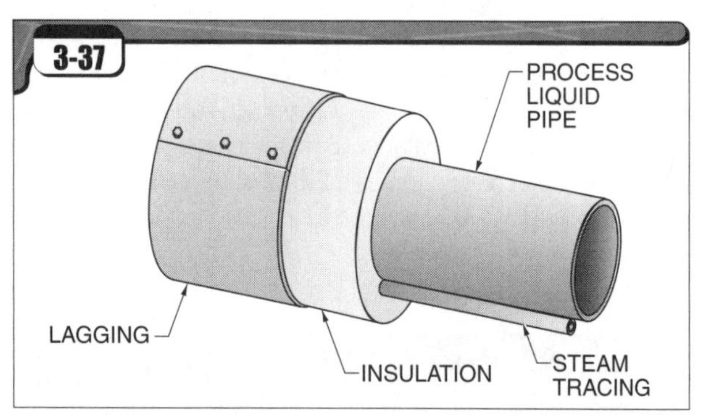

Figure 3-37. *Steam tracing is a small steam tube run beside another pipe or piece of equipment for freeze protection or to maintain a minimum process fluid temperature.*

101 Describe an in-line steam separator.

An *in-line steam separator* is a cylindrically shaped vessel that is installed in a steam pipe to remove moisture droplets after the steam has left the boiler. The steam passes through a steam separator on its way to a turbine or other equipment. The in-line steam separator contains baffles or stationary vanes that swirl or abruptly change the direction of the steam. **See Figure 3-38.** Centrifugal force and the rapid change of direction help remove moisture droplets from the steam before the steam enters an important piece of equipment like a turbine.

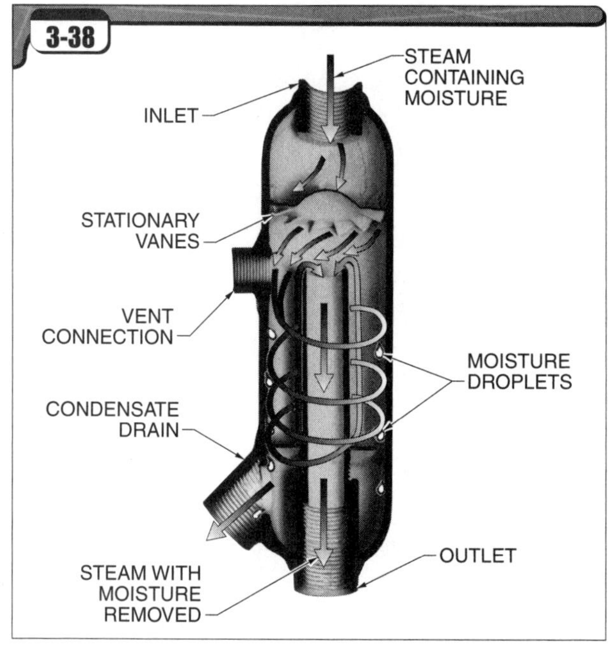

Clark-Reliance Co.

Figure 3-38. *In-line steam separators remove moisture from steam.*

102 What is an open heat exchanger?

An *open heat exchanger* is a heating unit in which steam or another heating medium and the fluid being heated come into direct contact. Cooking equipment that injects steam directly into a food product and basic percolation-type feedwater heaters are examples of open heat exchangers.

103 What is a closed heat exchanger?

A *closed heat exchanger* is a heating unit in which the heating medium and the fluid being heated do not mix but are separated by tube walls or other heating surfaces. Closed heat exchangers are made in many configurations. One of the most common is the shell-and-tube heat exchanger. **See Figure 3-39.** A shell-and-tube heat exchanger has a tube bundle contained within an outer shell. The fluid being heated usually flows through the tubes. The heating medium flows across the outside surfaces of the tubes. Baffles provide the most efficient contact between the heating medium and the tubes.

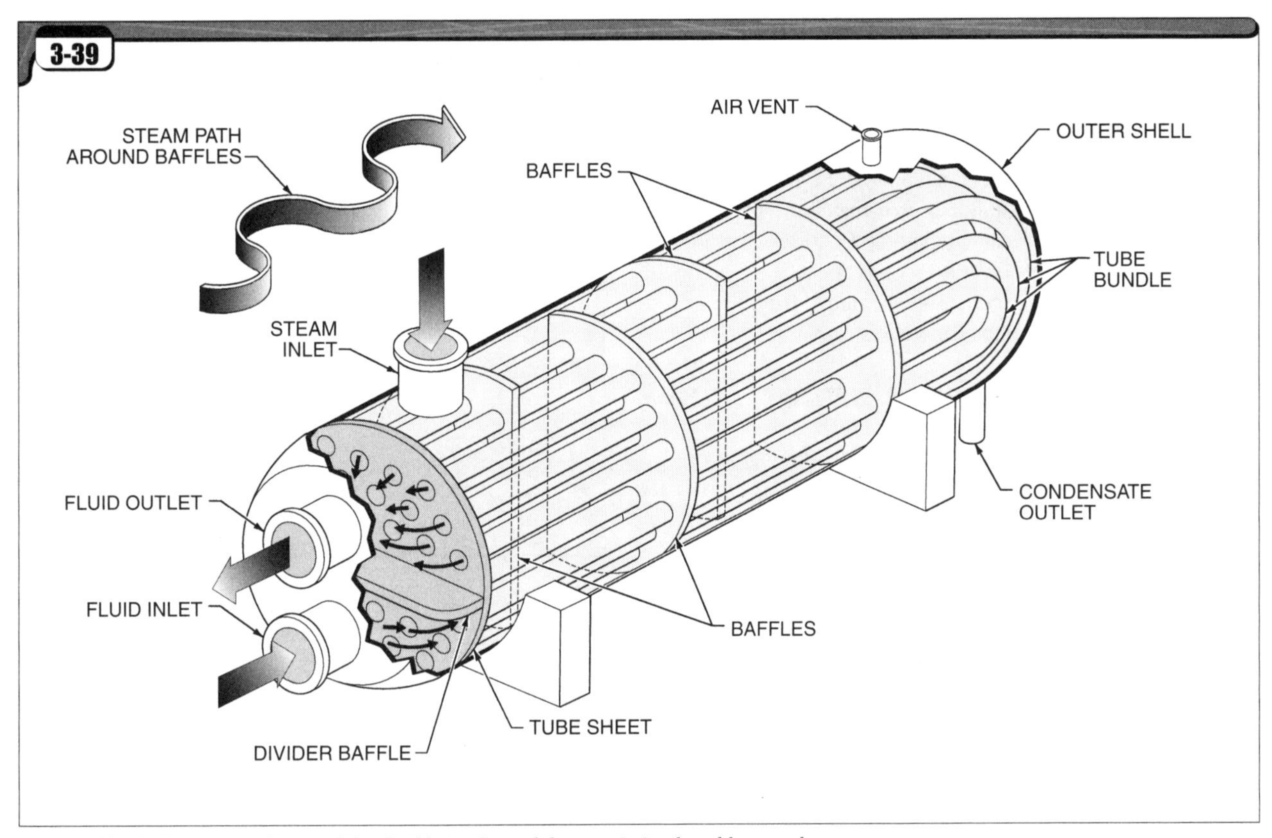

Figure 3-39. The heating medium and the fluid being heated do not mix in closed heat exchangers.

Condensate and Gas Removal

104 Why should condensate be removed from the steam piping and steam-using equipment?

Condensate interferes with heat transfer. Since condensate has already given up its latent heat, it contains far less energy available to do useful work in the steam-using equipment. Because the condensate occupies space that could be filled by steam, it interferes with the ability of the equipment to achieve maximum heat transfer.

Condensate is much more corrosive than steam. Condensate may contain carbonic acid, a compound that corrodes the piping and heat exchange equipment. This acid causes loss of metal, and eventually the piping and heat exchange equipment spring leaks. Steam is much less corrosive than condensate.

Condensate may freeze and damage the steam-using equipment, even when steam is present. Condensate that is allowed to accumulate and remain in low points in heat exchange equipment, such as steam coils used for building heat, cools as it dissipates heat to the surrounding atmosphere. The condensate may freeze and cause the heat exchange equipment to break, even though the steam-using equipment may still be under steam pressure. To prevent freezing and bursting, the heat exchange equipment must be kept drained of condensate.

Condensate accumulation in the steam-using equipment or steam piping may lead to water hammer. Slugs of water can be propelled ahead of the steam, causing damage when they encounter an elbow, tee, control valve, or other obstruction.

105 Why should condensate be saved for reuse?

Condensate is already hot. Less fuel must be burned to change it to steam again. The use of hot condensate instead of cold makeup water in the boiler results in fuel savings.

Condensate is already paid for. The condensate, if city water is used, has been through the water meter. If well water is used, the reuse of condensate saves wear on the well pumps and the energy required to run them.

Condensate has been chemically treated. Condensate is almost pure after having been distilled by the boiler. Use of additional cold makeup water requires further expense for water treatment chemicals.

Hot condensate contains less oxygen. Oxygen corrodes the boiler and the rest of the system.

Factoid

The reuse of hot condensate greatly increases boiler plant efficiency and reduces the cost of operation.

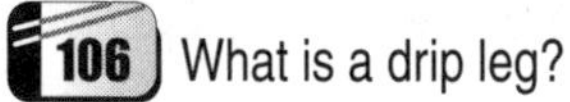

106 What is a drip leg?

A *drip leg* is a downward extension from a steam distribution line or piece of heat exchange equipment where condensate is allowed to drain. Drip legs are installed in steam systems wherever condensate can accumulate. **See Figure 3-40.** For example, drip legs are installed approximately every 150′ to 300′ along long runs of steam piping, at locations where the piping turns upward, at the bottom of heat exchange equipment, and other places. Condensate is drained from the drip leg through a runoff pipe that leads to a steam trap.

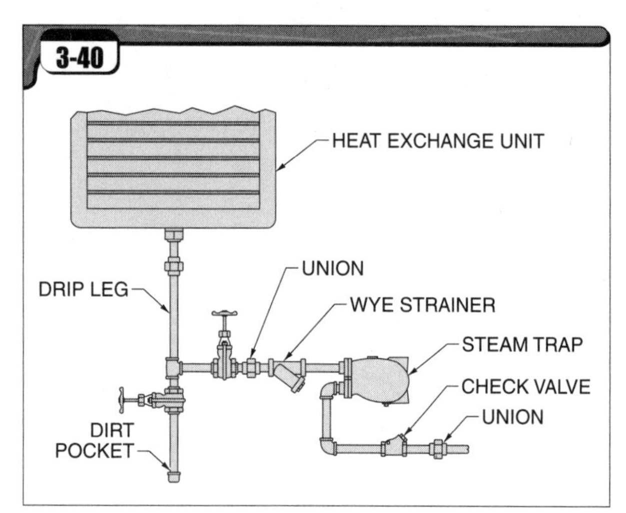

Figure 3-40. Condensate is drained from the drip leg.

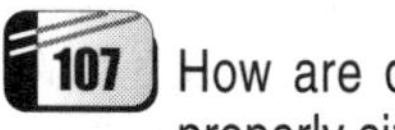

107 How are drip legs on steam distribution piping properly sized?

The drip leg must be large enough in diameter to allow the condensate, which may be flowing fairly rapidly through the pipe, to fall out before reaching the other side of the drip leg. For this reason it is common practice to make drip legs on steam piping the same diameter as the piping, up to 4″ in diameter. For example, a 3″ steam pipe requires a 3″ drip leg. **See Figure 3-41.**

For piping above 4″ in diameter, it is accepted practice to make the drip leg one-half the diameter of the steam pipe, but not less than 4″. For example, a 12″ diameter steam pipe would have 6″ diameter drip legs, but a 6″ diameter steam pipe would have 4″ diameter drip legs.

The length of a drip leg should be considered as well. Where feasible, the drip leg length should be approximately 28″ from the bottom of the steam pipe to the level of the steam trap inlet. This helps establish drainage from the steam system. For example, when this 28″ long drip leg fills with water, this creates approximately 1 psig of pressure to help force the condensate through the steam trap even before any pressure is established on the steam system. If the steam piping is drained manually by the boiler operators during system startup, the drip leg may be somewhat shorter.

STEAM PIPE SIZE*	RECOMMENDED DRIP LEG DIAMETER*	MINIMUM DRIP LEG LENGTH	
		Supervised Warm-up	Automatic Warm-up
½	½	10	28
¾	¾	10	28
1	1	10	28
1¼	1¼	10	28
1½	1½	10	28
2	2	10	28
2½	2½	10	28
3	3	10	28
4	4	10	28
6	4	10	28
8	4	12	28
10	6	15	28
12	6	18	28
14	8	21	28
16	8	24	28
18	10	27	28
20	10	30	30
24	12	36	36

* in in.

Figure 3-41. Both the diameter and length of drip legs are important to ensure proper condensate drainage.

108 What is a dirt pocket?

A *dirt pocket* is the pipe nipple installed on the bottom of the drip leg that catches rust and weld slag. This helps keep such debris from entering the steam trap where it could cause the small orifice in the trap to become plugged.

Trade Tip

It is a good idea to install valves on the bottom of dirt pockets. This provides a boiler operator a means of draining condensate from the steam lines when necessary.

109 What is a steam trap?

A *steam trap* is a mechanical device used for removing condensate and/or air from steam piping and heat exchange equipment. A steam trap is essentially an automatic valve that allows condensate to pass through but closes in the presence of live steam. All steam traps consist of an orifice through which condensate passes, and some means of opening and closing the orifice when necessary to drain condensate as it accumulates. Since condensate has already given up its latent heat, it is inefficient to have condensate in the steam system.

Most steam traps also remove air and other noncondensable gases from the steam distribution system. A *noncondensable gas* is a gas that does not change into a liquid when its temperature is reduced to room temperature. These gases cause operational problems if allowed to remain in the steam and condensate systems.

Note that all steam traps require differential pressure from the inlet side of the trap to the outlet side in order for the condensate and/or air to be removed from the steam system. If the pressure at the inlet side of the trap is not higher than the pressure at the outlet side, no steam trap works properly.

110 Why is it important to vent air and noncondensable gases from the steam system?

Air contains gases that are corrosive to steam and condensate systems. Gases are less soluble in hot water than in cold water. Therefore, when water is heated in a boiler, it begins to release the dissolved air. Air in the system contains some oxygen and some carbon dioxide. When oxygen is combined with moisture, especially under high temperatures such as in a condensate or steam line, it causes aggressive corrosion and pitting of the piping. Removal of the air helps prevent this corrosive attack.

Carbon dioxide in the system also creates acidic conditions that deteriorate the condensate return piping. After the condensate is discharged by the steam traps and begins to cool slightly, it begins to reabsorb air. If the air that is available to be reabsorbed contains carbon dioxide, the carbon dioxide chemically reacts with the condensate to form carbonic acid. Carbonic acid causes corrosion in the condensate piping.

Air insulates the heat transfer surfaces and interferes with heat exchange. Air is an excellent insulator. The fiberglass insulation used widely in modern homes and buildings takes advantage of this fact. The fiberglass contains pockets of trapped air. The air provides the insulation.

Air tends to "plate" on the heating surfaces in the process heat exchange equipment. This means that the air tends to form an insulating layer, inhibiting efficient heat transfer. By helping to remove the air accumulation from the steam system, steam traps improve system efficiency.

111 How does oxygen get into the boiler and pipes?

Oxygen is introduced into the system with the makeup water. Cold city water or well water contains dissolved oxygen. For this reason, makeup water is generally preheated to the boiling point in order to help drive out the air before the water is pumped into the boiler.

Oxygen may be pulled into the piping and equipment when vacuum conditions develop inside. When a piece of heat exchange equipment is shut down (i.e., valved off), the steam inside collapses into condensate as it gives up its heat. Because the volume of the condensate is much less than the volume of steam, a strong vacuum can form as the steam condenses. Air can be drawn into the equipment through vacuum breakers, small leaks in valve packings, tiny leaks in pipe fittings, etc. In many cases, it is important to break the vacuum to prevent damage to the equipment. However, it is equally important to vent the air back out as the system is heated up again.

Oxygen is introduced when the piping or equipment is opened up for maintenance work. For example, flanged connections may have to be opened to replace a leaking gasket.

112 Describe three common types of steam traps.

A *thermostatic steam trap* is a steam trap that contains a temperature-operated device, such as a corrugated bellows, that controls a small discharge valve. **See Figure 3-42.** The bellows contains a fluid that boils when heated. When condensate surrounds the bellows the fluid inside the bellows condenses and causes the bellows to contract as the condensate cools and the temperature drops below the steam temperature. This opens the discharge valve on the trap and allows the condensate to drain from the trap. When live steam reaches the trap, the heat of the steam causes the fluid in the bellows to boil and raises the pressure inside. The pressure rise causes the bellows to expand, which closes the discharge valve. Air is also cooler than steam, so air accumulation inside the trap causes contraction of the bellows as well and the air is discharged.

A *float and thermostatic steam trap* contains a thermostatic bellows or other thermostatic element and also contains a steel ball float connected to a discharge valve by a linkage. Condensate entering the trap makes the float rise, which opens the lower valve and allows the condensate to drain out. The bellows or other thermostatic device is installed at the top of the trap to remove air from the steam system. The bellows in a float and thermostatic steam trap operates identically to the bellows in a thermostatic steam trap.

An *inverted bucket steam trap* contains an upside-down steel cup, called a bucket, that is attached to a linkage that opens and closes a discharge valve as the cup rises and falls inside the trap. Condensate in steam enters through a standpipe inside the cup.

Steam is a gas and, like air, it tends to keep the inverted bucket afloat. When the bucket is about two-thirds filled with steam, the steam provides enough buoyancy to overcome the weight of the bucket and the bucket rises. With the bucket in the up position, the discharge valve is held closed. When condensate enters the trap and the bucket becomes about two-thirds filled with condensate, the weight of the bucket overcomes the buoyancy provided by the steam and the bucket sinks to the bottom of the trap. This opens the discharge valve by the linkage attached to the sinking bucket. The differential pressure across the trap pushes the condensate out of the trap until the condensate level drops to the point where the bucket is again about two-thirds filled with steam and regains its buoyancy. Then the bucket rises to the top position again and closes the discharge valve.

Air and other noncondensable gases that enter the trap do not pass around the bottom of the inverted bucket to exit the trap. However, the air and gases must be removed. Because air is a noncondensable gas, a bubble of air inside the bucket makes the bucket permanently buoyant and the trap does not function properly. For this reason, the bucket has a small air vent hole at the top to allow air and other gases to pass through. The air and other gases accumulate in the top of the trap body until the trap valve is opened. The gases then pass out with the condensate.

There are many variations of each type of steam trap. For example, two inverted bucket steam traps may look radically different due to different materials of construction, piping attachment locations, differences in condensate-handling capacity, and attached accessories such as strainers.

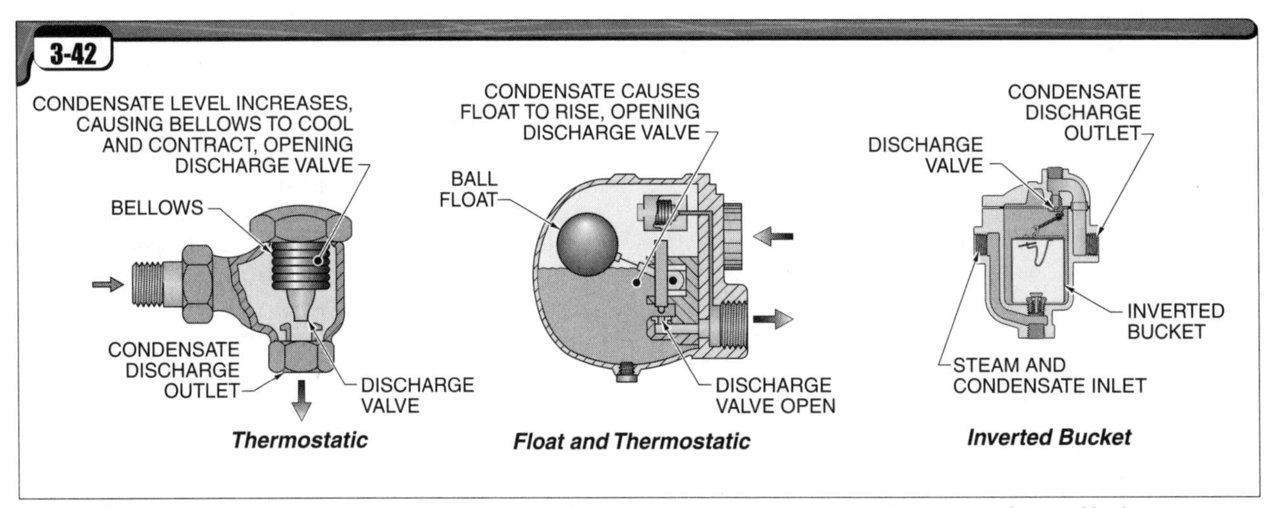

Figure 3-42. *Three common types of steam traps are the thermostatic steam trap, float and thermostatic steam trap, and inverted bucket steam trap.*

113 What is a disc steam trap?

A *disc steam trap* is a comparatively small steam trap that uses a flat, round disc as the means of opening and closing the outlet orifice. **See Figure 3-43.** Disc traps are sometimes called thermodynamic disc traps or thermo-disc traps.

The round disc in a disc trap covers both the inlet and outlet passages. The disc is very flat and smooth and closes against two flat, circular machined surfaces called seats. The disc is held in place by a cap that is attached to the trap by machined threads. The outside diameter of the disc is somewhat smaller than the inside diameter of the cap, so the disc can move freely up and down inside the cap.

When condensate enters the trap under adequate differential pressure, the condensate lifts the disc from the seats and condensate passes under the disc from the inlet port to the outlet ports. The condensate is thus discharged into the condensate return piping.

When very hot flashing condensate and steam enter the trap and begin to discharge, a small amount of steam is allowed to migrate to the back side (top) of the disc through the space between the outer rim of the disc and the inside of the cap. The steam quickly accumulates in the space above the disc. The top of the disc has much more surface area than the inlet orifice side of the disc. Therefore, the lower pressure steam on the top of the disc exerts more total force than the higher pressure steam on the bottom of the disc, and the disc is snapped onto the seats by the steam pressure above the disc. This stops the discharge of the trap. The disc remains on the seats until the steam above the disc condenses and thus releases the pressure above the disc. Then the trap opens and discharges until the steam above the disc is replaced.

A trap that is very similar in operation to the disc trap is the impulse trap. The impulse trap contains a piston that lifts vertically to open the trap orifice. The piston is surrounded by a round lip, configured much like the brim of a hat, that causes the piston to lift and reseat.

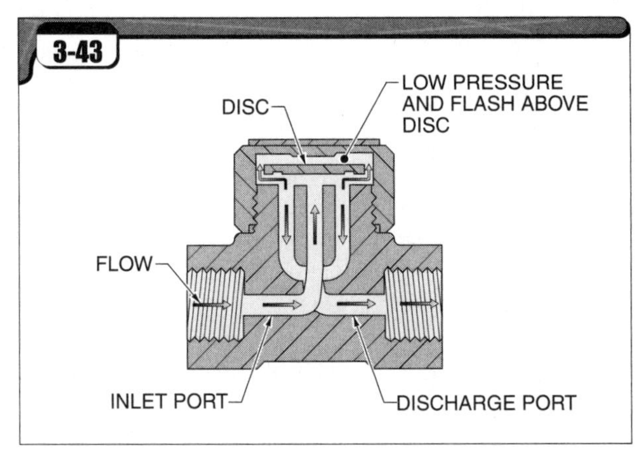

Figure 3-43. *Disc steam traps use a round, flat steel disc as the means of opening and closing the orifice.*

114 What care should be exercised when selecting a steam trap for a specific application?

Certain types of steam traps work well on certain applications and poorly on others. For example, some steam traps are very susceptible to damage from water hammer. Some have very limited ability to withstand the effects of dirt or rust entrained in the condensate. Some are susceptible to poor operation when installed in cold or wet environments. Some steam traps discharge air from a steam system much better than others. There is no universal steam trap that works well on all possible applications. There are many factors to consider when selecting a steam trap for a certain job, and a reliable steam trap supplier should be consulted for assistance.

The size of the piping connections on a steam trap is irrelevant when selecting a steam trap. Steam traps are made available with various size piping connections only for the convenience of the person installing the trap. For example, if all the condensate passing through the trap must pass through a ⅛″ orifice, it makes no difference whether the piping connections are 1″ or 1¼″.

115 List six places where drip legs and steam traps should be installed.

- at the ends of the main steam header
- at low points in piping where condensate may collect
- at points where the steam piping turns upward
- at specified spacing in long steam piping
- after steam radiators in a heating system or after any process equipment using steam for heating of a material
- before restrictions, such as pressure regulating valves

116 Why should a strainer be used upstream of a steam trap?

A strainer should be used upstream of a steam trap because the orifice through which the condensate flows inside most steam traps is quite small and easily clogged with debris. A strainer is used to catch these impurities before they can plug the orifice. **See Figure 3-44.** The strainer contains a screen that allows steam and condensate to pass through but catches debris that is larger than the openings in the screen.

Care should be taken when selecting a strainer for installation before a steam trap. Strainers are available with many different types of screens inside. The openings in the strainer screen (the mesh size) must be smaller than the orifice inside the steam trap for the strainer to be effective.

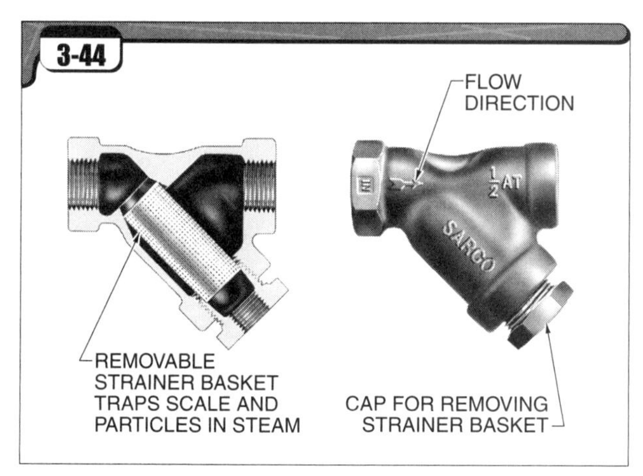

Spirax Sarco, Inc.

Figure 3-44. *A strainer should be used before a steam trap to catch dirt and impurities.*

117 What happens if a steam trap fails while open?

If a steam trap fails in the open position, live steam blows through the trap. This is a waste of steam energy and can cause water hammer in the condensate return piping.

118 What happens if a steam trap fails while closed?

If the steam trap fails while closed, the steam piping and the nearby equipment that use the steam become flooded with condensate. This results in lost efficiency of the equipment or damage because of water hammer, or freezing if the equipment is outdoors.

119 How may a steam trap be tested for proper operation?

The difference in temperature across the steam trap helps reveal whether the steam trap is working properly. When checking the difference in temperature across the steam trap, there typically is a drop of 10°F to 20°F from the supply side to the discharge side. This is an easy test to perform with an inexpensive probe-type digital thermometer, infrared thermometer, or inexpensive magnetic or strap-on thermometer. **See Figure 3-45.**

Trap test valves should be installed. Install a tee in the downstream piping from the steam trap with the branch directed downward. Install a valve on the branch of the tee, with the piping after the valve routed to a safe place. Also install a valve after the tee in the condensate return line. Alternatively, install a three-way ball valve for this purpose. Open the valve in the branch of the tee and close the valve in the condensate return line to visually observe the steam trap discharge. This requires some cost for valves and pipe fittings, but provides a very clear means of determining whether the trap is working correctly.

One limitation of this method is that the differential pressure conditions on the trap may not be exactly the same with the condensate diverted to the atmosphere as with the condensate discharging normally into the condensate return line. For example, the trap may operate correctly when discharging to the atmosphere but not when discharging against excessive pressure (back pressure) in the condensate return line.

An ultrasonic stethoscope is a very reliable method for testing steam traps. It detects sounds that are inaudible to the human ear and amplifies them so that the user can listen to the internal workings of the steam trap. One limitation of this method is that the user must understand what the various types of steam traps should sound like. For example, an inverted bucket steam trap should have an intermittent discharge, while a float and thermostatic steam trap should have a continuous light discharge.

The action of some steam traps is audible without the aid of listening devices. Sometimes, placing the blade tip of a long screwdriver against the steam trap and listening through the handle is sufficient if the surrounding area is relatively quiet.

Temperature crayons that have a specific melting point are used in the welding industry. The downstream side of the steam trap is marked with a calibrated crayon. If the downstream temperature gets too hot (the steam trap is blowing through), the crayon melts. This method can detect a steam trap that has failed in the open position, but not a trap that has failed in the closed position.

Flow indicators are installed on the downstream side of critical steam traps. Flow indicators are special pipe fittings that have glass ports that allow the boiler operator to look into the piping.

The simplest way to tell if a steam trap is functioning properly is by being familiar with the equipment or piping served by the steam trap and knowing what operating conditions are normal for that equipment. If the supply steam is at a normal temperature and pressure but the equipment is not maintaining its normal temperature or is experiencing water hammer, the steam trap could be the reason for the problem.

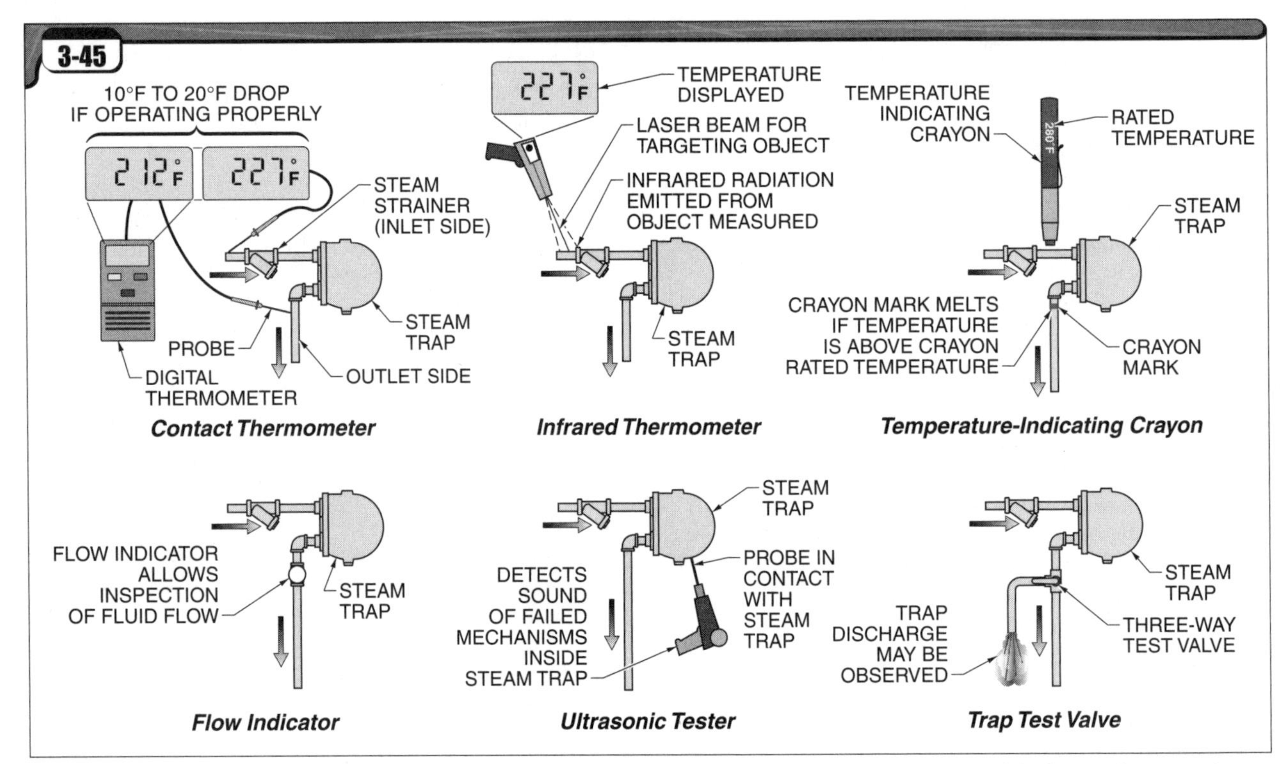

Figure 3-45. Steam traps should be routinely tested and repaired as needed to ensure trouble-free operation of the heat exchange equipment and avoid expensive steam losses.

120 How may a steam trap malfunction?

Dirt, scale, or other debris may clog the orifice or interfere with mechanical movements inside the steam trap. Water hammer can damage delicate floats and bellows assemblies. **See Figure 3-46.** The steam trap may freeze and crack if it is installed outdoors and not allowed to drain properly when the steam is shut off. The air vent orifice may become clogged with debris and cause the steam trap to become airbound.

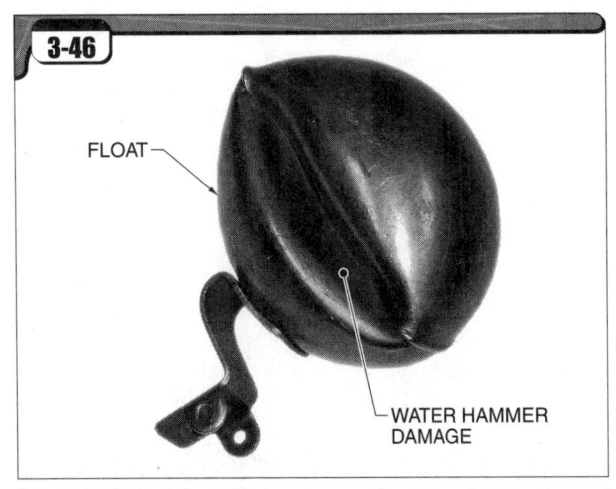

Figure 3-46. Water hammer can damage the internal components of steam traps.

121 What is a flash tank?

A *flash tank* is a pressure vessel in which condensate or other very hot water under a high temperature and pressure is allowed to partially flash into steam. **See Figure 3-47.** The flash steam may be used in a lower-pressure manufacturing process, feedwater heater, building heating or humidification system, or other process.

If the amount of flash steam formed is not adequate to meet the entire steam requirement of the low-pressure process, the flash steam can be supplemented by live steam provided through a pressure reducing valve. The temperature of the remaining condensate corresponds to the new, lower saturation pressure. The condensate is usually directed to a vented condensate return system.

Plants that use very high pressure steam may use a cascading flash steam system. A *cascading flash steam system* is a piping and tank arrangement where condensate is allowed to flash several times in progressively lower pressure flash tanks.

The use of a flash tank can result in greatly improved energy efficiency. The flash steam can be directed to another low-pressure process and used for heating. This results in substantial savings in fuel because it reduces the amount of steam that must be generated.

Factoid

There is no universal steam trap. Steam traps should be carefully selected and sized for each individual application.

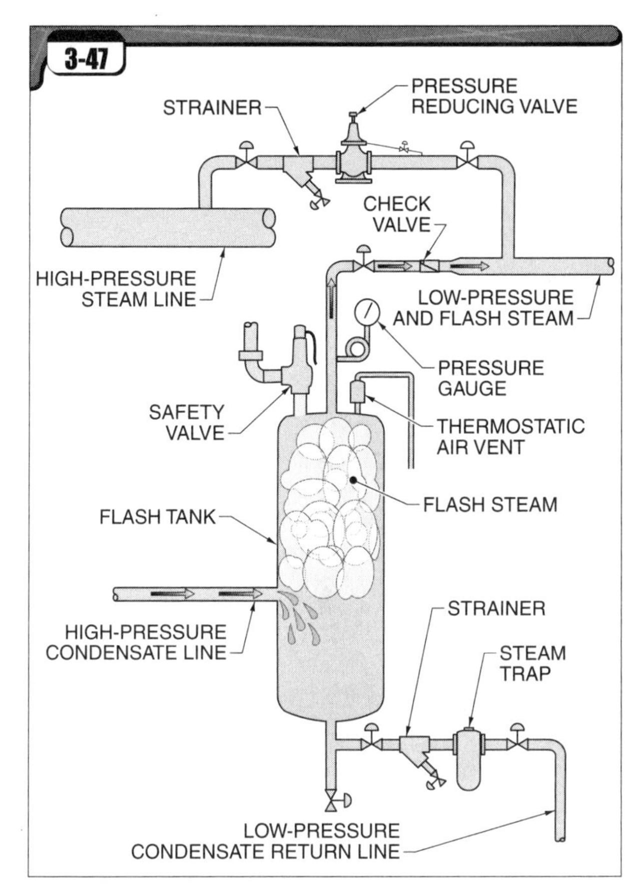

Figure 3-47. *Flash tanks significantly improve steam system efficiency by allowing flash steam to be reused in a lower-pressure process.*

122 Why is condensate return piping sometimes pitched?

Pitched means inclined or sloped. Condensate return piping should be pitched in a gravity return system to help the condensate flow more easily through the piping to a collection point. Piping is commonly pitched about ¼″ to 1″ per 10′-0″. Once the condensate accumulates in a condensate receiver, a pump is usually used to return the condensate to the boiler system. Condensate that is forced through the piping by a pump does not need to have the piping pitched, since the condensate does not rely on gravity for movement.

123 List two ways of returning condensate from a low-pressure steam heating system to the boiler.

Condensate is usually returned to the boiler system by a pump. For example, a condensate receiver tank, installed to collect the condensate, is mounted over a pump so that the water level in the receiver tank maintains a positive pressure on the suction side of the pump. The pump moves the condensate back into the boiler.

If a heating system operates at a very low pressure, for example less than about 5 psi to 8 psi, condensate may be made to return to the boiler by gravity. This works if the water column in the vertical condensate return main is high enough to create enough pressure to overcome the boiler pressure.

Why are condensate receivers usually vented to the atmosphere?

Many condensate receivers are not constructed to withstand any internal pressure. A vent pipe is provided so that, if steam traps in the system fail in the open position and pressurize the condensate return system, the condensate receiver vent vents the steam to the atmosphere. This eliminates the danger of the condensate receiver ever becoming overpressurized.

Additionally, most condensate receivers use motor-driven pumps to return the condensate to the boiler system. If the water is hotter than 212°F, this causes operational difficulties because the hot water may cause flashing in the pump. Venting the condensate receiver helps ensure that the water in the condensate receiver cannot exceed 212°F.

Superheaters

What is a superheater?

A *superheater* is a bank of tubes through which only steam passes, not water. **See Figure 3-48.** Steam flows through this bank of tubes after leaving the steam and water drum. The steam picks up additional heat in the superheater so that its temperature rises significantly above the saturated steam temperature. The superheater does not raise the pressure of the steam, only the temperature of the steam.

What determines whether a superheater is radiant or convective?

Whether a superheater is radiant or convective is determined by its location in the furnace area. For example, if the superheater is directly exposed to the radiant heat of the furnace, it is a radiant superheater. If it is not exposed to the radiant heat, but instead receives its heat by convection, it is a convective superheater. Many superheaters are constructed with more than one stage, or section, and may be combination radiant/convective designs.

How does superheating the steam improve the operation of a steam turbine?

Superheating evaporates any remaining moisture droplets in the steam. Moisture droplets in the steam may damage turbine blades by erosion or may leave deposits of impurities on the blades. Moisture also reduces the efficiency of a steam turbine by creating friction as the steam flows through the turbine. Superheating helps ensure that the steam flows completely through the steam turbine before the steam begins to condense.

The superheater also adds substantial heat energy in Btu to the steam, giving it the ability to do more useful work in the steam turbine. The savings in steam is about 1% for each 10°F of superheat up to about 100°F of superheat. Above 100°F of superheat, the savings are slightly less.

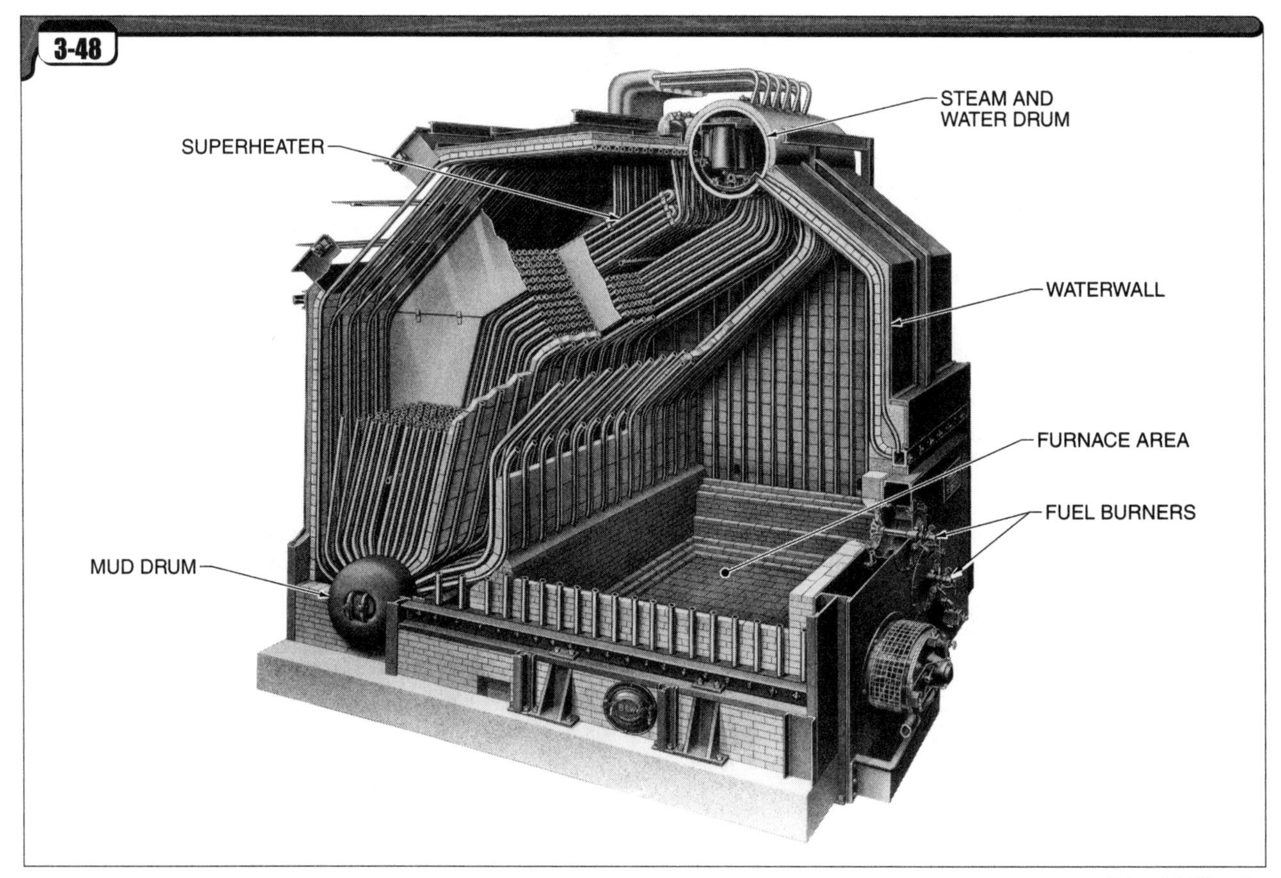

Figure 3-48. A superheater adds substantial heat to the steam leaving the steam and water drum in a watertube boiler.

128 What protects the superheater tubes from melting or being damaged while the boiler is on-line?

The flow of steam through the tubes is the only thing that prevents damage from overheating of the superheater tubes at any time. The steam is much lower in temperature than the gases in the furnace. A flow of steam must always be maintained through the superheater to prevent tube damage.

Factoid

Superheaters raise the temperature of the steam, but not the pressure. This helps ensure very dry steam for use in steam turbines. The presence of water droplets in the steam going to a turbine can cause significant damage from blade erosion.

129 How is the superheater protected from damage while the boiler is being heated before being put on-line?

A vent to the atmosphere is provided at the superheater outlet so that a flow of steam may pass through the superheater and out the roof of the plant. **See Figure 3-49.** The vent is placed before the nonreturn valve. The vent is left open until the boiler pressure becomes high enough to open the nonreturn valve and establish a flow of steam to the header. When the instrumentation detects that steam is flowing from the boiler through the superheater and into the steam header, the superheater vent may be closed.

If the superheater has a drain to remove condensate prior to startup, and the drain is installed at an appropriate location, the drain may also serve as the vent.

130 List two common causes of superheater tube failure.

Damage from overheating is a common cause of superheater tube failure. Overheating can happen when there is inadequate steam flow through the tubes during startup or low-load periods. Overheating may also occur because of scale deposits in the superheater, resulting from boiler carryover in the superheater.

Superheater tubes can fail because of erosion from fly ash in the flue gases and erosion by steam from misaligned soot blowers. *Fly ash* is ash from the combustion of coal or other solid fuels that is carried along with the draft through a boiler furnace and ductwork. Utility boilers often have tube shields constructed of high-temperature alloys attached to the superheater tubes to protect them from erosion.

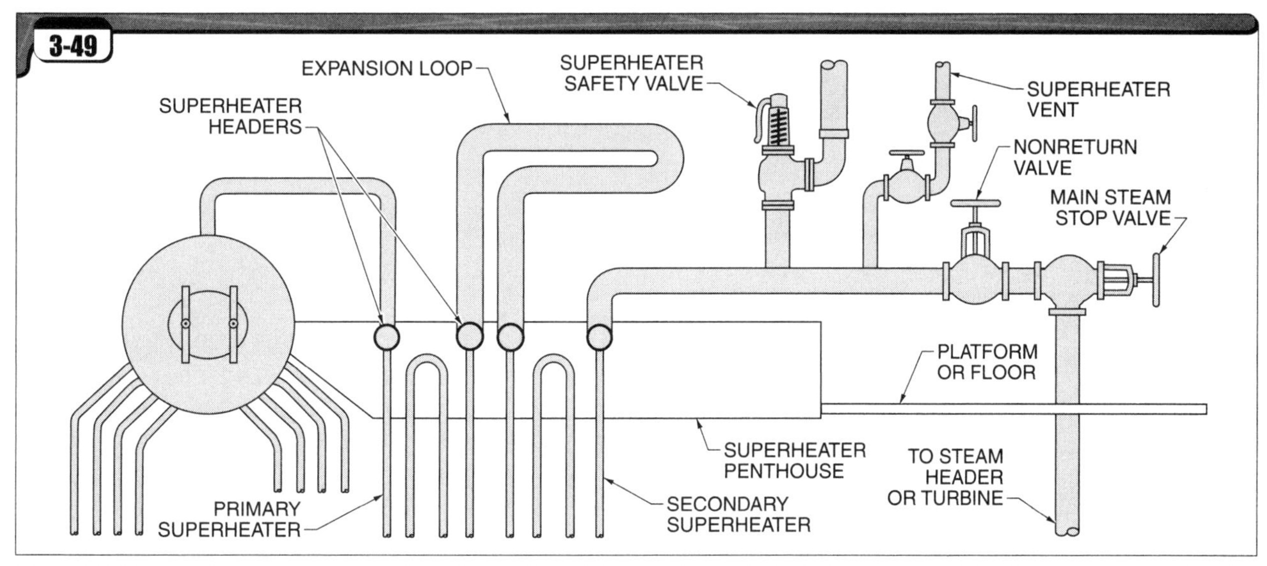

Figure 3-49. *The superheater vent ensures a flow of steam through the superheater during boiler startup and shutdown to prevent overheating of the superheater tubes.*

131 How is the temperature of superheated steam controlled?

An attemperator (also known as a desuperheater) is used to cool the superheated steam slightly when required. One type of attemperator injects a small, well-dispersed mist of water into one of the superheater passes. **See Figure 3-50.** This water is instantly vaporized and does not cause water hammer.

As the temperature of the superheated steam rises above the desired temperature, the attemperator opens to spray a small flow of water mist in proportion to the temperature rise. Heat from the superheated steam is used in evaporating the water, so the steam temperature drops. When the steam temperature falls lower than the temperature desired, the attemperator flow decreases or shuts off completely.

A *drum desuperheater* is an attemperator that diverts part of the superheated steam through a heat exchanger in the boiler mud drum. This portion of the steam is thus desuperheated, giving up its excess heat to the boiler water. After passing through the heat exchanger, the steam is remixed with the rest of the superheated steam.

Factoid

The terms desuperheater and attemperator are often used synonymously. Some engineering authorities say a desuperheater reduces the temperature of the steam to within 10°F of the saturation temperature while an attemperator does not reduce the temperature as much.

132 What is a line desuperheater?

A *line desuperheater* is a device that automatically removes superheat from superheated steam so that the steam becomes saturated steam. For example, large boilers in a utility plant, steel mill, paper mill, etc. may produce only superheated steam. However, some processes in the plant may require saturated steam at a certain temperature and pressure. A line desuperheater is used to inject a fine mist of feedwater or other very pure water into the steam line going to the process. The superheat in the steam is consumed in evaporating the mist of water, and the temperature of the steam is thus reduced.

A line desuperheater consists of a temperature-measuring device, a temperature controller, and a special water valve and nozzle that produce the water mist. The temperature-measuring device is installed downstream of the misting nozzle. It communicates the temperature of the steam to the temperature controller. The temperature controller compares the actual steam temperature to the desired steam temperature. It then causes a change in the position of the water valve to supply more or less water mist in response to the steam temperature.

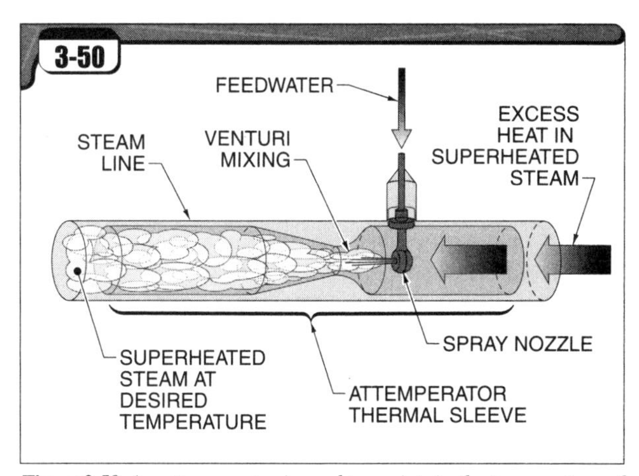

Figure 3-50. *An attemperator is used to maintain the temperature of the superheated steam at a specific temperature, or to reduce the superheated steam to saturated steam.*

133 How are superheater tubes supported?

Superheaters vary considerably in design, but fall into three basic classifications with regard to their configuration. All superheaters utilize inlet and outlet headers that serve as manifolds to supply steam to the individual superheater tubes and collect the steam as it exits the tubes. **See Figure 3-51.**

Pendant superheaters are those that hang from overhead supports. These superheaters are suspended in the flue gas stream, while their supports are protected from the heat of the furnace. Thus, these superheaters expand downward when heated.

Horizontal superheaters have horizontal tube banks. These tube banks are supported by brackets attached to the waterwall tubes or boiler tubes. The water inside these support tubes helps to keep the supports from overheating. The superheater tubes expand horizontally when heated, and the supports must allow for this movement. Sometimes these supports are located outside the flue gas stream to prevent overheating.

Bottom-supported superheaters are supported by a steel structure below the superheater. This structure supports the superheater's inlet and outlet headers, which in turn support the superheater tubes. The tubes expand upward in this design.

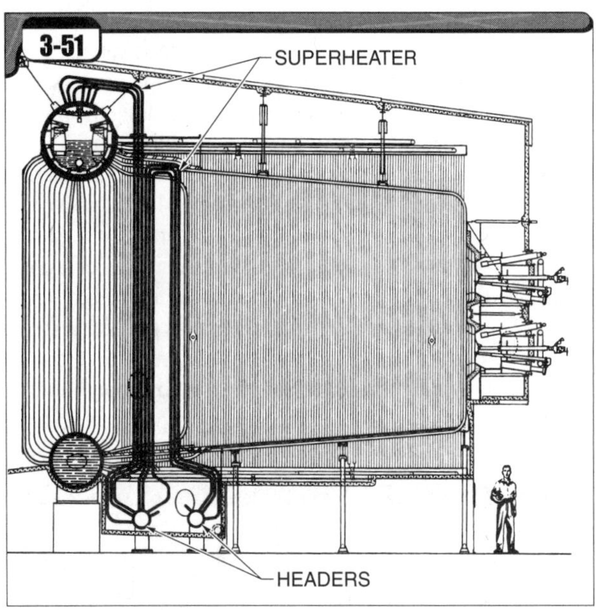

Babcock & Wilcox Co.

Figure 3-51. The inlet and outlet headers are at the bottom of a bottom-supported superheater.

134 How do drainable and nondrainable superheaters differ?

A drainable superheater is a superheater configured so that condensate in the superheater tubes migrates to a low point from which it can be drained. Very little if any condensate appears in the superheater during normal operation, but it may accumulate during shutdown as the superheater cools. Prior to startup of the boiler, the boiler operator may be required to manually cycle the drains to remove any accumulated condensate.

A *nondrainable superheater* is a superheater that does not have condensate drain connections. Pendant superheaters are generally nondrainable. As in the case of drainable superheaters, condensate may accumulate in the superheater as the boiler is shut down and while the unit is out of service. During startup, the condensate is simply boiled off as steam. Nondrainable superheater designs require that the boiler operators bring the temperature of the furnace up very gradually so that the condensate slowly evaporates and does not get carried over into the steam piping.

135 Do all superheaters require a safety valve?

A safety valve is required. The safety valve is installed on the superheater outlet piping from the boiler. It is provided to maintain steam flow through the superheater if the boiler pressure should become excessive. The superheater safety valve is installed after the superheater but before the nonreturn valve.

136 How should the superheater safety valve be set?

The superheater safety valve should be set so that it lifts before the drum safety valves lift. The superheater safety valve also reseats after the drum safety valves reseat. This ensures that a flow of steam through the superheater tubes is maintained at all times.

If the safety valves on the steam drum lifted before or reseated after the superheater safety valve, the flow of steam through the superheater would be shortcircuited. Overheating of the superheater tubes could result.

137 What is a slag screen?

A *slag screen,* or *screen wall,* is a loosely spaced bank, or several rows, of water tubes placed between the superheater and the combustion area of the furnace.

Some of the fly ash in the furnace, especially when firing coal or other solid fuels, becomes so hot that it begins to melt and become sticky. When semimolten fly ash comes into contact with a relatively cool boiler tube, it cools quickly and sticks on the tube, causing a buildup of slag.

Because the superheater is so vital to the efficient operation of the steam boiler/turbine installation, slag screen tubes are installed in front of the superheater to catch as much of the fly ash as possible before the flue gases enter the superheater. These tubes help maintain the efficiency of the superheater.

138 Why must extremely pure water be used in a boiler that is equipped with a superheater?

The boiler water must be extremely pure because any impurities contained in water droplets carried over from the boiler separate from the water and deposit on the superheater tubes. These deposits insulate the tubes, which causes the tubes to overheat and burst.

The impurities, especially silica, may also pass through the superheater and settle out on steam turbine blades, which results in loss of turbine efficiency. The water mist introduced into the superheater by the attemperator spray must also be pure.

How is boiler steam pressure affected as steam passes through a superheater?

The steam pressure at the superheater outlet is slightly lower than the pressure in the steam and water drum. The lower pressure is due to friction of the high-velocity steam against the tube walls in the superheater. For example, if the pressure in the steam and water drum is 750 psi, the pressure as the steam leaves the superheater may be 740 psi. This differential pressure is due to friction.

What is a reheater?

A reheater is a superheater through which steam is diverted after the steam has already passed through a section of a steam turbine. Reheaters are often used in electric power generating plant boilers, but rarely anywhere else. In cases where very large steam turbines are used, designers are often concerned that the steam may begin to condense before it passes completely through the turbine. To prevent this, the steam passes through a portion of the turbine and is then diverted to the reheater. The reheater is installed in the same furnace area as the superheater. The steam flows through the reheater, which raises the temperature of the steam again. The steam is then piped back to the remaining portion of the steam turbine.

Factoid

In some plants, a small flow of steam directly from the boiler, prior to the superheater, is condensed specifically for use in the attemperator. That condensate is called sweetwater when used for this purpose. Sweetwater is a source of very pure water.

What is the critical pressure of steam?

The *critical pressure* of steam is the pressure at which the density of the water and the density of the steam are the same. This pressure is 3206 psia. Some utility plant boilers operate at supercritical pressures, or pressures higher than the critical pressure. Because the density of the steam and the water are the same and there is not a definite water level, these boilers have no steam drums. The water in these boilers generally only passes through the tubes once. By the time the water reaches the outlet, it is all evaporated and there is no recirculation. For this reason, these boilers are also often referred to as once-through boilers. Such boilers may operate at up to 3850 psig.

Is the steam superheated after it flows through a pressure reducing valve?

Pressure reducing valves reduce the pressure of the steam, but not the temperature of the steam. Therefore, the temperature of the steam after a pressure reducing valve is higher than that which corresponds to the steam pressure, so technically the steam is superheated.

However, the superheat available in the steam after a pressure reducing valve is of little practical use. The small amount of superheat available in the steam after the pressure reducing valve is very quickly used up in re-evaporating any condensate that may be present in the bottom of the steam piping. For this reason, the superheat is dissipated within just a few feet of the pressure reducing valve, and is of no practical use except to make the steam slightly drier.

Factoid

Superheated steam is a relatively inefficient means of heat transfer. The steam must cool to the saturation temperature before condensing and releasing the heat of evaporation. The temperature of the superheated steam is not uniform and this can cause uneven heat transfer.

Why must leaks of very high pressure or superheated steam be quickly repaired?

Very high pressure or superheated steam is erosive. It cuts or enlarges the area surrounding the leak. For example, superheated steam leaking from between two flanges can cut a groove in the faces of the flanges along the path of the leakage. After a short time, this cutting action could ruin the machined faces of the flanges.

High-pressure or superheated steam leaks also create safety hazards in the boiler plant, due to the fact that the steam blowing out of the leak location may be totally invisible. **See Figure 3-52.** Live steam is a clear gas, and steam at a very high temperature may not begin to change back into a visible vapor until it has blown several feet from the leak location. The area in which the loud hiss from the leak is heard must be approached with extreme caution to avoid entering the path of the leakage. Severe burns would result from exposure to the steam.

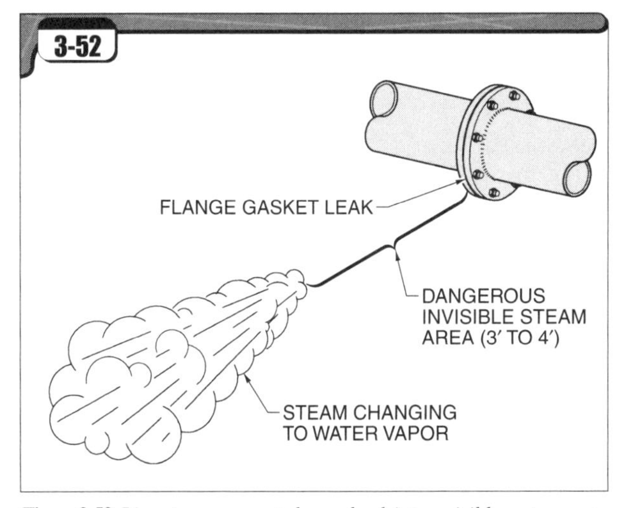

Figure 3-52. *Live steam may not change back into a visible water vapor until it has blown several feet from the leak locations.*

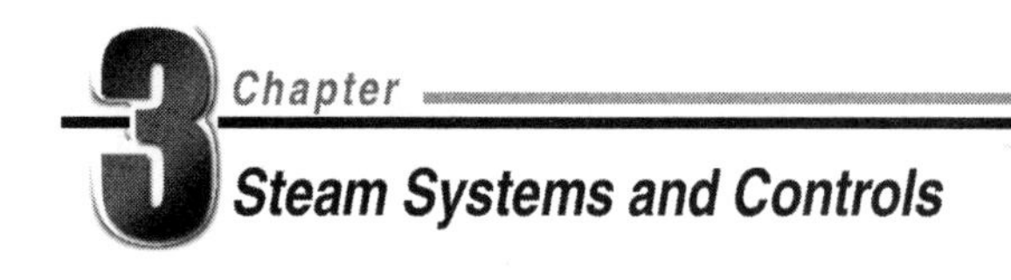

Name: ____________________ Date: ____________________

_______ **1.** The total force on a 3″ diameter safety valve disc installed on a boiler that carries a pressure of 150 psi is ___ lb.

_______ **2.** Boilers that have more than ___ sq ft of heating surface must have two or more safety valves.

_______ **3.** A boiler has a pressure of 275 psi and a 2 ½″ diameter safety valve. The total force acting on the safety valve is ___ lb.

_______ **4.** The blowdown for a safety valve set to pop at 240 psi and reseat at 232 psi is ___ psi.

_______ **5.** The safety valve keeps a boiler from exceeding its ___.

_______ **6.** The percent of blowdown for a safety valve set to pop at 260 psi and reseat at 251 psi is ___%.

7. List five items found on the safety valve data plate.

_______ **8.** The steam-venting capacity of a safety valve is confirmed by applying a(n) ___ test.

_______ **9.** A common pressure gauge uses a ___ tube to create movement of the pointer when pressure is applied.

A. siphon
B. Bourdon
C. mercury
D. thermostatic bellows

_______ **10.** ___ is the amount the pressure in the boiler must be reduced before the safety valve reseats.

T F **11.** A nonreturn valve can be opened manually.

_______ **12.** The range of a pressure gauge on a steam boiler should be ___ the MAWP.

A. 1 ½ to 2 times
B. not greater than
C. not less than
D. 2.31 times

13. What must be done to prevent damage from vacuum when draining a boiler?

T F **14.** Blowdown of a safety valve is expressed as a percentage of the steam pressure at which the boiler normally operates.

______________ **15.** A compound gauge can be used to read ___ and ___.

T F **16.** The total force exerted by steam equals the pressure of the steam multiplied by the area exposed to the steam.

T F **17.** Superheaters raise the pressure of steam.

______________ **18.** A ___ valve should be used for throttling purposes.

A. globe
B. gate
C. ball
D. check

______________ **19.** Vacuum breakers prevent damage to a pressure vessel by allowing ___ to be drawn in if a vacuum condition develops.

20. Why should condensate be removed from steam lines?

______________ **21.** The blowdown of the safety valve can be altered by changing the setting of the ___ ring.

A. throttling
B. accumulation
C. huddling
D. reaction

______________ **22.** Quality of steam refers to___.

A. how wet the steam is
B. how dry the steam is
C. how much the steam weighs
D. the pressure of the steam

23. What is the difference between an os&y valve and a rising stem valve?

T F **24.** The lift of a safety valve is expressed in psi.

T F **25.** Superheater tubes must be kept cool by maintaining a sufficient flow of water through the tubes.

______________ **26.** If a low-pressure and a high-pressure boiler have identical ratings in pounds of steam per hour, the ___ boiler would require the larger safety valve.

T F **27.** Gate valves used for throttling purposes are subject to deterioration from wire drawing.

28. What purpose does the actuator serve on a control valve?

T F 29. A superheater adds Btu to the steam and gives it the capacity to do more work.

_______________ 30. ___ is a small steam tube or pipe which runs alongside the piping for thick liquids to keep the liquids warm and pumpable.
A. Lagging
B. Jacketing
C. Siphon tubing
D. Steam tracing

31. What is a battery of boilers?

T F 32. Steam scrubbers remove moisture from steam as the steam leaves the boiler.

_______________ 33. A(n) ___ gas is a gas that does not change into a liquid when its temperature is reduced.

_______________ 34. Oxygen combines with ___ to cause corrosion and pitting of steam piping.

_______________ 35. A(n) ___ test is used to confirm whether the safety valves have sufficient relieving capacity to vent all excess steam that the boiler can produce.
A. hydrostatic
B. throttling
C. huddling
D. accumulation

_______________ 36. A constant steam pressure should be maintained on the boilers to ensure that ___.
A. steam consumption is minimized
B. the steam does not become superheated
C. expansion of the piping is maximized
D. the steam temperature and volume remain constant

T F 37. A safety valve pops open fully and immediately because the valve disc that closes against the seat of the safety valve is made in a two-stage pattern.

_______________ 38. A safety valve should be manually tested when the pressure is about ___% of the normal popping pressure.

T F 39. The capacity of a safety valve must be great enough to vent all of the excess steam while never allowing boiler pressure to build to more than 8% above the MAWP.

T F 40. An accumulation test should be performed monthly on the safety valve.

_______________ 41. ___ is the amount of steam, in pounds per hour, that the safety valve is capable of venting at the rated pressure of the valve.

_______________ 42. A(n) ___ tester exerts exact hydraulic pressure on the steam pressure gauge, which can then be recalibrated.

_______________ 43. ___ settings on the pressure controls are normally staggered so boilers can lead and lag.

T F 44. A check valve is a one-way flow valve for fluids.

_______________ 45. Condensate return piping is commonly pitched approximately ___ per 10′-0″.
A. 1⁄16″
B. 1⁄8″ to 1⁄4″
C. 1⁄8″ to 1⁄2″
D. 1⁄4″ to 1″

________________ **46.** ___ is the development of froth on the surface of the water in a boiler.

________________ **47.** ___ is a severe form of carryover in which large slugs of water leave the boiler with the steam.

A. Entrainment B. Priming
C. Foaming D. Capacity

T F 48. Hot condensate contains less oxygen than cool makeup water.

________________ **49.** Water hammer first damages a steam line at ___ in the piping.

A. elbows B. tees
C. restrictions D. all of the above

T F 50. A boiler is fired at its maximum firing rate during an accumulation test.

T F 51. Boiler vents vent air from a boiler as it fills with water.

________________ **52.** A hydrostatic test uses ___ pressure to check for leaks.

A. air B. water
C. steam D. none of the above

________________ **53.** Steam pressure gauges should be accurate to within ___% of the working pressure.

A. 2 B. 5
C. 6 D. 10

________________ **54.** The flow of steam through a superheater should never be less than ___% of the capacity of the boiler.

________________ **55.** A ___ is a manifold that feeds several small branch pipes or takes in steam or water from several smaller pipes.

A. vacuum breaker B. primer
C. superheater D. header

________________ **56.** A(n) ___ pressure control is a control device that regulates the burner for a higher or lower fuel-burning rate depending on the steam pressure in the boiler.

________________ **57.** The differential setting of an automatic pressure control that turns the burner ON at 84 psi and OFF at 110 psi is ___ psi.

A. 26 B. 84
C. 110 D. 194

T F 58. A gate valve is used to shut off or admit flow.

________________ **59.** ___ is the erosion that occurs as steam or another high-velocity fluid streaks through a small opening.

A. Throttling B. Blowdown
C. Steam tracing D. Wire drawing

________________ **60.** A(n) ___ may be formed inside a piece of heat exchange equipment when the steam supply is shut off and the steam inside the equipment collapses into condensate.

T F 61. Standard globe valves erode badly if used on high-pressure steam systems.

________________ **62.** The main steam stop valve is used for isolating the ___ side of a boiler.

T F 63. A ball valve is a quick-acting, two-position shutoff valve.

______________ **64.** Cyclone separators are found in many ___ boilers.
A. firetube
B. watertube
C. cast iron
D. all of the above

T F **65.** The use of hot condensate instead of cold makeup water results in fuel savings.

______________ **66.** A ___ is the pipe nipple and cap installed on the bottom of the drip leg.
A. steam trap
B. steam scrubber
C. dirt pocket
D. water hammer

T F **67.** The superheater safety valve should be set so that it lifts after the drum safety valves lift.

______________ **68.** Changing a control device to a degree that is proportional to the need is called ___.
A. regulating
B. equalization
C. wire drawing
D. modulation

T F **69.** Priming can cause water hammer.

______________ **70.** ___ is the hydraulic shock that results from water buildup in steam piping.
A. Gravity
B. Slag screen
C. Water hammer
D. Condensation

______________ **71.** ___ water is water that is added to the system to replace water that is lost or drained.
A. Added
B. Extra
C. New
D. Makeup

______________ **72.** When checking the difference in temperature across a steam trap, there typically is a drop of ___°F to ___°F.
A. 5; 10
B. 10; 20
C. 20; 30
D. 30; 35

T F **73.** A strainer should be used upstream of a steam trap.

______________ **74.** The total force acting on the safety valve disc of a boiler with a pressure of 300 psig and a 3½″ diameter safety valve disc is ___ lb.
A. 525
B. 1050
C. 2885
D. 3299

______________ **75.** A pressure gauge that reads 0 psi to 120 psi is the most accurate between approximately ___ psi and ___ psi.
A. 0; 40
B. 0; 80
C. 40; 80
D. 80; 120

______________ **76.** Automatic pressure controls maintain a constant pressure by ___.
A. throttling the steam to achieve the highest feasible pressure
B. superheating the steam
C. adjusting the fuel-burning rate of the combustion equipment
D. removing moisture from the steam

______________ **77.** A control system in which the flows of fuel and air are precisely measured and adjusted so as to always be in the correct proportions is called ___ control.
A. positioning
B. metering
C. operating limit
D. differential

______ **78.** A(n) ___ heat exchanger is a heating unit in which steam or another heating medium and the fluid being heated come into direct contact.

______ **79.** Steam is ___ for a short distance after passing through a pressure reducing valve.
A. superheated
B. supersaturated
C. wire-drawn
D. reheated

______ **80.** A modulating pressure control adjusts the firing rate of the burner by operating the ___.
A. equalizing line
B. solenoid valve
C. modulating motor
D. reheater

______ **81.** A(n) ___ heat exchanger is a heating unit in which the heating medium and the fluid being heated do not mix but are separated by tube walls or other heating surfaces.

______ **82.** If a boiler has multiple safety valves, the valve set to lift at the highest pressure must not lift at greater than ___% above the MAWP.
A. 3
B. 6
C. 10
D. 20

______ **83.** A(n) ___ is a pressure vessel in which condensate or other very hot water under high temperature and pressure is allowed to partially flash into steam.
A. expansion tank
B. steam trap
C. steam separator
D. flash tank

______ **84.** A safety valve that is marked with the letters "UV" inside the ASME symbol stamp is suitable for use only on ___.
A. hot water boilers
B. steam boilers
C. natural gas- or fuel oil-fired boilers
D. unfired pressure vessels

______ **85.** A boiler vent is located on the ___ of the boiler.
A. side
B. end
C. bottom
D. top

______ **86.** Copper and brass piping may be used for piping systems as long as the temperature does not exceed ___°F.
A. 212
B. 250
C. 406
D. 500

______ **87.** A low-pressure boiler's safety valve(s) must not allow the boiler pressure to build to greater than ___ psig during an accumulation test.
A. 15.9
B. 16.5
C. 20
D. 22.5

______ **88.** Carbon dioxide can chemically react with condensate to form ___ acid.

______ **89.** ___ is an alloy of chromium, cobalt, and tungsten used to treat the interior components of valves to make them very hard and erosion resistant.

T F **90.** Condensate in heat exchange equipment can freeze even if the steam to the equipment is turned on.

______ **91.** The stem on a(n) ___ valve rises through the center of the handwheel, allowing the operator to visually determine whether the valve is open or closed.

______________ **92.** The minimum pressure at which a low-pressure boiler's safety valve should be manually tested is ___ psig.

______________ **93.** A temperature control valve uses a temperature-sensing device connected to the valve actuator by a ___ tube.

A. capillary
B. Bourdon
C. equalizing
D. siphon

______________ **94.** A ___ valve consists of a circular disc that is rotated by the valve stem so that the disc is parallel to the flow through the valve, perpendicular to the flow, or somewhere between.

A. piston
B. diaphragm
C. plug
D. butterfly

T F **95.** The size of the piping connections determines the amount of condensate that a steam trap can discharge.

______________ **96.** A ___ is an actuator that consists of an iron plunger surrounded by an encased coil of wire.

A. solenoid
B. modulator
C. pressurestat
D. diaphragm

______________ **97.** Condensate receivers are often vented to the atmosphere in order to ___.

A. prevent overpressurization of the receiver
B. prevent condensate spills
C. promote energy efficiency
D. all of the above

______________ **98.** Every steam trap must have adequate ___ across the steam trap in order for the condensate and/or air to be removed from the steam system.

A. differential pressure
B. steam pressure
C. back pressure
D. pressure equalization

T F **99.** Solenoid valves are desirable as actuators for steam valves because they bring the temperature of the steam-using equipment up quickly.

______________ **100.** ___ is a benefit from the use of pressure reducing valves.

A. Smaller steam distribution piping
B. More precise temperature control
C. Lower pressure ratings for steam-using equipment
D. all of the above

______________ **101.** The pressure at the outlet of a superheater is ___ the pressure in the steam and water drum of the boiler during normal operation.

A. substantially higher than
B. slightly higher than
C. the same as
D. slightly lower than

______________ **102.** A steam pipe is to be run for a distance of 112′. The coldest temperature expected for this pipe is 40°F. During operation, the pipe reaches a temperature of 406°F. If the coefficient of expansion for the piping material is 0.0000073, the pipe expands ___″ when heated.

A. 3.59
B. 4
C. 35.9
D. 159.3

Chapter 4

Water Supply Systems and Controls

Maintaining a constant, safe water level in a steam boiler is the boiler operator's most important duty. Boilers are damaged or destroyed by overheating because of inadequate water level much more frequently than by any other cause. Therefore, it is essential that the boiler operator have a clear understanding of the pump systems that provide the feedwater to the boiler, and the level indicators and controls that maintain the water level. These components require continual care in order to maintain optimum reliability.

Equipment that is well maintained and kept in proper adjustment will provide many years of satisfactory service. Boiler operators protect the facility's substantial investment in equipment by performing such tasks as adjusting pump packing, following proper lubrication schedules, and testing safety devices. Especially in larger facilities, boiler operators may perform only first-line maintenance functions. However, a thorough understanding of the mechanical principles of each component helps the boiler operator evaluate and respond to upset conditions much more quickly and competently.

Level Indicating Devices

1 Why is it important to maintain a constant water level in a steam boiler?

The water level in a steam boiler should be maintained within a narrow range as determined by the boiler manufacturer. Operation of the boiler with the water level outside this range is dangerous.

A low water condition is a lower-than-acceptable water level in a boiler. It is dangerous because the heating surfaces in the boiler can be damaged by overheating. The steel and cast iron that are used in boilers melt at a considerably lower temperature than the temperature of the combustion gases in the furnace area. The circulating water inside the boiler protects the metal by carrying away the heat as quickly as it is conducted through the metal. The boiler tubes may melt and the possibility of a boiler explosion exists if there is inadequate water to remove the heat. **See Figure 4-1.**

A high water condition is a higher-than-acceptable water level in a boiler. It is dangerous because water can be carried over with the steam into steam lines. This can result in water hammer, which can damage steam-using equipment or burst piping.

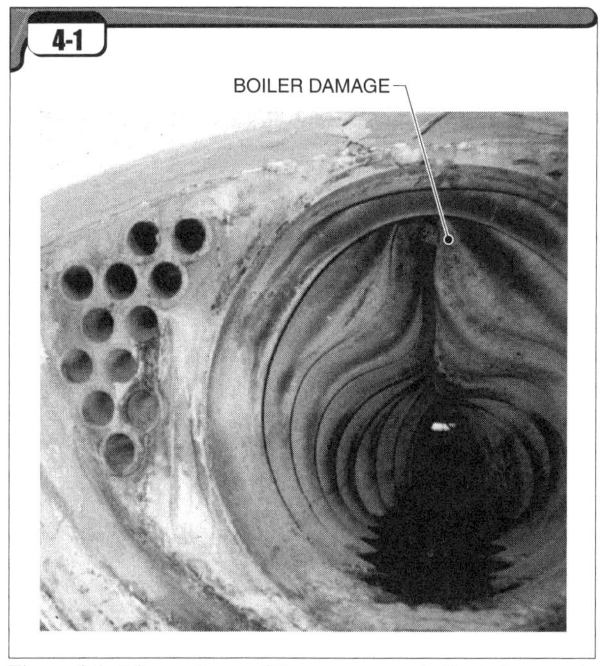

Figure 4-1. *A low water condition can cause a boiler to be severely damaged by overheating.*

2 What is a water column?

A *water column* is a metal vessel installed on the outside of the boiler shell or drum at the normal operating water level (NOWL) for the purpose of determining the location of the water level. **See Figure 4-2.** Water columns range from approximately 12″ to 36″ high. The exact height of a water column is based on the size of the boiler.

In a boiler that is firing, the surface of the water is fairly turbulent because of the large number of steam bubbles breaking through the water surface. If there were a simple window on the boiler drum or shell itself, the movement of the water would make determining the actual water level difficult. The water column provides an area where the movement of the boiler water can calm down so that the operator can determine the water level. Water columns and the gauge glasses attached to them work on the principle that water seeks its own level.

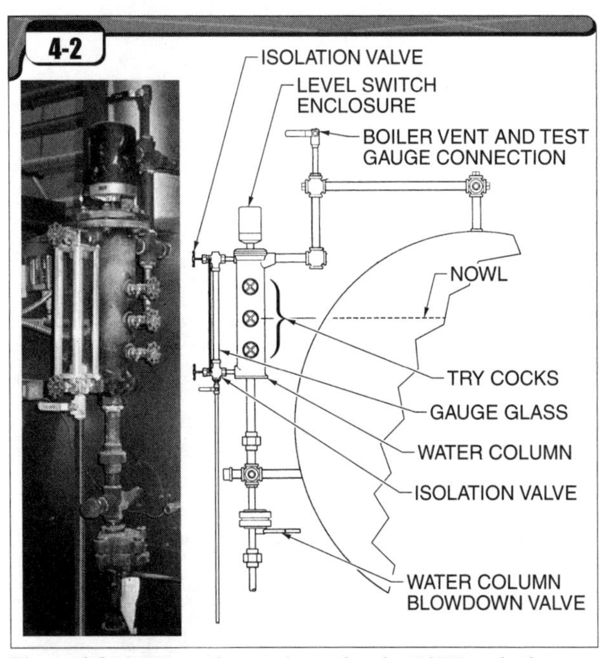

Figure 4-2. *A water column is located at the NOWL on boilers.*

3 What is a gauge glass?

A *gauge glass* is a tubular or flat glass connected to a water column that allows an operator to see the water level in the water column, and thus in the boiler, at a glance. The gauge glass is the primary water level indicator. The gauge glass may be connected directly to some small boilers.

Factoid

The gauge glass is considered the primary boiler water level indicator, even if remote or electronic indicating devices are provided.

4 What are try cocks?

A *try cock* is a valve used as a secondary water level indicator. There are usually three try cocks installed on the side of a water column. Opening each try cock should help the operator determine the water level in the boiler. The top try cock should blow pure steam. The bottom try cock should blow water, some of which may flash into steam. The middle try cock may blow steam, water, or a mixture of steam and water. This is because the water level in a boiler fluctuates slightly during normal operation.

Try cocks are referred to as gauge cocks in the ASME Code. The ASME Code does not require try cocks to be installed.

5 Describe how to connect a water column to a boiler.

A water column is installed so that the middle of the gauge glass is at the NOWL, which also places the middle try cock at the NOWL. **See Figure 4-3.** The top piping to a water column is connected to the boiler shell at a point that is higher than the highest visible point in the gauge glass. The bottom piping to a water column is connected to the boiler shell at a point that is lower than the lowest visible point in the gauge glass. The piping between the boiler and the water column must not be configured such that any low points are created that will permit the accumulation of sludge or debris. Generally, this means that the water column piping should be plumb and level, not pitched.

The water column for firetube boilers is piped so that the lowest visible point in the gauge glass is at least 3″ above the highest point of the tubes, flues, or crown sheets. For watertube boilers, the lowest visible point in the gauge glass must be at least 2″ above the lowest permissible water level, as determined by the boiler manufacturer. This ensures that an adequate blanket of water will exist to prevent overheating of the tubes while the boiler cools. For example, when a boiler is shut down, it will continue to produce steam for a period of time. This occurs because the boiler contains a large amount of hot metal and hot refractory (firebrick) materials, and these materials will continue to provide residual heat for a short time. In addition, if there are no other operating boilers on the same steam header (in battery), there will be no steam pressure from other boilers to close the nonreturn valve on boiler shutdown. The water in the boiler will continue to flash into steam as the system pressure drops. These conditions result in a continued conversion of boiler water to steam after boiler shutdown. The blanket of water above the tubes ensures that the tubes are cooled during this period.

Pipe with an ID of at least 1″ is used for connections from the boiler to the water column. This piping should be at least extra heavy pipe.

Cross fittings are used at directional changes in the piping instead of elbows or tees, to allow easy inspection and cleanout while the boiler is shut down. Blowdown piping from a water column is at least ¾″ ID.

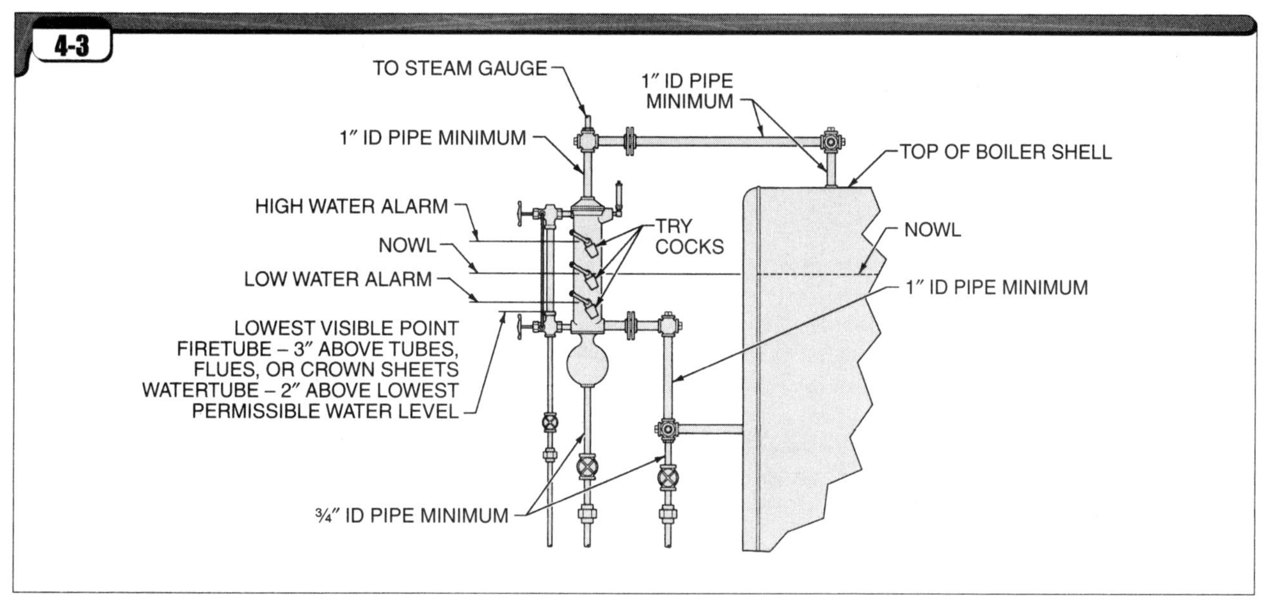

Figure 4-3. *The middle try cock of a water column is located at the NOWL.*

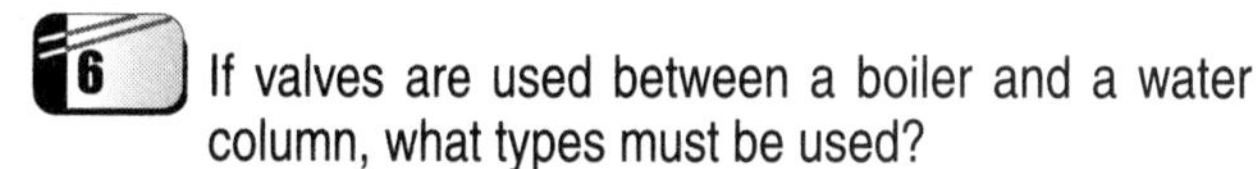

6 If valves are used between a boiler and a water column, what types must be used?

Valves at this location, if left closed, can be perceived by the boiler operator that a safe water level exists in the boiler, while the level may actually be unsafe. Therefore, while the ASME Code does not prohibit the use of valves between the boiler and the water column, it does specify the permissible types of valves and how they should be operated.

Stopcocks with permanent position-indicating levers or os&y valves are permitted between the water column and boiler. The boiler operator can easily observe the valve position to determine whether they are open or closed. Per the ASME Code, they shall be sealed or locked in the open position while the boiler is in service. This is intended to prevent unauthorized closing of the valves.

7 What could happen if the valves in water column piping were left closed?

On many boilers, the feedwater controls are contained inside the water column. In this case, the controls would not properly call for water to be added to the boiler. This is because the water could be low in the boiler but satisfactory in the water column. Thus, if the valves in water column piping are left closed, the chance of boiler damage or a boiler explosion from low water is increased.

8 What materials of construction are acceptable for water columns?

Water columns are made of cast iron, ductile iron, or steel. Cast iron water columns are used for pressures up to 250 psi. Ductile iron water columns are used for pressures through 350 psi. Steel water columns are used for pressures above 350 psi.

9 What types of gauge glasses are used on boilers?

Tubular gauge glasses for steam boilers are used up to about 250 psig. Tubular gauge glasses are provided in a variety of lengths and diameters, with various pressure ratings. Tubular gauge glasses are made stronger by increasing the wall thickness. In this case, the outside diameter (OD) stays the same but the ID is reduced. **See Figure 4-4.**

Most tubular gauge glasses are provided with a vertical painted red stripe flanked by white stripes. When viewing the stripes through a tubular gauge glass, the width of the red stripe is magnified by the water. This makes the red stripe appear wider, and makes the water level easier to see.

Flat gauge glasses may be used for pressures up to about 2200 psi. They are considerably thicker than tubular gauge glasses. Flat gauge glasses for boilers are usually between approximately ⅜″ thick and approximately ⅝″ thick. In addition, a metal cover plate is added to provide additional support and strength.

The great majority of flat gauge glasses for boilers are of either the round port type or the elongated port type. The round port type is often referred to as the bull's-eye type. The use of small individual ports minimizes the surface area of the glass and thus adds strength against very high pressures. Both types form windows into a hollow cast iron or steel chamber. The metal chamber often has glass ports on both sides, allowing the boiler operator to look completely through the assembly. This makes the water easier to see. Many operators install a light behind flat gauge glasses of this type to illuminate the water level.

Some elongated-type flat gauge glasses have V-shaped grooves cut lengthwise into the glass. These grooves bend the light rays passing through the glass like a prism, making the water in the gauge glass appear dark. This makes the water level much easier to see.

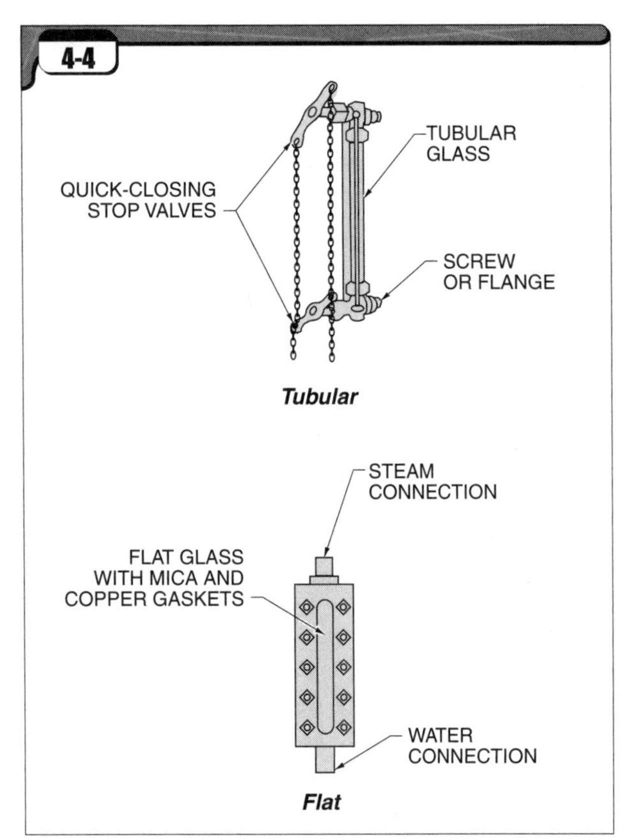

Figure 4-4. *Tubular gauge glasses are available for pressures up to 250 psi. Flat gauge glasses may be used for pressures up to 2200 psi.*

10 On what principle do bicolor gauge glasses work?

Bicolor gauge glasses work on the principle of light refraction. An external high-intensity light shines through a red/green glass filter and then through the round port type flat gauge glass. **See Figure 4-5.** The gauge glass is configured so that either the red or the green light rays are bent out of view, depending on whether steam or water is in the gauge glass. If steam is in the gauge glass, the light rays are slightly bent and make the light look red on the other side of the gauge glass. If water is in the gauge glass, the light rays are bent and make the light look green on the other side of the gauge glass.

Bicolor gauge glasses are very convenient, especially if the boiler is large and the upper drum is far above the operating floor. The two colors make the water level much easier to see. Sets of mirrors that are configured in a periscope fashion are often installed to allow the boiler operator to view the bicolor gauge glass from several floors below.

Bicolor gauge glasses are also often provided with a bonnet that uses fiber optic cable to transmit the light image to a remote control room. Separate fiber optic cables are provided for each red or green port, and these cables transmit the red or green image to a panel-mounted indicator, sometimes several hundred feet away.

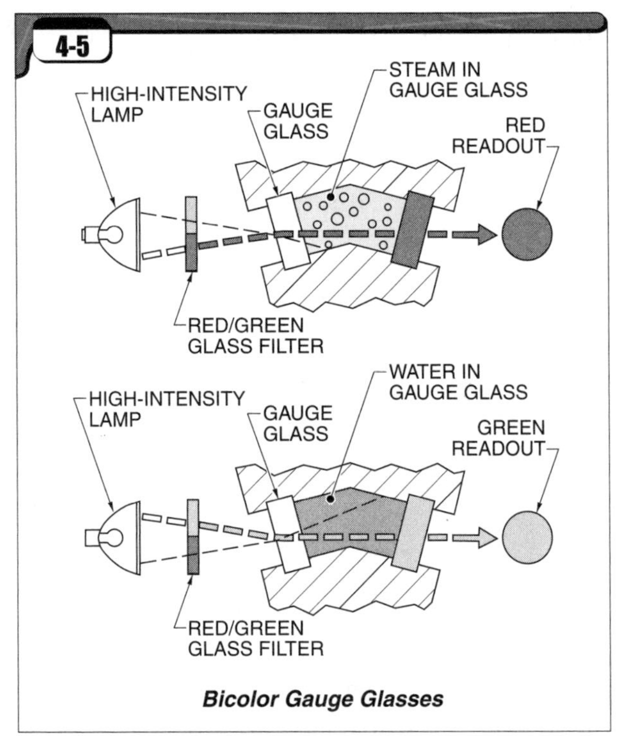

Figure 4-5. *Bicolor gauge glasses work on the principle of light refraction.*

11 Do all steam boilers require a gauge glass?

Under the ASME Code, all steam boilers except forced-flow boilers that have no fixed steam and water line require at least one gauge glass. Boilers that are operated at over 400 psig (except electric boilers of the electrode type) must have two gauge glasses. These may be connected to a single water column or directly to the steam and water drum.

For very high-pressure boilers (those with all drum safety valves set at or above 900 psig), two independent remote water level indicators may be provided in place of one of the two gauge glasses.

12 What provisions must be made if the boiler operator does not have direct visual contact with the gauge glass on the boiler when making control changes?

Many boiler plants have a control room from which the boiler operators cannot directly see the gauge glass. If a bicolor gauge glass is used, the red and green light images from the glass ports can be transmitted through fiber optic cables and displayed in the control area. The ASME Code considers this the same as direct visual contact with the gauge glass. If a bicolor gauge glass image is not available, the ASME Code provides that two dependable remote level indications may be substituted for direct visual contact with the gauge glass. These may consist of transmission of the gauge glass image via a camera and monitoring screen, or another type of remote indicator. For example, electronic measurements of the boiler water level may be made and transmitted to a panel-mounted indicator or computer terminal. **See Figure 4-6.**

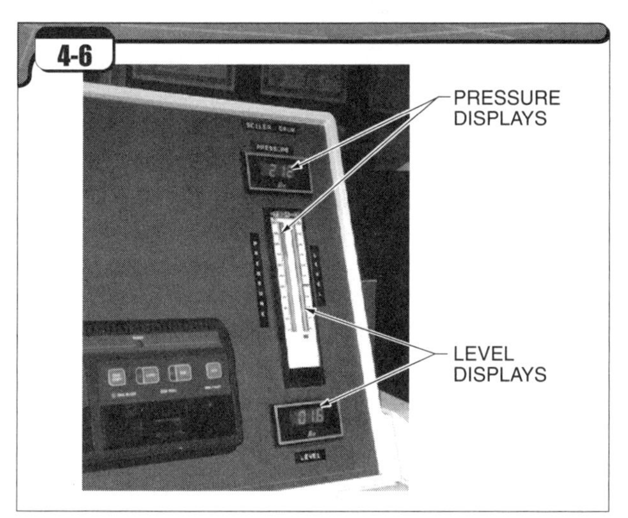

Figure 4-6. *Remote water level indicators are used when direct visible contact with the gauge glass is not feasible.*

13 What is the purpose of sheet mica in flat, high-pressure gauge glasses?

Sheet mica is a flat, clear mineral sheet. Sheet mica is placed between the gauge glass and the steam and/or water. The mica protects the flat, high-pressure gauge glass from the etching and eroding action of the steam.

14 Why are gauge glasses fitted with hand valves at the top and bottom?

Hand valves are used to isolate a gauge glass if it breaks so that it may be replaced. If these isolating valves are located 7′ or higher from the operating floor level, they must be a type that indicates by their position whether they are open or closed.

If the valves attached to the water column and gauge glass cannot be reached from the operating floor level, most boiler plants install chain operators on the valves. For example, if the gauge glass should break, this allows the boiler operator to quickly pull a chain and isolate the broken gauge glass.

15 How is a ball check valve used in conjunction with a gauge glass?

A *ball check valve* is an automatic self-closing gauge glass valve. A ball check valve acts as a safety device to automatically isolate a gauge glass if the glass breaks. Each of these valves contains a brass or stainless steel ball about ½″ in diameter. **See Figure 4-7.** If the glass tube should break while in service, the rush of steam and water causes these metal balls to be pushed against the openings in the valves, isolating the glass automatically so that the dangerous spray of steam and hot water is minimized.

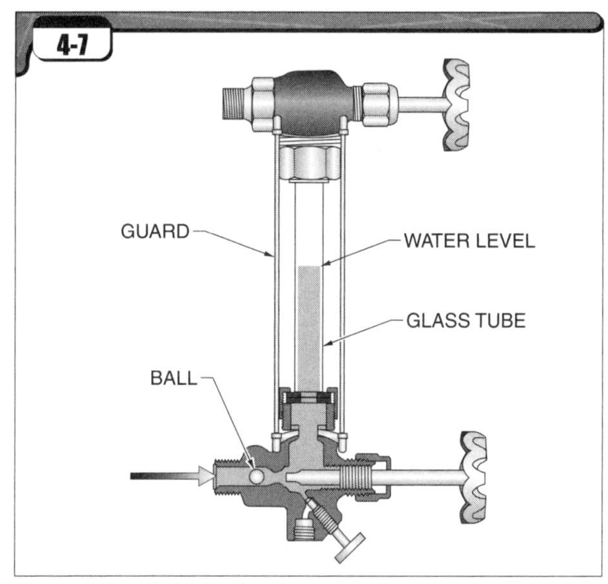

Figure 4-7. *Ball check valves automatically isolate the gauge glass if it breaks.*

16 Can a boiler be operated with a broken gauge glass?

A boiler can be operated temporarily with a broken gauge glass. The gauge glass can be isolated by closing the top and bottom valves. Then the boiler operator can frequently check the water level with the try cocks while replacing the gauge glass.

17 What should be closed first if a gauge glass breaks?

If a gauge glass breaks, the bottom valve should be closed first because of the greater danger of burns. If a person is burned by steam, the burning is greatly reduced if the person pulls out of the path of the steam. However, if hot water gets on a person's clothes, it keeps burning until the clothes are removed or until the water cools below about 140°F.

Factoid

Ball check valves on steam boiler gauge glasses are an excellent safety investment.

18 What would happen if the top or bottom valve to a gauge glass were left closed?

If the top or bottom valve to a gauge glass were left closed, the water level in the gauge glass would rise and show a higher reading than the actual water level in the boiler. If the top valve to a gauge glass were left closed, the collapsing steam would form a vacuum in the top of the gauge glass. The water level would rise to fill the vacuum.

If the bottom valve to a gauge glass were left closed, the steam vapors condensing in the gauge glass would eventually fill the gauge glass since there would be no way for the water level in the gauge glass to equalize with the level in the boiler.

19 Why might a gauge glass show a false level?

- One or both of the valves to the water column or gauge glass may be closed.
- One or both of the piping connections to the water column or gauge glass may be plugged with scale or debris.
- The water column may have been piped to the boiler in such a manner as to allow pockets of water to settle in the piping. These pockets of water would not allow the water column to show the actual water level.
- The boiler framework or brick setting may have settled and allowed the boiler to tilt slightly.

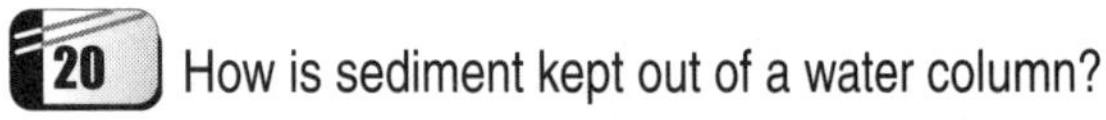

20 How is sediment kept out of a water column?

A water column should be blown down frequently to ensure that debris does not clog up the piping to the column. When a water column is blown down, the gauge glass should also be blown down, if piped separately. This ensures that the piping connections from the water column to the gauge glass are clear as well.

21 How are a water column and gauge glass blown down safely?

The blowdown pipe at the bottom of a water column should run to a safe drain area. Opening the valve in this pipe causes the pressure and velocity of the steam and water in the boiler to blow impurities from the water column.

The water column should be blown down before the gauge glass is blown down to prevent sludge and impurities from the water column from being blown into the small piping connections to the gauge glass. The water column blowdown valve is opened for 5 sec to 10 sec. Then the valve is closed.

A carefully planned sequence should be followed for the gauge glass blowdown. **See Figure 4-8.** The valve on top of the gauge glass is closed and the gauge glass blowdown valve is opened. This clears out any sludge or debris from the lower part of the piping to the gauge glass. Next, the valve on the bottom of the gauge glass is closed and the valve on top of the gauge glass is reopened. This clears out any sludge or debris from the top part of the piping to the gauge glass. Then the valve on the bottom of the gauge glass is reopened. Next, the gauge glass blowdown valve is closed to allow the boiler water to enter the gauge glass again. It should be verified that the water level returns to the NOWL.

It is critical that the boiler operator watch the water return to the gauge glass when the blowdown valve is closed. Water should quickly and freely reenter the gauge glass. If it does, the piping connections to the water column and gauge glass are clear of debris. If the water returning to the gauge glass is very dirty, the water column and gauge glass should be blown down again until the water is reasonably clear.

If the gauge glass is fitted with ball check valves, these valves will close as the gauge glass is blown down. In order to prevent this, the boiler operator may close the valves about halfway. In this position, the valve stems will hold the check ball off the seat inside the valve and allow water to pass. The operator must be sure to reopen the ball check valves fully after blowing down the gauge glass, in order to restore the functionality of the valves.

22 What is the minimum size pipe used for a gauge glass blowdown line?

The minimum size pipe used for a gauge glass blowdown line is ¼″ ID, according to the ASME Code. When the boiler operating pressure exceeds 100 psi, the gauge glass must be provided with a valved drain to a safe discharge point.

23 How often should a water column and gauge glass be blown down?

A water column and gauge glass should be blown down once per shift if the boiler is in continuous 24-hour operation. If the boiler is not in continuous 24-hour operation, the water column and gauge glass should be blown down once per day.

24 What are the first duties of the boiler operator when taking over a shift?

The boiler operator should discuss the operation with the operator going off duty and then check the log book for notes about any abnormal conditions. As quickly as feasible thereafter, the water columns and gauge glasses of all boilers in service should be blown down to confirm that all boiler water levels are satisfactory.

25 Can connections that use a flow of steam or water be made to a water column?

Lines that would allow a flow of steam or water cannot be connected to a water column as a false water level reading would result. The flow of steam or water from a water column would cause an inaccurate water level reading because of the slight pressure drop at the point where the flow of steam or water leaves the water column.

26 What appliances may be attached to a water column?

Only appliances that do not use a flow of steam or water may be attached to a water column. Permissible attachments include automatic pressure controls, level alarms, pressure gauges, and feedwater regulating devices.

27 Why do most high-pressure boilers that operate at 500 psi and above not have try cocks?

In high-pressure boilers operating at about 500 psi and above, much of the water instantly flashes to steam as it blows from the try cock. This occurs because the water temperature is so high. This makes it difficult to tell whether steam or water is coming out of the try cock; thus try cocks are not useful on very high-pressure boilers.

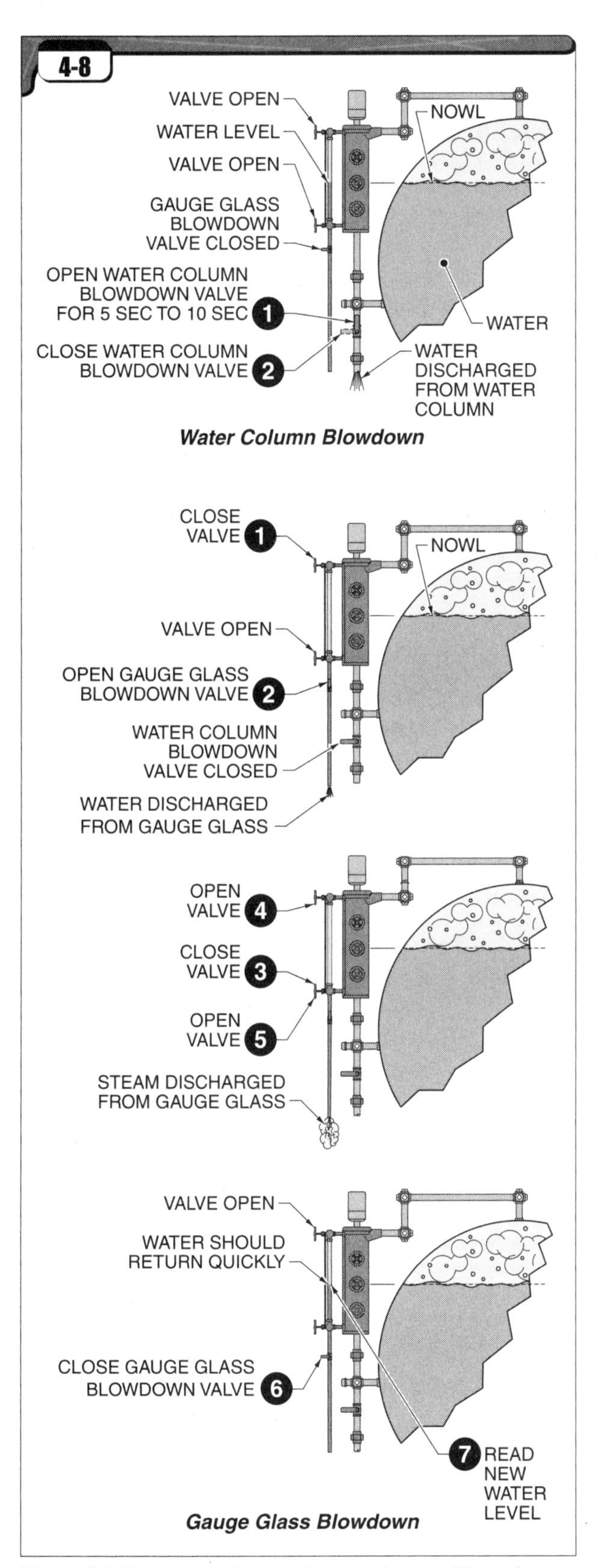

Figure 4-8. *The water column and gauge glass are blown down to remove any accumulated sludge and/or sediment.*

Level Controls

What is the purpose of an automatic feedwater regulating system (feedwater regulator)?

A feedwater regulating system (feedwater regulator) automatically maintains a constant safe water level in a boiler. This automatic control relieves the boiler operator of the tedious duty of having to constantly watch the water level in the gauge glass and manually manipulate the water valves. The feedwater regulating system consists of a means of automatically detecting the boiler water level, and a means of automatically increasing or reducing the feedwater supply as needed to keep the water level constant.

Describe the operation of an ON/OFF feedwater regulating system.

An *ON/OFF feedwater regulating system* is a level control system that uses a water level-detecting device to turn the feedwater pump ON when the boiler water level drops to a preset point, and turn the pump OFF when the water level has risen to an upper setting. This type of feedwater regulating system requires a small range, or differential, between the lower level and the upper level. Usually this range is about an inch to two, depending on the physical size of the boiler. The ON/OFF method of control is usually used for packaged boilers. The level-detecting device is most commonly either a float-operated switch such as a mercury switch, or an electric probe. **See Figure 4-9.**

A variation of this system is used when multiple boilers are fed water from a single feedwater pump. This is very often the case in small to medium-size hospitals, manufacturing plants, and similar size facilities. In this variation, the feedwater pump runs continuously and feedwater pressure is always maintained in a feedwater header. The feedwater lines from the feedwater header to each boiler are equipped with a valve that is opened and closed by the level-detecting device on that boiler. Usually the valve has a solenoid actuator or electric motor actuator.

The ON/OFF feedwater regulating system is the simplest and least expensive system. However, it has the disadvantage of putting all the comparatively cool water in the boiler at one time. For example, when the lower setting on the level-detecting device is reached, the pump is started or the feedwater valve is opened, depending on the system configuration. In most cases, the feedwater is considerably cooler than the water that is already inside the boiler. When the large volume of cooler feedwater is added to the boiler at one time, it cools the body of water that is already in the boiler to some degree. The new water must then be heated. The end result is that the steam production from the boiler drops each time the boiler takes on water, and the steam header pressure may fluctuate somewhat. In larger facilities that require a very constant steam header pressure, this may not be acceptable and a more sophisticated modulating feedwater regulating system must be used.

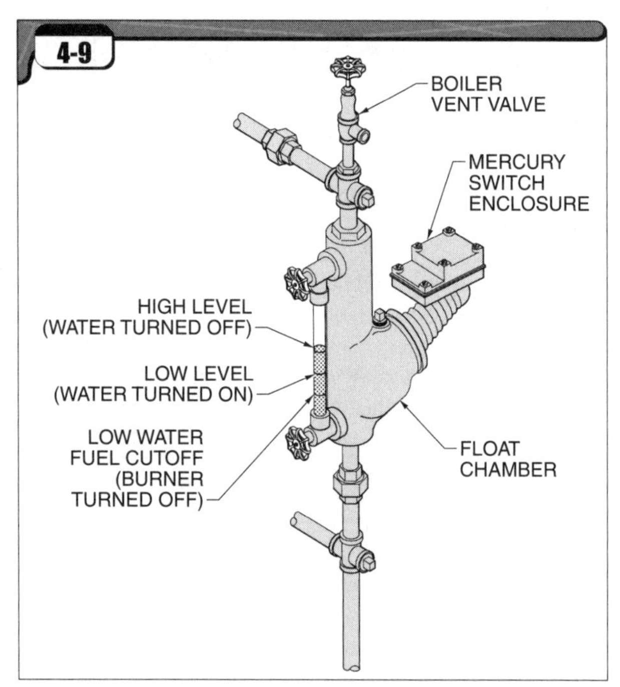

Figure 4-9. *The ON/OFF feedwater regulator turns the feedwater OFF at high level and ON at low level. It turns the burner OFF at the low water cutoff point.*

30 What is a modulating feedwater regulating system?

A *modulating feedwater regulating system* is one that continually adjusts the position of a feedwater regulating valve as needed in an effort to maintain a constant boiler water level by matching the position of the valve to the change in the boiler water level. There are a number of different modulating feedwater regulating systems. Some use electrical power to position the regulating valve, some use compressed air, and some are self-contained, purely mechanical systems. The self-contained systems are commonly referred to as feedwater regulators.

Factoid

All automatic controls will eventually fail to work properly. Boiler operators should periodically review procedures and be prepared to manually take control of various parts of the steam system in the event of a control failure.

31 Describe the operation of a thermohydraulic feedwater regulator.

A *thermohydraulic feedwater regulator* is a modulating control that controls feedwater flow in direct response to changes in the boiler water level. As the term thermohydraulic implies, a change in temperature generates a change in hydraulic pressure, and the hydraulic pressure is used to operate the feedwater regulating valve.

This regulator consists of a radiator chamber, a tube that passes through the middle of the radiator chamber, and a feedwater regulating valve with a diaphragm actuator. **See Figure 4-10.** The tube that passes through the middle of the radiator chamber is connected to the drum high in the steam space at one end and well below the NOWL at the other end. The water level in this tube is the same as the water level in the boiler drum.

The radiator chamber is full of distilled water, which is confined and does not circulate. The radiator chamber is connected to the top of the diaphragm actuator on the control valve with a piece of copper or stainless steel tubing. The radiator chamber has fins that slowly dissipate heat from the steam and hot water.

If there is a drop in the boiler water level, the water level in the tube that passes through the center of the radiator chamber drops by the same amount. This drop causes more of the water in the radiator chamber to be exposed to the greater heat transfer rate of the steam. Heat is conducted into the water faster than it is radiated from the fins, so the pressure and temperature of the distilled water in the radiator chamber rise and the water expands. The increase in pressure and expansion is applied to the top of the diaphragm actuator through the tubing, deforming the diaphragm downward and opening the feedwater valve farther. This results in more feedwater flow, and the water level in the boiler is brought back up to normal. If the level rises above the NOWL, the process reverses.

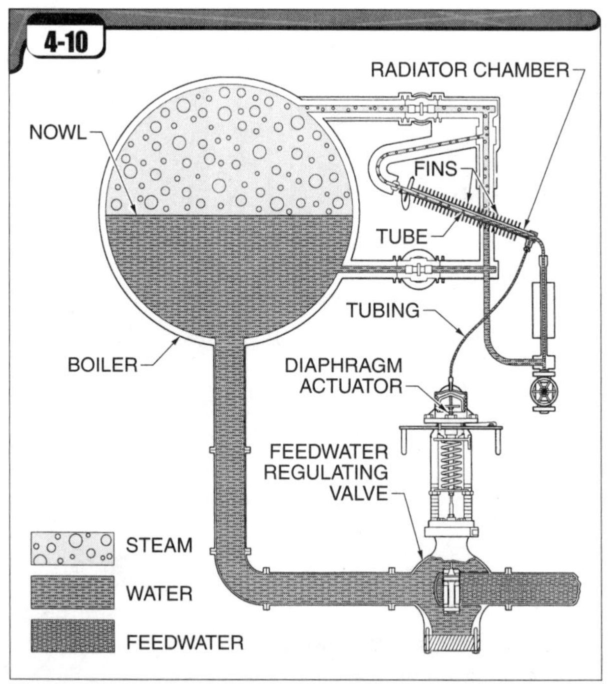

Figure 4-10. *The thermohydraulic feedwater regulator uses the heat of steam to create the hydraulic pressure needed to control feedwater flow.*

32 Describe the operation of a thermostatic expansion tube feedwater regulator.

A *thermostatic expansion tube feedwater regulator* is a modulating control that controls feedwater flow in direct response to changes in the boiler water level. With this regulator, changes in temperature result in a proportional movement of the feedwater regulating valve.

This unit has a long, slightly tilted expansion tube that is anchored at the lower end. **See Figure 4-11.** The expansion tube is the main functioning part of the regulator. The upper end of the expansion tube is connected to the drum high in the steam space and the lower end is connected well below the NOWL. The system also has a lever-actuated valve in the feedwater piping. A linkage at the upper end of the expansion tube uses an offset pivot point. One side of the linkage is connected to the expansion tube and the other side is connected to a rod that extends down to the lever actuator on the valve.

A drop in the water level in the boiler also results in a drop in the level inside the expansion tube. The material of construction of the expansion tube has a high coefficient of expansion. This means that the expansion tube material expands and contracts significantly with changes in temperature as compared to other materials. When more of the tube is exposed to the heat of the steam, the tube expands. This expansion moves the pivoted linkage on the end of the tube and allows the attached rod to drop slightly. The rod moves the lever actuator attached to the feedwater valve and causes the valve to open farther. The feedwater flow to the drum increases and the drum level is brought back to normal. As the water level rises, the process reverses.

33 List an advantage and a disadvantage of using an automatic feedwater regulating system.

An advantage of an automatic feedwater regulator is that the operator is relieved of the monotonous duty of constantly watching the water level in a boiler and can perform other tasks in the plant. Therefore, the chances of a low water level due to the operator being distracted or inattentive are minimized.

A disadvantage of an automatic feedwater regulator is that operators often become complacent and too dependent on the automatic feedwater regulating system and may not be prepared to respond in the event of a system failure.

34 What is a metering feedwater regulating system?

A *metering feedwater regulating system* is a control that continually measures the boiler conditions and adjusts the feedwater control valve. The measured conditions may include the steam flow from the boiler, the feedwater flow going to the boiler, the boiler water level, or all three.

A pound of steam is produced from a pound of water. In an ideal case, if the feedwater flow could be kept perfectly matched to the steam flow, the boiler water level would stay constant. However, the boiler water level will shrink and swell slightly with changes in the steam flow rate and the heat input from the combustion equipment. In addition, some water is lost to boiler blowdown. *Blowdown* is removal of impurities from the boiler water by draining some of the water. Therefore, the actual boiler water level is also measured and is used to fine-tune the feedwater regulating valve position.

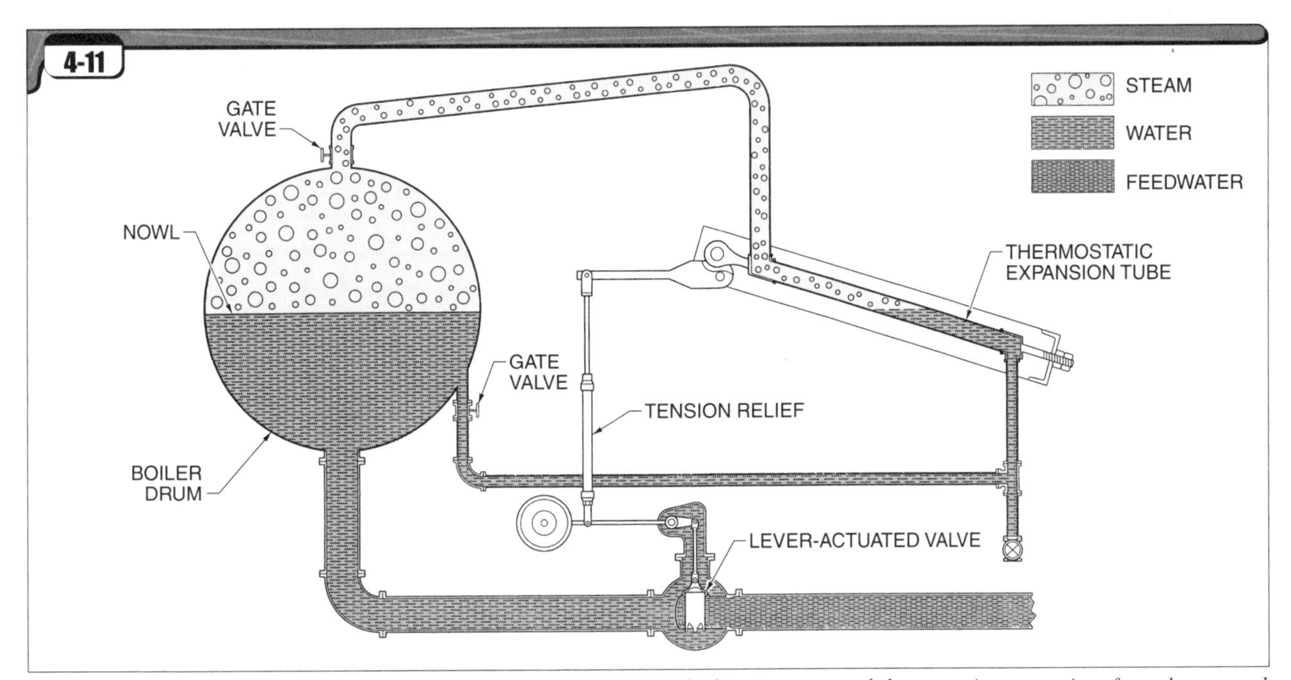

Figure 4-11. *The thermostatic expansion tube feedwater regulator uses the heat in steam and the expansive properties of metal to control feedwater flow.*

Metering feedwater regulating systems are the most expensive and complicated, but have the advantage of being able to maintain the boiler water level much more precisely than ON/OFF or modulating systems.

35 What is the advantage of using a two- or three-element feedwater regulating system?

In instrumentation terminology, the word "element" refers to a device used to measure something in a process. For example, a float-operated device, or element, may be used to measure the water level in a tank. A *two-element feedwater regulating system* is a water level control system that measures the steam flow from the boiler in addition to the water level. A *three-element feedwater regulating system* is a water level control system that measures the steam flow from the boiler and the feedwater flow into the boiler in addition to the water level.

Two- and three-element feedwater regulators have the advantage of being able to anticipate changes in the water level of a boiler before the changes actually occur. For example, if the steam flow increases significantly because of additional steam demand in the plant, the water level will rise slightly in the boiler. This occurs because the momentary drop in pressure as more steam leaves the boiler causes additional boiler water to flash into steam bubbles. These bubbles displace water, thus the entire body of water in the boiler will swell slightly. If the feedwater regulator only senses the water level, it can be deceived by this slight momentary swell. When the flash steam bubbles in the boiler water dissipate, the water level can drop too far the other way.

In such a scenario, the two-element or three-element feedwater regulating system will maintain a more stable water level. It will sense that more water is needed because more steam is leaving the boiler. Thus, the drop in the boiler water level is avoided because the regulator starts adding water even before the level drops.

36 What is a double-seated valve?

A *double-seated valve* is a control valve that has two discs on one stem and two seats in the body. **See Figure 4-12.** Double-seated valves are also referred to as balanced valves. In a typical single-seated control valve, the pressure of the fluid flowing through the valve presses against one side of the valve disc. This means that a strong spring must oppose the force of the flowing fluid. In a double-seated valve, the pressure of the fluid flowing through the valve presses against the bottom of one disc and the top of the other disc. This means that the forces almost balance each other and the force required to change the valve position is small. This type of valve is often used for feedwater regulators such as the thermohydraulic type because the force available to move the valve is fairly slight. Double-seated valves are used for other automatic process control valve applications as well, such as speed governors on small steam turbines.

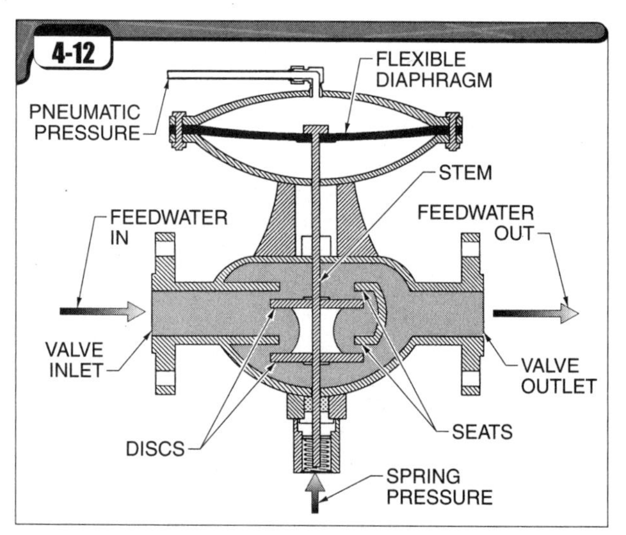

Figure 4-12. *Double-seated valves are often used for feedwater regulation.*

37 Why should the boiler operator be concerned with the amount of feedwater that a boiler uses?

The amount of feedwater that a boiler uses is important because a pound of steam produced by a boiler requires a pound of water to replenish the steam. If enough water is not supplied to replenish the steam leaving a boiler, the water level will drop. Conversely, if much more water is being supplied than is needed to replenish the steam leaving the boiler, the blowdown may be excessive or a leak may be present.

The feedwater flow to the boiler is usually measured and expressed in gallons per minute (gpm). The steam flow is expressed in pounds per hour (lb/hr). The boiler operator may compare the two. This is done by multiplying gallons of feedwater by 8.33 to obtain lb/min, and then multiplying by 60 (min/hr) to obtain lb/hr.

38 What is a condensate receiver? Describe a typical modern package unit.

A *condensate receiver* consists of a small vessel for receiving condensate from steam-using equipment and a pump to return the condensate to the boiler system. A typical modern package unit contains a small tank, level control switches for the pump, a sight glass on the tank for determining the tank level, and the condensate pump. **See Figure 4-13.** The feedwater that is added to the boilers should consist of as much condensate as possible and enough makeup water to satisfy the remainder of the feedwater demand.

Trade Tip

To quickly compare lb/hr of steam flow to gpm of feedwater flow, drop three zeros from the steam flow and double the resulting number. For example, 25,000 lb/hr steam flow equals 25 × 2 or 50 gpm.

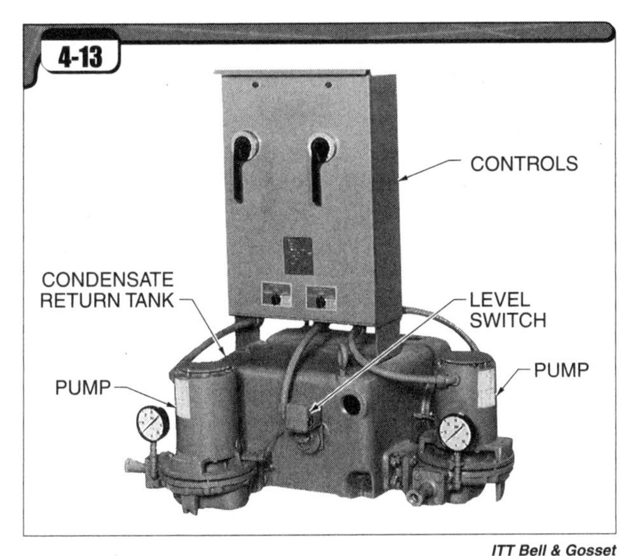

ITT Bell & Gosset

Figure 4-13. *A condensate receiver receives condensate from steam-using equipment and returns it to the boiler system.*

39 What should an operator suspect if large quantities of makeup water are used?

There may be condensate leaks or steam leaks somewhere in the plant, or steam traps may be failed in the open position. Steam traps that fail open allow live steam to go out the vents at the condensate receivers.

In most boiler systems, the makeup water passes through a water softener prior to going into the boiler system. Most commercial and industrial water softeners are supplied with flowmeters that include a totalizer. The totalizer continually adds up the total amount of water that has passed through the softeners. This reading is usually in gallons. Boiler operators should read this meter daily and become familiar with the normal amount of makeup water used on a daily basis. A particularly high reading serves as an indicator of steam or condensate losses.

40 What is a makeup water feeder?

A *makeup water feeder* is an automatic float-operated valve that feeds makeup water to a low-pressure heating boiler to replace condensate that has been lost from the system or water that has been lost in the form of steam leaks. **See Figure 4-14.** In a low-pressure heating system, the great majority of the condensate is usually returned to the boiler. The amount of water that is lost in the form of steam leaks, condensate leaks, and blowdown is relatively small. For this reason, the quantity of oxygen introduced into the system is small. A deaerator is therefore usually not justified and air that is liberated from the boiler water is vented at the radiator vents and high-point vents in the distribution system. The makeup water is therefore often added directly into the boiler by the makeup water feeder. Sometimes the makeup water feeder feeds the makeup water to a condensate receiver rather than to the boiler itself.

The makeup water feeder is not intended to take the place of a feedwater regulating system. The makeup water feeder is only intended to replace water that is lost. The float of the makeup water feeder is located at a slightly lower point than the NOWL in the boiler or condensate receiver.

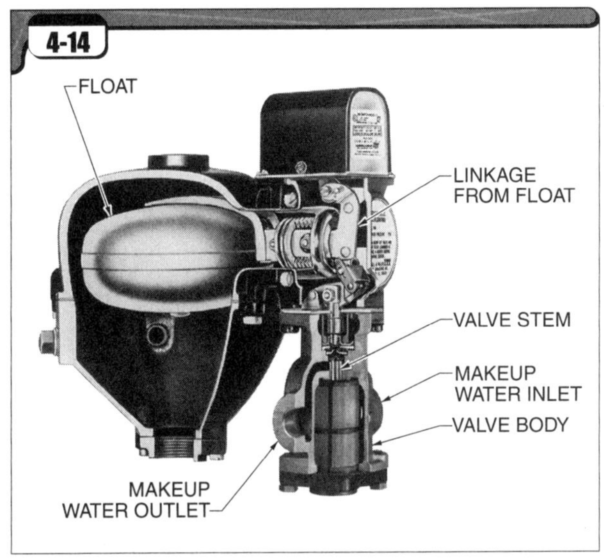

McDonnell & Miller

Figure 4-14. *Makeup water feeders replace water that is lost to leaks in a low-pressure boiler system.*

41 What care should a makeup water feeder be given?

Like any automatic device that contains water, the makeup water feeder's reliability may be reduced by the accumulation of sludge or scale, either in the float chamber or in the connecting piping. For this reason, the makeup water feeder should be blown down regularly to flush out any debris.

Since the feedwater provided to low-pressure heating boilers almost always consists of mostly clean condensate, it may be wasteful to blow down the makeup water feeder too frequently. A weekly blowdown is usually adequate, but this frequency should be influenced by experience.

The makeup water feeder also contains a strainer that is intended to catch any debris that may exist in the makeup water piping and prevent it from interfering with proper operation. For example, particles of rust or dirt may prevent the valve in the makeup water feeder from closing properly. If the strainer becomes plugged with debris, this could interfere with the makeup water feeder's ability to supply adequate makeup water. For this reason, the strainer should be blown down regularly by using the supplied blowdown valve. A monthly blowdown of the strainer should be adequate.

Factoid

Bicolor gauge glasses assist the boiler operator in seeing the boiler water level from a considerable distance.

Level Alarms and Safety Devices

What does a blowing whistle on top of a water column indicate?

A blowing whistle on top of a water column indicates either a high or a low water level in the boiler. The whistle is caused by floats inside the water column that open a small steam valve to the whistle if a high or low water level is reached. If the whistle blows, check the gauge glass to determine the water level.

Factoid

The piping between the boiler proper and any attachment to the boiler falls under applicable sections of the ASME Boiler and Pressure Vessel Code and/or the ASME Code for Pressure Piping, B31.1. A competent engineering source should be consulted for specific questions regarding allowable pipe types, joining methods, qualifications for welding personnel, and any other questions about pressure vessel design.

List three mechanisms used to signal boiler water level alarms.

Floats may be used inside a water column to open a steam valve to a whistle as an alarm, or the float may be used to trip a switch that activates an alarm. In water columns for boilers that fire coal on grates, two floats are commonly provided. The upper float controls a high water alarm and the lower float controls a low water alarm. Floats are commonly made of thin-gauge steel.

Electric probes utilize the conductive qualities of water to either open or close an electric circuit to an alarm annunciator. For example, the water can conduct a small electrical signal between two immersed probes. If the water level falls to a level low enough to break the circuit between the probes, an annunciator sounds a low water alarm. An *annunciator* is an audible alarm that is created electrically or electronically. For example, a beeper or buzzer accompanied by a light on a control panel serves as an annunciator. Whether the alarm circuit is opened or closed depends on whether the alarm is for high or low water and the wiring scheme.

A differential pressure transmitter may also be used to signal a boiler water level alarm. **See Figure 4-15.** A *transmitter* is an instrument used to send information about the condition of a process to a control device. Differential pressure transmitters have two sensing pipes. One pipe is used to sense the pressure exerted by the weight of the water in a boiler pushing down on one side of the transmitter. The other pipe is used to sense the pressure exerted by the weight of a full column of water on the other side of the transmitter. The height of the full column of water is calibrated to produce pressure equal to the full range of the transmitter. The transmitter converts the difference between the weights of the two columns of water into a proportional electrical signal. For example, most instruments of this type use a 4 mA to 20 mA signal. A milliamp (mA) is 1/1000 of an ampere.

As the boiler water level rises or falls, the electrical signal produced by the transmitter rises or falls proportionally. Annunciators, which receive the electrical signal from the transmitter, may be set to signal alarms at preset values. Differential pressure transmitters are also used to provide continuous remote water level indication.

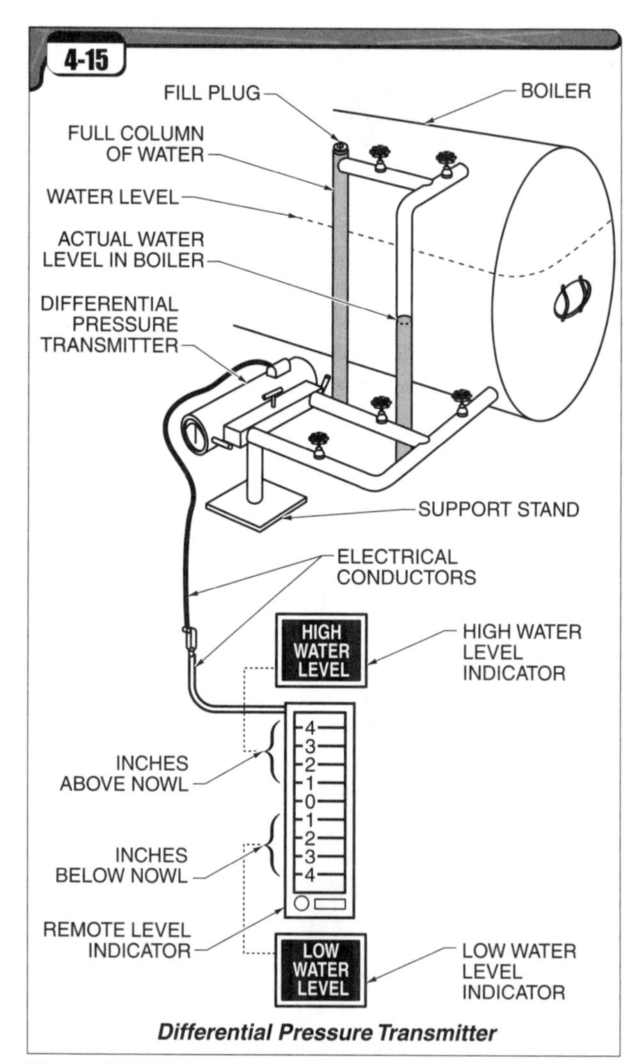

Figure 4-15. *Differential pressure transmitters may be used to signal alarms.*

Factoid

Boilers are damaged by low water approximately four to five times more frequently than any other cause, even though two separate low water fuel cutoff controls are usually provided. This is because these critical safety controls are often neglected.

Describe a low water fuel cutoff.

A *low water fuel cutoff* is a device located slightly below the NOWL that turns off the boiler burner in the event of low water. **See Figure 4-16.** The device prevents damage to the boiler tubes and the possibility of a boiler explosion.

The same types of devices that are used to create low water alarms are used as low water fuel cutoff controls. For example, a float in a water column or in a separate float chamber is often used in conjunction with a pair of mercury switches. When the float drops to a preset level, the first mercury switch operates the feedwater pump or valve to admit feedwater into the boiler. If the float continues to drop, the second mercury switch breaks the fuel valve circuit and the fuel valves fail closed. Electric probes and level-sensing transmitters are often configured to work the same way. The electric probe type of low water fuel cutoff uses the conductivity of water to activate an electrical relay that shuts off the boiler burner. If the water level falls below the probes, the water cannot conduct between the probes and the fuel valve circuit is broken. Transmitters can also be used to serve this function. When the signal from a transmitter falls below a predetermined point, the computer or panel-mounted controller can initiate a fuel system shutdown.

Combining the water column and low water fuel cutoff into one housing reduces initial cost. Additionally, the low water fuel cutoff can be tested when the water column is blown down. This configuration also reduces the number of connections that must be made to a boiler, which reduces the potential for leaks. A disadvantage to this arrangement is that if the piping to a combination water column and low water fuel cutoff should plug up, neither appliance would operate.

Because the potential boiler damage that can happen from low water is so great, a redundant (backup) low water fuel cutoff control is usually provided. For example, another low water fuel cutoff device that is separate from the water column may be attached to the boiler in the same manner as the water column. The redundant device is installed so that it shuts off the burner when the water level in the boiler drops to a level slightly below that of the primary low water fuel cutoff device.

In most cases, the redundant control is a manual-reset device. This means that the boiler operator must press a reset button in order to reset the controls and allow the boiler to be started again. The intent of this arrangement is to ensure that the operator is made aware that the primary low water fuel cutoff did not work properly.

Where should a low water fuel cutoff be placed on a boiler?

A low water fuel cutoff should be placed on a boiler so that the control causes the fuel valves to fail closed before the water level drops below the lowest safe level. **See Figure 4-17.** The exact level depends on the boiler, but is usually from 2″ to 6″ below the NOWL. This generally corresponds with the location of the lowest try cock. In no case should a low water fuel

Float-Operated Mercury Switch

Electric Probes in Water Column

Figure 4-16. *Float-operated switches and electric probes are used as low water fuel cutoffs.*

cutoff be set to operate at a level below the lowest visible level in the gauge glass. A low water fuel cutoff should be piped in the same manner as a water column, with a provision made for blowing down the piping and testing the device.

A redundant low water fuel cutoff is installed in the same way as the primary low water fuel cutoff, except that it is installed at a slightly lower level. This is usually about 1″ to 2″ below the level of the primary low water fuel cutoff.

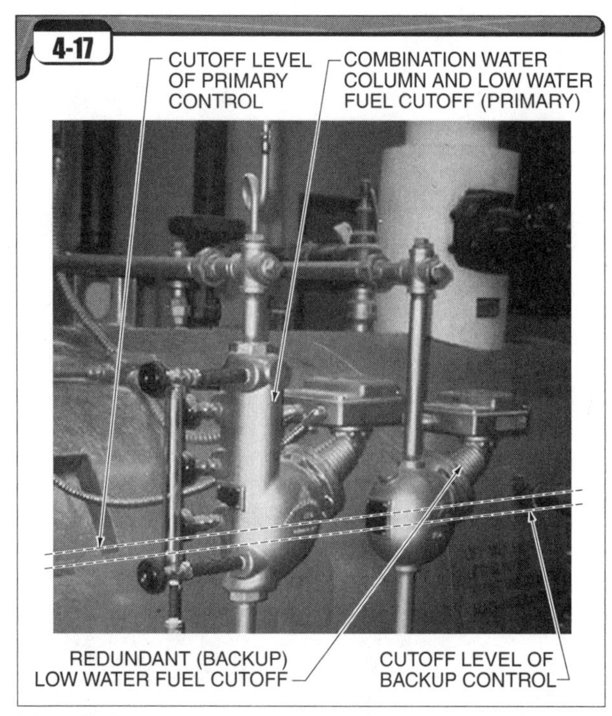

Figure 4-17. *A redundant low water fuel cutoff is installed at a slightly lower level than the primary low water fuel cutoff.*

46 What is a fusible plug? How does it function?

A *fusible plug* is a temperature sensitive device that causes an audible alarm when exposed to excessive temperature. Fusible plugs are most commonly found in firetube boilers. The fusible plug is a hollow brass plug that is filled with a core of tin solder that is designed to melt at 450°F. **See Figure 4-18.**

Fusible plugs are manufactured as either fire side or water side types. A fire side fusible plug screws in from the fire side toward the water side. A water side fusible plug screws in from the water side toward the fire side.

A fusible plug is installed at the lowest permissible water level (usually just above the highest heating surface) in smaller boilers. The fusible plug extends through the rear tube sheet in an HRT firetube boiler, the top of the flue furnace in a scotch marine firetube boiler, the crown sheet in a firebox-style firetube boiler, or other area of intense heat. The fusible plug is located in the direct path of hot gases of combustion on the fire side and is covered with water on the water side. Water in a boiler has a higher heat transfer rate than steam, and can carry heat away from the plug faster than steam can. If the water level drops below the plug, the solder melts out, and steam rushes through the opening in the plug. Because of the design of the opening, the steam causes a loud sound that alerts the operator that the water level in the boiler is very low.

Fusible plugs are most commonly found on low pressure to medium pressure coal-fired boilers. If the fusible plug melts, the boiler operator should stop the fuel and air supplies, dump the coal from the grates into the ash pit and thus remove the source of the heat. In coal-fired firebox firetube boilers or scotch marine boilers, the location of the fusible plug directly above the fire causes steam and water to blow into the furnace area, helping to put out the fire. These devices are much less common on gas-fired or oil-fired boilers. Low water fuel cutoff controls are used on these boilers to shut off the fuel if a low water condition occurs. Additionally, many package boilers are unattended. Therefore, no one is present to respond to the audible alarm created by the fusible plug.

A fusible plug is maintained by keeping it clean of soot on the fire side and clean of scale on the water side. Fusible plugs must be replaced annually because the fill material can oxidize over a period of time. This can cause the plug to not melt out when needed. When replacing a fusible plug each year, only those displaying an ASME International stamp should be used.

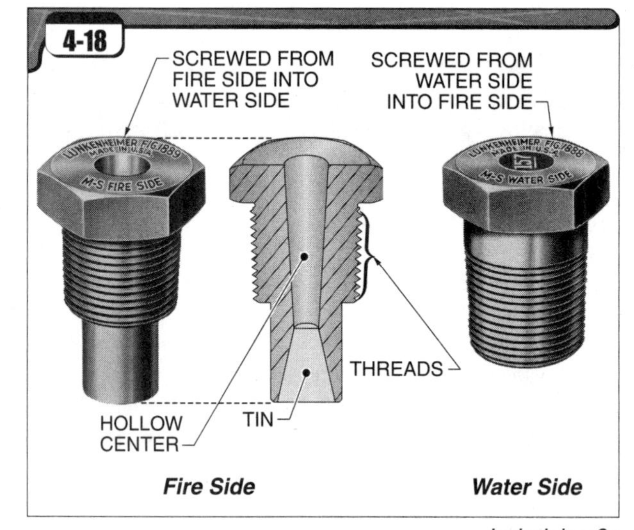

Figure 4-18. *Fusible plugs are temperature-sensitive devices that cause an audible alarm when exposed to high temperatures.*

47 How often should low water fuel cutoff testing be conducted with the quick-drain method?

A low water fuel cutoff should be tested daily by blowing it down in the same manner as blowing down a water column. **See Figure 4-19.** The burner should cut off. This is commonly referred to as the "quick-drain method." This test is feasible if there are several boilers and the momentary shutdown of one boiler will not cause a significant upset in the facility's processes.

If it is not feasible to have the boiler shut down during low water fuel cutoff testing, a test switch is often installed. Veteran operators often refer to this test switch as a "dead man switch." This switch is a spring-loaded pushbutton type switch that must be manually held in by the boiler operator while blowing down and testing the low water fuel cutoff devices. If the switch is released while the water level in the water column is low, the boiler will shut off. When the water level returns to normal, the switch can be released.

It is important that such testing devices not be configured to simply electrically bypass the low water fuel cutoff. For example, if the low water fuel cutoff does not operate properly while the pushbutton is held in, the boiler operator needs to know this. To facilitate a conclusive test, the test switch location should also be equipped with an indicating light or audible alarm.

The electrical circuit to the light or alarm is wired so that when the pushbutton is pressed, it bypasses the actual electrical relays that cause the fuel valves to close. The indicating light or audible alarm is energized when the same relays open. This system confirms with certainty that the low water fuel cutoff is in working condition. That is, if the boiler operator had not been holding the pushbutton, the fuel valves would have closed.

48 How often should low water fuel cutoff testing be conducted with the slow-drain method?

Float-operated low water fuel cutoffs should be tested monthly by actually allowing the boiler water level to drop and confirming that the low water fuel cutoff shuts the boiler off. This is commonly referred to as the "slow-drain method." This test is particularly appropriate for float-operated low water fuel cutoffs, because they are more susceptible to malfunction due to failure of the float to drop. This method more realistically simulates the development of an actual low water condition because the water level drops gradually.

To test the low water fuel cutoff using the slow-drain method, close the manual feedwater valve. **See Figure 4-20.** Then open the bottom blowdown valve to allow water to drain from the boiler. The boiler burner(s) should be firing during the test and should shut down when the float drops in the float chamber and reaches the trip point of the low water fuel cutoff. This test may also be performed without opening the bottom blowdown valve, because the water level in the boiler will drop as the water evaporates into steam. The sequence is reversed to return to normal operation.

Low water cutoff tests should be performed while the boiler is on low fire if possible, to minimize thermal strains on the boiler. The tests should be documented and records maintained for review by the boiler inspector.

49 What safety rules should be followed while testing a low water fuel cutoff by the slow-drain method?

The operator must not leave the boiler or otherwise be distracted until the test is complete and all valves are returned to normal position. The operator must also have a clear view of the gauge glass on the boiler while testing or have another operator present who can watch it.

50 What should be done if a low water fuel cutoff does not work during testing?

If feasible, the boiler should be shut down immediately and repairs made to the low water fuel cutoff. If this is not feasible, a boiler operator should be specifically assigned to constantly monitor the water level until the boiler can be shut down and the low water fuel cutoff controls repaired.

The great majority of all boiler accidents are due to low water fuel cutoff controls that do not work when needed. Most of these cases are due to lack of testing and maintenance or improper testing and maintenance.

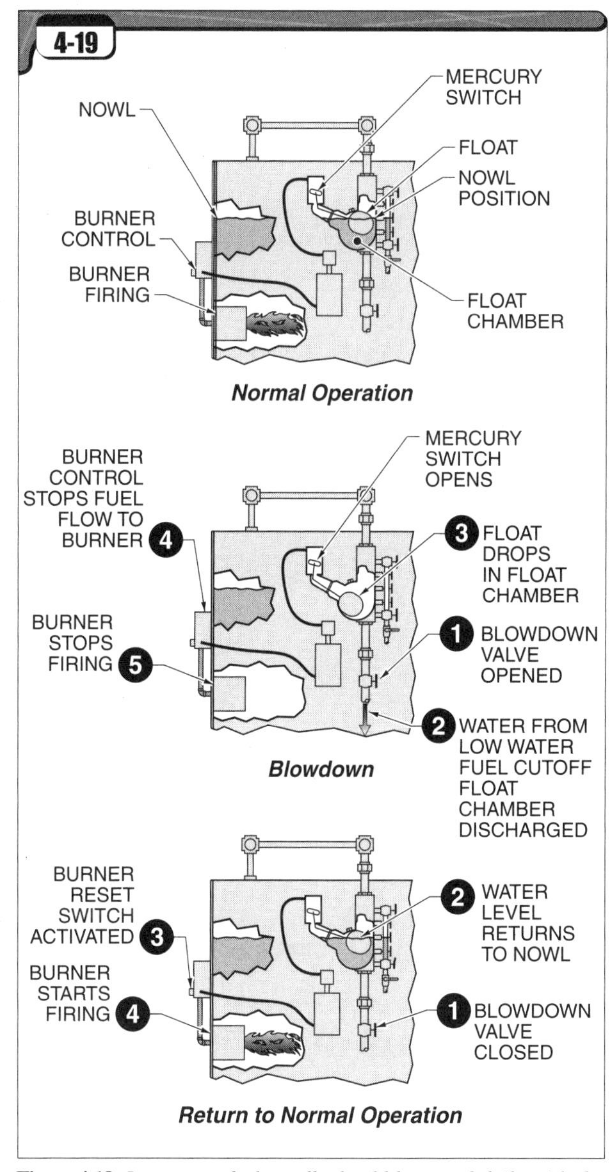

Figure 4-19. *Low water fuel cutoffs should be tested daily with the quick-drain test.*

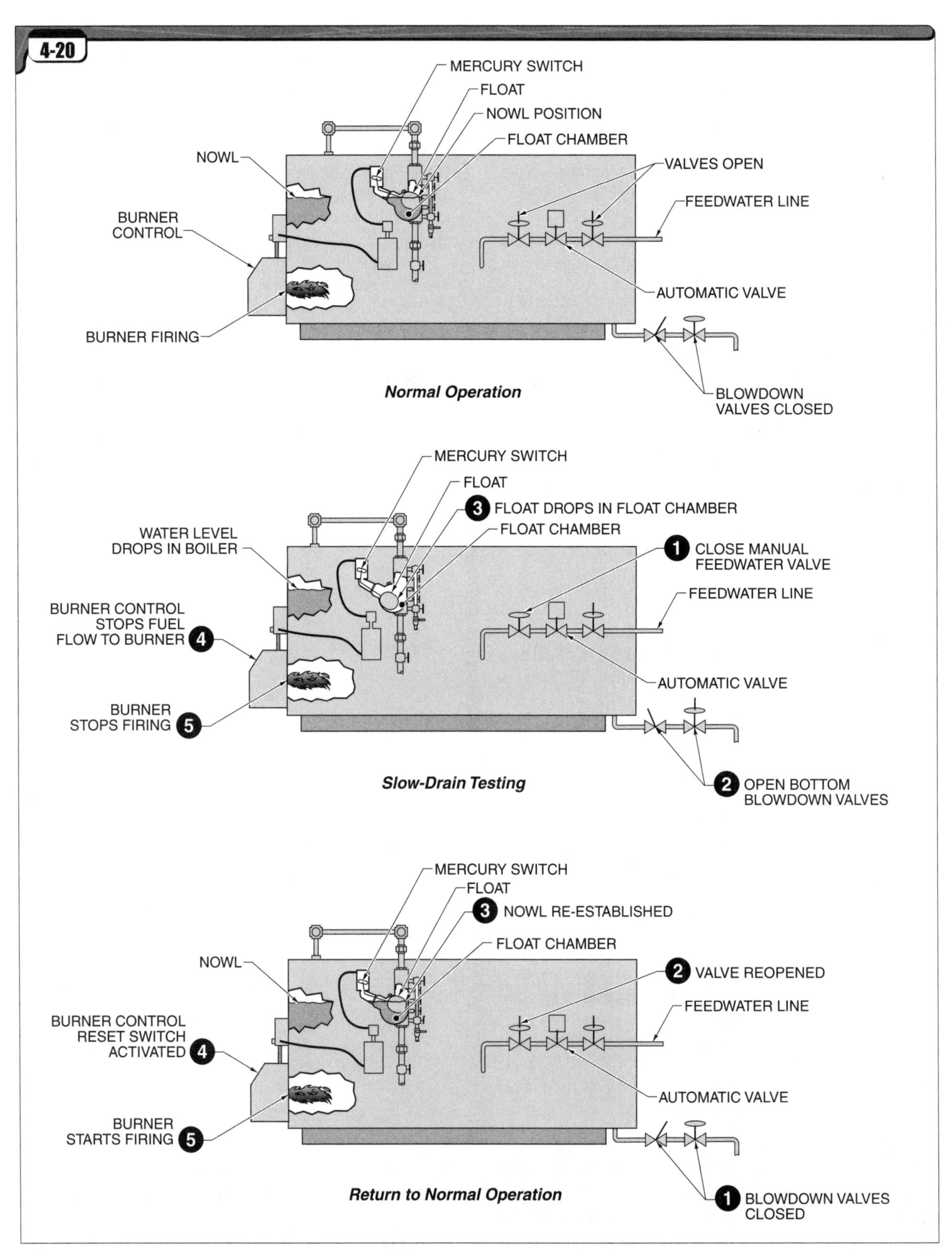

Figure 4-20. *Float-operated low water fuel cutoffs should also be tested monthly with the slow-drain test.*

Factoid

American industry is eliminating the use of mercury switches and other devices containing mercury because of environmental and safety concerns. Mercury is highly toxic if ingested.

Pump Theory and Principles

51 What is meant by vapor pressure of water?

All liquids evaporate to some extent. *Vapor pressure* is the equilibrium pressure where the number of molecules evaporating from a liquid surface equals the number of molecules condensing back to the liquid. The vapor pressure changes with changes in temperature. The equilibrium temperature is the boiling point. Water at 212°F has a vapor pressure of 14.7 psia (0 psig). This means that water boils at 212°F at standard atmospheric pressure. If the pressure on the water is increased, the equilibrium temperature is also increased. For example, water at 406.1°F has a vapor pressure of 250 psig. This is equivalent to saying that if the pressure on a closed vessel is 250 psig (264.7 psia), the boiling point is 406.1°F.

If water starts boiling and the steam is removed for use elsewhere, the number of molecules leaving the surface of the water is more than the number of molecules condensing back to the water. Therefore, steam is generated and water is being consumed.

52 What is specific gravity of a liquid?

Density is the weight of a material divided by its volume. Specific gravity is the density of a liquid divided by the density of water, when both are measured at the same temperature. In almost all cases, the standard temperature used is 60°F. For example, what is the specific gravity of a 1 gal. sample of fuel oil that weighs 7.58 lb? Water weighs 8.33 lb/gal. at the same temperature.

$SG = density\ liquid \div density\ water$
$SG = 7.58 \div 8.33$
$SG = \mathbf{0.910}$

53 What is the specific gravity of a salt brine solution that weighs 9.37 lb/gal. at 60°F?

$SG = density\ liquid \div density\ water$
$SG = 9.37 \div 8.33$
$SG = \mathbf{1.125}$

54 What is head?

Head is a vertical column of liquid that, due to its weight, exerts pressure on the bottom and sides of its container. The pressure at any particular elevation in the container of liquid will directly reflect the elevation (head) of the liquid above that point. The pressure due to a head of water is 0.433 psig/ft. For example, if a column of water is 10′ above the point of measurement, the pressure at the point of measurement will be 0.433 psig × 10, or 4.33 psig. The pressure created can be very slightly affected by the temperature of the water, but not enough to be of concern in normal boiler room operations.

The term "head" is often used in pump applications. If the water supply to a pump is taken from a tank or other supply source that is above the pump, the pump operates under suction head. This means that the water enters the suction side of the pump under a certain amount of pressure, which is due to the elevation of the water.

55 What is static suction head?

Static suction head is the vertical distance from the centerline of the pump up to the level of the liquid in the supply tank. **See Figure 4-21.** A static item is at rest. A dynamic item is in motion. The water being supplied to a pump is actually in motion, but the pressure exerted due to the head is approximately the same as if it were static. However, if the water level in the tank above the pump were to drop (creating low static head), the pump might encounter operational problems.

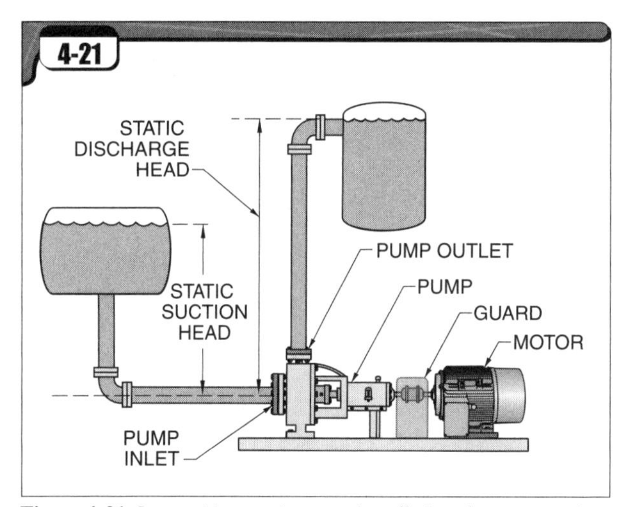

Figure 4-21. *In a positive-suction pump installation, the pump receives liquid with a static suction head from a source above the pump.*

Trade Tip

If a vacuum pump serving a condensate system runs continuously or for long periods without stopping, it is usually an indication of either air leaks into the condensate system or steam traps that have failed in the open position.

56 What is a positive-suction pump installation?

A *positive-suction pump installation* is any installation where the pump receives liquid on the suction side from a source above the pump—that is, under head. This configuration is advantageous because the liquid automatically flows into the suction side of the pump due to gravity.

57 What is lift?

Lift is the condition where the level of the liquid to be pumped is below the elevation of the pump. Lift is the opposite of head. The pump must expend energy not only in moving the liquid through the discharge pipe, but also in lifting the liquid from the lower elevation. In this case the pump operates under suction lift.

58 What is static suction lift?

Static suction lift is the vertical distance from the centerline of the pump down to the level of liquid in the supply source below. Again, static refers to a body that is not in motion; however, this measurement is adequate for normal plant operations.

59 What is a negative-suction pump installation?

A *negative-suction pump installation* is any installation where the pump must draw (lift) liquid up from a source below the pump. **See Figure 4-22.**

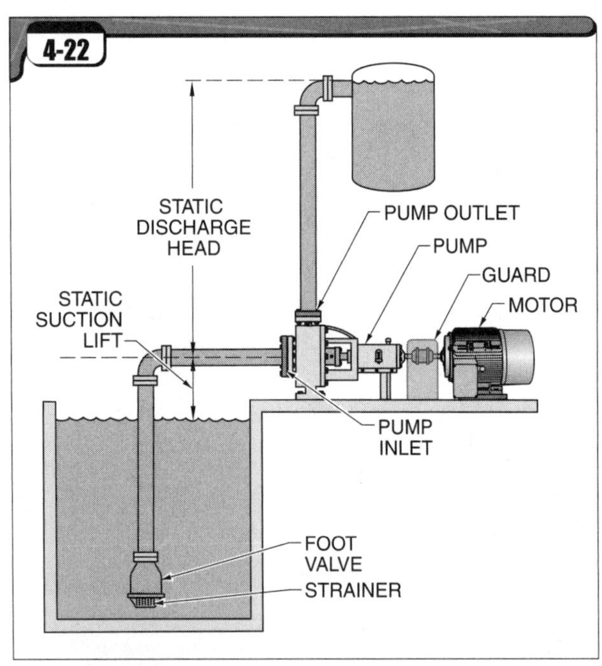

Figure 4-22. *In a negative-suction pump installation, the pump must draw liquid from a source below the pump.*

60 What is a foot valve? What function does it serve?

A *foot valve* is a check valve installed at the bottom of the suction line on a negative-suction pump that keeps the suction line primed when the pump shuts down. Usually, a foot valve has a screen or strainer installed on its suction side to keep debris from getting stuck in the valve. The pump can lose its prime if debris causes the valve to stick open.

61 What is static discharge head?

Static discharge head is the vertical distance from the centerline of the pump up to the surface of the liquid in the tank or vessel into which the piping discharges. Note that if that tank or vessel is under pressure, as a boiler is, the pressure must be converted to equivalent static discharge head when calculating the performance of the pump.

62 What is friction head?

Friction head is the pressure loss associated with friction in the pump piping and fittings, converted into equivalent feet of static head.

The movement of the water or other liquid through a pump's suction and discharge piping is impeded by friction. This friction occurs as the liquid flows against the walls of the piping, around elbows, and at directional changes at tees. Design engineers convert this friction to equivalent feet of vertical head so that this loss may be subtracted from the available work that the pump is capable of performing. Losses due to friction head may be substantial if the piping is undersized or if an excessive number of elbows or tees exist in the piping. Friction head may also become excessive if additional demand is placed on the pump and piping, such as an additional boiler that must be provided with water from an existing feedwater header.

63 How is boiler pressure changed to equivalent static head when calculating the output of a feedwater pump?

The boiler pressure in psig is multiplied by 2.31 to change pounds per square inch to equivalent feet of water. The multiplier 2.31 is used because the force exerted at the bottom of a column of water 2.31′ high is 1 psi. The boiler pressure can also be divided by 0.433. The divisor 0.433 is used because the pressure exerted at the bottom of a column of water 1′-0″ high is 0.433 psi.

Boiler pressure is changed to equivalent static head by applying the following formula:

$$SH = P \times 2.31$$

where

SH = static head (in ft WC)

P = boiler pressure (in psig)

2.31 = conversion factor (in ft/psig)

For example, what is the equivalent static head of a boiler operating at 225 psig?

$$SH = P \times 2.31$$

$$SH = 225 \times 2.31$$

$$SH = \mathbf{520' \ WC}$$

64 What is total dynamic head?

Total dynamic head is the total amount of head produced by the pump and available to perform useful work, after losses have been subtracted. In other words, it is the head created by the pump on the discharge side, minus friction head, and minus static suction head or plus static suction lift.

65 What is displacement?

Displacement is the volume of fluid forced out of a full container when another body is forced into the container. For example, when a piston in a reciprocating pump sweeps through the length of a cylinder, it forces out, or displaces, the volume of liquid that was contained in that cylinder.

66 What is capacity of a pump?

Capacity is the volume of fluid that can be delivered by a pump over a given unit of time. The most common expression of capacity in reference to pumps is gallons per minute (gpm).

67 What is a positive-displacement pump?

A *positive-displacement pump* is a pump that moves the same amount of liquid with every stroke or rotation. A positive-displacement pump may be either a reciprocating or rotary pump. The flow from a positive-displacement pump cannot be throttled while the pump is running. Positive-displacement pumps are often capable of building extreme discharge pressures if the discharge flow is shut off. This is because liquids are incompressible. That is, they cannot be made smaller through the application of pressure in the same way that gases can. In some cases, this extreme pressure may damage the pump, burst piping, blow fuses in the pump's motor drive circuit, or cause other damage. In cases where an overpressure situation is possible, a relief valve must be used to protect the discharge side of a positive-displacement pump.

Factoid

To keep positive-displacement pumps from developing excessive pressure, a relief valve should be provided on the discharge side of the pump. The relief valve will open to release any dangerous pressure before damage results.

68 What is a relief valve?

A relief valve is an automatic spring-loaded valve designed to open when excessive pressure develops in the system it is protecting. Safety valves are generally used for compressed gases, while relief valves are generally used for liquid service. Relief valves do not have blowdown (blowback) like safety valves, and they do not pop open fully and instantly. Relief valves open in proportion to the excess pressure.

69 Where are relief valves commonly located?

The most common locations for relief valves are on the discharge side of pumps and on systems where excessive pressure may develop because of the thermal expansion of liquids. For example, if a positive-displacement pump is operating against a closed valve, the pressure will need to be relieved before damage results. **See Figure 4-23.** If a full heat exchanger is valved off while still exposed to heat, the expansion of the liquid can result in tremendous pressure being developed inside. This extreme pressure may cause the exchanger to burst, resulting in serious damage and injuries. A relief valve, if properly selected and installed, will open and relieve the excess pressure before it can develop into a danger.

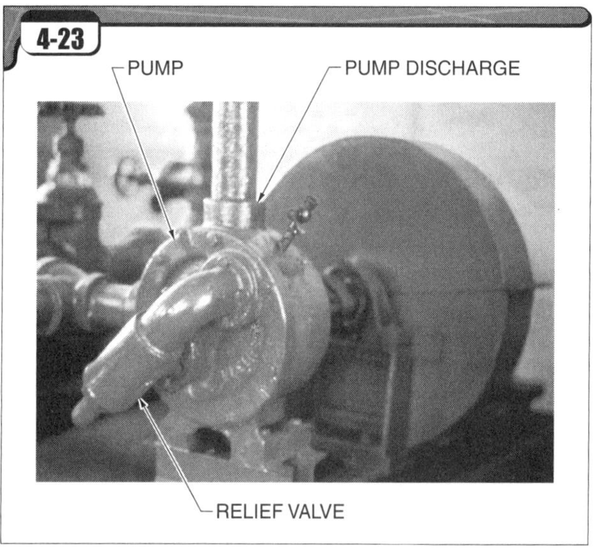

Figure 4-23. *Relief valves are often provided on the discharge side of a positive-displacement pump.*

70 How may pumps be classified?

The most basic groups into which all pumps may be classified are displacement and dynamic pumps. A displacement pump imparts energy to the fluid by using the force of a separate object, such as a piston or gear, to transfer a fluid through valves or ports of a sealed chamber. A dynamic pump adds momentum to the liquid while taking in and discharging fluids in a continuous flow across an unsealed chamber.

Pumps may further be classified as motive fluid, reciprocating, or rotary. A *motive fluid pump* is a pump that uses the force of a secondary fluid to pump the primary fluid. A motive fluid pump may be a displacement pump or a dynamic pump. The separate fluid may mix with the fluid being pumped. For example, steam may be used to force water through a nozzle of special design. A *reciprocating pump* is a displacement pump that uses a reciprocating piston or diaphragm to repeatedly displace fluid from a cylinder or chamber. A *rotary pump* is a pump with a rotating shaft. A rotary pump may be either a dynamic pump or a displacement pump.

A positive-displacement pump is a pump in which every stroke or revolution (depending on whether the pump is reciprocating or rotary) moves a predetermined amount of liquid. A typical configuration includes a piston that moves back and forth in a cylinder and a system of one-way valves that restrict the direction of flow.

Dynamic pumps produce a predictable outlet pressure as long as there is sufficient inlet flow and the outlet pressure remains below the capacity of the pump. A typical configuration includes a motor-driven rotating shaft with an impeller that adds momentum to move the fluid across the chamber and out of the pump.

Beyond these basic groupings, pumps in industry are segregated by many specific criteria, including the following:

- application (what is being pumped)
- materials of construction (metallic vs. nonmetallic)
- orientation (vertical vs. horizontal)
- suction configuration (positive suction vs. negative suction)
- casing design (horizontal split case, back pullout, or in-line)
- number of pumping stages (single stage vs. multistage)
- energy source (electric motor, steam turbine, or fuel combustion engine)
- submersible vs. nonsubmersible

Reciprocating Pumps

What is a single-acting pump?

In a reciprocating pump, the stroke is the distance the reciprocating component travels when sweeping through the liquid chamber from one end to the other. Thus, if a piston in a pump travels from one end of the cylinder to the other and back, it has traveled two strokes.

A *single-acting pump* is a reciprocating pump that moves fluid in only one direction of the stroke. For example, a bicycle air pump is a single-acting pump. It pumps air only on the downward stroke. The cylinder is refilled on the return stroke. In a steam plant, single-acting pumps are used for applications such as feeding water treatment chemicals to the boiler.

What is a double-acting pump?

A *double-acting pump* is a reciprocating pump that moves fluid in both directions of stroke. A double-acting pump is equipped with check valves on both sides of the piston so that the pump moves fluid when the reciprocating component is moving in either direction. For example, a steam-driven, double-acting pump may be used to pump boiler feedwater, condensate, or fuel oil.

What is a simplex pump?

A *simplex pump* is a steam-driven, reciprocating, positive-displacement, double-acting pump with one steam cylinder and one liquid cylinder. **See Figure 4-24.** A steam valve controls the amount of steam entering the steam cylinder. The position of this valve determines the side of the steam piston to which the steam is admitted. This determines the direction of movement of the steam piston. The pump has a reversing gear for changing the direction of travel at the end of each stroke. The liquid piston is on a common piston rod with the steam piston so that they move in unison. Check valves in the liquid end control movement of liquid through that end.

An air chamber is often located above the liquid valve chest. The air trapped in the air chamber alternately compresses and expands as the reciprocating piston in the liquid cylinder produces pulsations. This helps smooth out the flow to prevent water hammer in the discharge piping.

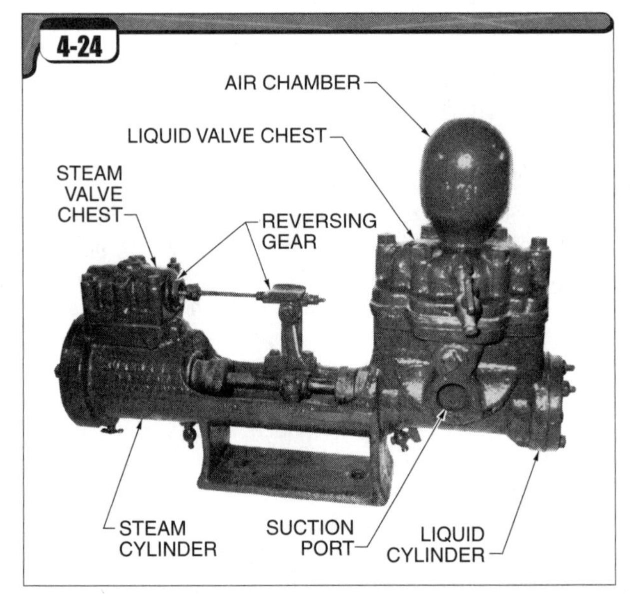

Figure 4-24. *A simplex pump has one steam cylinder and one liquid cylinder.*

How is the steam valve reversed at the end of the stroke in a simplex pump?

An auxiliary valve is provided that is shifted by a reversing gear when the piston reaches the end of the stroke. When the reversing gear causes the auxiliary valve to shift positions, the main steam valve is carried with it. This in turn opens the steam inlet port that provides steam to the opposite side of the piston. It also opens an exhaust port on the side of the piston where the steam pressure had been previously.

What is a duplex pump?

A *duplex pump* is a steam-driven, reciprocating, positive-displacement, double-acting pump with two steam cylinders and two liquid cylinders. **See Figure 4-25.** As in the simplex pump, the movement of the steam pistons, and therefore the liquid pistons, in a duplex pump is controlled by the positioning of the steam valves. The piston rod on one side operates the steam valve on the other side. The liquid end of the duplex pump is essentially the same as that of the simplex pump, except that there are two separate pumping sides that take in liquid from a common suction line and feed liquid into a common discharge line.

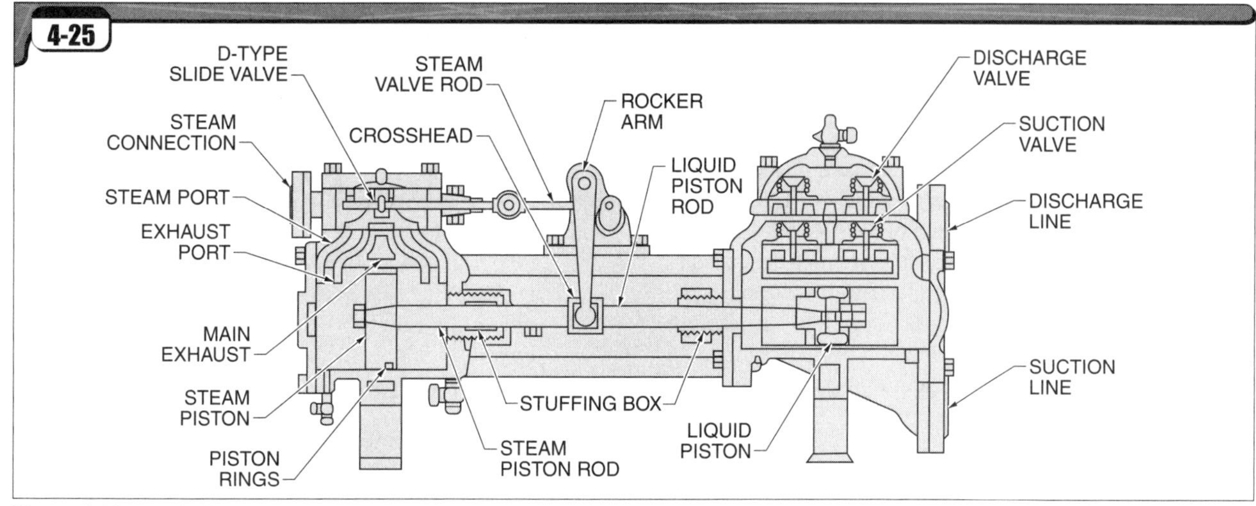

Figure 4-25. *In a duplex pump, the movement of the steam valves controls the movement of the steam pistons, and thus the liquid pistons.*

76 For what applications are simplex and duplex pumps typically used?

Simplex and duplex pumps are being used less and less often in modern boiler room operations. Both have significant deficiencies compared to other available pumps. For example, both are limited in capacity and are therefore used mainly in small boiler plant installations. These types of pumps require substantial labor to fabricate and therefore their initial cost is higher. Additionally, the exhaust steam from these pumps contains oil. Therefore, the resulting condensate cannot be directly returned to the boiler system without first removing the oil in an oil separation process. These pumps are also more expensive to operate when using fossil fuels to generate the steam because of the cost of the fuel.

Duplex pumps have largely replaced simplex pumps because of greater reliability. Simplex pumps were more likely to stall at the end of the stroke if the reversing gear did not function properly. Duplex pumps are much less susceptible to this, because one side or the other is always in motion. In fact, reliability and long life are the main features of duplex pumps. Many boiler systems use duplex pumps as backup pumps for this reason.

These pumps do have a few applications in which they still excel. They are primarily used in industries such as refineries and chemical plants for pumping flammable or combustible liquids. This is because these pumps do not use electric motors and therefore do not present ignition sources from potential sparks. In facilties such as refineries and steel mills, waste fuels are very often available, and this greatly reduces the cost of operation of duplex pumps.

77 What are D-slide valves as used on a reciprocating duplex pump?

A *D-slide valve* is a valve that controls the movement of steam into and out of the steam cylinder in a duplex pump. Two D-slide valves are contained within the steam chest of a duplex pump. **See Figure 4-26.** As they reciprocate, they alternately admit steam through the ports on each side of the duplex pump.

Steam is admitted through the ports and flows into the ends of the steam cylinders and drives the pistons. At the same time, exhausted steam flows up from the cylinders through the two exhaust ports. When this exhaust steam reaches the underside of the D-slide valve, it flows through the hollow cutout area on the underside of the valve and into the center exhaust port, which is connected to the pump exhaust steam piping.

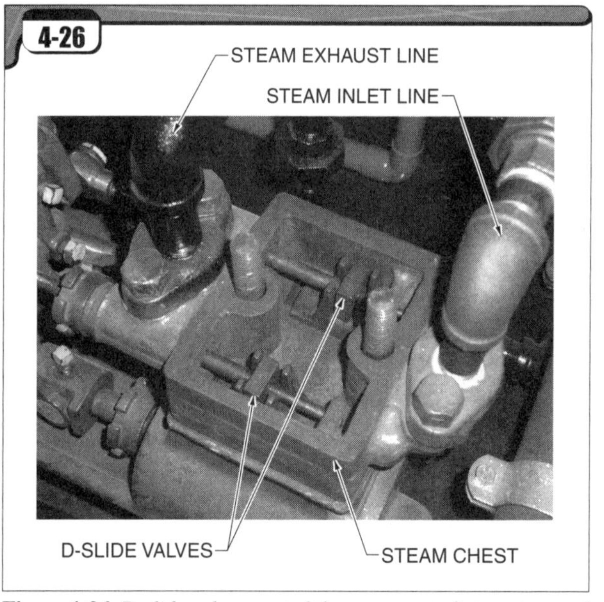

Figure 4-26. *D-slide valves control the movement of steam into and out of the steam pistons of a duplex pump.*

What do the numbers 8 × 4 × 10 mean in relation to a reciprocating pump?

The first number is the diameter of the steam piston in inches. The second number is the diameter of the liquid piston in inches. The last number is the length of the stroke in inches. In this example, the diameter of the steam piston is 8″. The diameter of the liquid piston is 4″. The length of the stroke is 10″.

These numbers are used to easily determine the capacity of a reciprocating pump when the speed in strokes per minute is known or can be estimated. Boiler operators can also use the steam pressure supplied to the pump in conjunction with these numbers to determine the liquid pressure that may be developed by the pump.

What is the slip of a pump?

Slip is the difference between calculated and actual displacement of a pump. A small portion of the liquid in a reciprocating pump may leak from the discharge side of the piston back to the suction side. This slip occurs because of imperfect contact between the piston and the cylinder wall. Another example of slip is leakage through the packing glands around the piston rods. Slip is expressed as a percentage of the theoretical flow and typically ranges from 5% to 10%.

What determines the quantity of liquid moved by a simplex or duplex pump?

Since these pumps are reciprocating, positive-displacement pumps, the quantity of liquid moved is determined by the diameter of the liquid end piston, the length of the stroke, and the number of strokes made by the piston(s) per minute. The calculated quantity of liquid is lessened slightly by slip. The capacity may be found by applying the following formula:

$gpm = (A_L \times L \times N) \times (1 - slip) \div 231$

where

gpm = gallons per minute

A_L = area of the liquid end piston

L = length of stroke

N = number of strokes per minute

231 = conversion factor, cu in./gal.

$slip$ = % difference

For example, a simplex pump is 6 × 4 × 5. At full capacity, it averages 60 strokes per minute. Neglecting slip and the volume of the piston rod, what will be the capacity of the pump in gpm?

$gpm = (A_L \times L \times N) \times (1 - slip) \div 231$

$gpm = (3.14 \times 2 \times 2) \times 5 \times 60 \times (1 - 0) \div 231$

$gpm = (12.56 \times 5 \times 60) \div 231$

$gpm = 3768 \div 231$

gpm = **16.3 gpm**

Note: The volume of the piston rod on the inboard side of the liquid piston occupies a portion of the cylinder volume; thus in reality the volume of the piston rod must be subtracted from the total cylinder volume to obtain the usable volume.

A duplex pump is 9 × 6 × 10. It makes an average of 30 strokes per minute when running at full capacity. Slip totals 6%. Neglecting the volume of the piston rod, what is the capacity of the pump?

Because the pump is a duplex pump, the volume of liquid pumped on each stroke must be multiplied by 2, as the duplex pump has two sides.

$gpm = 2 \times (A_L \times L \times N) \times (1 - slip) \div 231$

$gpm = 2 \times (3.14 \times 3 \times 3) \times 10 \times 30 \times (1 - 0.06) \div 231$

$gpm = 2 \times (28.26 \times 10 \times 30 \times 0.94) \div 231$

$gpm = 15{,}938 \div 231$

gpm = **69.0 gpm**

Why can a simplex or duplex pump develop enough force to put water into the same boiler that it takes its steam supply from?

The total force developed on each piston is equal, but the steam piston is usually 2 to 2½ times larger in area than the liquid piston. Therefore, the pressure developed on the liquid end is greater than the steam pressure. The maximum possible pressure developed on the liquid end may be found by applying the following formula:

$P_L = (D_S^2 \div D_L^2) \times P_S$

where

P_L = maximum liquid end pressure that develops

D_S^2 = diameter squared of the steam piston

D_L^2 = diameter squared of the liquid piston

P_S = pressure of the steam

For example, a duplex pump is 6 × 4 × 6. Steam is supplied to the pump at 100 psi. What maximum pressure may be produced at the liquid end?

$P_L = (D_S^2 \div D_L^2) \times P_S$

$P_L = (36 \div 16) \times 100$

$P_L = 2.25 \times 100$

P_L = **225 psig**

List several reasons why a simplex or duplex reciprocating pump may fail to provide a normal amount of water.

- Packing on the liquid end piston may be worn, allowing excessive slip.
- Steam piston rings may be worn or broken.
- There may be an inadequate supply of water to the pump. For example, the supply tank may be empty.
- The steam admission valve(s) may be worn or improperly adjusted.
- Suction or discharge valves on the liquid end may be throttled or closed.
- Suction or discharge check valves on either side may not be seating properly. Improper seating allows some of the flow to leak back through the valves.
- A strainer on the suction side of the pump may be clogged.

84 Can exhaust steam from a simplex or duplex pump be used to heat boiler feedwater?

Exhaust steam from a simplex or duplex pump can be used to heat boiler feedwater, but it must pass through an oil separation process first. Lubricating oil is injected into the steam supply of a steam-driven reciprocating pump to lubricate the internal moving parts. The exhaust steam and resulting condensate are therefore contaminated with oil.

85 Describe the procedure for starting a steam-driven pump and placing it under load.

The suction and discharge valves on the liquid end should be open. The lubricator for the steam valves and pistons is checked to make sure that it contains cylinder oil. Lubrication should be checked on the other moving parts of the pump as well. The steam cylinder drain cocks are opened to drain the condensate out, and the steam exhaust valve is opened. The steam inlet valve to the pump may then be cracked open to allow the pump to reciprocate slowly so that the mass of the pump casting heats evenly. The cylinder drains will blow considerable condensate for a short period, and then may be closed when live steam comes out of them. The pump is then brought up to speed by slowly opening the steam inlet valve. The lubricator should be checked once more to confirm that the lubricating oil has begun to feed.

When a steam-driven pump is used for a controlled application such as boiler feedwater, a control valve is provided to modulate the steam supply and thus control the output of the pump. The manual steam inlet valve is normally opened completely, since the control valve controls the pump output.

Factoid

Metering pumps are used where a small and precisely measured flow rate is desired, such as when feeding boiler water treatment chemicals.

86 What is a metering pump?

A *metering pump* is a small-capacity pump used to pump a closely measured amount of a liquid. **See Figure 4-27.** Metering pumps are commonly used for pumping water treatment chemicals into various parts of a boiler system. Metering pumps are usually reciprocating pumps of either the piston or diaphragm type. In a piston-type metering pump, a small reciprocating piston slides into and out of a chamber that has a suction check valve on one side and a discharge check valve on the other side. As the piston withdraws from the chamber, a vacuum is created. This allows liquid to flow into the chamber through the suction check valve. When the piston is forced back into the chamber, the liquid is displaced through the discharge check valve.

In a diaphragm-type metering pump, a reciprocating rod is attached to the center of a flexible diaphragm, or bladder. When the reciprocating rod withdraws from the diaphragm chamber, a vacuum is created under the diaphragm and liquid enters through the suction check valve. When the reciprocating rod extends, the diaphragm is pressed into the diaphragm chamber and the liquid is forced out through the discharge check valve.

The length of the stroke and/or the speed of the pump are usually variable, allowing the boiler operator to control the amount of liquid that is pumped.

Figure 4-27. *Metering pumps are small-capacity pumps commonly used for pumping water treatment chemicals.*

Centrifugal Pumps

87 What is a centrifugal pump?

A *centrifugal pump* is a pump in which a rotating impeller throws liquid from its vanes through centrifugal force. **See Figure 4-28.** In most centrifugal pumps, the casing around the impeller is in the shape of a volute. A *volute* is a spiral-shaped form. The volute casing directs the flow of liquid smoothly toward the pump discharge. A volute is also known as a diffuser.

In some centrifugal pumps, the casing is cylindrical, or pillbox-shaped, and diffusion rings are used to perform the job of directing the liquid toward the discharge port. A *diffusion ring* is a stationary vane in the pump casing. In either type of centrifugal pump, the passageways get progressively larger as the liquid approaches the discharge port. This causes the velocity energy in the liquid to be converted to pressure energy.

Centrifugal pump impellers are available in open, semi-closed, or closed configurations. An *open impeller* is an impeller that has vanes that are not enclosed or supported by a shroud (side wall) on either side. Open impellers with longer vanes, however, must have a partial shroud to provide structural integrity. A *semiclosed impeller* is an impeller that has a shroud on one side of the vanes. A *closed impeller* is an impeller that has shrouds on both sides of the vanes. The use of shrouds adds structural strength to the vanes. Closed impellers are almost always used for pumping clear liquids at high speeds. Open impellers are almost always used with small, inexpensive pumps. Semiclosed impellers use the clearance between the open side of the vanes and the pump casing to serve the same purpose as the shroud. This provides slightly more clearance to allow the pump to operate with some debris in the liquid.

Centrifugal pump impellers and casings are made from a variety of materials for various applications. For example, stainless steel or fiberglass may be used for chemical resistance.

88 What controls the quantity of water moved by a centrifugal pump?

The quantity of water moved by a centrifugal pump is a function of the diameter of the impeller, impeller speed, shape of the impeller, friction losses, slip, and size of the suction and discharge ports. During operation, the quantity of water moved by any given centrifugal pump is controlled by throttling the discharge flow. This can be done with a manual valve or an automatically controlled valve.

89 What results from a pressure increase on the suction side of a centrifugal pump?

A pressure increase on the suction side of a centrifugal pump results in the same pressure increase on the discharge side of the centrifugal pump.

90 What are the advantages of using a centrifugal pump instead of a steam-driven reciprocating pump?

Centrifugal pumps have fewer parts, have simpler construction, produce a more uniform flow, and can provide much higher capacities in gallons per minute (gpm). Centrifugal pumps are also easier to tear down and repair and are cheaper initially because a duplex pump requires complex casting and machining work.

91 What is a horizontal split-case pump?

A *horizontal split-case pump* is a pump that has a horizontally split pump casing where the top half of the pump casing can be lifted off for inspection and maintenance without disturbing the shaft, impeller, or bearings. **See Figure 4-29.**

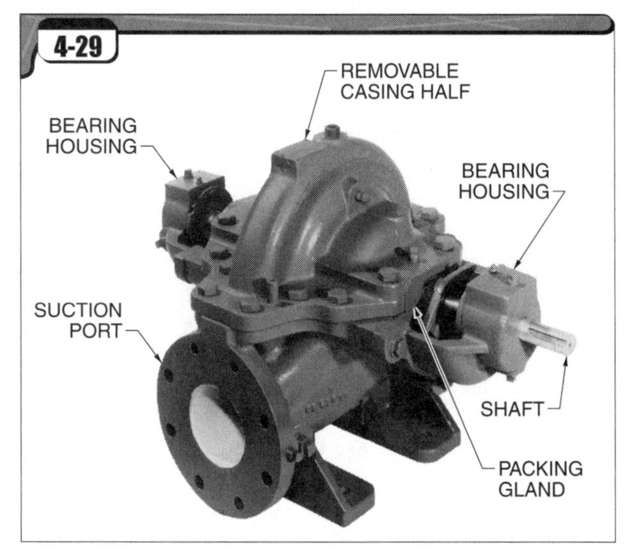

Dresser Industries, Inc.

Figure 4-29. *The top half of a horizontal split-case pump is removable.*

92 What is staging in reference to pump design?

Staging is the placement of more than one impeller on the same shaft in a centrifugal pump. For example, in a multistage pump, the discharge of one impeller will feed into the suction of the next impeller so that high pressures can develop. **See Figure 4-30.** Sometimes the impellers are opposed, or oriented in opposite directions, to minimize the axial thrust on the pump shaft. This allows designers to use smaller and lighter bearings.

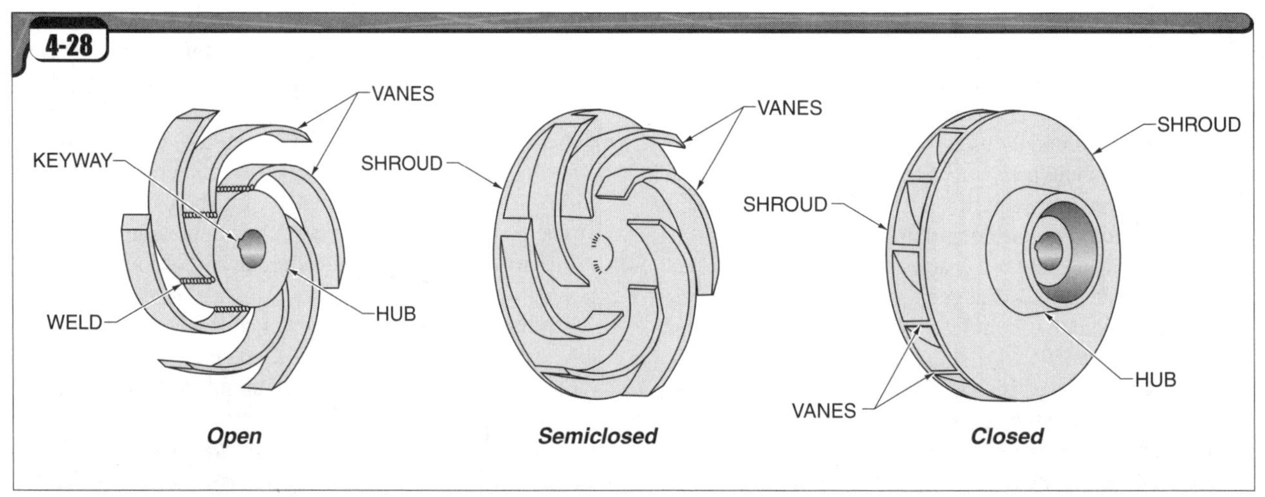

Figure 4-28. *Centrifugal pump impellers have either an open, semiclosed, or closed configuration.*

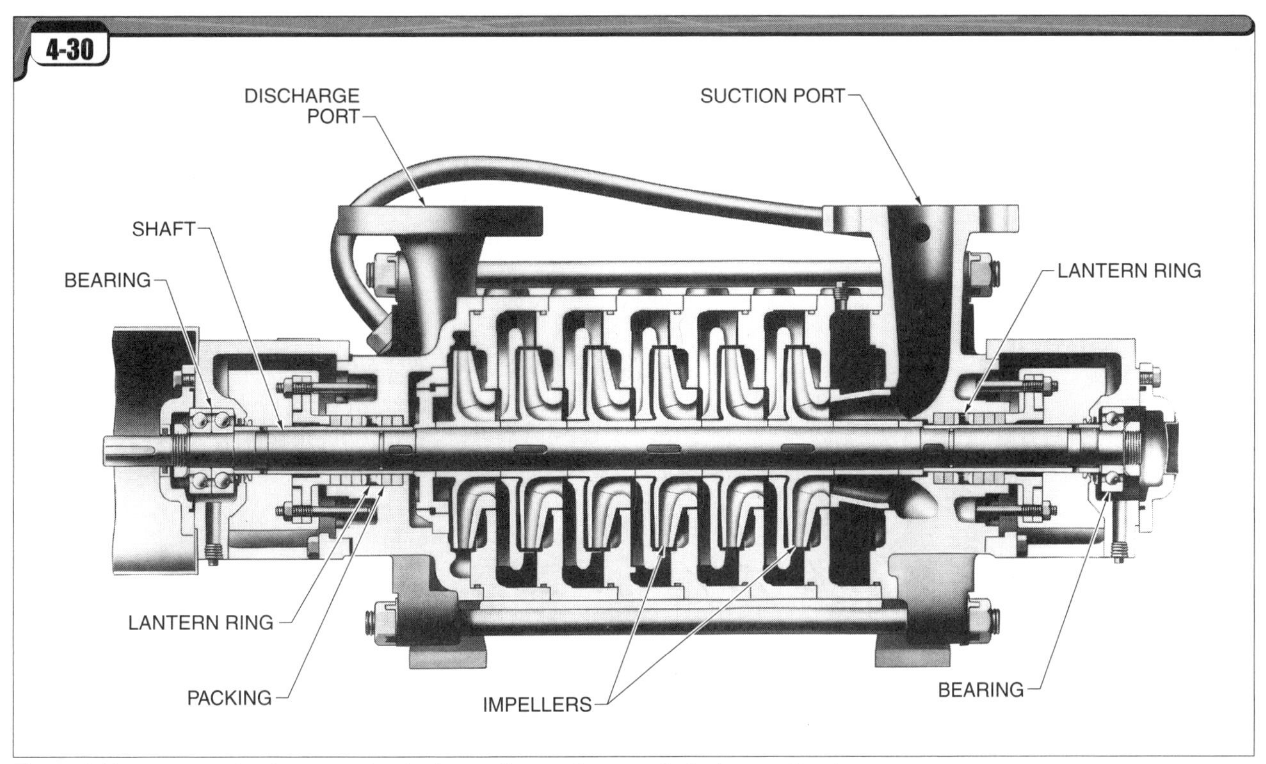

Figure 4-30. *Staging is the placement of multiple impellers on the same shaft of a centrifugal pump.* Dresser Industries, Inc.

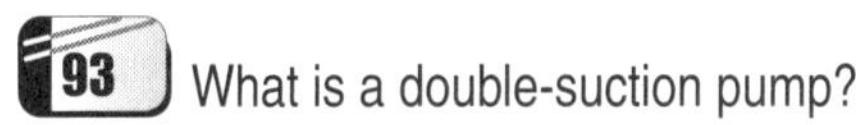

93 What is a double-suction pump?

A *double-suction pump* is a pump with a casing and impeller designed to allow a liquid to flow into both sides (eyes) of the impeller at once. The pump shaft passes completely through the impeller and is supported by a bearing on both ends. **See Figure 4-31.**

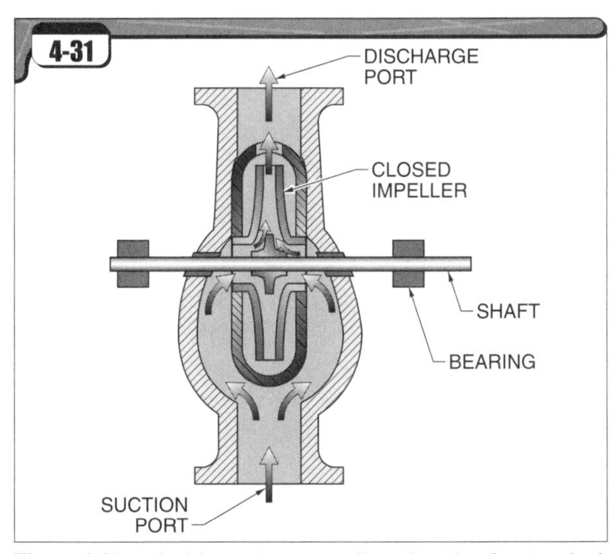

Figure 4-31. *A double-suction pump allows liquid to flow into both sides of the impeller at once.*

94 Why is a shaft sleeve often placed around the shaft of a rotary pump?

A *shaft sleeve* is a replaceable sacrificial part covering a pump shaft. If shaft damage should occur because of erosion, overheating, or friction, it is less expensive to replace a shaft sleeve than the shaft. O-rings are installed between the shaft and the shaft sleeve to prevent leakage of water or process liquid. The shaft sleeve is pinned or keyed to the shaft to keep it from turning on the shaft.

95 What prevents leakage of water or other process liquid from around the pump shaft?

Packing is installed around the pump shaft to keep water or other process liquid from leaking out of the pump. Packing is commonly made of graphite-impregnated rope which is square in cross section, or it may be made of Teflon® or other materials. **See Figure 4-32.**

Packing is installed in the packing gland area, or stuffing box. The packing is cut to length to create individual rings around the pump shaft. The locations of the cut ends of each ring should be staggered so that the liquid in the pump cannot easily leak through the cut locations. Packing in the packing gland area is adjusted by tightening the packing follower. The packing follower compresses the packing so that it throttles the leakage of liquid from around the shaft. A mechanical seal may be used instead of packing.

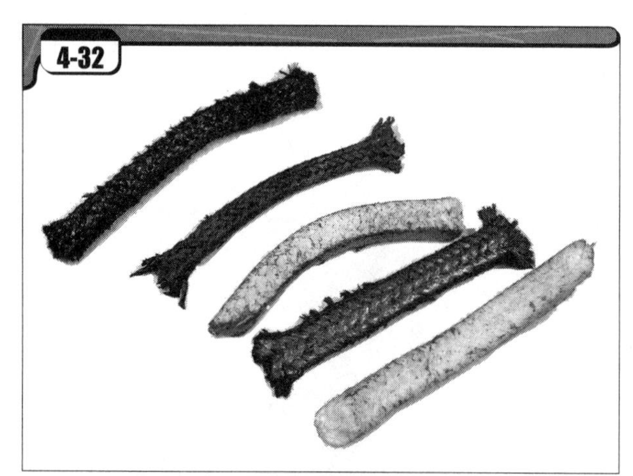

Figure 4-32. *The packing used to prevent leaks may be made of a number of materials, depending on the applications.*

How is the water seal of a centrifugal pump shaft accomplished?

Liquid (usually water) is supplied to the packing gland area from the high-pressure side of the pump through a small tube or pipe. **See Figure 4-33.** Generally, this liquid flows into the packing gland area through the lantern ring. A *lantern ring* is a spacer installed between two of the rings of packing in a pump. It creates an open space where seal flush liquid can freely enter the packing gland area.

The liquid cools the packing, shaft, and shaft sleeve. Additionally, the liquid flushes debris from the shaft and packing gland area and prevents debris from entering the packing gland area. The liquid is commonly known as seal flush. If a mechanical seal is used, the seal flush also cools the mechanical seal.

If the liquid being pumped is drawn from below the pump, the suction piping will be under a vacuum. Another purpose of the seal flush in this case is to prevent air from being drawn in through the packing, which could cause the pump to lose its prime.

What care should be used in adjusting the packing of a centrifugal pump?

The packing gland adjustment should never be drawn up so tightly as to completely stop the flow of liquid out of the packing. A small leakage is designed into the system to keep the shaft and the packing from overheating.

Packing should be tightened about one half of a flat on each packing gland nut at a time, alternating from side to side, until the leakage out of the packing is slowed to the desired rate. This rate is specified in the manufacturer's literature. A few minutes should be allowed between adjustments, especially with new packing, which may swell slightly as it becomes soaked.

If for any reason packing is tightened so much that smoke comes from the packing gland, the packing is very likely ruined. Once it has overheated to this degree, the packing should be replaced because the heat will have hardened it. Continued use of this packing may cause damage to the shaft or shaft sleeve.

Figure 4-33. *Water supplied to the packing glands from the high-pressure section of the pump cools the packing and shaft sleeve.*

What is a mechanical seal?

A *mechanical seal* is an assembly installed around a pump shaft that prevents leakage of the pumped liquid along the shaft. **See Figure 4-34.** A mechanical seal is used instead of packing around a pump shaft or shaft sleeve. Mechanical seals are used in modern boiler plants for everyday applications such as pumping feedwater or condensate. They are particularly useful, though, when leakage of the material could create a fire hazard or noxious vapors, when the liquid is valuable, or in any other case where leakage is unacceptable.

The two main parts of a mechanical seal are the stationary ring and the seal assembly. The stationary ring is placed around the shaft in the gland cover. It is often made of relatively soft material such as carbon or graphite. The seal assembly is attached to the pump shaft or shaft sleeve and rotates with the shaft. The seal assembly supports a ring made of much harder material such as tungsten carbide or silicon carbide. In many mechanical seals, the stationary and rotating seal faces are made of the same material.

O-rings prevent leakage between the seal assembly and the shaft or shaft sleeve, and between the stationary portion and gland cover. Springs in the seal assembly maintain proper compression between the two seal faces.

The faces of the softer material and harder material are machined perfectly smooth. The tolerances for these faces are in millionths of an inch. These two faces are manufactured as a matched pair so that they mate together exactly. The mating faces form the main sealing surfaces. As one side rotates against the other, a small amount of the pumped liquid infiltrates between the mating surfaces and forms a film of liquid. This film lubricates the surfaces without leaking. Mechanical seals are often provided with a small flow of very clean water, or sometimes other liquids, from an external source. This keeps any dirt and debris in the pumped liquid from infiltrating between the seal faces. Dirt or debris could scratch the faces and cause leakage.

There should be no leakage from a mechanical seal. Leakage greater than a few drops per day from a mechanical seal indicates that the seal is about to fail. A mechanical seal should not be adjusted after it is installed. The springs in the seal make the adjustments automatically. Attempts to adjust a mechanical seal will cause premature failure.

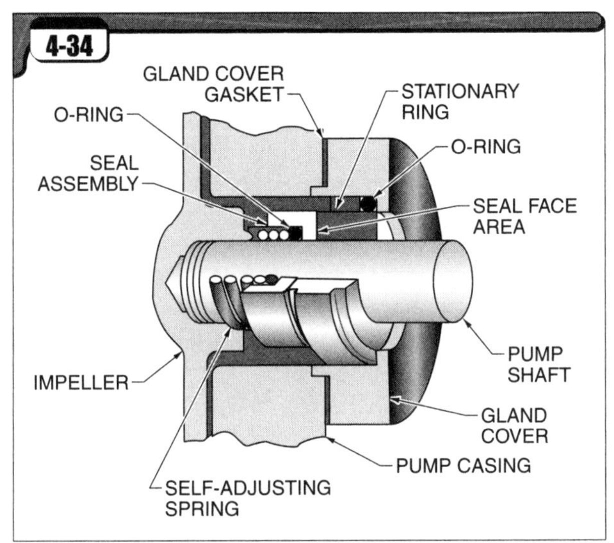

Figure 4-34. *Mechanical seals are self-adjusting after installation and allow no leakage.*

99 Should a centrifugal pump be started with the discharge valve open or closed?

A centrifugal pump should be started with the discharge valve closed. The advantage of starting a centrifugal pump with the discharge valve closed is that, under these conditions, the pump is performing almost no work. This minimizes the surge of electrical current that is experienced when the motor is energized. This surge, known as starting current, may be as much as ten times the normal continuous current required while the pump is running. If the pump is driven by a steam turbine, this also minimizes the starting torque on the turbine and allows the turbine to heat up smoothly and evenly before placing it under load.

The suction valve should be opened completely before starting the pump. With the discharge valve still closed, the impeller then simply churns in the liquid inside the pump casing. A centrifugal pump impeller can spin in the liquid inside the pump for a brief period with no negative effects. Once the pump is started, the boiler operator can open the discharge valve.

The need to start a centrifugal pump with the discharge valve closed often takes care of itself. For example, if the pump operates in parallel with another pump sharing a common discharge line, a check valve on the discharge side of the pump being started will remain closed until the pump comes up to speed and generates discharge pressure.

100 Why is an air vent valve often installed on top of a centrifugal pump?

The air vent valve allows the boiler operator to vent all air from the inside of the pump before the pump is started. This prevents air bubbles from creating voids in the liquid as it flows through the pump. Such voids can cause turbulence, noisy operation, and reduced capacity of the pump.

101 What do the letters NPSH represent?

The letters NPSH represent "net positive suction head." NPSH is a numerical value that equals the minimum suction pressure that must be maintained to prevent cavitation or flashing in a pump. Required NPSH is supplied with the pump documentation.

Available NPSH is the pressure, in feet of liquid, available at the pump datum exerted by the weight of the column of liquid pushing down on the pump and the process pressure, minus the vapor pressure of the liquid in feet of liquid. The pump datum is almost always the centerline of the impeller.

Available NPSH is a function of the piping arrangement on the suction side of the pump. For example, NPSH is affected by whether the pump has a positive or negative suction, the suction piping size, and the number and type of pipefittings and valves in the suction piping.

More simply stated, available NPSH is the net amount of pressure available to the suction side of the pump after subtracting the amount of pressure required to keep the liquid from boiling. NPSH is expressed in feet of liquid (for example, water) absolute, rather than psi. The available NPSH must always exceed the required NPSH. Calculations for available or required NPSH are affected by the specific gravity of the liquid.

102 What is cavitation of a pump?

Cavitation is the condition caused when a portion of the water or other liquid entering the eye of a pump impeller flashes into steam bubbles. For example, cavitation can occur when the water being pumped is too hot for the amount of head available at the pump suction inlet. As the water enters the eye of the pump impeller, the velocity of the water increases and the pressure of the water drops slightly. This lower pressure may be below the vapor pressure. If so, flash steam bubbles will form in the water. These bubbles are referred to as cavities, thus the term "cavitation."

Factoid

Motor-driven centrifugal pumps should be started with the discharge valve closed to minimize the starting current draw on the motor.

How does cavitation affect the pump?

When flash steam bubbles begin to pass through the pump impeller vanes and into the high pressure on the discharge side of the impeller, they collapse, or implode, back into water. The collapse of these bubbles can occur with thousands of pounds of force, and it occurs against the surface of the impeller. This force may be enough to cause pitting of the impeller.

Other detrimental effects of cavitation include vibration, bearing wear or failure, damage to mechanical seals, noisy operation, and shaft damage. More immediately, the capacity of the pump drops, and the amount of water delivered may be insufficient.

Why may some centrifugal pumps have a balancing valve in the discharge piping?

Suppliers of packaged equipment, such as condensate receiver/ pump packages, often attempt to select a pump for the particular application from a fairly limited range of available pump sizes. This is done to reduce the initial capital cost by using a pump that is already stocked in inventory. For example, a pump may be slightly too large for its intended application. In such cases, a globe valve is installed on the discharge side of the pump, and this valve is used to throttle the flow from the pump. This is done in order to create backpressure on the pump and reduce the flow slightly in an effort to match the actual conditions to the design requirements of the pump. This helps to eliminate the cavitation that would otherwise occur. Once the globe valve is adjusted to the correct position, it is left in that position and not changed. This is often the simplest and lowest-cost approach to matching a pump to its intended application, although it results in lower operating efficiency.

What is required at the suction side of the pump to allow pumping of feedwater at 225°F?

Enough head of the water on the suction side of the pump is required to prevent it from flashing to steam at the eye of the impeller. This condition causes cavitation of the pump. The hotter the water to be pumped, the more head will be required. In large steam plants that use very hot feedwater, the supply tank may be several floors above the feedwater pumps in order to provide enough head at the pump suction.

What limits the lift a negative-suction pump can create on the suction side?

Water can boil at a fairly low temperature (less than 212°F) when under a vacuum condition. Boiling prevents water from being lifted up by the pump. The boiling temperature of water increases as the pressure increases. Therefore, the boiling temperature decreases as the pressure decreases. If the pressure decreases to the point of being a vacuum, then the boiling temperature may drop substantially below 212°F.

Wear of the casing, impeller, and wearing rings causes slip, which results in a loss of lift. Loose or worn packing allows air leakage into the suction side of the pump, limiting lift. Air leaks may also occur at corroded or loose pipefittings. Other factors include friction losses due to excessive length or inadequate diameter of the suction line and excessive numbers or incorrect types of valves and elbows. For these reasons, the practical suction lift attained by centrifugal pumps is usually limited to 10′ to 15′, although 20′ may be attained by pumps of special design.

List several reasons why a centrifugal pump may fail to deliver an adequate amount of water.

- There may be air leaks in the suction line (negative suction).
- The impeller may be damaged or worn.
- A suction or discharge valve may be throttled or closed, or there may be an obstruction in the suction or discharge line.
- Suction supply tank may be empty.
- One fuse in a three-phase motor control circuit may be blown and thus the motor is single phasing.
- Pump running in the wrong direction of rotation (e.g., two electrical leads in a three-phase circuit were reversed).
- Check valve may be sticking on the parallel pump, causing part of the water to backflow.
- Suction-side strainer may be clogged.
- Water being pumped may be too hot, causing cavitation.
- There may be excessive clearances between the impeller and casing because of erosion or wear.
- Pump not primed.

What is the purpose of a recirculation line in a centrifugal pump installation?

A *recirculation line* is a line that provides a minimum flow through a pump to prevent overheating. If a centrifugal pump runs for an extended period of time against a closed discharge valve, friction created by the impeller churning in the trapped water causes the pump to get hot. The extra heat eventually causes cavitation and damage to the pump. This problem is prevented by supplying a recirculation line back to the process from which the pump takes its suction. **See Figure 4-35.**

Recirculation lines are used in applications where a pump runs continuously but supplies a varying demand. For example, a number of boilers may be supplied with feedwater from a common pump. If none of the boilers are calling for water, the recirculation line provides a small flow of water through the pump to keep it cool.

The quantity of recirculating liquid should be limited to only that amount which is necessary because this amounts to work done by the pump with no productive result. Special self-adjusting recirculation valves are made for use with larger pumps having recirculation lines. In the absence of a recirculation valve, a small orifice may be inserted in a pipe union or between two flanges in the recirculation line. In such applications, the orifice is simply a small metal plate with a hole of predetermined size drilled through it.

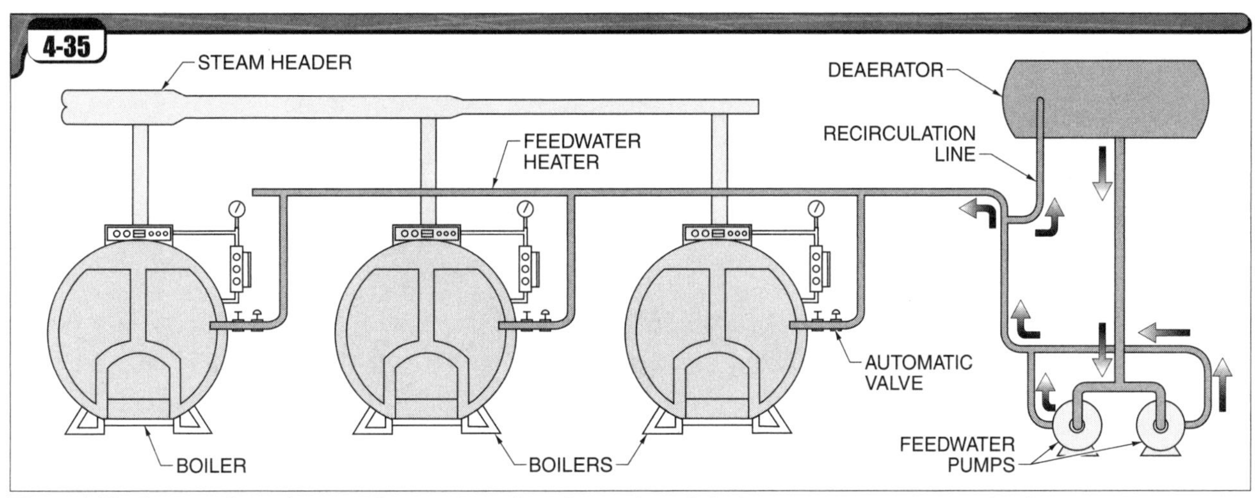

Figure 4-35. *The recirculation line provides a minimum flow of water through feedwater pumps to prevent overheating.*

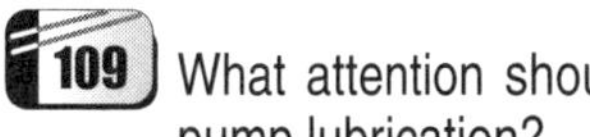

What attention should the boiler operator give to pump lubrication?

Lubrication of the rotating equipment is one of the boiler operator's most important duties. If the bearings in the rotating equipment are starved for oil or grease, the equipment can be seriously damaged. Damage may also occur if the lubricant is contaminated or the wrong type. This results in very high maintenance costs and may interrupt production if the damaged equipment has no backup.

There are several means by which rotating equipment is lubricated. **See Figure 4-36.** Some bearings are sealed and require no attention. Greasing with a grease gun lubricates others. The type and amount of grease used is important. The wrong type of grease may not provide adequate lubrication or may break down under excessive temperatures. On the other hand, excessive greasing may cause bearing seals to be pushed out of place, allowing dirt and contaminants to enter the bearing. Some bearing housings have a zerk fitting (grease gun fitting) on the top and a plug on the bottom. In order to prevent excessive pressure in the bearing housing, the bottom plug should be removed before greasing. The new grease should be pumped into the bearing housing with the grease gun until clean grease comes out the bottom hole. Then the plug may be replaced in the bottom hole.

Lubricating oil may be supplied to bearings under pressure from an internal or external oil pump, or it may be carried to the bearing by slinger rings. A *slinger ring* is a metallic ring that hangs on the rotating shaft and is considerably larger in diameter than the shaft. As the shaft spins, the small amount of friction between the inside of the slinger ring and the surface of the shaft causes the slinger ring to rotate around the shaft. As the ring turns, the bottom of the slinger ring passes through an oil sump below the shaft bearing and the ring picks up oil. As the ring rotates over the top of the shaft, the oil is deposited in the bearing.

External lubricators are sometimes used to provide oil to bearings in rotating equipment as needed to maintain the level in an oil sump. The lubricator, or oiler, is filled with the proper oil and then placed upside down so that the level of oil in the lubricator is higher than the level of oil in the oil sump. The oil dispenses into the oil sump until the level of oil in the oil sump is above the lubricator connection. At this point, a vacuum forms in the top of the oil bulb and the oil flow stops. When the level in the oil sump falls below the level of the lubricator connection, the vacuum in the oil bulb is broken and the lubricator dispenses oil.

While automatic oiling systems are convenient, the boiler operator should not assume that they are working properly. For example, external bulb type lubricators may show a proper oil level, but will not provide adequate lubrication if the connection to the oil sump is blocked by dirt or congealed oil. Part of the boiler operator's normal equipment inspections includes checking for proper lubrication. This consists of feeling the various bearings to see that they are not excessively warm, opening inspection plugs over the top of slinger rings to see that they are turning, observing flow indicators in pumped oil systems to confirm that oil is circulating, etc.

Larger pumps, steam turbines, etc. may have external oil conditioning systems such as oil coolers and filters. Oil coolers usually consist of a small closed heat exchanger, generally of the shell-and-tube type. The boiler operator should frequently check and adjust the flow of cooling water to the oil coolers so that the oil is neither too hot nor too cool. Excessive cooling water use results in wasted water. It can also result in condensation of water vapor in the bearing housings and can change the viscosity of the oil. *Viscosity* is a measurement of a liquid's resistance to flow. High viscosity liquids have a high resistance to flow. Thus, if the oil is cooled excessively, the oil may not flow adequately to provide proper lubrication. The equipment manufacturer's guidelines for oil temperature should be closely followed.

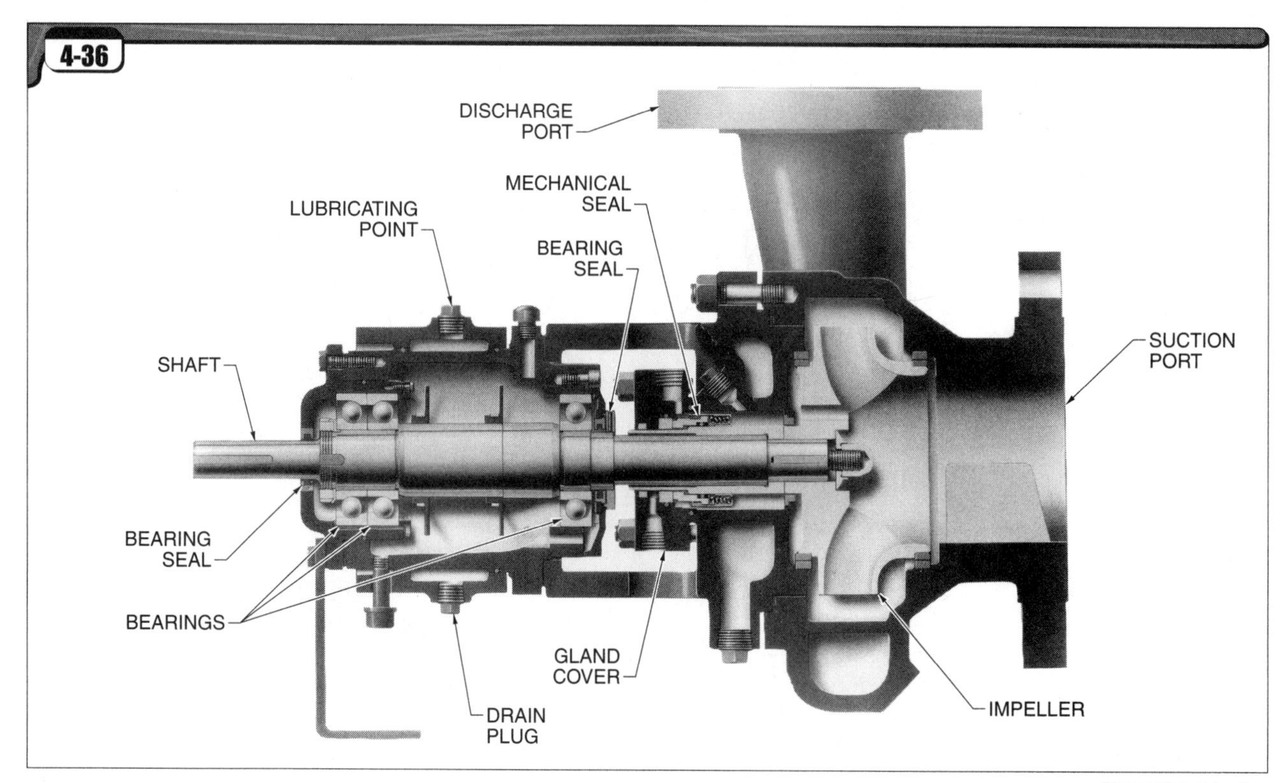

Dresser Industries, Inc.

Figure 4-36. *Care must be taken when lubricating rotating equipment such as pumps to prevent seals from being dislodged and dirt from entering bearings.*

110 Why is the size of the piping connected to a pump important?

If the piping is too small, the flow velocity through the piping becomes excessive. When this occurs, friction along the walls of the piping results in significant pressure losses. This is particularly important on the suction side of a pump. If the suction piping is too small, excessive friction along the pipe walls can cause the pump to be starved for flow. The pressure drop due to the excessive friction may cause the liquid, if very hot, to begin flashing. These two conditions can lead to pump cavitation. For this reason, the suction piping is often sized 1 to 2 pipe sizes larger than the pump suction connection.

If the discharge piping from the pump is too small, friction along the walls of the pipe may lead to excessive pressure drop. This may result in inadequate pressure by the time the liquid reaches its destination. For example, if the pump is a feedwater pump, an inadequate discharge pipe size may result in the feedwater pressure at the boiler being too low to force the water into the boiler.

The primary drawback of the suction or discharge piping being too large is simply the cost of the piping. Piping costs, along with the cost of pipe fittings, flanges, gaskets, pipe hangers and supports, etc. increase substantially as the pipe size is increased.

Rotary Positive-Displacement Pumps

111 For what applications are rotary positive-displacement pumps used?

Rotary positive-displacement pumps are used in applications where a constant flow rate without regard to the pump discharge pressure is desired. For example, the fuel oil supply to a fuel oil burner should be provided at a constant flow rate.

With centrifugal pumps, the discharge flow rate will vary somewhat as the discharge pressure changes. This is not acceptable in applications where a very constant flow rate is required. For this reason, rotary positive-displacement pumps are normally used for these applications.

Rotary positive-displacement pumps may develop extreme discharge pressures if run deadheaded. Deadheading a pump means running the pump against a closed discharge valve. Rotary positive-displacement pumps should be equipped with a pressure relief valve that will relieve excessive pressure on the discharge side of the pump back to the suction side.

Factoid

By monitoring lubrication conditions, boiler operators play a critical role in maximizing the life of rotating equipment.

112 What is a turbine pump?

A *turbine pump* is a rotary positive-displacement pump that uses a flat impeller with small flat perpendicular fins machined into the impeller rim. **See Figure 4-37.** The fins sweep the liquid along from the suction port to the discharge port. Turbine pumps are often used as boiler feedwater pumps for smaller packaged boilers of less than about 150 HP. These pumps can produce several times the discharge pressure of a centrifugal pump with the same size impeller, but have relatively low capacity.

The paddle-like blades on the rim of the impeller pass through a channel in the pump casing that is slightly wider than the blades of the impeller. The effect of the impeller is to cause the liquid to flow through the channel in a spiral fashion. Each impeller blade that sweeps the liquid along adds more velocity. Turbine pumps are often also referred to as regenerative turbine pumps.

113 What is a lobe pump?

A *lobe pump* is a rotary positive-displacement pump in which the liquid being pumped fills the open spaces between the lobes of matched rotors and the pump housing. A lobe pump is very similar to a gear pump, but has rounded lobes instead of gear teeth. The lobes in a lobe pump perform the same function and in basically the same way as the gear teeth in a gear pump. Unlike a gear pump, however, the shape of the lobes makes it infeasible for one lobe to drive the other. For this reason, lobe pumps require timing gears.

Factoid

When a new motor is installed, it should be "bumped," or momentarily energized, to confirm the correct direction of rotation. A pump that is rotating backward will have a greatly reduced capacity.

114 What is a vane pump?

A *vane pump* is a rotary positive-displacement pump that uses a rotating drum located eccentrically inside a cylindrical pump casing. The rotating drum is equipped with sliding vanes that slide radially into and out of the drum. Centrifugal force or springs cause the vanes to slide outward from the rotating drum as it spins, and decreasing clearance between the rotating drum and the pump casing causes the vanes to slide inward.

As the vanes in the rotating drum pass across the pump suction port, the clearance between the rotating drum and the pump casing increases from minimum to maximum. This occurs because the interior volume of the pump casing is not symmetrical because of the eccentric mounting of the drum. As the clearance increases, the sliding vanes extend and continue to contact the casing wall. Liquid entering the pump fills the voids between the sliding vanes and is swept along by the vanes. As the vanes pass across the discharge port, the clearance between the rotating drum and the casing wall decreases from maximum to minimum. This causes the liquid to be displaced into the discharge port.

115 What is a gear pump?

A *gear pump* is a rotary positive-displacement pump in which the liquid being pumped fills the open spaces between the teeth of rotating cylindrical gears and the pump housing. **See Figure 4-38.** The liquid is forced out through the discharge port as the gear teeth mesh together.

In a typical gear pump, two meshing gears rotate in opposite directions. In most cases, one of the gears drives the other. As the meshed teeth of the two gears separate, liquid fills the spaces between the gear teeth and the pump casing. As the gears again mesh together, the liquid is displaced and forced out the pump discharge.

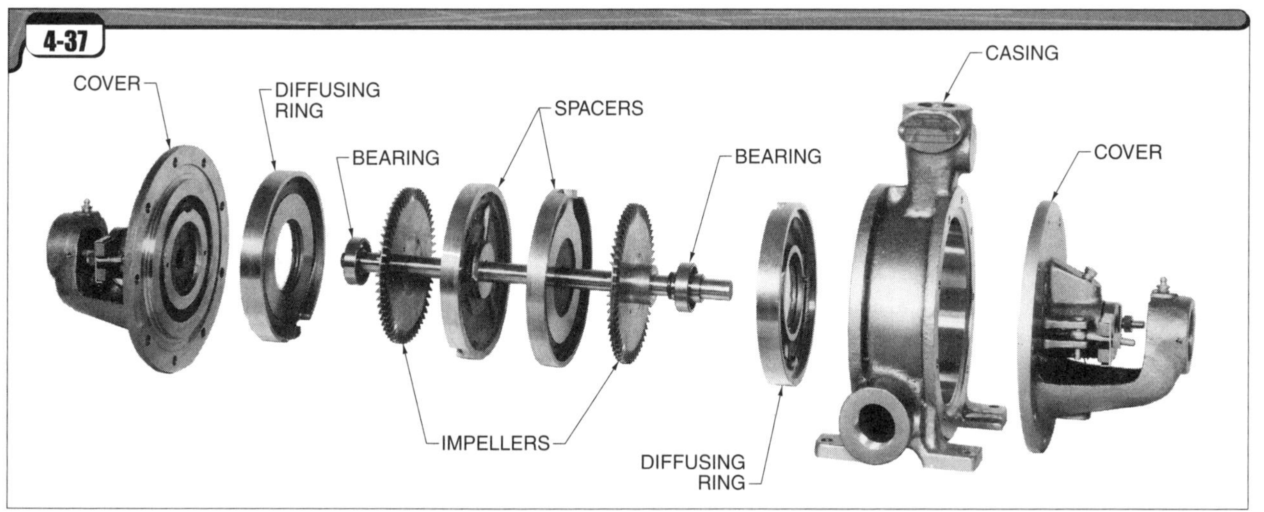

Figure 4-37. *Turbine feedwater pumps are positive-displacement pumps and require an open discharge valve when starting.*

The gears in a gear pump are configured in a number of different variations. Spur gears are fairly simple cylindrical gears with the gear teeth machined parallel to the pump shaft. Helical gears are machined such that the teeth are not parallel with the pump shaft, but angled slightly. This produces a slightly more uniform flow. Internal gear pumps use a smaller offset spur gear that turns inside of and is driven by a larger toothed rotor. The larger rotor is driven by the pump motor.

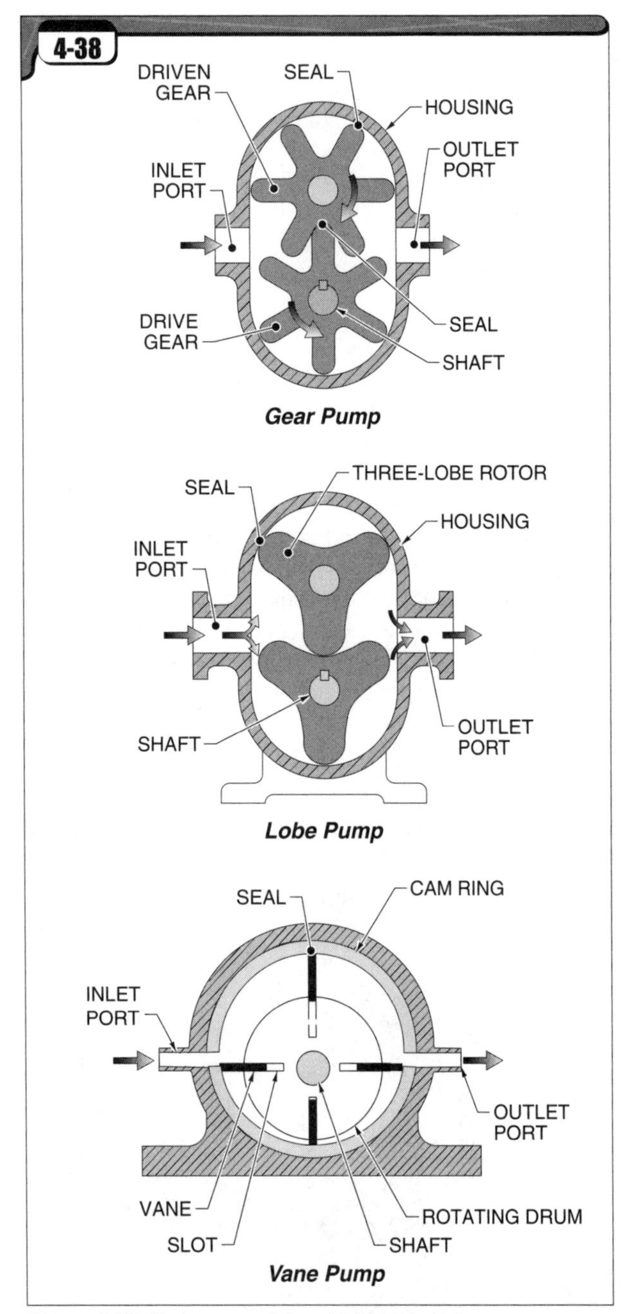

Figure 4-38. *Gear pumps, lobe pumps, and vane pumps are rotary positive-displacement pumps.*

Motive Fluid Pumps

116 What is a motive fluid pump?

A *motive fluid pump* is a pump that uses the force of a secondary fluid to pump the primary fluid. The secondary fluid is usually steam or compressed air.

117 Describe the operations of a typical motive fluid condensate return pump.

A typical motive fluid condensate return pump consists of a condensate collection chamber, a condensate inlet check valve, a motive fluid inlet valve, a vent valve, a condensate discharge check valve, and a ball float to cause the system to function at the proper time. **See Figure 4-39.** The condensate discharge check valve is spring-loaded and requires a specific amount of differential pressure to cause it to open.

Condensate enters the collection chamber through the condensate inlet check valve. As the condensate accumulates in the collection chamber, the ball float is lifted. When the ball float rises to a certain point, it causes the vent valve to close and the motive fluid inlet valve (steam or air valve) to open. The motive fluid pressurizes the condensate collection chamber. This causes the condensate inlet check valve to close and the condensate discharge check valve to open. The motive fluid's pressure forces the condensate out of the collection chamber and into the discharge line. When the ball float drops to a lower preset point, it causes the motive fluid valve to close and the vent valve to open. When the vent valve opens, it releases the pressure in the condensate collection chamber. This releases the backpressure on the condensate inlet check valve and allows condensate to enter the collection chamber again through the condensate inlet valve.

Though motive fluid condensate return pumps cost considerably more than simple packaged condensate receivers with motor-driven pumps, they have the advantages of being able to handle hotter condensate without cavitation, and requiring no electricity.

118 What is a venturi?

A *venturi* is a nozzle with a slight hourglass-shaped taper. It is wide on both ends and narrow in the middle, with an open space around the outside of the inlet. A motive fluid, most commonly steam, is blown through a tapered nozzle and then through the throat of the venturi. **See Figure 4-40.** This causes a substantial increase in the steam's flow velocity and a subsequent decrease in pressure. A vacuum is created around the inlet of the venturi throat, and this vacuum is used to draw the gas or liquid to be pumped toward the inlet throat. The gas or liquid to be pumped is entrained with the motive fluid and forced through the venturi. Thus, a venturi serves as a pump with no moving parts.

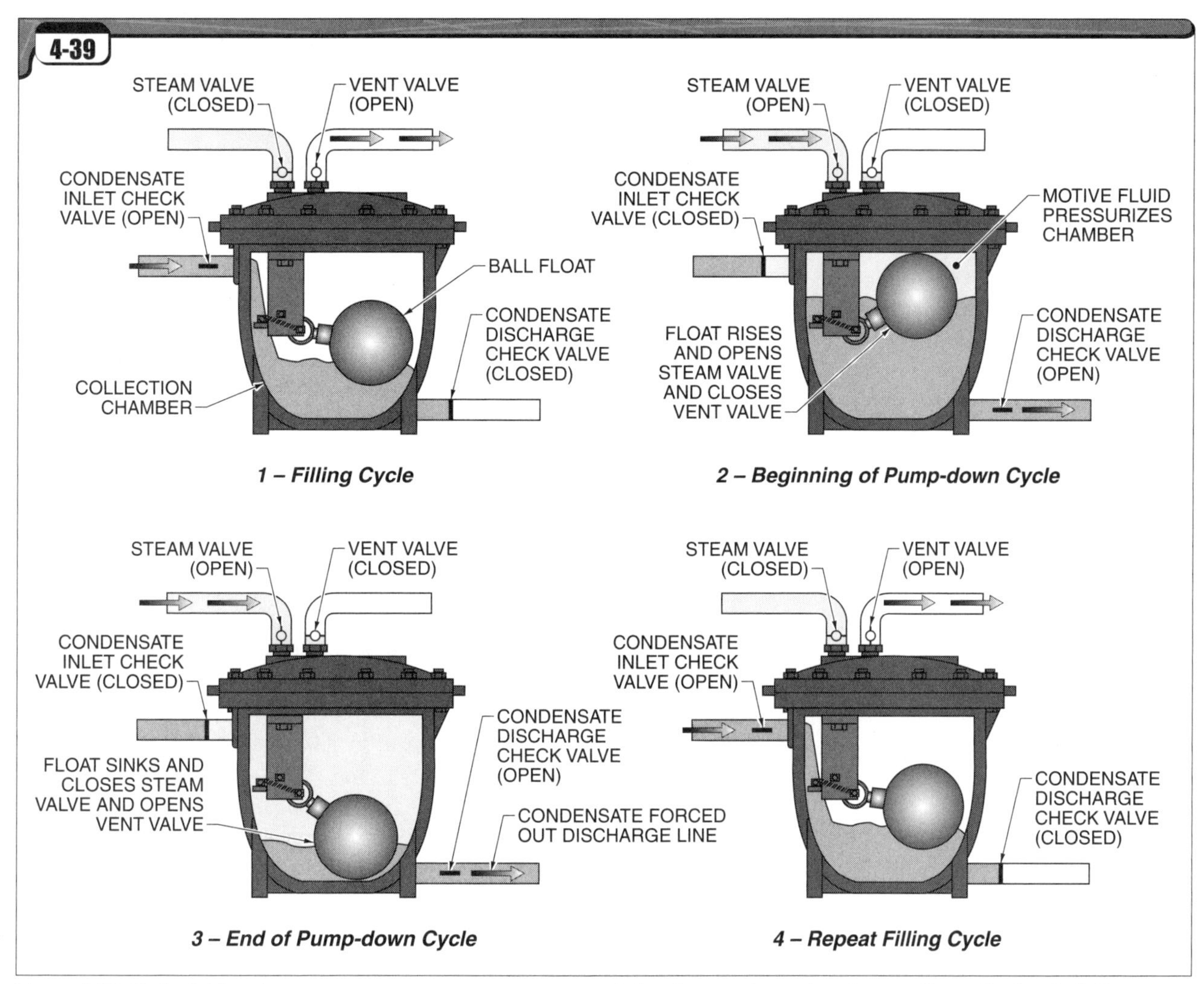

Figure 4-39. *Motive fluid condensate return pumps use steam or compressed air to force condensate into the return line and back to the boiler system.*

The venturi principle is used in a variety of applications in boiler plants. For example, venturis are used in water softening equipment, vacuum condensate pumps, and air ejectors on steam condensers. Motive fluid pumps that use venturis have few moving parts and require no power source other than the motive fluid.

119 What is an injector?

An *injector* is a motive fluid pump that uses the velocity of steam to draw water and pump it into a boiler. **See Figure 4-41.** Injectors are most commonly used as standby feedwater pumps. Steam passes through a venturi, which creates a vacuum around the outside of the venturi. This in turn causes the liquid to flow up the suction line and mix with the steam. The momentum from the high velocity of the steam is transferred to the water, which is forced out the injector discharge line. The steam condenses in the process and becomes part of the feedwater. Injectors usually force water into the same boiler that the steam supply was taken from.

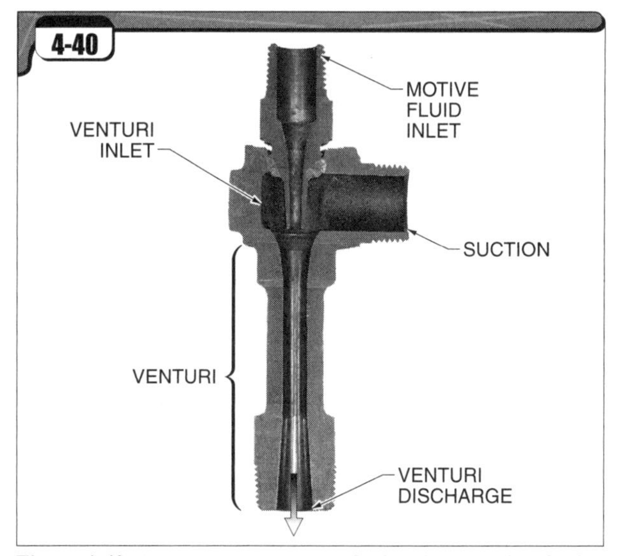

Figure 4-40. *A venturi uses a motive fluid to draw another fluid to the inlet and force it through the discharge.*

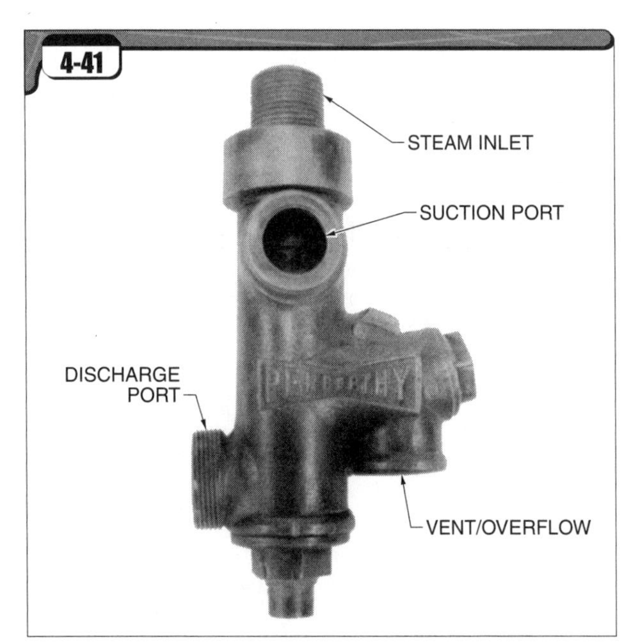

Figure 4-41. *An injector commonly uses steam as the motive fluid to draw water and pump it into a boiler.*

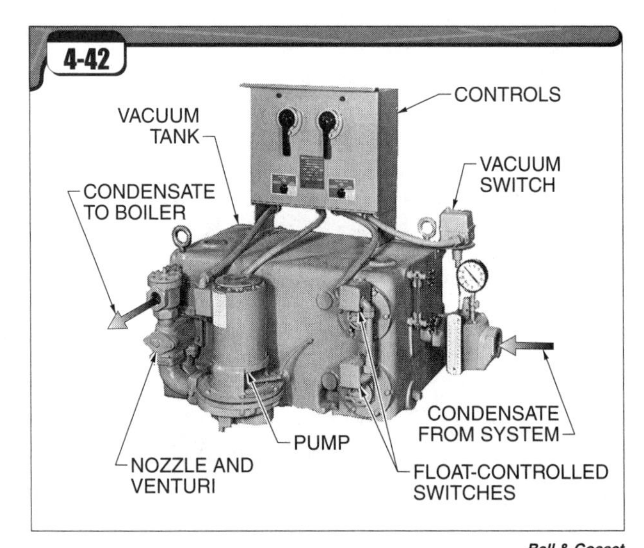

Bell & Gosset

Figure 4-42. *The vacuum pump of a condensate system draws condensate back to the tank and pumps it to the boiler system while it vents air to the atmosphere.*

Vacuum Pumps

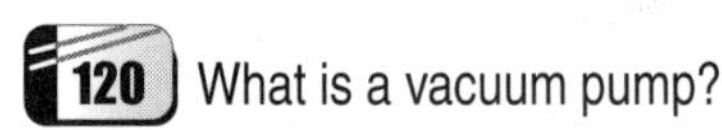

120 What is a vacuum pump?

A *vacuum pump* is a pump that withdraws gases or vapors from a closed container and creates a vacuum in the container. **See Figure 4-42.** Vacuum pumps are used to create a vacuum in process heat exchangers, steam turbine condensers, and many other vessels. Vacuum pumps are also used as priming pumps. A *priming pump* is a vacuum pump that ejects air from the suction line of a larger negative-suction pump installation. The vacuum created causes liquid to flow up from below and into the larger pump, thus priming the larger pump.

A typical vacuum pump serving a condensate system is a two-compartment condensate vessel, divided horizontally. One compartment acts as a condensate accumulation area and is under the same vacuum as the condensate piping system. The other compartment is vented to the atmosphere.

As condensate accumulates in the lower compartment, a float switch energizes the condensate pump. The pump may also be energized if the vacuum in the condensate system drops below a preset level. The pump draws condensate from the upper compartment and forces it through a venturi or set of venturis and back into the upper compartment. This causes condensate and air to be drawn from the lower compartment and into the venturi. The condensate and air are discharged through the venturi and into the upper compartment. Air separates from the condensate and vents to the atmosphere. As the condensate level rises in the upper compartment, a float switch opens a valve on the discharge side of the pump and condensate is discharged from the upper compartment to the boiler.

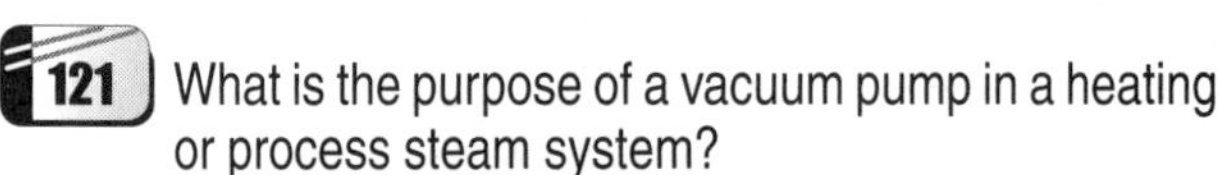

121 What is the purpose of a vacuum pump in a heating or process steam system?

A vacuum pump creates a vacuum on condensate lines, which helps condensate flow toward the condensate tank. This also creates additional differential pressure across the heating system steam traps, allowing the steam pressure on the heating system to be lowered. The lower pressure steam contains more latent heat, and this makes the heating system more efficient. The condensate tank and vacuum pump are typically a package unit.

122 List three things that a vacuum pump serving a condensate system does.

A vacuum pump creates a vacuum in the condensate piping. This causes condensate, air, and flash steam to be drawn toward the condensate receiver to which the vacuum pump is connected.

A vacuum pump sends condensate back to the boiler system. As condensate accumulates to a preset level in the condensate receiver, the pump discharges the condensate through the discharge valve and forces it back to the boiler.

A vacuum pump ejects air that is present in the condensate to the atmosphere. Air that is discharged by the steam traps is carried through the condensate return piping to the vacuum pump. There the air is allowed to separate from the condensate and vent to the atmosphere.

123 What is an aspirating pump?

An aspirating pump is a pump that removes air from the process liquid and discharges the air to the atmosphere. A vacuum condensate pump usually separates air bubbles from the condensate and vents air to the atmosphere.

Factoid

In the past, mechanical seals were expensive and used only for critical applications. Modern mechanical seals are used even for mundane applications such as pumping condensate and makeup water.

Pump Power Requirements

What is horsepower?

In the early days of the steam engine, Scottish inventor James Watt required a standard measurement for units of work and power with which steam engines could be compared. At that time, horses were the primary source of power. Through experimentation, Watt determined that a typical draft horse could perform the equivalent work of lifting a 33,000 lb weight to a height of 1 ft in 1 min. This led to the modern definition of power, the horsepower.

Power is the rate at which work is done. A *horsepower (HP)* is a unit of power equal to 33,000 foot-pounds (ft-lb) of work done in 1 minute. A *foot-pound* is a unit of work equal to the movement of a 1-lb object over a distance of 1 ft.

In pump applications, a force is applied to a liquid that results in the movement of the liquid. The amount of work done is the product of the weight of the liquid and the distance the liquid is moved. The weight of the liquid moved is usually expressed in lb/min or gpm, where 1 gallon of water equals 8.33 pounds. One gallon of other liquids will weigh different amounts, depending on the specific gravity of the liquid. The head of the pump in feet is the vertical distance or equivalent vertical distance to which the water or other liquid is pumped.

How is the theoretical horsepower requirement found for a pump?

The theoretical horsepower requirement is found by applying the following formula:

$HP = Q \times 8.33 \times H \times sp\ gr \div 33{,}000$

where

HP = horsepower

Q = capacity of the pump (in gpm)

8.33 = conversion constant (in lb/gal.)

H = head (in ft)

$sp\ gr$ = specific gravity of the pumped liquid

33,000 = conversion constant (in ft-lb/min/HP)

What is the theoretical horsepower requirement of a pump that delivers water at 100 gpm against a head of 578′? Disregard friction and other losses. Water has a specific gravity of 1.0.

$HP = Q \times 8.33 \times H \times sp\ gr \div 33{,}000$

$HP = 100 \times 8.33 \times 578 \times 1.0 \div 33{,}000$

$HP = 481{,}474 \div 33000$

$HP =$ **14.6 HP**

How may friction losses and slip be accounted for in pump horsepower calculations?

Design engineers use charts that assign values to each resistance in a piping system. For example, friction losses occur in pipe, pipe fittings, and valves. The sum of these losses, converted to feet of head, must be added to the required pump discharge head when selecting the pump in order to ensure that the pump has sufficient capacity. If the pump has a negative suction, the suction lift must be added to the total.

Pump slip and internal losses such as bearing friction must also be added to the mechanical losses in a pump installation. Of course, losses will increase with wear and when certain adjustments such as the impeller clearance are not kept within the manufacturer's recommended tolerances.

The total amount of losses in the pump determines pump efficiency. *Efficiency* is a ratio of the energy output to the energy input of a piece of equipment. Pump efficiency is expressed as a fraction or as a percentage. The pump manufacturer can provide the expected pump efficiency. The effects of friction losses and slip may be considered by modifying the formula as follows:

$HP = Q \times 8.33 \times H \times sp\ gr \div (33{,}000 \times E_p)$

where

HP = horsepower

Q = capacity of the pump (in gpm)

8.33 = conversion constant (in lb/gal.)

H = head (in ft)

$sp\ gr$ = specific gravity of the pumped liquid

33,000 = conversion constant (in ft-lb/min/HP)

E_p = pump efficiency

What horsepower is required to drive a pump that is running at 68% efficiency and delivering 450 gpm against a head of 822′?

$HP = Q \times 8.33 \times H \times sp\ gr \div (33{,}000 \times E_p)$

$HP = 450 \times 8.33 \times 822 \times 1 \div (33{,}000 \times 0.68)$

$HP = 3748.5 \times 822 \times 1 \div 22{,}440$

$HP =$ **137 HP**

Feedwater Piping and Valves

How is the pump affected if the check valve in the boiler feedwater piping sticks open?

If the pump control scheme is such that the pump starts and stops as the boiler requires water, then the pressurized water from the boiler will reverse direction and flow back through the pump when the pump stops. The pump will start and stop more often because the boiler will be draining back through the feedwater line. Some of the very hot boiler water may flash into steam inside the pump.

In larger boiler installations, a pair of boiler feedwater pumps are provided and one pump runs continuously. If this is the case and the faulty check valve is on the idle pump, the pump that is running will cause the idle pump to spin backward. This happens because at least part of the boiler feedwater flows back through the idle pump. This condition will likely result in insufficient feedwater being supplied to the boiler. The problem may be diagnosed by determining which pump motor is running. If only one motor is running, the other motor and pump will spin backward because of the faulty check valve.

⚠ WARNING

A pump that is spinning backward must not be started. The correct response is to close the discharge valve on the pump that is spinning backward. This will stop the backflow through the idle pump and restore the feedwater supply to the boiler. The pump should then be locked out and the check valve repaired.

130 How should a stop valve and a check valve be installed in the feedwater line?

The stop valve should be installed as close to the boiler as is practical. The check valve should be installed between the feedwater pump and the stop valve. This is the minimum number of valves that are provided. The stop valve can be closed for working on the check valve without having to drain the boiler.

In most cases, a stop valve and check valve are commonly provided at the pump location and another stop valve and check valve are provided at the boiler.

131 How should the piping be arranged when a battery of boilers is fed from one feedwater pump?

If one feedwater pump is to supply water to multiple boilers, the pump supplies the water to a feedwater header. Each boiler receives its water supply from the feedwater header. **See Figure 4-43.** The feedwater line from the feedwater header to each boiler should contain a stop valve close to the feedwater header, a check valve to prevent water from flowing back out of the boiler, and another stop valve as close to the boiler as practical. Each boiler feedwater line also contains an automatic feedwater regulating valve to control the boiler water level. A bypass line should be provided around the automatic feedwater regulating valve so that the boiler water level may be maintained manually if the feedwater regulating valve malfunctions.

The feedwater header should taper down in size so as to be appropriately sized to provide roughly the same flow velocity in feet per minute to each boiler. This helps ensure that boiler water treatment chemicals are transported evenly to the boilers and that there are no areas where air bubbles or dirt may collect. The branches to each boiler ideally should be taken using wye fittings rather than tees, so that friction losses will be minimized.

Feedwater Preheating

132 What is the advantage of preheating the feedwater before it enters the boiler?

The hotter the feedwater is as it enters the boiler, the faster the conversion of the water to steam can occur because less energy is needed to heat up the water in the boiler. Higher feedwater

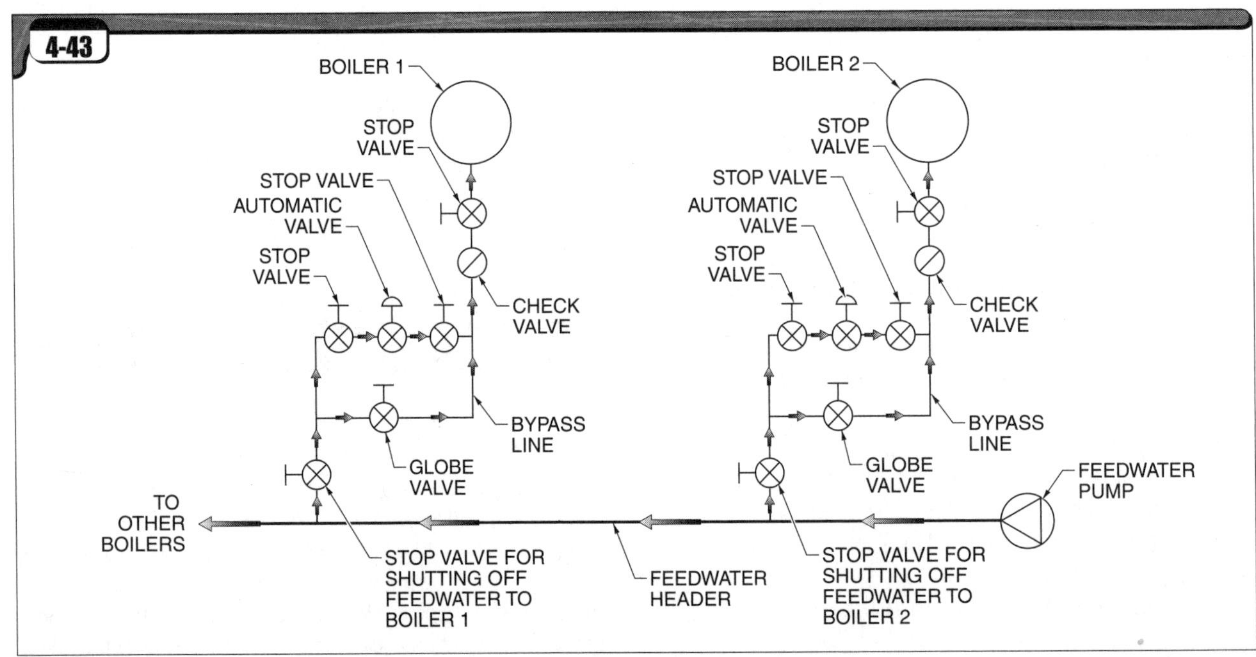

Figure 4-43. *A feedwater pump may supply water to several boilers through a feedwater header.*

temperatures also minimize thermal strains in the boiler. If the feedwater can be preheated with a source of heat that has already performed some work or that would otherwise have been wasted, this significantly improves the overall boiler plant efficiency.

133 What is an open feedwater heater?

An open feedwater heater is an open heat exchanger in which the boiler feedwater and the steam used as the heating medium come into direct contact with each other. The steam is condensed in the process of heating the feedwater and becomes part of the feedwater. Open feedwater heaters are also referred to as deaerators and are used to remove oxygen from the feedwater to prevent corrosion.

134 What is a closed feedwater heater?

A closed feedwater heater is a closed heat exchanger that is used specifically for heating the feedwater before the feedwater enters the boiler. Most closed feedwater heaters are shell and tube-type heat exchangers. The feedwater is directed through tubes in a tube bundle that is installed inside a cylindrical shell. Steam is piped to the shell and transfers heat to the water inside the tubes. The tubes are generally configured such that the feedwater passes through two to four banks of tubes before leaving the heater. The tubes pass through intermediate baffles that support the tubes and keep them straight. The baffle plates also promote heat transfer by maximizing contact of the steam with the tubes.

Closed feedwater heaters are normally classified as either high-pressure feedwater heaters or low-pressure feedwater heaters. High-pressure feedwater heaters are installed between the feedwater pump and the boiler, and therefore must withstand the full feedwater pressure delivered by the feedwater pump. Low-pressure feedwater heaters are installed before the feedwater pump.

135 What items should be connected to the shell of a closed feedwater heater?

- A safety valve should be provided on the steam side of a closed feedwater heater to keep the vessel from exceeding its rated steam pressure.
- A relief valve should be provided on the water side of a closed feedwater heater to prevent damage if the feedwater pressure increases beyond the rated pressure of the liquid side of the vessel. This could happen because of expansion of the water if the water side were full and the water were heated before the feedwater inlet and outlet valves were opened.
- Air can become trapped in the top of a closed feedwater heater. Therefore, a small air vent should be provided on top of larger closed feedwater heaters to vent this air.
- A properly-sized drip leg and steam trap should be provided to drain the condensate as fast as it accumulates.
- A gauge glass is often provided on larger closed feedwater heaters so that the operator can confirm that the drainage system is working properly.
- A condensate level switch may be provided to set off an alarm if the closed feedwater heater becomes flooded.

136 How can a closed feedwater heater be tested for tube leaks?

A fairly simple test to determine whether any tubes are leaking in a closed feedwater heater can be conducted by isolating the feedwater heater from the steam, feedwater, and condensate piping, and then disconnecting the condensate drain at a pipe union. The feedwater valve is then cracked open so that the feedwater side of the heater is pressurized. If water comes out of the condensate drain, at least one tube is leaking. This simple test only determines whether a tube leak exists, however. It does not identify which particular tube is leaking.

To isolate the specific location of the leak, the steam side of the heater is pressurized with water while observing the tube ends at the tube sheet(s). If water comes out of one of the tubes, that tube is defective. If water comes out around a tube end, the tube end is loose in the tube sheet or cracked where it is rolled into the tube sheet.

137 What is an economizer?

An *economizer* is a series, or bank, of boiler tubes used to recover heat from the boiler flue gas. Feedwater flows through an economizer en route to a boiler drum. An economizer is placed in the breeching between the boiler flue-gas outlet and the stack. **See Figure 4-44.** An economizer recovers heat that would otherwise be lost to the stack and uses this heat to preheat water flowing to the boiler. This typically increases the overall steam system efficiency in large industrial and utility boilers about 6% to 10%.

138 On what type of boilers are economizers most frequently found?

Economizers are much more common on industrial and utility watertube boilers than on firetube boilers, although some firetube boilers are equipped with them as well. An economizer becomes justified when it can absorb heat more economically than other types of heating surface. The design of modern firetube boilers lends itself well to absorption of heat when there is a comparatively low differential temperature between the flue gases and the boiler water. The addition of tubes to capture more available heat is simpler in firetube boilers, as the tubes are straight. In the design of larger watertube boilers, a point is reached where it becomes impractical to install enough boiler tube surface to extract additional heat from the flue gases.

Heat transfer takes place more quickly and efficiently when there is a larger temperature difference between the heating medium and the fluid to be heated. Therefore, as the flue gas temperature falls to within a few hundred degrees of the boiler water temperature, the heat transfer process slows. Increasing the heat transfer surface proportionally can compensate for this, but soon becomes impractical and prohibitively expensive in a watertube boiler. Since the feedwater is cooler than the water that is already in the boiler, there is a greater differential temperature between the flue gases and the feedwater. Thus, heat exchange between these two fluids will take place more readily.

139 How are economizers constructed?

Economizers for large boilers are almost always configured in stacked rows of horizontal tubes with headers and/or return bends on each end. **See Figure 4-45.** The overall construction of large economizers is in a rectangular configuration designed to become part of a breeching duct of the same size. The feedwater follows a serpentine path from the inlet header, through the many rows of horizontal tubes, and out the outlet header. Most large economizers are *counterflow* designs, where the flue gases contact the tubes closest to the economizer outlet first. Thus, the hottest flue gases encounter the hottest feedwater first. In a *parallel-flow* economizer, the opposite is true. Economizers often have bypass lines installed, allowing the economizer to be bypassed completely if one of the economizer tubes should fail.

Economizers are sometimes constructed of cast iron, but modern economizers are much more commonly made of steel. Cast iron economizers frequently have fin-tube construction, where the outer surfaces of the economizer tubes have extended fins to provide better heat absorption.

Packaged economizers are available for use with smaller watertube and firetube boilers. Many of these are configured as a coiled tube with extended surface fins for additional heat absorption. This design lends itself well to use as an add-on installation with existing older boilers. If the boilers are absorbing the heat of the flue gases as well as possible but the exiting flue gas temperature is high enough to make additional heat recovery feasible, an add-on economizer may be appropriate. This can sometimes increase the boiler plant efficiency considerably and have a short payback period.

Figure 4-45. *Feedwater follows a serpentine path through the rows of tubes in an economizer.*

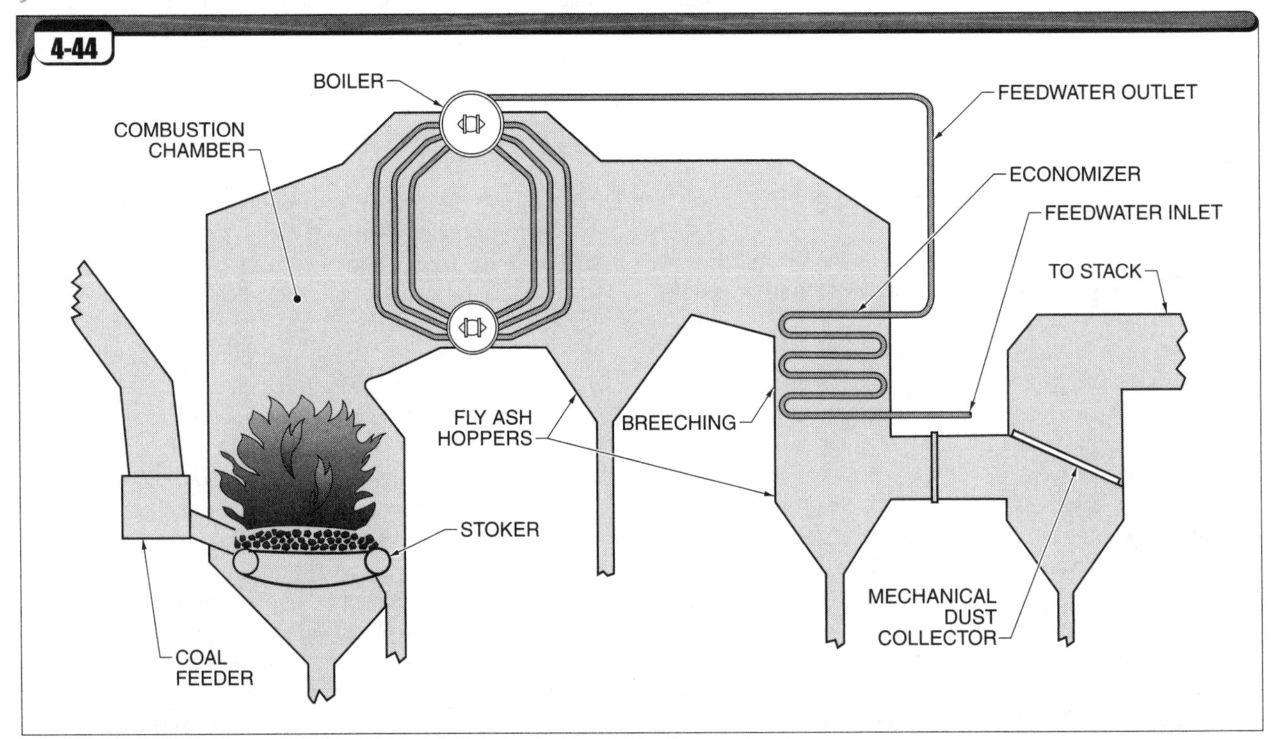

Figure 4-44. *Economizers recover heat that would otherwise be lost to the atmosphere.*

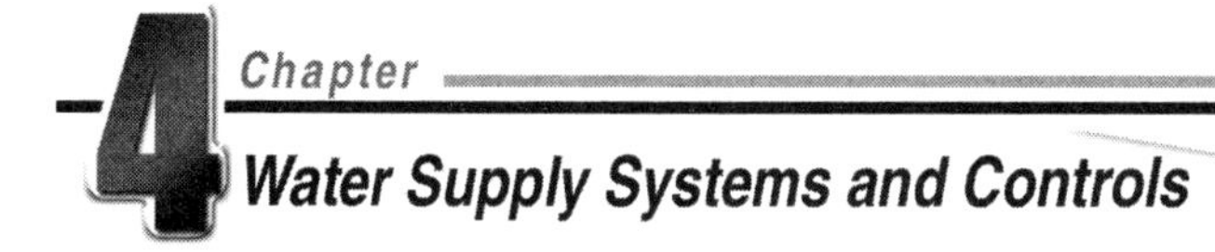

4 Chapter — Water Supply Systems and Controls

Name: ______________________________ Date: ______________

T F **1.** Circulating water is the only thing that keeps boiler heating surfaces from being damaged by intense heat.

______________ **2.** A ___ is a metal vessel installed on the outside of the boiler shell or drum at the NOWL.
A. try cock B. stopcock
C. gauge glass D. none of the above

______________ **3.** The letters NOWL represent ___.

T F **4.** The exact height of a water column is based on the size of the boiler.

______________ **5.** ___ water is dangerous because boiler heating surfaces could be damaged by overheating.
A. High B. Low
C. Pure D. none of the above

6. What is the purpose of a water column?

______________ **7.** Assuming a normal water level in the boiler, the top try cock should blow ___ when opened.
A. pure steam B. a mixture of steam and water
C. water with some flash steam D. pure water

______________ **8.** There are usually ___ try cocks installed on the side of the water column.
A. two B. three
C. four D. five

T F **9.** A flat gauge glass is normally used with boilers operating at 100 psi or less.

T F **10.** Try cocks are the primary water level indicator.

______________ **11.** Water columns are made of ___.
A. cast iron, ductile iron, or copper B. cast iron, steel, or copper
C. malleable iron, steel, or copper D. cast iron, ductile iron, or steel

T F **12.** Economizers recover heat from flue gases that would otherwise be lost up the stack.

______________ **13.** The capacity of a pump is normally given in ___.
A. cubic feet per minute (cfm) B. pounds per minute (lb/min)
C. pounds per hour (lb/hr) D. gallons per minute (gpm)

______________ **14.** A water column and gauge glass should be blown down ___ on a boiler in continuous 24-hour operation.
A. once per shift B. every 6 hours
C. once per day D. once per week

T F 15. A boiler must be taken out of service to repair a broken gauge glass.

______________ **16.** ___ is the weight of a given volume of a liquid divided by the weight of an equal volume of water, when both are measured at the same temperature.
A. Static head B. Specific gravity
C. Specific density D. Net positive suction head

T F 17. The minimum size pipe used for gauge glass blowdown is ¼″, as specified by the ASME Code.

T F 18. Connections that use a flow of steam or water may be made to a water column.

______________ **19.** A blowing whistle on top of a water column indicates a ___ water level in the boiler.
A. high B. low
C. either A or B D. neither A nor B

T F 20. Floats may be used inside a water column to open a steam valve to a whistle as an alarm.

______________ **21.** A(n) ___ is an audible alarm that is created electronically.
A. capacitor B. thyristor
C. annunciator D. none of the above

T F 22. Differential pressure transmitters use electricity to trip alarms.

______________ **23.** Bicolor gauge glasses work on the principle of light ___.
A. diffusion B. reflection
C. refraction D. none of the above

T F 24. Makeup water is necessary in steam heating systems because small amounts of water are lost through system leaks and boiler blowdown.

T F 25. A low water fuel cutoff must be located at least 4″ above the NOWL.

______________ **26.** The low water fuel cutoff should be tested ___ by blowing it down in the same manner as blowing down a water column.
A. daily or once per shift B. weekly
C. monthly D. none of the above

______________ **27.** The ___ regulator automatically maintains a constant safe water level in a boiler.

T F 28. Steam in a boiler carries heat away from the fusible plug as readily as water does.

______________ **29.** Water at 14.7 psia begins to boil at ___°F.
A. 100 B. 114.7
C. 212 D. 226.7

______________ **30.** A ___ should be placed at the discharge of a positive-displacement pump.
A. safety valve
B. foot valve
C. relief valve
D. ball check valve

______________ **31.** A negative-suction pump installation draws liquid ___ from a source ___ the pump.
A. down; above
B. over; beside
C. up; below
D. none of the above

T F **32.** The boiling temperature of water increases as pressure decreases.

T F **33.** A duplex pump is a steam-driven, reciprocating, positive-displacement, double-acting pump with two steam cylinders and two liquid cylinders.

______________ **34.** A simplex pump has the numbers $4 \times 2.5 \times 6$ on its side. The 4 indicates the ___ in inches.
A. diameter of the steam piston
B. diameter of the liquid piston
C. length of the stroke
D. none of the above

T F **35.** The flow from a positive-displacement pump can be throttled while the pump is running.

T F **36.** A simplex pump is a steam-driven, reciprocating, positive-displacement, double-acting pump with one steam cylinder and one liquid cylinder.

______________ **37.** A simplex pump has the numbers $4 \times 2.5 \times 6$ on its side. The 2.5 indicates the ___ in inches.
A. diameter of the steam piston
B. diameter of the liquid piston
C. length of the stroke
D. none of the above

______________ **38.** A simplex pump has the numbers $4 \times 2.5 \times 6$ on its side. The 6 indicates the___ in inches.
A. diameter of the steam piston
B. diameter of the liquid piston
C. length of the stroke
D. none of the above

______________ **39.** The slip of a pump typically ranges from ___% to ___%.
A. 0; 5
B. 0; 10
C. 5; 10
D. 5; 15

______________ **40.** What is the purpose of using ball check valves in a gauge glass installation?
A. To automatically bleed the pressure from the boiler if the gauge glass should break.
B. To permit blowing down the gauge glass to remove impurities.
C. To automatically shut off flow to the gauge glass if the gauge glass should break.
D. To permit removal of the gauge glass for cleaning while the boiler is in operation.

______________ **41.** A duplex pump is $6 \times 3 \times 6$. Steam is supplied to the pump at 100 psi. The maximum pressure that may be produced at the liquid end is ___ psi.
A. 100
B. 300
C. 400
D. 500

T F **42.** The impeller in a centrifugal pump throws liquid from its vanes through centrifugal force.

______________ **43.** The packing gland area of a rotary pump is also known as the ___ area.
A. seal
B. packing box
C. compression box
D. stuffing box

__________ **44.** A ___ is a spacer installed between two of the rings of packing in a pump.

A. packing follower
B. seal flush
C. key
D. lantern ring

T F **45.** Aspirating pumps remove air from the process liquid and discharge it into the boiler.

T F **46.** Mechanical seals require manual adjustment after installation.

__________ **47.** A pressure increase on the suction side of a centrifugal pump produces ___ on the discharge side.

A. a pressure drop
B. the same pressure increase
C. no pressure change
D. 5% to 10% pressure change

T F **48.** The pressure at the bottom of a column of water 2.31′ high is 1 psi.

__________ **49.** The equivalent static head of a boiler operating at 250 psi is ___′.

A. 108.2
B. 247.7
C. 252.3
D. 577.5

T F **50.** The venturi in an injector causes a drop in the steam velocity.

T F **51.** One horsepower equals 33,000 ft-lb of work in 1 minute.

__________ **52.** The letters NPSH represent ___.

__________ **53.** The formula for determining the horsepower requirement for a pump is ___.

A. $\frac{Q \times H}{sp\ gr \times 33{,}000 \times E_p}$

B. $\frac{Q \times H \times sp\ gr}{33{,}000 \times E_p}$

C. $\frac{Q \times 8.33 \times H \times sp\ gr}{33{,}000 \times E_p}$

D. $\frac{Q \times 0.433 \times H \times sp\ gr}{33{,}000 \times E_p}$

__________ **54.** What is the horsepower requirement of a pump that moves 2500 gpm of water against a head of 475′ if the pump is operating at 58% efficiency?

__________ **55.** ___ pumps are small-capacity pumps normally used in boiler rooms for pumping water treatment chemicals.

56. How should the packing gland around a pump shaft be adjusted if the leakage of water is excessive?

T F 57. High water in a boiler can result in water hammer, which can damage steam-using equipment.

______________ **58.** A ___ is a nozzle with a slight hourglass-shaped taper.

A. regulator
B. venturi
C. fusible plug
D. none of the above

T F 59. Try cocks are not useful on high-pressure boilers operating at about 500 psi and above.

______________ **60.** A low water fuel cutoff is usually placed on the boiler from ___″ to ___″ below the NOWL.

A. 2; 3
B. 2; 4
C. 2; 6
D. none of the above

T F 61. The majority of firetube boiler accidents are due to low water fuel cutoffs that did not work when needed.

______________ **62.** A high-pressure feedwater heater is located ___.

A. between the low-pressure feedwater heater and the feedwater pump
B. between the feedwater pump and the boiler
C. between the boiler and the steam header
D. in the breeching between the boiler and the stack

T F 63. Water can boil at less than 212°F when under a vacuum condition.

______________ **64.** A(n) ___ impeller has no shroud to support the vanes, or only a partial shroud on one side.

A. open
B. semiclosed
C. closed
D. none of the above

T F 65. Only appliances that use a flow of steam or water may be attached to a water column.

T F 66. The gauge glass should be blown down after blowing down the water column.

T F 67. The bottom valve should be closed first if a gauge glass breaks.

______________ **68.** A(n) ___ is $1/1000$ of an ampere.

______________ **69.** A ___ pump is preferred for boiler feed.

A. reciprocating
B. centrifugal
C. pump is never required for boiler feed
D. none of the above

______________ **70.** ___ may occur when a portion of the water or other liquid entering the eye of a pump impeller flashes to steam bubbles.

______________ **71.** A thermostatic expansion tube feedwater regulator is a type of ___ feedwater regulating system.

A. modulating
B. ON/OFF
C. metering
D. recirculating

T F 72. Economizers are only used with watertube boilers.

T F 73. The makeup water feeder on a low-pressure heating boiler is intended to serve as the feedwater regulating system.

______________ **74.** What is the horsepower required to drive a pump that develops 68% efficiency at full capacity if it moves 4900 gpm of water against a head of 326′.

______________ **75.** How should a motor-driven centrifugal pump be started to minimize the starting current draw on the motor?

A. With the suction and discharge valves wide open
B. With the suction valve closed and the discharge valve open
C. With the suction and discharge valves closed tightly
D. With the suction valve wide open and the discharge valve closed

76. Why would a redundant (backup) low water fuel cutoff control be a manual reset type control?

______________ **77.** A duplex pump is 6 × 3 × 6. It runs at an average speed of 68 strokes per minute and slip in the pump totals 10%. Neglecting the volume of the piston rod, what is the capacity of the pump?

A. 11.2 gpm
B. 22.5 gpm
C. 25.0 gpm
D. 29.9 gpm

T F **78.** A gear pump is a positive-displacement pump.

______________ **79.** In a three-element feedwater regulating system, ___ are measured.

A. steam flow, fuel flow, and feedwater flow
B. feedwater flow, water level, and condensate return
C. makeup water, condensate return, and water level
D. steam flow, feedwater flow, and water level

______________ **80.** If valves are installed between the boiler and the water column, the valves must be ___.

A. a type that may be repacked while the boiler is in service
B. combination stop and check type valves
C. a ball check type
D. sealed or locked in the open position while the boiler is in service

T F **81.** A motive fluid pump uses the force of a secondary fluid to pump the primary fluid.

Chapter 5

Water Treatment Systems and Controls

The proper preparation of the feedwater for a steam boiler system minimizes the inherent problems associated with having water and steel in the same environment. For example, removal of oxygen from the water prevents pitting. Pitting can seriously degrade the mechanical integrity of piping and pressure vessels. Removal of scale-forming minerals ensures efficient heat transfer in steam boilers and helps prevent overheating of tube surfaces. Proper chemical treatment of the steam and condensate prevents corrosion due to acidic conditions.

The impurities contained in water supplies can vary considerably within the same city or geographical area. Therefore, there can be no "one size fits all" solution to water treatment challenges. Boiler operators should develop a clear understanding of the behavior of the various impurities in the system water, the equipment used to remove them, and the function of the chemicals used to counteract them. A proactive approach helps ensure the best possible water quality in the boilers and this helps minimize expensive blowdown.

By interacting knowledgeably with the facility's water treatment specialist, the boiler operator helps maximize the steam system efficiency, longevity, and reliability while minimizing expensive downtime and repairs.

Water Treatment Objectives

What is the hydrologic cycle?

The hydrologic cycle is the collection of continually repeated processes that natural water goes through on the earth. It has a great deal to do with the sources of contaminants that cause problems in boilers, steam distribution systems, and condensate return systems.

For example, as rain falls from the clouds, it absorbs atmospheric gases such as carbon dioxide and oxygen. It also picks up industrial pollutant gases and smoke, as well as natural impurities such as dust and pollen. When the rain contacts the earth, part of it soaks into the ground. As the water percolates through the ground, it absorbs impurities such as iron, calcium, magnesium, decaying organic matter, silica, and others.

Part of the rainwater also flows across the surface of the earth as the first few inches of the ground become saturated. This water flows downhill into lakes and rivers. Some water accumulates in municipal reservoirs. As the water flows across the ground, it absorbs and carries other types of natural and man-made impurities, such as agricultural chemicals, silt, oily residues, and other pollutants.

The natural water that is used in boilers generally consists of groundwater and surface water. Groundwater is drawn from sources below the surface of the earth, such as wells and springs. Surface water is obtained from bodies of water that lie on the surface of the earth, such as rivers, lakes, and reservoirs. The types and quantities of impurities in a particular water supply are largely a function of the source from which the water was obtained.

What are some classifications of impurities found in natural water?

Dissolved gases are gases that have gone into solution in water. Examples of gases that may be dissolved in natural water include oxygen, carbon dioxide, and hydrogen sulfide. *Dissolved solids* are solid impurities that have gone into solution in water. Dissolved solids exist in natural water even though they are not visible. This is similar to the way that sugar dissolves in a cup of coffee. Examples of solids that dissolve in natural water include some calcium and iron compounds. *Suspended solids* are solid impurities that are suspended in water. Suspended solids are visible in natural water. Examples of solids that may be suspended in natural water include algae and silt.

What are the objectives of a good boiler system water treatment program?

- prevention of corrosion and pitting
- prevention of scale
- prevention of steam contamination

Dissolved gases are undesirable in a boiler system because they contribute to corrosion and pitting of the boilers, piping, and vessels. **See Figure 5-1.** The two dissolved gases that cause the most damage are dissolved oxygen and dissolved carbon dioxide.

Scale interferes with the ability of the boiler to transfer heat from the hot combustion gases to the water. Scale can result in overheating of tubes and other heating surfaces. Dissolved solids, especially compounds containing calcium, become crystalline and can form cement-like boiler scale when water is boiled away. Compounds containing magnesium and iron can also contribute to significant scale problems.

Boiler water can contain significant amounts of dissolved and suspended solids, oily residues, or other contaminants. When the boiler water is dirty, the water can foam and the water level can become very erratic. This can result in carryover and priming. Dissolved gases that are released from the boiler water and pass into the steam have insulating properties and interfere with the ability of the steam to transfer heat properly.

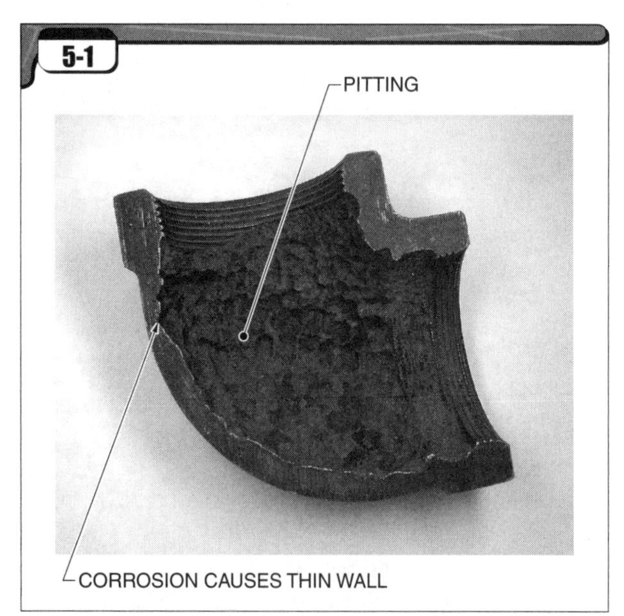

Figure 5-1. *Dissolved gases cause corrosion and pitting of boiler metal and piping.*

Oxygen Pitting

What is a deaerator?

A *deaerator* is an open heat exchanger that removes dissolved gases from the feedwater going to a boiler. **See Figure 5-2.** Feedwater is less corrosive when the dissolved gases are removed. Most deaerators are pressure vessels. Some small units operate at atmospheric pressure. The great majority of pressurized deaerators operate at between 5 psig and 15 psig, but some larger units operate at up to 100 psig.

Solubility is the ability of a material to dissolve in water. As water is heated, the solubility of oxygen and carbon dioxide decreases. At the boiling point, the solubility of oxygen and carbon dioxide decreases to almost zero. However, the remaining small amount of oxygen is difficult to remove. Therefore, three operational objectives must be met in order for a deaerator to achieve the best possible oxygen-removal efficiency.

The deaerator must heat the water to as near the temperature of the steam as possible. This minimizes the solubility of the gases in the water.

The deaerator must maximize the surface area of the water. Since oxygen must separate at the water's surface, maximizing the surface area improves the efficiency of oxygen removal. The flow of water going through a deaerator is broken into many small droplets and/or thin films, depending on the deaerator design.

The deaerator must provide agitation to scrub the gases from the water. The last two to three percent of the dissolved oxygen in water is not readily released, even when the water temperature approaches the boiling point. For this reason, deaerators are designed to create agitation to scrub as much of the remaining oxygen from the water as possible.

Condensate is already heated and therefore contains less oxygen than cold makeup water. However, condensate may absorb oxygen and carbon dioxide as it accumulates and cools in condensate receivers and piping. Therefore, both condensate and makeup water are deaerated before being pumped to the boilers.

Figure 5-2. *Deaerators remove dissolved gases from the boiler feedwater to make the feedwater less corrosive.*

In what cases might a deaerator not be used?

Deaerators are justified where the amount of cold makeup water is more than a few percent of the total steam load. The cold makeup water presents a greater corrosion potential because of its comparatively high oxygen content. In closed, low-pressure heating systems, almost all of the condensate is typically returned to the boiler. Therefore, a deaerator is not usually included in these systems.

What are the comparative advantages of pressurized and atmospheric deaerators?

Pressurized deaerators have several advantages over atmospheric deaerators. Deaerators for larger facilities are very sophisticated and may have several successive sections that the steam must flow through. Therefore, it takes considerable pressure to force steam to flow through the unit. The pressure in the deaerator also helps the unit vent the gases to the atmosphere more positively, with less chance of reabsorption. Condensate from high-pressure steam heat exchange equipment may flash when entering a deaerator, and the pressurized design is better for capturing this flash steam and using it to heat the incoming water. In addition, pressurized units are able to preheat the water to higher temperatures.

The primary advantage of atmospheric deaerators is cost. Since the vessel has a large vent to atmosphere, it is very unlikely that it can become overpressurized. Therefore, the deaerator does not need to be constructed to pressure vessel code requirements. Atmospheric units are less sophisticated in design as well. These features result in considerably lower initial cost. However, atmospheric deaerators are less efficient at oxygen removal.

Describe three types of pressurized deaerators.

The main differences between various deaerators are their approaches to creation of surface area and agitation. The three major types are the tray deaerator, spray deaerator, and combination tray-spray deaerator. **See Figure 5-3.** Manufacturers make many variations on these major designs.

A tray deaerator causes water to cascade downward in a zigzag pattern over small, staggered trays. Depending on the design, steam flows either upward or downward through the trays and contacts the water. The water is heated and agitated and oxygen is released as a result.

A spray deaerator uses self-adjusting spray nozzles in the top section to break the water flow into a spray of small water droplets. Steam entering a spray deaerator is directed into the water sprays. The water is heated and agitated and oxygen is released as a result.

A combination tray-spray deaerator uses trays and spray nozzles to break the water flow into small droplets. The trays and spray nozzles are usually in separate sections. Baffles control the flow of steam and water through the sections of a tray-spray deaerator. The water is heated and agitated and oxygen is released as a result.

The heat and the flow of steam release oxygen from the water and carry it to the top of the deaerator. Oxygen, along with other gases, is vented out the top of the deaerator to the atmosphere. A small amount of steam passes out of this vent with the oxygen.

Water flowing through a deaerator is heated to within a few degrees of the steam temperature. The water accumulates in a feedwater storage section or storage tank in the bottom of the vessel. The feedwater storage tank holds a reserve of water for periods when a boiler demands feedwater faster than the deaerator can provide it. This tank is typically sized to provide about 20 min to 30 min of boiler feedwater in case the water supply should be lost. This ensures that adequate water is always available to the boilers.

The operation of deaerators is significantly better with a steady flow of incoming water. Unsteady flow can occur, for example, if a condensate pump provides a large on and off flow to the deaerator. A *deaerator surge tank* is a tank or pressure vessel used to even out flow into a deaerator and provide a continuous, modulated flow. Condensate from the condensate receivers is pumped into a surge tank. Very often, the makeup water is also added to the surge tank to help provide a more constant water temperature to the deaerator. The blended water is then pumped in a smaller, more continuous stream from the surge tank to the deaerator.

Some manufacturers of deaerators use the term "surge tank" to describe the feedwater storage tank located below the deaerator. Used in this context, the term "surge" implies the periodic short-duration demands of the boiler(s) for feedwater at a faster rate than the deaerator can prepare it. The tank provides a ready volume of deaerated feedwater during these periods.

Can boiler feedwater be heated above 212°F in a vented heater?

Boiler feedwater can be heated above 212°F if the vent is small enough that pressure can be maintained in the heater. For example, the steam in a deaerator can be supplied faster than the small vent can relieve pressure from the vessel. This allows the pressure to build up slightly in the vented heater.

What should the water temperature be when the water leaves the deaerator?

The water temperature should be within 3°F to 5°F of the temperature of the steam entering the deaerator. This temperature difference should be monitored daily by the operator. If the temperature difference begins to increase, a problem in the operation of the deaerator is indicated. For example, the spray nozzles may be partially plugged or damaged.

Factoid

Deaerators remove most dissolved gases from feedwater by heating the feedwater. The gases are vented from the deaerator along with a small amount of steam.

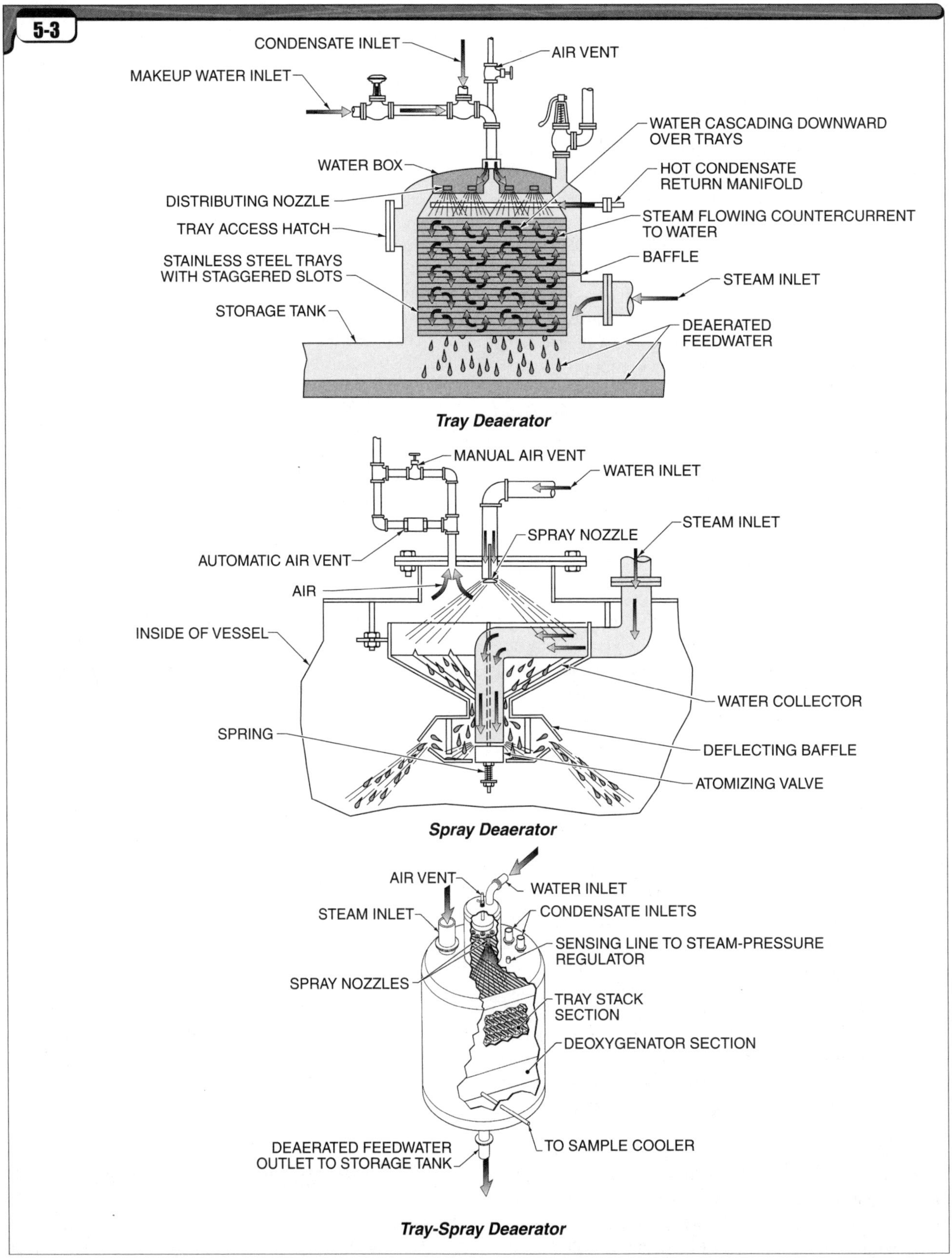

Tray Deaerator

Spray Deaerator

Tray-Spray Deaerator

Figure 5-3. *Deaerators remove oxygen from the feedwater going to boilers.*

How is the supply of steam to a deaerator controlled?

Most deaerators use a pressure controller that senses the steam pressure in the deaerator and modulates a control valve in the steam supply line to maintain a constant pressure. The controller may be a stand-alone unit mounted near the deaerator or it may be part of a centralized computer control system. Small deaerators frequently use a simple pressure-reducing valve to control the steam pressure.

Atmospheric deaerators are maintained at atmospheric pressure inside the vessel. Therefore, the temperature of the deaerated feedwater is measured instead. A temperature regulator is used to vary the steam supply depending on the feedwater temperature. This is something of a disadvantage, because temperature controls normally do not respond to changes in the operation as quickly as pressure controls.

How is the water level controlled in a deaerator?

The water level in a deaerator is usually controlled by a float-operated pneumatic controller. The controller senses the water level in the storage tank and operates a control valve that modulates the amount of water entering the deaerator. On some deaerators, the float that senses the water level is connected by a direct mechanical linkage to the control valve. The water level in a deaerator may also be controlled electronically.

An overflow line is provided on most deaerators to prevent backup of water into the steam supply line if the level-regulating controls fail. The overflow line on a pressurized deaerator must contain a float-operated valve or other means to prevent steam from being wasted through the overflow.

How are boiler feedwater pumps and a deaerator placed in relation to each other?

A deaerator is placed before the boiler feedwater pumps because the storage section of the deaerator supplies feedwater to the pumps. The deaerator is also placed above the feedwater pumps so that the required NPSH at the inlet of the pumps is maintained to keep the pumps from cavitating. In large power plants, the deaerators may be 50′ or more above the feedwater pumps.

What is a vent condenser?

A *vent condenser* is an in-line heat exchanger installed in the vent from a deaerator to the atmosphere. **See Figure 5-4.** A vent condenser minimizes steam loss to the atmosphere. The vent condenser usually consists of a shell-and-tube heat exchanger. The makeup water flows through the tubes and the steam and air mixture passes over the outside surfaces of the tubes. Most of the heat in the steam is given up to the makeup water flowing through the tubes. The steam condenses and the condensate is returned to the deaerator. The oxygen and other gases are vented to the atmosphere. Some vent condensers are installed inside the top dome of the deaerator.

What is the purpose of a vacuum breaker placed on a deaerator?

If steam flow to a deaerator is lost or shut off, water entering the vessel condenses the steam remaining in the vessel and causes a strong vacuum. When steam changes to condensate, the volume decreases dramatically. For example, steam at 212°F collapses to 1/1600 its volume when it condenses back into water. The collapsing steam can cause a powerful vacuum.

A deaerator is a pressure vessel designed to withstand pressure on the inside pushing out and not the atmospheric pressure outside pushing in. If a vacuum develops inside a deaerator, the vacuum breaker opens and allows air to flow into the vessel to relieve the vacuum and keep atmospheric pressure from crushing the vessel.

A vacuum breaker on a deaerator is usually a check valve connected to the top of the vessel. It is oriented so that pressure in the vessel holds it in the closed position. Vacuum in the vessel allows the atmospheric pressure outside to open the check valve into the deaerator.

Why is a deaerator equipped with a safety valve?

A deaerator, like any other pressure vessel, is designed for a maximum pressure rating. Operating above the maximum pressure rating could cause the vessel to explode. A safety valve prevents an explosion.

How is the small quantity of oxygen treated that is not removed from the feedwater by the deaerator?

The small quantity of oxygen remaining in the feedwater is treated with oxygen scavengers. An *oxygen scavenger* is a chemical that reacts with any oxygen remaining in boiler feedwater and changes it into a form that does not cause corrosion. Oxygen scavengers are added to feedwater as it leaves the deaerator. Commonly used oxygen scavengers include sodium sulfite and hydrazine.

Most impurities in water are measured by the concentration of the specific impurity in the water in parts per million. A *part per million (ppm)* is the concentration of a solution equal to one part of a chemical in one million parts of the solution (1/1,000,000). For example, if calcium impurity is present at 1 ppm, there is 1 lb of the calcium impurity for every 1,000,000 lb of the calcium and water solution. *Note:* Some iron testing apparatus indicate measurements in milligrams per liter (mg/l). One mg/l is the same as 1 ppm.

Most modern deaerators are extremely efficient at oxygen removal. Oxygen remaining in feedwater is typically as low as 0.03 ppm for atmospheric deaerators and 0.005 ppm for pressurized deaerators. Another way of stating these amounts is to use parts per billion. A *part per billion (ppb)* is the concentration of a solution equal to one part of a chemical in one billion parts of the solution (1/1,000,000,000). A concentration of 0.03 ppm equals 30 ppb and 0.005 ppm equals 5 ppb.

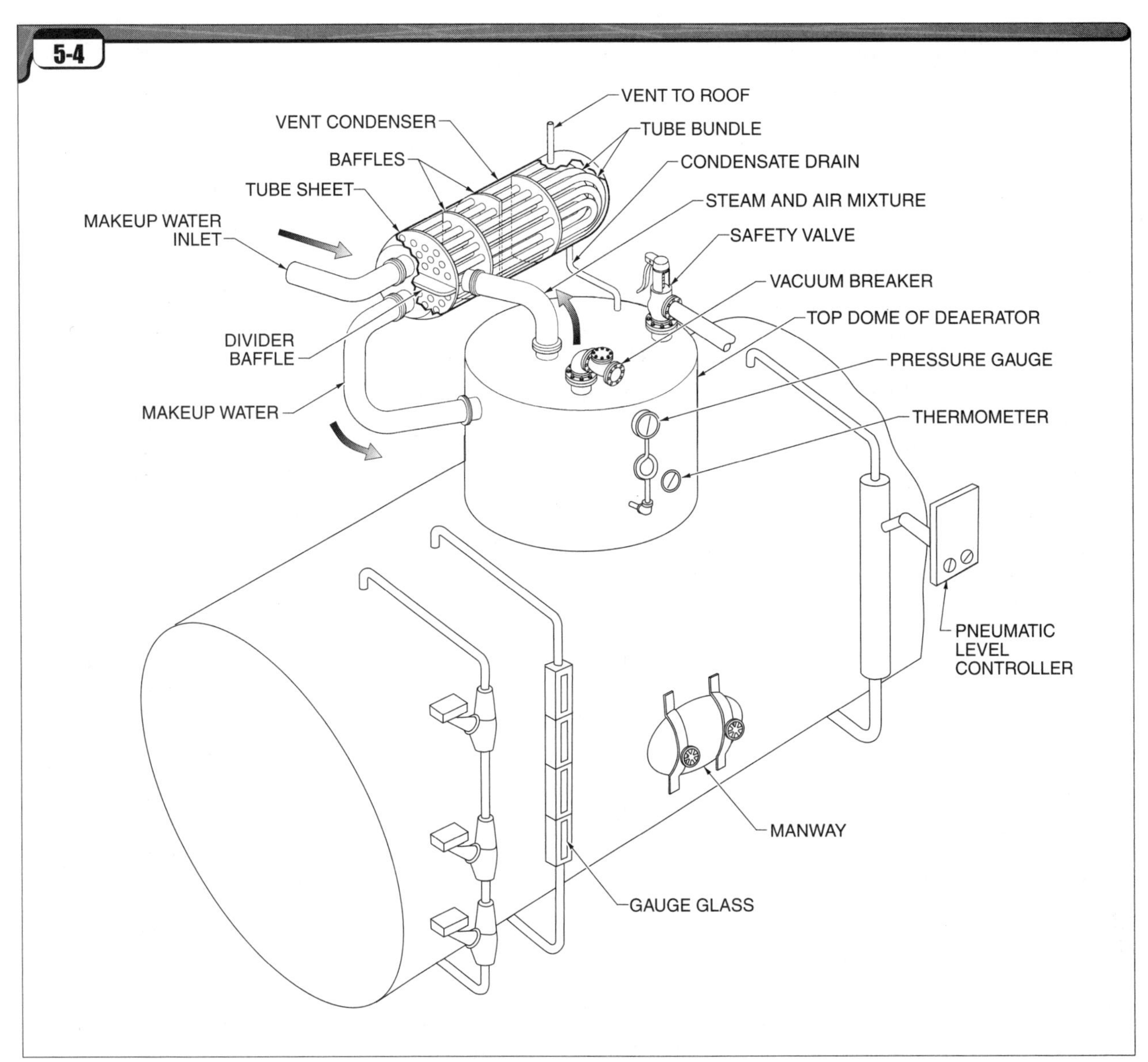

Figure 5-4. *Vent condensers minimize steam loss to the atmosphere.*

Factoid

Oxygen scavengers do not remove oxygen from the feedwater, but change it to a noncorrosive form. A residual of the scavenger is always maintained in the boiler water.

17 What determines the quantity of oxygen scavenger that should be used?

The amount of oxygen to be treated determines the amount of oxygen scavenger that should be used. The amount of oxygen to be treated depends directly on the overall quantity of water contained in the steam system, the percentage of makeup water used, and the concentration of oxygen in the water. Therefore, if a large amount of hot condensate is returned to the boiler system, less makeup water is needed, and less chemical treatment must be used. A water treatment specialist should be consulted for help in determining the expected quantities of chemical treatments to be used.

The treatment chemical is fed in an adequate quantity to treat all the remaining oxygen in the feedwater. Enough extra treatment chemical is fed to maintain a predetermined residual in the water. A residual is a small amount of the chemical that is left over after all the particular contaminant has been treated. This small residual of the treatment chemical serves two purposes. First, it provides evidence that enough of the treatment chemical was fed to treat all the contaminant present. Second, it provides a small amount of treatment chemical that remains immediately available in the water in case any of the contaminant material should get into the system. For example, enough sodium sulfite is typically fed to boiler feedwater to chemically react with all the oxygen in the feedwater and maintain a residual of 3 ppm to 6 ppm.

Whether the impurity being treated is oxygen or another troublesome contaminant, it is almost always less expensive, more efficient, and less problematic to remove the impurity with mechanical means first. For example, deaerators perform the job of oxygen removal more efficiently and more cheaply than the use of sodium sulfite or hydrazine. Although sodium sulfite performs a needed function, it is itself an impurity in the boiler water. In large amounts, the sodium sulfite can contribute to foaming and other problems.

Scale

What is hardness of water?

Hardness of water is the measurement, in ppm, of calcium and magnesium in water. Compounds containing calcium and magnesium account for the majority of boiler scale. These compounds exist in a number of possible forms such as calcium carbonate and magnesium silicate. In boiler water, a level of 0 ppm of calcium and magnesium hardness is desirable but difficult to attain. The feedwater entering a typical industrial boiler may have 0.005 ppm to 1 ppm of hardness.

Hard water is treated in several ways in an effort to minimize scale formation in boilers. It is important to remove the hardness before the water enters the boiler, because once scale is formed, it is difficult to remove and causes substantial efficiency losses. **See Figure 5-5.** Some other natural salts and minerals, such as silica, also contribute to scale.

Figure 5-5. *Water that contains excessive hardness causes scale in boilers.*

How does temporary hardness of water differ from permanent hardness of water?

Temporary hardness of water is a type of hardness that can be reduced by heating the water. Temporary hardness of water consists primarily of calcium carbonate and calcium bicarbonate. These salt compounds exist in natural water as the result of the action of acidic rainwater on natural minerals in the earth such as limestone. Temporary hardness is also known as alkaline hardness because the salts dissolve to form an alkaline solution.

Permanent hardness of water is a type of hardness that can be reduced only by the use of chemicals or distillation. Permanent hardness of water consists of noncarbonate salts that are the result of naturally occurring acidic reactions. Seawater is an example of water containing permanent hardness.

How is scale detrimental to a boiler?

Scale insulates heating surfaces so that less heat passes through the boiler tubes into the water. Even a very light accumulation of scale in a boiler causes boiler efficiency to drop dramatically. **See Figure 5-6.** Scale results in higher fuel costs because larger quantities of fuel must be burned to satisfy the same steam demand.

Heavy accumulations of scale can be dangerous. Heat transfer can be so severely impeded that water can no longer adequately remove the heat from the metal tubes. The tubes can then overheat, soften, and burst.

EFFECT OF BOILER SCALE ON FUEL CONSUMPTION				
Thickness of Scale*	Efficiency Loss	Fuel Consumption per 100 Boiler Horsepower		
		Coal†	Nat. Gas‡	No. 6 Oil§
0	0%	335	28.0	4200
1/64	4%	348	29.1	4370
1/32	7%	358	30.0	4490
1/16	11%	372	31.1	4660
1/8	18%	395	33.0	4960
3/16	27%	425	35.6	5330
1/4	38%	462	41.2	5800
3/8	48%	496	41.4	6220

* in in. † in lb/hr ‡ in cu ft/hr § in gal/hr

Figure 5-6. *The accumulation of scale in boilers causes drastic efficiency losses.*

What causes hardness in the water to form scale in a steam boiler?

Calcium and magnesium hardness in boiler water crystallizes and becomes boiler scale in two primary ways. Calcium and magnesium are inversely soluble. *Inverse solubility* is the tendency of certain impurities in water to crystallize and precipitate as the temperature of the boiler water increases. Since the hottest water is against the heating surfaces, compounds containing calcium and magnesium form scale on the heating surfaces first.

When steam is formed, only pure water is boiled off. The scale-forming impurities that are dissolved or suspended in the boiler water are left behind in the remaining water. This may be compared to a drinking glass that is filled with tap water and then left to evaporate on a windowsill. The dissolved solids in the tap water are not visible because they are dissolved into solution. After the water evaporates from the glass, the solids that were in the water may be seen as residue in the empty glass.

If large amounts of water are boiled off, larger quantities of these impurities are left behind. Boilers can operate with some concentration of these salts and impurities in the water without significant problems. However, if the impurities in the boiler water become too concentrated, scale and other problems result.

22 Why must boiler water treatment be more closely monitored when operating a high-pressure boiler than when operating a low-pressure boiler?

Boiler water treatment must be more closely monitored when operating a high-pressure boiler because the higher temperature of the water results in a greater tendency to form scale.

23 Why do boilers used only for heating require less attention to water treatment than other boilers?

Almost all of the condensate from boilers used for heating is returned to the boiler. This condensate is basically distilled water. Therefore, after the original concentrations of scale-forming impurities are removed, only small amounts of makeup water introduce additional impurities into the system. In addition, the condensate from a steam heating system is rarely subject to contamination.

The quantity of returning condensate in any boiler plant should be monitored. If the quantity of returning condensate is inadequate because of leaks or other problems in the system, larger amounts of makeup water will be required. This brings larger quantities of impurities into the boiler.

24 What are pretreatment and after-treatment of boiler system water?

Pretreatment of boiler system water consists of the water treatment processes that occur before the water enters the boiler. Deaeration and softening are examples of pretreatment.

After-treatment of boiler water consists of the water treatment processes that occur during or after steam generation. After-treatment occurs in the boiler or steam distribution system. For example, scale-control chemicals are added to boiler water to prevent scaling due to the small quantities of impurities not removed by pretreatment.

25 What is ion exchange?

An *ion* is an atom or molecule with an electrical charge. A *cation* is an ion that has a positive electrical charge. For example, the calcium ions (Ca^{+2}) in calcium carbonate ($CaCO_3$) and the magnesium ions (Mg^{+2}) in magnesium sulfate ($MgSO_4$) are cations. An *anion* is an ion that has a negative electrical charge. For example, the carbonate (CO_3^{-2}) in calcium carbonate and the sulfate (SO_4^{-2}) in magnesium sulfate are anions.

Ion exchange is an exchange of one ion for another ion in a solution. Ion exchange is the process by which most modern water softeners operate. Boiler water ion exchange typically occurs when feedwater is passed through an ion-exchange bed of resin particles. A resin is a man-made substance composed of tiny plastic-like beads. Resins used in water treatment equipment are composed of different molecular compounds depending on what kind of impurities need to be removed from the water.

Ion exchange occurs because the resin has a stronger attraction to one ion than another. For example, a typical sodium zeolite water softener is regenerated and the surface of the resin beads is coated with sodium ions. When hard water is sent through the softener, the calcium cations in calcium bicarbonate replace the sodium ions on the resin. **See Figure 5-7.** The sodium ions are free to form soluble sodium bicarbonate. Another common exchange is to replace the magnesium cations in magnesium sulfate with sodium atoms from the resin to form soluble sodium sulfate.

Common water softeners are generally capable of removing some amount of dissolved iron from the water as well. However, large amounts of iron in the water being softened can cause operational problems with the softener. The iron is more difficult to remove from the resin than the hardness. After a period of time, the iron can interfere with the ability of the resin to perform properly. Separate treatment processes designed specifically to remove iron may be required for some incoming water.

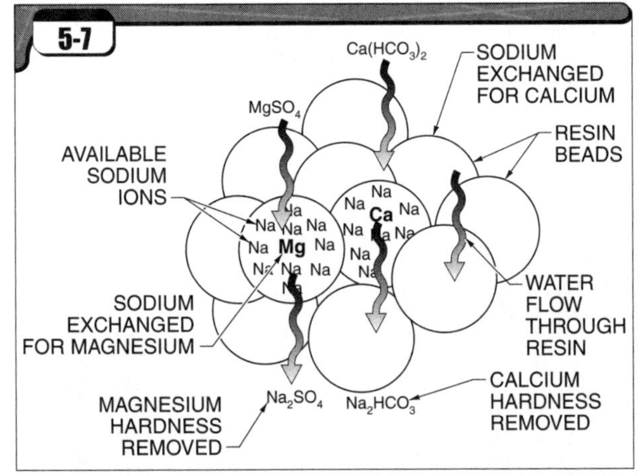

Figure 5-7. *Sodium ions are exchanged for calcium and magnesium ions on the surface of water softener resin.*

26 Describe a sodium zeolite water softener.

A *sodium zeolite water softener* is an ion-exchange water softener that uses resin beads and a brine solution to soften water. A *zeolite* is a synthetic sodium aluminosilicate cation-exchange material based on naturally occurring clays called zeolites. All ion-exchange materials used during the early development of ion-exchange water softeners were zeolites or materials closely resembling zeolites. Therefore, the term sodium zeolite became synonymous with all cation exchange materials. Modern water softeners use synthetic resins rather than zeolites, but the term sodium zeolite is still commonly used.

A water softener is a vertical tank that is about two-thirds filled with resin material. **See Figure 5-8.** The water softener has a separate open tank that contains brine. *Brine* is a solution of salt and water. The brine regenerates the water softener's resin by flushing the captured calcium and magnesium hardness ions from the resin and replacing the calcium and magnesium with sodium ions. During each regeneration, the resin beads are coated with sodium cations from the salt brine.

An automatic timer controls the period of time that each step of the regeneration process takes. At the end of each step, the timer energizes a stager supplied with the water softener. The stager is a motorized unit that controls a multiport valve or group of valves (valve nest). The multiport valve or the group of valves directs the water flow through the water softener. The direction of water flow depends on the stage of regeneration of the water softener.

When the water softener is in service, the inlet water flows downward through the resin beads. The resin has a slightly stronger attraction to calcium and magnesium ions than it does to sodium ions. As the hard water comes in contact with the resin, the resin gives up sodium ions. Sodium does not contribute to scale formation. The resin attracts the calcium and magnesium ions. The resulting water leaving the water softener has greatly reduced scale-forming potential.

27 How does a sodium zeolite water softener regenerate?

Some water softener systems have two softener tanks in parallel. As one softener tank regenerates, the other tank is used to soften water for use. Other water softener systems use a storage tank to supply soft water while the softener regenerates. The regeneration of the water softener takes place in three steps. The timer control on the softener is set to control the duration of each step. As each step completes, the timer positions the stager to redirect the flow through the softener for the next step.

The first step is backwash. The water softener is removed from service and the automatic valve stager starts a flow of water upward through the resin beads. This reversed flow flushes out impurities and solids, and fluffs up the resin beads. This helps separate the resin beads and prepares them for the next step. Water and impurities pass out of the top of the water softener vessel and down the drain.

For the next step, brine is drawn from the brine tank and diluted with softened water so that the diluted solution contains about 10% to 12% salt. The diluted brine passes downward through the resin beads. Because of the very high concentration of sodium in the brine, the brine drives the calcium and magnesium ions from the surface of the resin and recoats the resin beads with sodium ions. The calcium and magnesium ions are flushed down the drain.

A two-stage rinse finishes the regeneration and prepares the softener for service again. First, a slow rinse of softened water at a low flow rate is directed downward through the bed, removing the majority of the unused brine. Then a fast rinse of softened water at a high flow rate removes the final traces of brine and packs the resin down. The sodium zeolite water softener is then ready to return to service.

Factoid

The great majority of modern water softeners work by ion exchange.

5-8

Figure 5-8. *Sodium zeolite water softeners use a brine solution and resin beads to soften water.*

28 What process occurs in a hot process water softener?

A *hot process water softener* is a pressure vessel that uses steam to heat makeup water by direct contact and uses chemical injection to precipitate the hardness out of the water. **See Figure 5-9.**

The calcium and magnesium hardness and any dissolved oxygen in the water become less soluble as the steam heats the makeup water. Oxygen released from the makeup water is vented to the atmosphere. Thus the hot process water softener also serves to at least partially deaerate the makeup water. The pressure maintained in a hot process water softener is usually about 3 psig to 10 psig, and the temperature of the water is generally between about 218°F and 239°F.

Lime is used to make the hardness ions precipitate from the water as a sludge that can be removed from the water softener. The sludge settles to the bottom of the softener and is removed by blowing down the softener tank. A portion of the sludge is continually recirculated through the main vessel by a sludge recirculation pump. This ensures maximum contact of the hardness ions with the lime to increase the effectiveness of hardness removal.

Lime is injected in proportion to the flow rate of the water entering the water softener. Coagulant chemicals are added to help precipitate the hardness. Soda ash may also be added to help remove hardness.

Hot process water softeners were common in large industries as the primary means of hardness removal for many years. Resin-type ion-exchange water softeners have replaced hot process water softeners to a large degree. However, in industries that require a very large amount of makeup water for the boilers, hot process water softeners are sometimes used to remove the majority of the hardness in the water before softening in an ion-exchange water softener. Air released from the water is vented through a vent condenser on top of the water softener. The softened water leaving the unit is filtered through anthracite coal filters to remove remaining suspended sludge before the water is used.

29 What is the purpose of a vacuum breaker valve on a hot process water softener?

Hot process water softeners are designed to withstand internal pressure. An internal vacuum can cause the vessel to implode. When the steam supply to a hot process water softener is lost, any remaining steam condenses in the vessel. This can result in the formation of a vacuum in the vessel. A vacuum breaker opens and allows air into the vessel to relieve the vacuum.

30 What is a condensate polisher?

A *condensate polisher* is an ion-exchange water softener similar to a sodium zeolite water softener but with a resin that can withstand the high temperatures encountered with condensate. A condensate polisher takes out iron that returns with condensate. It also takes out hardness that may return in the condensate because of leaking heat exchangers.

Condensate polishers used for utility boiler systems (over about 1500 psig) may be very sophisticated units that also remove silica that could deposit on steam turbine blades.

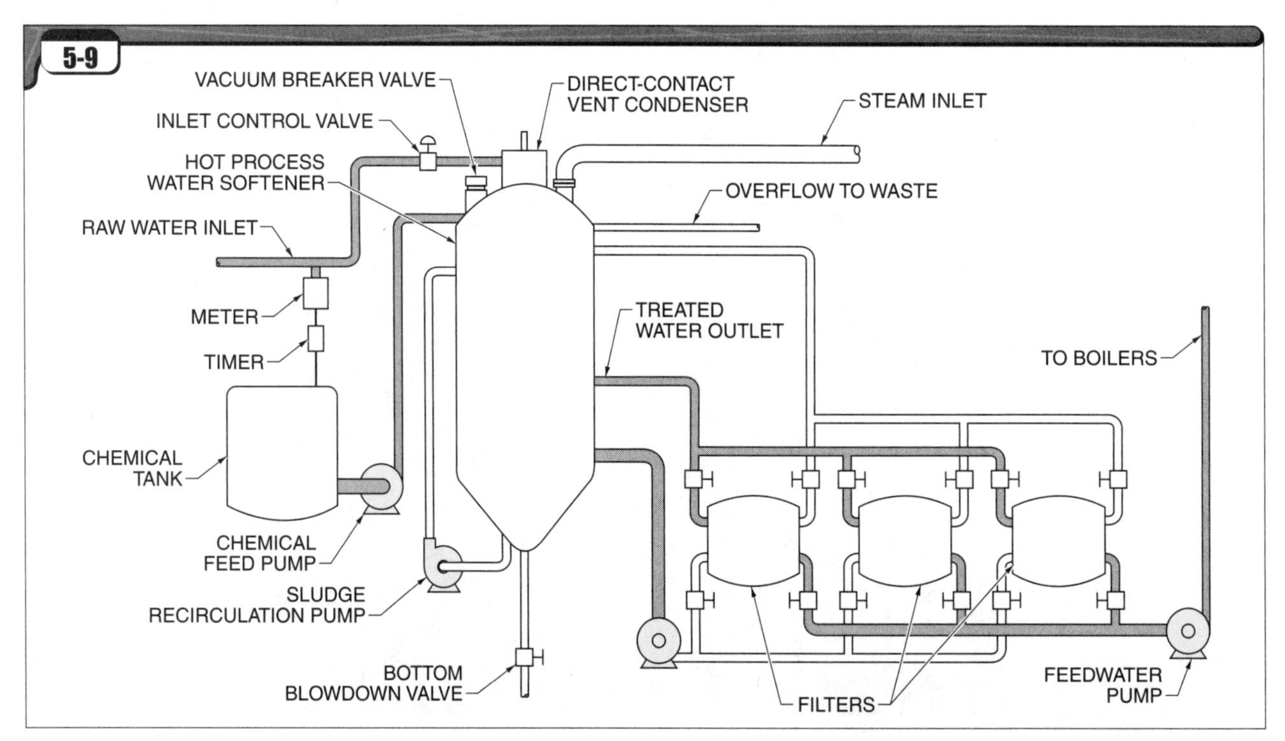

Figure 5-9. *Hot process water softeners use steam to heat makeup water so that impurities in the water become less soluble.*

31 How is the small amount of hardness treated that remains in the water after softening?

A number of chemical products are used to prevent the small remaining amount of hardness from adhering and becoming scale in the boiler. The particular combination of treatment chemicals used depends on the types and quantities of scale-forming impurities in the makeup water, the boiler configuration, the pressure and temperature of the boiler water, and other factors.

Most water treatment specialists prescribe scale control agents that are based on one of four general families of chemicals. The families of chemicals are phosphates, chelants, polymer dispersants, and phosphonates.

A *phosphate* is a chemical that causes hardness in boiler water to precipitate and settle out as a heavy sludge. The sludge tends to settle to the bottom of the boiler and is usually removed through bottom blowdown. Phosphates are used in several forms, including disodium phosphate and trisodium phosphate. Phosphates are usually fed in large enough quantity to maintain a small residual amount in the boiler water.

Boiler operators should be alert to softener problems when using phosphate treatments. If the softener allows significant amounts of hardness to pass through, a large amount of sludge can be formed in the boilers. In this scenario, the boiler operator needs to blow down the boiler more frequently.

A *chelant* is a chemical that helps keep the hardness in the water dissolved so that it does not crystallize on heating surfaces. Chelants, or chelating agents, are known in the industry as solubilizing agents. The impurities can then be removed by the continuous blowdown system. Chelants are based on a natural acid called ethylenediamine tetraacetic acid (EDTA).

Chelants can become highly corrosive to the boiler steel in the presence of oxygen. Therefore, chelants are only used in steam plants that have good deaeration. In addition, chelants are used only where the softening equipment produces water with nearly zero hardness. This is because chelants are very expensive. Because chelants are difficult to test for directly, they are fed in response to the amount of hardness detected in the feedwater.

A *polymer dispersant* is a synthetic compound that prevents scale deposits by dispersing the scale before it deposits on boiler surfaces. The polymers essentially coat the individual microscopic hardness crystals in the early stages of formation.

The polymer dispersant treatment provides a negative electrical charge on the surface of each crystal. Since charges of the same polarity ("like" charges) repel, the crystals repel each other and do not adhere to form larger crystals. The polymer dispersants apply the same surface charge to the heating surfaces in the boiler. The microscopic hardness crystals in the water are repelled from the heating surfaces as well.

Polymer dispersants also tend to prevent the spiny growth that is typical of a growing scale crystal, and this helps keep the crystals from interlocking and linking together. This attribute of the polymer dispersant treatments is known as "crystal distortion." Polymer dispersants are fed in response to the amount of hardness or iron detected in the feedwater.

Water treatment specialists frequently use polymer dispersants in conjunction with other scale control agents to take advantage of the desirable qualities of each. For example, when phosphates are used to cause the hardness in the boiler water to settle out in the boiler water, dispersants keep the settled sludge from compacting. The dispersants keep the sludge in a fluffy consistency so that it will go down the blowdown lines easily. For this reason, some water treatment specialists refer to polymer dispersants as sludge conditioners.

A *phosphonate* is an organic phosphate that provides multiple functions in water treatment. The most common phosphonate is hydroxyethylidene diphosphonic acid (HEDP). Phosphonates have phosphates as their root component, but also include other organic compounds in their molecular structure.

In many cases, phosphonate treatments offer an attractive blend of the attributes of the other scale control agents. For example, phosphonates provide some of the crystal distortion characteristics of polymer dispersants, so they help minimize the growth of scale crystals.

Phosphonates also have some solubilizing ability, although weaker than chelant treatments. In addition, they provide some precipitation characteristics to help larger, heavier solids in the boiler water settle out and be removed by blowdown.

However, phosphonates do not disperse hardness particles in the boiler water. Therefore, phosphonates are blended with polymer dispersants to help precipitated solids flow easily down the bottom blowdown line.

32 Why should iron be removed from makeup water?

Iron can deposit in a boiler at high temperatures and form iron scale. Iron is also referred to in the water treatment industry as a deposit binder. Iron helps other contaminants stick together and form deposits inside a boiler.

33 What is an iron filter and how does it work?

An *iron filter* is a pressure vessel through which raw water flows on its way to the water softeners. Potassium permanganate ($KMnO_4$), an oxidizer, is injected into the water piping before the water enters the iron filter.

The potassium permanganate changes the ferric form of the iron that dissolves in water to the ferrous form that settles out of the water. After the iron settles out, it can then be filtered out of the flow by greensand in the bottom half of the filter. *Greensand* is a dark, coarse, sandy material. The filter is backwashed when the greensand begins to plug up from too much iron.

34 How are polymer dispersants used in controlling problems related to iron in boiler water?

Polymer dispersants are used with chelants, phosphates, and phosphonates to optimize scale control. The same polymer dispersants are used for iron control. Polymer dispersants surround the iron particles in the water and coat the boiler heating surfaces with a large number of negative charges. When the particles come near each other and near the heating surfaces, they are repelled because they have like charges. This prevents the particles from depositing on the heating surfaces.

35 What is an evaporator?

An *evaporator* is a set of heat exchangers that produces water suitable for boiler use from water that contains large quantities of impurities. An evaporator heats contaminated water in the first heat exchanger to generate steam. When the steam is driven off, the solids in the water are left behind. The steam is then condensed in the second heat exchanger. This condensed steam is distilled water.

Evaporators often operate under a vacuum so that less energy is used and a lower boiling temperature is possible. Evaporators are most commonly used on ships where the quantity of fresh water available for boiler use is limited.

Acidic Corrosion

36 What is pH of water?

In the context of boiler systems, pH provides an index of how likely the water is to cause corrosion and other damage to steel and iron. Water always has a very small concentration of free ions present because the water molecules dissociate into hydrogen ions (H^+) and hydroxide ions (OH^-). The pH of a solution is a value on a scale that represents the relative available amounts of these two types of ions. The pH scale is 0 to 14 with 0 being the most acidic and 14 the most alkaline. At a pH of 7, the concentrations of the H^+ ions and the OH^- ions are the same and the solution is balanced, or neutral. **See Figure 5-10.**

The presence of other chemicals dissolved in water can change the relative proportions of the H^+ and the OH^- ions. If there are more H^+ ions than OH^- ions, the solution is acidic. If there are fewer H^+ ions than OH^- ions, the solution is alkaline, or basic. A simple rule to remember is that a low pH is acidic and has a high concentration of H^+ ions, and a high pH is alkaline and has a low concentration of H^+ ions.

37 What is caustic embrittlement?

Boiler water that has a concentration of H^+ that is too low (concentration of OH^- ions that is too high) may cause a problem known as caustic embrittlement. *Caustic embrittlement* is a problem in which boiler metal becomes brittle and weak because of cracks in the crystalline structure of the metal (crystalline cracking). Boiler water that has a very high pH (over about 13) for a long period of time can cause caustic embrittlement of boiler metal. The metal is mainly prone to caustic embrittlement in areas of high stress, such as welds and rolled tube ends. Caustic embrittlement can lead to failure of the metal by cracking and has resulted in boiler explosions. Caustic embrittlement is sometimes known as stress corrosion cracking. It is rare today as a result of advances in boiler water treatment technologies.

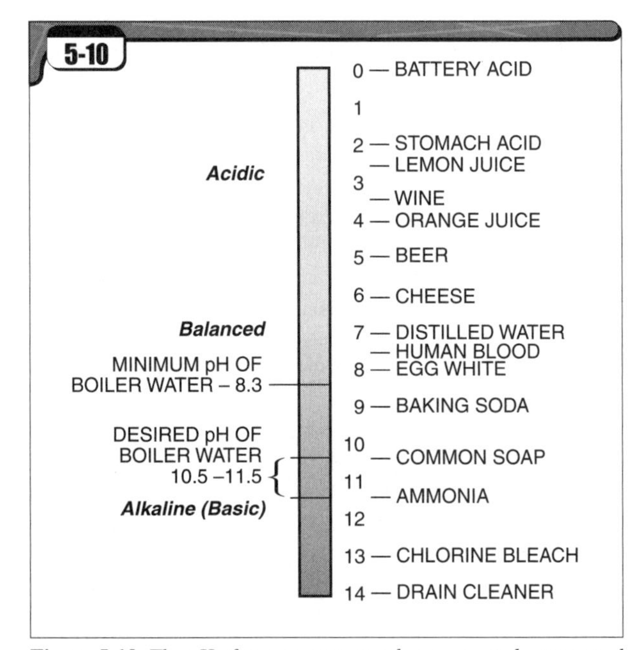

Figure 5-10. *The pH of many common substances can be measured on a scale ranging from 0 to 14.*

38 What is the acceptable pH range of boiler water?

The H^+ ions in water can often cause corrosion of metal that is in contact with the water. The minimum acceptable pH of boiler water is 8.3. **See Figure 5-11.** If the pH of the boiler water is lower than 8.3, significant acidic corrosion can occur. The acidity in water with a pH of 8.3 or above is high enough that there is very little or no acidic corrosion. Acidic boiler water corrodes boiler metal in such a way that the metal is dissolved, or eaten away. Boiler metal plate thickness is reduced by corrosion and cannot be restored.

Alkaline boiler water can be more conducive to scale formation, foaming, and carryover. However, these conditions can be controlled through proper water treatment techniques. The lower corrosion rate associated with alkaline boiler water more than justifies these potential problems. The pH specifications for most steam boilers are therefore set between approximately 10.5 and 11.5. In addition, a pH in this range facilitates desirable reactions of the chemicals used in treating the boiler water.

39 How is the pH measured in a water-based solution?

The pH of a water-based solution is most simply measured with litmus paper, which changes color depending on the pH of the solution. The color of the litmus paper is then compared to a graduated color chart. **See Figure 5-12.** The pH reading is listed next to the matching color.

The pH may also be read with an instrument that uses an electrode probe and provides readouts on a meter or digital display. A pH meter is more accurate than litmus paper but it requires the purchase of a pH meter and electrode probe as well as calibration solutions.

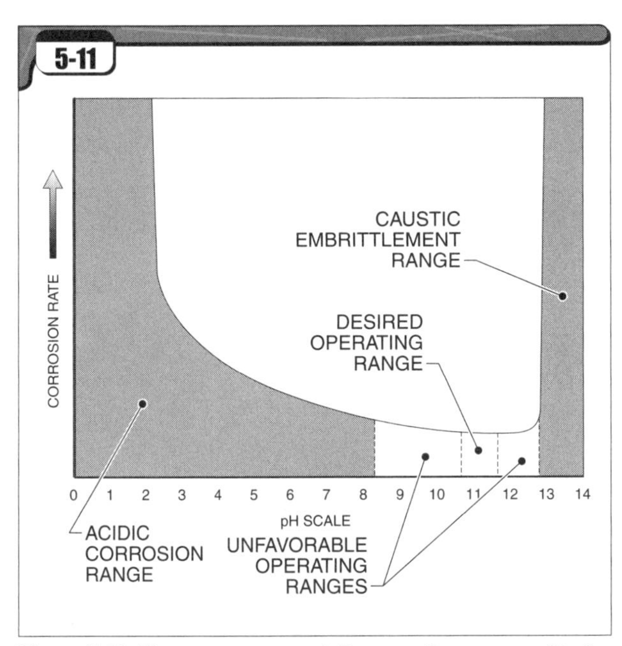

Figure 5-11. *Water treatment specialists usually recommend boiler water pH of 10.5 to 11.5 to minimize acidic corrosion while also minimizing caustic embrittlement.*

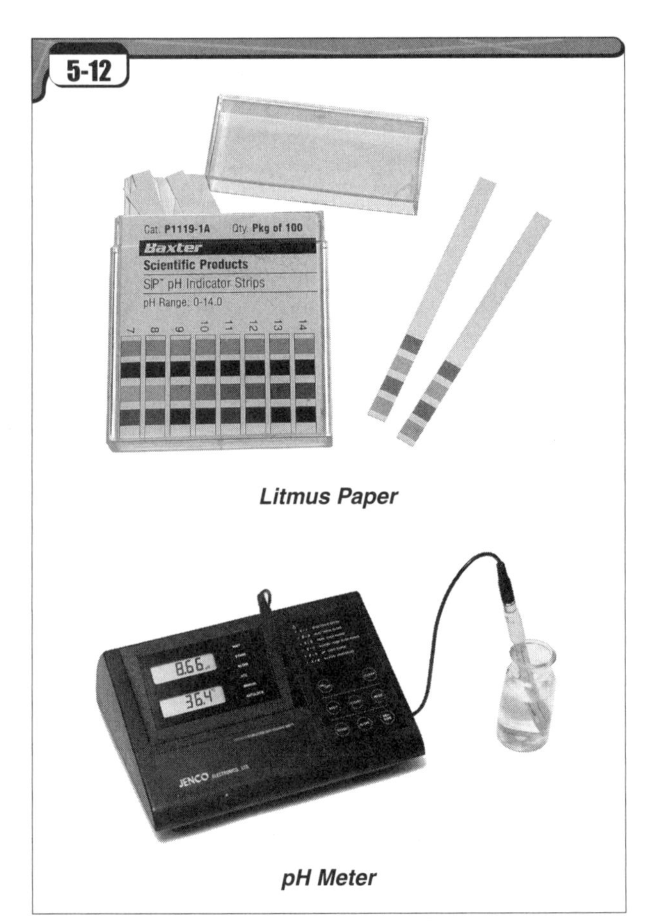

Figure 5-12. *The pH of boiler water can be measured with litmus paper or with a pH meter.*

40 How can boiler water pH be changed if it is too high or too low?

Caustic soda (sodium hydroxide) is used to adjust boiler water pH. Caustic soda solution has a pH of about 14. The pH of feedwater is around 6.5 to 7.5 prior to the addition of any caustic soda. Blowing down the boiler and adding feedwater containing no caustic soda therefore lowers boiler water pH. Increasing the amount of caustic soda fed to the boiler raises boiler water pH. The caustic soda is usually mixed with other treatment chemicals being fed to the boiler.

41 What is acidity?

The term "acidity" has two common meanings in the context of boiler water. The first meaning of acidity is the condition where water has a pH below 7 and the concentration of H^+ ions is greater than the concentration of OH^- ions. The second meaning of acidity is the presence of the H^+ ions themselves. Chemists refer to the acidity present in water in terms of chemical reactions that may occur due to the H^+ ions. This is the case even when the water is alkaline and the concentration of H^+ ions is less than the concentration of the OH^- ions.

42 What is alkalinity?

The term "alkalinity" can be quite confusing to the boiler operator because the term is used in different contexts when referring to water conditions in a boiler system.

The first meaning of "alkalinity" is the condition where water has a pH above 7 and the concentration of H^+ ions is less than the concentration of OH^- ions. When referring to the pH scale, alkalinity is used to represent the opposite of acidity. Therefore, alkaline is the opposite of acidic.

The second meaning of "alkalinity" is the presence of materials dissolved in water that make the water alkaline. The forms of alkalinity in a water supply that are of primary relevance to boiler operation are carbonates, bicarbonates, and hydroxides.

Carbonates and bicarbonates in the water make up almost all of the naturally occurring alkalinity. These most commonly result from contact of natural water with limestone on and below the Earth's surface. Natural water is slightly acidic, and slowly dissolves limestone. Limestone is calcium carbonate and calcium bicarbonate, and as it is dissolved, these solids mix with the water.

The quantity of carbonate and bicarbonate alkalinity in the makeup water is determined through the M alkalinity test. The M stands for the test chemicals commonly used in this test, methyl orange or methyl purple. The M alkalinity test is used to measure the amount of alkalinity in water with a pH of above 4.3. M alkalinity is also known as total alkalinity.

Another test chemical, phenolphthalein, is used to measure the quantity of alkalinity that exists with a pH of above 8.3. This test is known as the P alkalinity test. The P stands for phenolphthalein.

Hydroxide alkalinity is not naturally occuring in water supplies, but is formed as bicarbonate (HCO_3) alkalinity from limestone breaks down in the boiler under heat and pressure to form carbon dioxide and hydroxides. Hydroxide alkalinity is often referred to as OH alkalinity, or simply O alkalinity. Hydroxide alkalinity is a desirable form of alkalinity because it helps minimize the water's corrosivity to steel. It also helps facilitate desirable chemical reactions of the water treatment chemicals. For example, in the presence of hydroxide alkalinity, scale-forming calcium may be more readily precipitated as a nonadhering sludge. If the amount of hydroxide alkalinity formed in the boiler water is less than optimum, boiler operators often add small quantities of sodium hydroxide to the boiler water as a means of boosting the hydroxide alkalinity into the desired range. Sodium hydroxide (NaOH) is commonly known as caustic soda.

It is desirable for the pH of the boiler water to be about 10.5 to 11.5. The amount of hydroxide alkalinity in the boiler water is often used as an indirect way of confirming that the pH of the boiler water is in the desired range. A residual of 300 ppm to 600 ppm of hydroxide alkalinity in the boiler water indirectly proves this pH range, because unless the pH is in this range, this residual cannot exist.

Once the quantities of alkalinity from the M alkalinity and P alkalinity tests are determined, the quantity of OH alkalinity may be determined by applying the formula $2P - M = OH$.

43 What problems are caused by high carbonate and bicarbonate alkalinity of boiler water?

High alkalinity from carbonates and bicarbonates in boiler water contributes to the formation of carbon dioxide (CO_2). The CO_2 is released from the boiler as a gas when the alkalinity decomposes in the boiler under the high temperature and pressure conditions. The CO_2 passes out of the boiler with the steam. When the steam condenses, the cooling condensate absorbs the CO_2. The combination of condensate and CO_2 forms carbonic acid, which attacks the condensate return lines. **See Figure 5-13.**

Very high alkalinity also causes foaming of the boiler water. Foaming makes it difficult to determine the water level and may lead to carryover of boiler water into the steam piping. High alkalinity also creates conditions that are conducive to scale formation in the boiler. Municipal water treatment plants generally remove most of the carbonate and bicarbonate alkalinity before the water is supplied to commercial and industrial customers.

In cases where a substantial amount of this alkalinity still exists and would create problems in the boilers, a dealkalizer or other mechanical method is used to remove this alkalinity from the makeup water. If a lesser amount is present, many water treatment specialists prescribe the use of an antifoam agent to reduce moderate foaming of the boiler water. Antifoam agents usually consist of a polyglycol solution that reduces the surface tension of the boiler water. Finally, the buildup of carbonate and bicarbonate alkalinity in the boilers is managed by draining off (blowing down) a portion of the concentrated boiler water and replacing it with makeup water containing a much lower concentration of alkalinity.

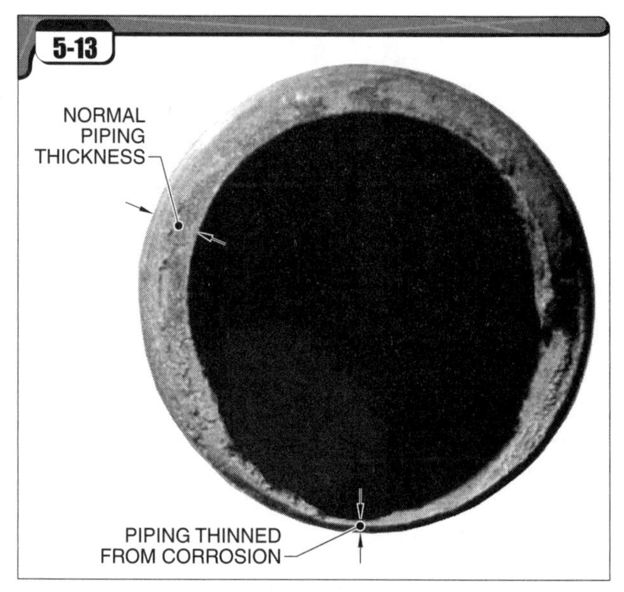

Figure 5-13. *Condensate and carbon dioxide combine to form carbonic acid, which corrodes condensate piping.*

44 What is a dealkalizer?

A *dealkalizer* is an ion-exchange unit that works in a manner similar to a sodium zeolite water softener, but exchanges anions rather than cations. Anions such as carbonates and bicarbonates are removed by ion exchange and replaced with chloride from the salt brine used in dealkalizer regeneration. The chloride does not contribute to CO_2 formation or to conditions that are conducive to scale formation.

45 What are amines? Name and explain the function of two classes of amines.

An *amine* is a chemical that prevents corrosion in condensate and steam piping. The primary purpose of amine treatments is to counteract the effect of carbonic acid in the condensate piping. The two common classes of amines are neutralizing amines and filming amines.

A *neutralizing amine* is a chemical that neutralizes the pH of the condensate. Neutralizing amines are continually injected in small amounts into a boiler while the boiler is operating. The amine vaporizes and passes out of the boiler with the steam.

When the amine condenses, it blends and reacts with the condensate. The amine raises the pH of the condensate, which prevents the corrosion caused by carbonic acid. Common neutralizing amines include morpholine, cyclohexylamine and diethylaminoethanol.

A *filming amine* is a chemical that prevents corrosion of piping by providing a protective barrier. Filming amines are continually injected in small amounts directly into the steam. Common locations for the injection point are in the main steam line between the boiler and the steam header or in the header itself. The vaporized filming amine flows with the steam to the steam-using equipment. When the filming amine condenses, it forms a protective film on the walls of the condensate piping that prevents the condensate from corroding the steel piping. The result is a coating that acts like automobile wax. It prevents the condensate from coming into direct contact with the steel piping. The most common filming amine is octadecylamine.

A given boiler plant may use more than one type of amine treatment. Filming amines and neutralizing amines may both be used in one system. Water treatment specialists may blend the various amines to provide better protection for a specific system. For example, morpholine tends to drop out of the steam system sooner than cyclohexylamine. Therefore, the water treatment specialist may use morpholine to protect parts of the system closer to the boiler room and cyclohexylamine to protect more remote areas.

Sample coupons are often installed in the condensate system at several locations so that the water treatment specialist may monitor the distribution of amines and the resulting corrosion rate of the piping. A *sample coupon* is a small, flat strip of steel that is inserted into an elbow or tee fitting in a piping network. The weight and dimensions of the sample coupon are logged when the coupon is installed, and then checked at intervals. A loss of weight or size of the coupon indicates corrosion occurring in the system, and the amount of weight or dimensional loss helps determine the corrosion rate.

Boiler operators monitor the pH of the returned condensate as an indicator of the presence of carbonic acid. The feed of amines is adjusted in response to the condensate pH.

Combined Approaches

What is a demineralizer?

A *demineralizer* is a highly efficient ion-exchange process generally used for high-pressure boilers. A demineralizer looks similar to a pair of sodium zeolite water softeners arranged in series. A demineralizer is also known as a deionizer.

The first vessel is a softening process that removes calcium, magnesium, iron, and other cations from the makeup water. The resin in this vessel is regenerated with sulfuric acid instead of salt, and hydrogen ions (H) from the sulfuric acid are exchanged for the calcium, magnesium, and other cations.

The second vessel is a dealkalization process that removes carbonates, bicarbonates, sulfates, and other anions from the makeup water. The resin in this vessel is regenerated with caustic soda instead of salt, and hydroxide ions from the caustic soda are exchanged for the carbonates, bicarbonates, sulfates, and other anions.

Because a demineralizer replaces all the cations in the inlet water with hydrogen (H^+) ions and all the anions with hydroxide (OH^-) ions, the H^+ and OH^- ions combine in the effluent to form water.

What is reverse osmosis?

Reverse osmosis (RO) is a water-purification process in which the water to be treated is pressurized and applied against the surface of a semipermeable membrane. The membrane material contains billions of extremely tiny pores and functions much like the wall of a human cell. The water molecule is one of the smallest molecules, while water contaminants of various kinds are generally larger molecules. When applied against the surface of the membrane under pressure, pure water will be forced through the membrane while contaminants in the water will not. Thus, the membrane essentially serves as a filter with microscopic pores.

The pores in the membrane are very small and many of the pores plug with contaminants. Therefore, a very large membrane surface area is needed in order to pass a reasonable amount of water. For this reason, the membranes are supplied in a rolled fashion, similar to a rolled-up rug. Passageways between the layers of the rolled membrane allow untreated water to have access to the membrane. Many of these rolls or membrane elements are used to achieve the volume of treated water needed.

RO systems are usually only used as a pretreatment process for very high pressure and utility boilers that require extremely high quality makeup water. For example, an RO system may be used instead of a demineralizer. However, RO systems are frequently used to prepare boiler makeup water for industrial boilers when a manufacturing process such as pharmaceutical products or foods requires water of extreme purity. In such plants, the RO system is sized so that a portion of the purified water may be used for boiler makeup water.

Water Quality Monitoring

Why is a sample cooler necessary when gathering condensate or steam for testing?

A *sample cooler* is a small, closed heat exchanger that cools condensate or other hot water to a temperature below about 130°F to 140°F before the water emerges from the cooler and into a sample container. **See Figure 5-14.** This prevents part of the sample from flashing to steam. If part of the sample flashes off, the impurities in the water will remain. Therefore, the water that remains in the sample container is more concentrated in impurities than the liquid in the actual process and is not a representative sample. A *representative sample* is a sample that is exactly the same as the item being tested. If a steam sample is required, a sample cooler condenses it so that the steam may be tested for impurities as a liquid.

Sample coolers are also important for safety reasons. Sample coolers allow the boiler operator to obtain condensate, feedwater, and boiler water samples without the risk of steam burns.

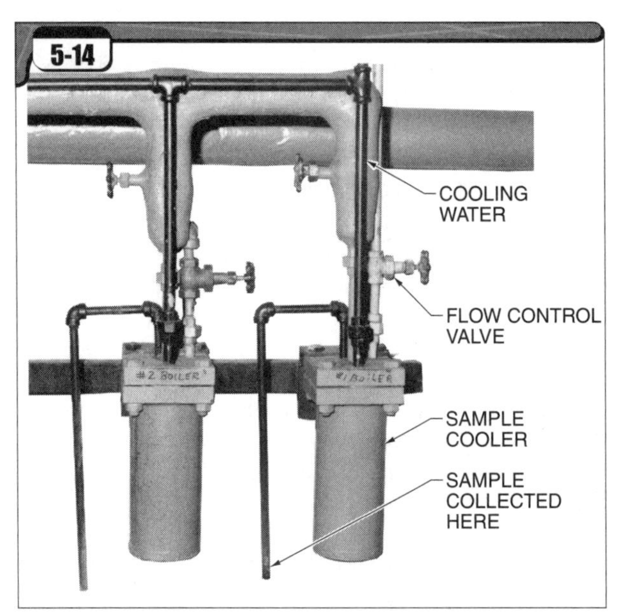

Figure 5-14. *Sample coolers cool hot water to below the boiling point to prevent part of the sample from flashing to steam.*

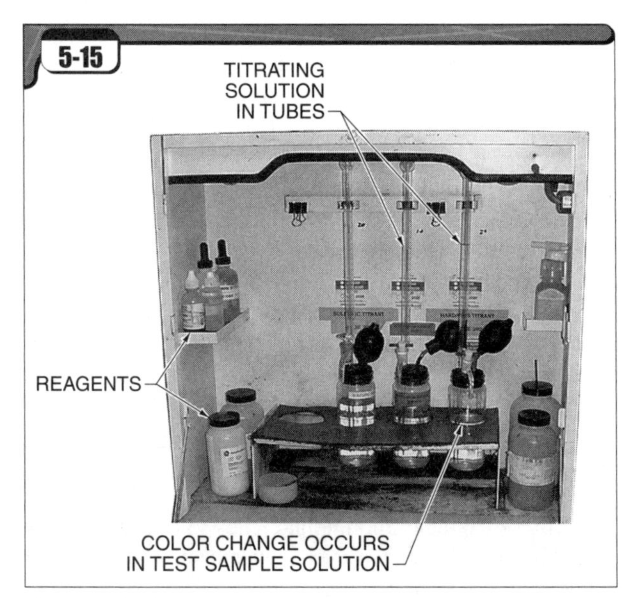

Figure 5-15. *Titration tests use reagents to measure the concentration of a specific dissolved substance.*

What is a reagent?

A *reagent* is a chemical used in a water treatment test to show the presence of a specific substance, such as hardness. Reagents cause a color change or facilitate a color change by another chemical known as an indicator. For example, a titration test uses reagents. **See Figure 5-15.** A *titration test* is a test that determines the concentration of a specific substance dissolved in water. A small, predetermined amount of a reagent is added to a specified volume of a test sample. For example, the test may call for 10 drops of the reagent to be added to a 50 ml sample. The sample is then titrated. This means that titrating solution is gradually added, one drop at a time, until a specific color change occurs in the test sample. The chemicals used to make the titrating solution may also be called reagents.

In a test for hardness, the test sample turns pinkish-red if there is any hardness present in the sample. A titrating solution is gradually added until the sample turns to a sky blue. The amount of titrating solution used is converted, using a multiplier, to a value that indicates the amount of the specific impurity in the sample. For example, the amount of titrating solution used in a hardness test may be converted to parts per million of hardness content of the sample by multiplying by 20.

An older test for hardness used a standard soap solution. Since hard water interferes with development of lather, the quantity of standard soap solution used before lather would develop was used as a means of determining the amount of hardness in the water. Titration testing is much more precise.

Reagents are also used in several other water tests as well. For example, the reagents used in a phosphate test cause the boiler water sample to turn a shade of blue, and that shade is matched with a chart showing many progressive shades of blue. The particular shade indicates the phosphate residual in ppm.

What do the letters TDS represent?

Total dissolved solids (TDS) is a measurement of the concentration of impurities in boiler water. TDS is used as a very general statement about the overall quality of the water. A TDS test does not identify specific impurities in the water. It only identifies the quantity of all dissolved solid impurities. Because the TDS test is somewhat prone to errors, the conductivity of the water is often used as a substitute measure of the relative purity of the water.

What is conductivity of water?

Conductivity is the ability of a conductor to allow current flow. *Conductance* is the quantitative measure of the ability of an electric circuit to allow current flow. For boiler water, the water sample is the conductor. Pure water with no impurities is a poor conductor of electricity. Therefore, the conductance of pure water is very low. The conductivity increases as the concentration of impurities, such as TDS, increases. Therefore, by determining how well the water conducts an electrical current, the concentration of impurities in the water can be determined. The conductance of the water does not indicate which specific impurities are contained in the water, only the relative amount of impurities. The conductance measurement is made between the poles in a conductivity meter. **See Figure 5-16.** The reading obtained indicates how well the water conducts the electrical current in a straight line between the poles.

ASME International and the American Boiler Manufacturers Association (ABMA) have guidelines that show the amount of conductivity that boilers operating at certain pressures and temperatures can typically tolerate before scale and other problems are likely to develop. In addition, water treatment specialists recommend acceptable conductivity levels in specific boilers based on actual water conditions. Scale, carryover, foaming, or other problems may develop if the water conductivity in a boiler is too high. Blowing down the boiler and adding fresh feedwater reduces the conductivity of the boiler water. However, water, heat, and expensive treatment chemicals may be wasted because of excessive blowdown.

In addition, boiler operators may periodically test the conductivity before and after a softener. The conductivity should be very nearly the same on both sides since the total charges removed from the water in the softener are the same as the total charges replaced. If the conductivity after the softener is higher than the conductivity before the softener, this probably means that the excess brine left after regenerating the softener is not being thoroughly rinsed out.

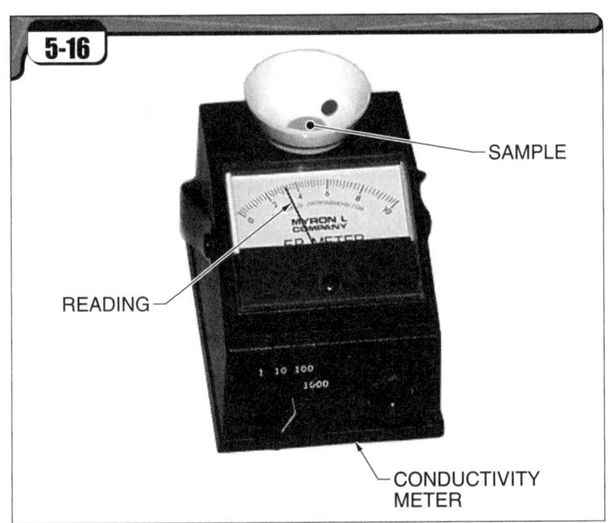

Figure 5-16. *A conductivity meter measures the ability of the water sample to conduct electrical current as an indirect method of measuring TDS.*

52 What is the unit of measure of electrical conductance in water?

Resistance is the measure of the ability of an electric circuit to oppose current flow. Resistance is measured in ohms (Ω). Since conductance is the opposite of resistance, people started using a unit called the mho (ohm spelled backward) as the unit of conductance. Since the actual conductance of boiler water is very low, the unit micromhos is used. A micromho is one-millionth (1/1,000,000) of a mho.

Factoid

The modern unit of measure of conductance is the siemens and the use of the term mho is being discouraged. However, common usage in boiler plants is still to use the term mho and conductivity meters still use the unit micromho.

53 How is the return condensate percentage in feedwater determined?

The return condensate percentage in feedwater is determined by applying the following formula:

$$RC = \frac{MC - FC}{MC - CC} \times 100$$

where

RC = return condensate percentage (in %)

MC = makeup conductivity

FC = feedwater conductivity

CC = condensate conductivity

100 = conversion constant to percent

This calculation determines the percentage of condensate return only at the time the particular samples are drawn. Therefore, the calculation must be run frequently over a period of time to determine a good average. Also, a composite (blended) sample of all condensate streams is needed for accurate results. For example, the condensate sample should be taken from a main condensate tank to which all satellite condensate receivers return condensate, or from the main condensate line just before the deaerator.

54 What is the return condensate percentage in feedwater if the makeup conductivity is 880 micromhos, the feedwater conductivity is 180 micromhos, and the condensate conductivity is 66 micromhos?

$$RC = \frac{MC - FC}{MC - CC} \times 100$$

$$RC = \frac{880 - 180}{880 - 66} \times 100$$

$$RC = \frac{700}{814} \times 100$$

$$RC = \mathbf{86\%}$$

55 What are the cycles of concentration?

Cycles of concentration measure the concentration of the solids in the boiler water compared to the concentration of the solids in the feedwater. As water in a boiler evaporates, solid impurities (dissolved solids and suspended solids) in the water are left in the boiler. Therefore, when large amounts of water are evaporated into steam, large amounts of impurities are left behind. For example, if the concentration of impurities in the boiler water is 20 times more than the concentration of impurities in the feedwater, the boiler is operating at 20 cycles of concentration.

Determining the cycles of concentration is a useful tool for the boiler operator and the water treatment specialist. For example, the water supply can be analyzed to identify which impurities in the water are likely to be the most troublesome.

A water treatment specialist can recommend the maximum cycles of concentration based on the maximum acceptable quantity of that impurity in the boiler water and the normal quantity found in the feedwater.

The boiler operator can use cycles of concentration to make equipment adjustments and changes to blowdown procedures. Cycles of concentration may also be a useful index in gauging effectiveness of improvement projects. For example, improvements in the condensate return system may result in more condensate being returned to the boiler room. The conductivity of the condensate is lower than the conductivity of makeup water. Therefore, an increase in the returned condensate results in lower conductivity of the feedwater. This allows the boiler to operate at higher cycles of concentration and reduced blowdown.

56 What methods are commonly used to determine the amount of dissolved iron in water?

A *color wheel comparator test* is a relatively simple test used to determine the quantity of iron in the condensate or makeup water. **See Figure 5-17.** The color wheel comparator test uses a graduated color wheel and viewer (comparator). In this test, a reagent is added to a sample of the water to be tested. Over a period of 2 min to 5 min, the reagent causes the sample to turn a yellow or orange tint if iron is present in the sample. The shade may vary from an almost undetectable yellow to a pronounced orange. The shade of the actual sample is matched to the tint on the color wheel and the corresponding amount of iron in the sample is read from the wheel. This method is somewhat subjective because of variations in the way people see color, but is sufficiently accurate for many boiler plant operations.

A *spectrophotometer* is an instrument that measures the ability of different frequencies of light to pass through a sample of liquid. Specific reagents are used to test for the presence of various impurities in a given liquid sample. The reagent used when testing for the specific impurity, such as dissolved iron, causes a color change in the sample. Because of the color change, certain frequencies in the light spectrum are partially blocked. The amount of change in the light spectrum of the sample compared to pure water is detected by the spectrophotometer and converted into a corresponding quantity of iron present in the sample.

57 What methods are used to feed water treatment chemicals into a boiler system?

Metering pumps are commonly used to feed a continuous, metered flow of various chemicals to the boiler, the deaerator, or the main steam outlet line. **See Figure 5-18.** In a typical metering pump application, a small reciprocating pump injects a very small amount of a water treatment chemical into the process fluid with each stroke. The stroke length, the number of strokes per minute, or both may be adjusted to control the amount of water treatment chemical injected.

The chemical is supplied in drums or in bulk. It is then mixed with condensate or softened water in a mixing tank by an agitator. The agitator ensures complete mixing of the solution.

Bypass chemical feed tanks are used in some boiler installations to feed water treatment chemicals into the system. They are most common with closed, low-pressure steam heating boilers and hot water boilers. A bypass chemical feed tank consists of a vessel into which the premixed water treatment chemical is poured. Valves in the vessel piping are changed so that the flow of water going to the boiler passes through the vessel, carrying the water treatment chemicals along with it. Bypass chemical feed tanks are often referred to as shot feeders or slug feeders, because the entire contents of the tank are carried into the boiler in a very short time.

The metering pump method is preferred because metering pumps keep concentrations of various water treatment chemicals in a boiler or other system at a more constant level. The chemical concentration increases when a bypass chemical feed tank releases the water treatment chemicals. The concentration is then allowed to dissipate. By maintaining a constant concentration of chemicals in a boiler, an operator may more closely control the results of the impurities in the water.

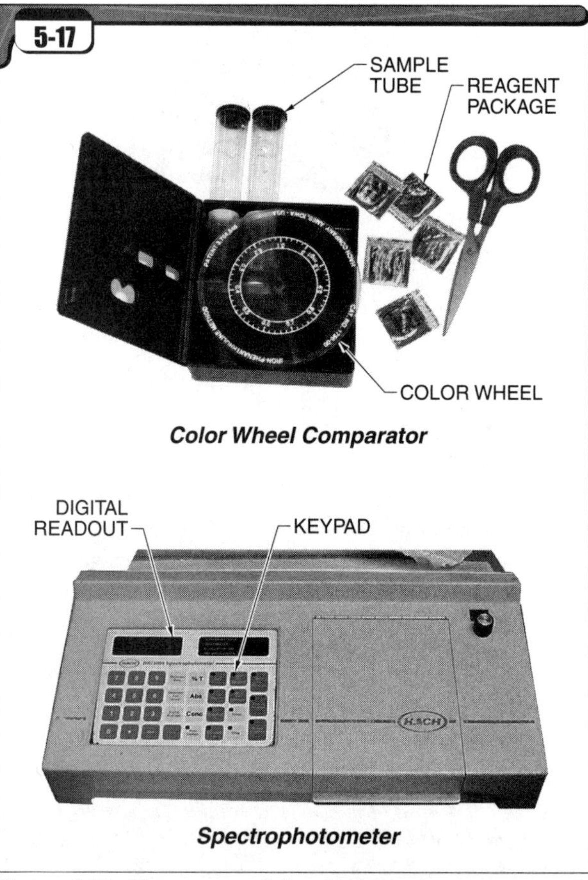

Figure 5-17. *The amount of dissolved iron in a water sample may be measured with a color wheel comparator or with a spectrophotometer.*

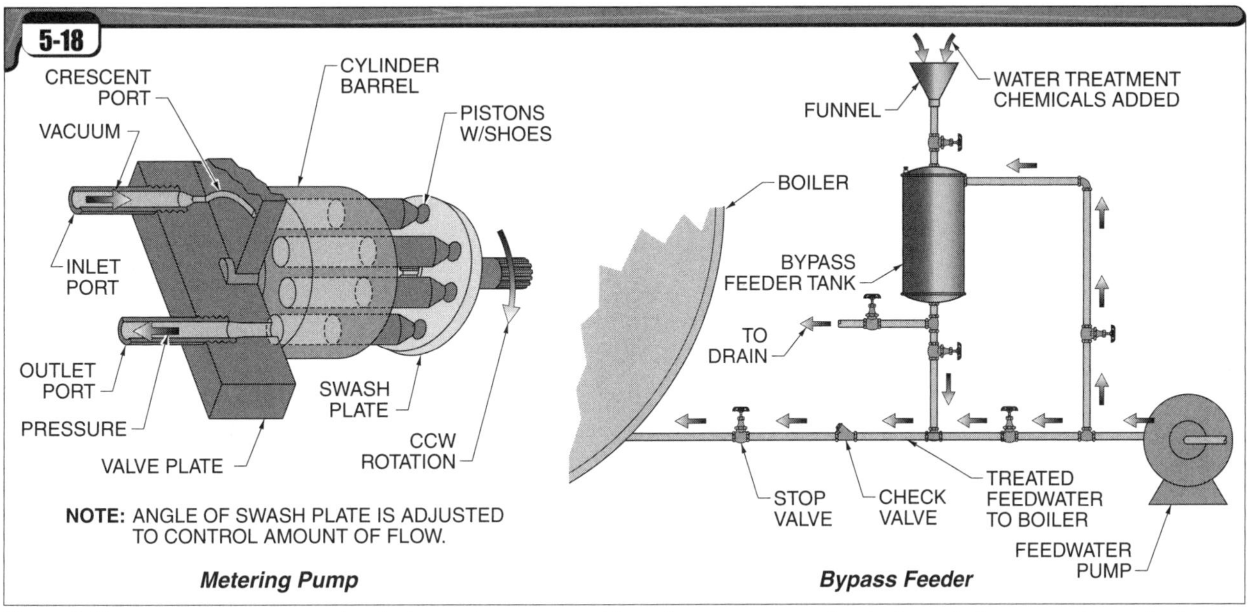

Figure 5-18. *Metering pumps are used to feed a continuous metered flow of water treatment chemicals to a boiler system. Bypass feeders are used to add chemicals all at once.*

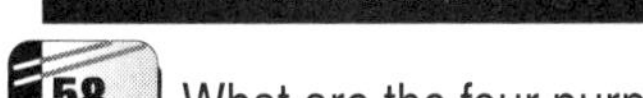

Boiler Blowdown

58 What are the four purposes of boiler blowdowns?

- Blowdown removes sludge and impurities from the boiler water.
- Blowdown is used to completely drain the boiler. This is also known as dumping the boiler.
- Blowdown is used to control the concentration of chemicals in the water. If the treatment chemicals become too heavily concentrated, blowing down the boiler and adding fresh feedwater can reduce the chemical concentration.
- Blowdown can be used to lower a high boiler water level.

59 What types of blowdown are commonly used?

Bottom blowdown is the process of periodically draining part of the boiler water to remove heavy sludge that settles to the bottom of the boiler. **See Figure 5-19.** The amount and frequency of bottom blowdown depend on the type and amount of impurities in the water, the type of water treatment program, and the practice that produces the best results. Bottom blowdown in modern boilers is not as necessary as in years past because of significant advances in pretreatment processes and chemical technologies.

Continuous blowdown is the process of continuously draining water from a boiler to control the quantity of impurities in the remaining water. Because impurities in the boiler water are left behind when the steam separates from the water, the greatest concentration of these impurities is a few inches below the NOWL. A continuous blowdown system uses a blowdown manifold that is located about 2″ to 6″ below the NOWL and runs most of the length of the boiler drum or shell. The manifold is a pipe, usually ¾″ or 1″, that is capped at the end and has small ¼″ holes drilled at spaced intervals along the length. This permits removal of impurities from the whole length of the drum or shell at once. Continuous blowdown is normally controlled with an automatic valve and conductivity controller or with a manual valve with a position-indicating scale on the side.

Surface blowdown is the process of intermittently removing water from a boiler to control the quantity of impurities in the remaining water or to remove a film of impurities on the water. The term "surface blowdown" is used in two contexts in modern boiler technology. In most cases, surface blowdown is exactly the same in configuration as continuous blowdown, but it is used intermittently rather than continuously. In some small packaged boilers, however, surface blowdown is located at the NOWL rather than a few inches below. This location helps remove the film of impurities that may float on the surface of the water inside a boiler. This type of surface blowdown is often referred to as a skimmer for this reason. This type of blowdown is also intermittent and manually controlled.

60 What factors affect the amount and frequency of blowdown?

The amount and frequency of blowdown depend on the specific impurities in the water, the quantity of those impurities, the chemicals used to treat the water, the boiler design, and the practices that produce the best results.

Factoid

Amines are used to protect condensate piping and steam piping from acidic corrosion.

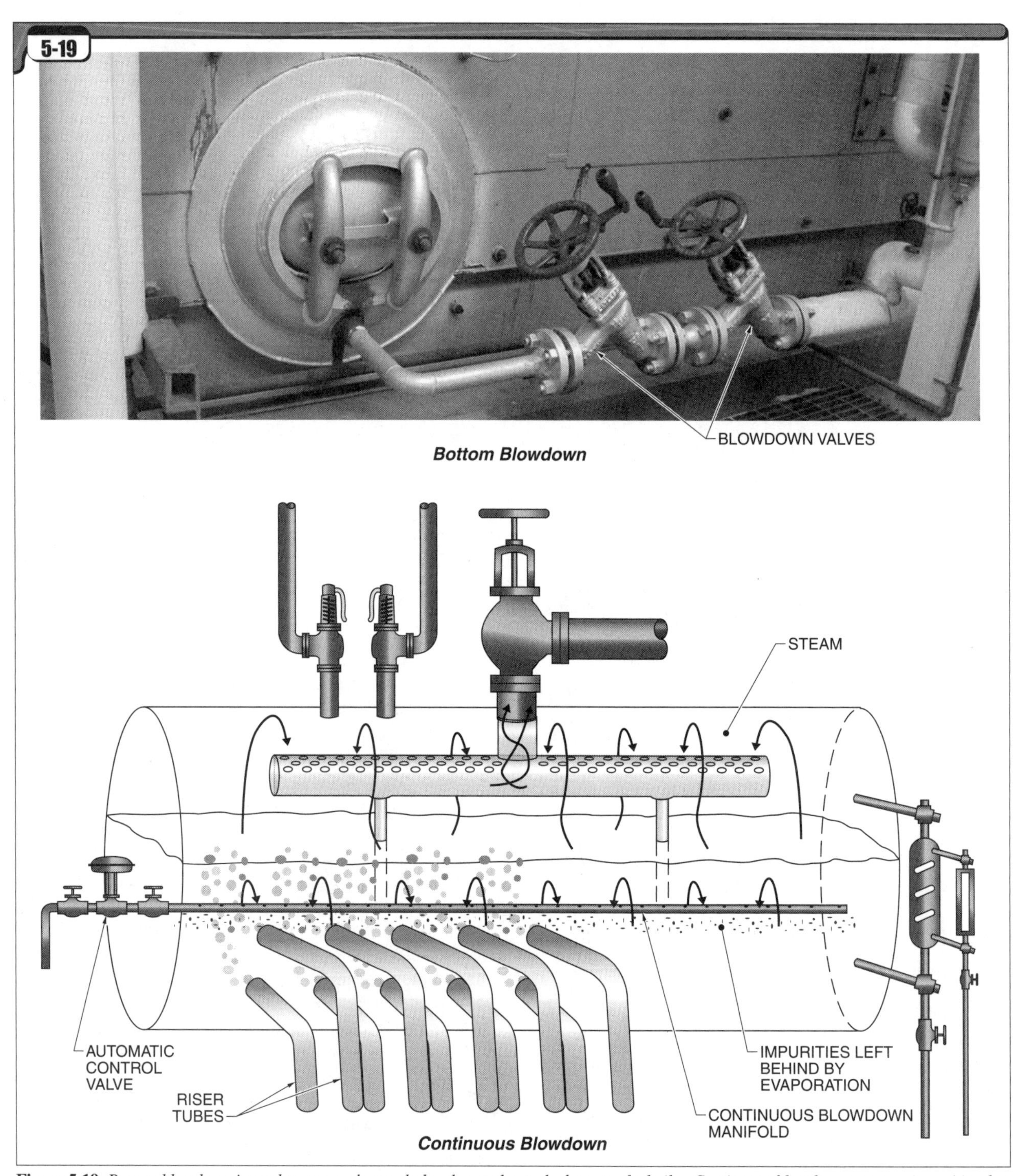

Figure 5-19. *Bottom blowdown is used to remove heavy sludge that settles to the bottom of a boiler. Continuous blowdown removes impurities that are left behind when boiler water evaporates into steam.*

What type of piping and valve arrangement is required for the bottom blowdown of a high-pressure boiler? How are these valves operated?

Two bottom blowdown valves in series must be provided when a high-pressure boiler operates at 100 psig or more. **See Figure 5-20.** A high-pressure boiler is usually equipped with a quick-opening gate valve and a slow-opening globe or plug valve. In some applications, two slow-opening valves are used. A *quick-opening valve* is a valve that requires only a 90° change in the position of a lever arm to move the valve from fully closed to fully open. The quick-opening valve is typically a lever-operated gate valve. This valve should be installed closest to the boiler and should be opened first and closed last. A *slow-opening valve* is a valve that requires five or more full turns of a handwheel to move the valve from fully closed to fully open.

The slow-opening valve is often configured in a wye (Y) pattern in which the disc or plug closes at a 45° angle. This valve configuration is less likely to hold debris and is less prone to erosion than a standard globe or plug valve design.

In general, the valve farthest from the boiler should be the one that is throttled open and closed (the slow-opening valve). This valve will incur the wear associated with blowing down the boiler. The valve closest to the boiler is intended to act as a sealing valve. If the slow-opening valve tends to leak because of erosion, it can be replaced without having to drain the boiler.

Slow-opening valves help to minimize thermal stress on the ambient temperature blowdown piping by controlling the flow of hot water into the piping. This allows the piping to heat up slowly and minimizes water hammer due to the water in the piping flashing to steam.

Bottom blowdown piping for boilers should be Schedule 80 steel or heavier, as appropriate for the pressure of the boiler. The blowdown piping must not be smaller than 1″ in diameter or larger than 2½″ in diameter. However, when the boiler has fewer than 100 sq ft of heating surface, the bottom blowdown line may be ¾″.

In some high-pressure installations, tandem valves are used. A *tandem valve* is a blowdown valve configuration with two valves in series machined into a common valve body. Tandem valves are constructed to handle high pressures. In contrast to most blowdown valve installations, the manufacturers of these valves recommend that the valve farthest from the boiler be opened first and closed last. This causes the inner valve to take the wear from the blowdown. There are several variations in the ASME Code requirements for very high pressure blowdown valves, and the code should be consulted for guidance in specific cases.

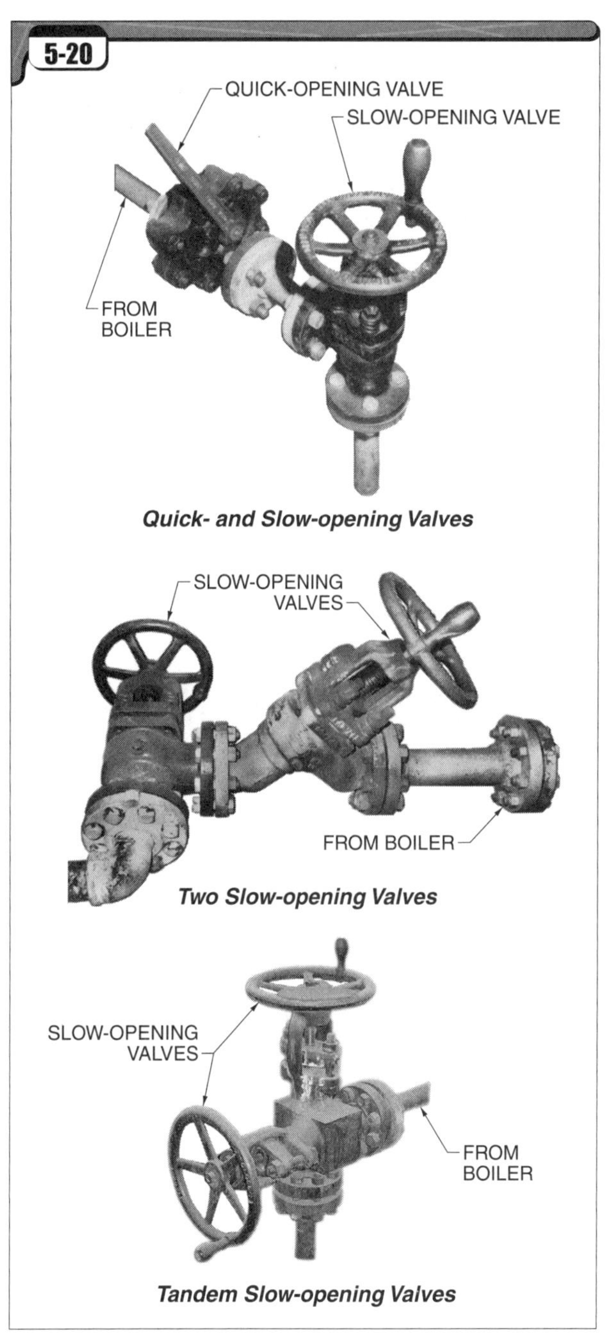

Figure 5-20. *Bottom blowdown valves for high-pressure boilers may consist of quick-opening and slow-opening valves or of two slow-opening valves. Very high pressure boilers often use tandem slow-opening valves.*

62 Is it more effective to perform a bottom blowdown during light-load periods or during heavy steaming?

Bottom blowdown is much more effective when it is used during light-load periods. Solids in a boiler settle to the bottom of the vessel more efficiently when there is less agitation of the water.

Continuous blowdown is more effective during heavy steaming. As the steam separates from the water, the impurities are left behind in the water. The continuous blowdown manifold is located just below where the steam and water separate. This is where the impurities are the most concentrated.

63 How is continuous blowdown controlled?

Continuous blowdown is either controlled by a manual valve or by an automatic analyzer/controller in conjunction with an automatic control valve. **See Figure 5-21.** The manual valve used for this purpose includes a graduated scale on the side that assists the boiler operator in gauging the valve opening.

When using automatic control, a conductivity probe continuously measures the conductivity of the water going through the blowdown line and sends this information to the controller. The controller compares the water conductivity reading received from the conductivity probe with the setpoint in the controller. A *setpoint* is the desired point at which an automatic controller maintains a variable condition within a process. The controller calculates the amount of change needed and sends an output signal to the automatic control valve. The automatic control valve opens or closes as required to bring the water conductivity reading back to the setpoint. A bypass line with an orifice is piped around the automatic control valve to maintain a small continuous flow across the sensing probe.

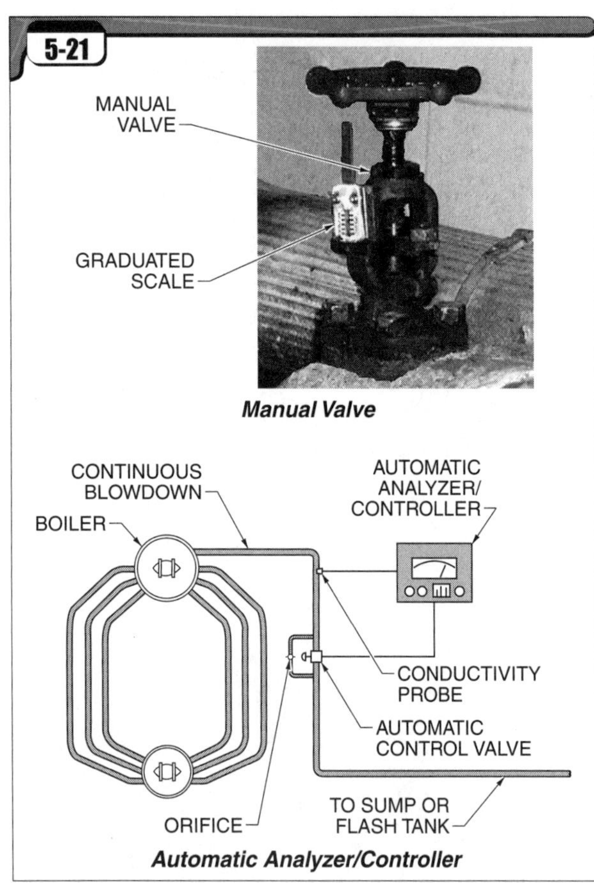

Figure 5-21. *Manual continuous blowdown is controlled with a hand valve with a position-indicating scale. Automatic continuous blowdown is controlled with an automatic valve and an automatic analyzer/controller.*

Factoid

In addition to impurities, expensive water treatment chemicals, heat, and water go down the drain during boiler blowdown. Therefore, the amount of boiler blowdown should be limited to only that amount necessary. However, too little blowdown allows impurities to build up to unacceptable levels.

64 Why should a sump or blowdown tank be provided before allowing blowdown water to enter the sewer?

A sump or blowdown tank provides a place where blowdown water can safely flash into steam. **See Figure 5-22.** For example, when a boiler operating at 100 psig is blown down, the water in the blowdown piping is approximately 338°F. When this water is released from the boiler at 100 psig to atmospheric pressure, a portion of it flashes into steam until the water temperature falls to 212°F.

If a sump or blowdown tank is not provided, the flash steam could pressurize the drain piping in the building. This could cause steam, hot water, and any other process fluid that may be connected to the drain system to flow backward and up through any drains. Additionally, the steam and hot water suddenly entering the piping could cause damage. If the drain piping is buried in the ground, this results in substantial time, expense, and difficulty in making repairs.

Blowdown tanks vary somewhat in configuration. Some blowdown tanks hold a volume of water and the incoming boiler blowdown causes this reservoir to overflow to the drain. The reservoir of water thus helps to raise the temperature of the water entering the sewer piping gradually. Other blowdown tanks drain directly to the sewer.

Local codes usually prohibit the introduction of water above about 140°F into city sewer lines. Most blowdown tanks also have a source of cooling water attached. Most commonly, this is city water that is controlled by an automatic temperature control valve. As the boiler is blown down into the blowdown tank, a temperature probe senses the high temperature of the blowdown and causes a control valve in a city water line to open and admit cool water into the blowdown tank drain to mix with and cool the blowdown water as it enters the sewer.

65 How is the heat of blowdowns typically recovered?

The continuous blowdown system is usually routed into a flash tank. The flash tank is a pressure vessel. **See Figure 5-23.** Because the continuous blowdown system flow is fairly continuous, the flow of flash steam that results is also fairly continuous. This flash steam, at some lower pressure than the boiler, can be piped to another process such as a deaerator or heating system for use. This steam is essentially free since the boiler must be blown down anyway. The use of this steam increases overall plant efficiency.

In many plants, the remaining blowdown water also passes through a closed shell-and-tube heat exchanger before going down the drain. In this heat exchanger, the remaining heat from the blowdown water is given up to the makeup water that is flowing to the deaerator. This process cools the blowdown water sufficiently to prevent damage to the sewer piping and recovers the heat from the blowdowns in heating the makeup water.

It is rarely economically feasible to recover heat from the bottom blowdown, because it only occurs for about a minute or less per day.

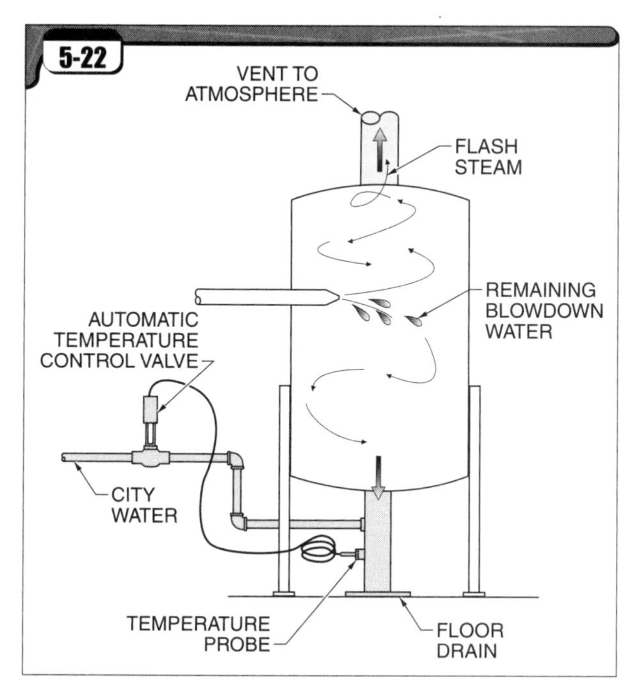

Figure 5-22. *Blowdown tanks serve to relieve flash steam to the atmosphere and cool blowdown water before the blowdown water enters the sewer piping.*

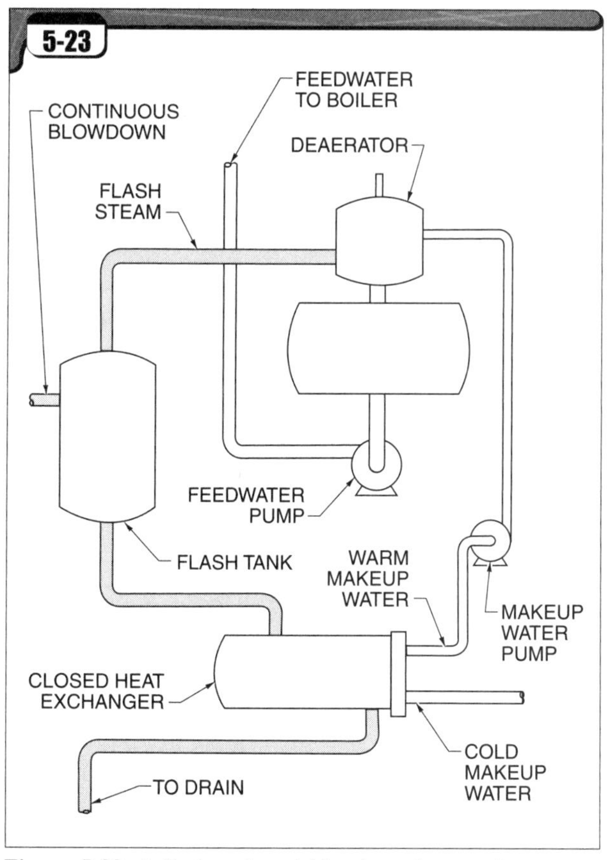

Figure 5-23. *A flash tank and blowdown heat exchanger are commonly used to recover heat from continuous blowdown.*

Boiler Protection during Outages

66 What is laying up a boiler for storage?

Laying up is the procedure used to take a boiler out of service for a longer than normal period of time. Extra care must be taken when putting a boiler in storage to prevent deterioration in the form of rusting, oxygen pitting, or acidic corrosion.

67 What are the two methods of laying up a boiler for storage?

The wet layup method is used for storage periods of up to four to six weeks, or during periods when the boiler may be needed for service on short notice. Corrosion is the main concern during the storage period. To prevent corrosion, the pH in the boiler water must be maintained at the desired alkaline value to avoid acidic conditions.

The boiler should be filled completely with deaerated water. This water should have extra oxygen scavenger added to it. The chemical will react with any remaining oxygen in the water to prevent oxygen corrosion. Sodium sulfite is commonly used as the scavenger, and a residual of 250 ppm should be maintained. In order to control the pH of the water, caustic soda should be added to maintain a concentration of approximately 400 ppm.

All outlets from the boiler must be closed to prevent loss of the treated water. When filling, water is allowed to overflow out of the drum vent to ensure that the boiler is completely full. By filling the boiler completely, corrosion is prevented in the upper part of the drum as well.

An alternate wet layup method uses a nitrogen blanket to prevent corrosion. **See Figure 5-24.** A nitrogen tank and regulator are attached to the drum vent or other suitable location and a 3 psig to 10 psig blanket of nitrogen is maintained in the space over the water in the boiler. Any leakage that occurs will be nitrogen leaking out and not oxygen leaking in. The best method of applying the nitrogen blanket is to connect the nitrogen supply to the boiler vent while there is still a slight steam pressure on the boiler. The nitrogen pressure regulator is set for about 3 psig to 10 psig, so that when the remaining steam in the boiler collapses and begins to form a vacuum, the nitrogen regulator opens and allows nitrogen to fill the space formerly occupied by the steam.

The nitrogen blanket method also allows for shrinking and swelling of the water with slight changes in temperature. The water conditions must be checked periodically and chemicals replenished as required to maintain the original specifications.

The dry layup method is used for long-term boiler storage. The dry method involves keeping the inside of the boiler totally dry to prevent corrosion. Oxygen needs moisture to cause corrosion of the boiler metal surfaces. The inside of the boiler is thoroughly cleaned and dried. Trays of quicklime or other desiccants such as silica gel are placed inside the boiler to absorb moisture from the air. *Quicklime* is limestone that has been thoroughly dried. A *desiccant* is a drying agent. When using silica gel as the desiccant, approximately 5 lb per 30 cu ft of boiler interior volume should be used. If using quicklime, 2 lb per 30 cu ft of boiler interior volume should be used.

The boiler is tightly sealed to prevent infiltration of atmospheric air. It is critical that the nonreturn valve, the feedwater valves, and the blowdown valves do not leak through. Leaking valves could allow water to enter the boiler. Plastic or stainless steel trays should be used to hold the desiccant to keep the trays from corroding. The trays of desiccant are checked every two to three months to see if they have absorbed their limit of moisture. Some desiccants change color when they have absorbed their limit. The color change is a convenient signal of the need for replacement. When the desiccant is saturated, it must be regenerated or replaced. Some types of desiccants can be regenerated by heating them in an oven.

When using either the dry layup or wet layup method, the fire side of the boiler should be thoroughly cleaned of soot and ash to prevent corrosion from the acidic elements, especially sulfur, in the soot. While the boiler is stored, the circulation of warm, moist air from the boiler room through the fire side of the boiler should be prevented. This circulation may occur naturally in some cases as air is drawn through the furnace, through the boiler tubes, and up the stack. In order to prevent this circulation, the furnace doors below the breeching connection to the boiler can be opened. Air can then flow directly to the stack without first flowing through the boiler. Portable kerosene heaters may be used initially in the fire side to thoroughly dry the unit.

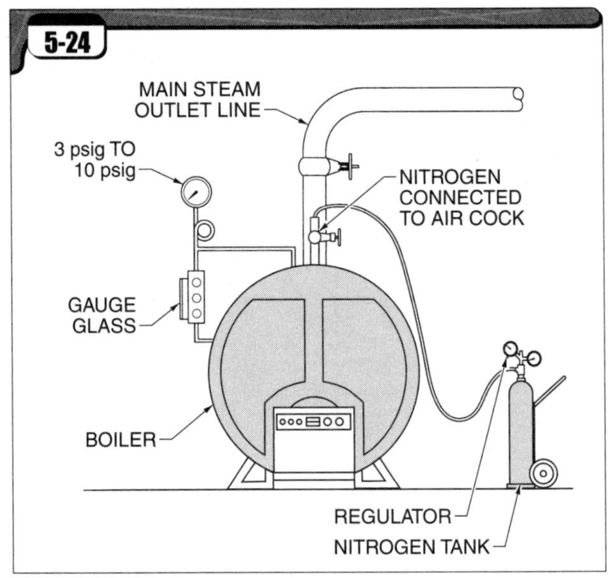

Figure 5-24. *An alternate wet layup method uses a nitrogen blanket to prevent corrosion in the boiler.*

Chapter

Water Treatment Systems and Controls

Trade Test

Name: Date:

T F **1.** Dissolved oxygen can be converted to free oxygen in a deaerator.

__________ **2.** The three major types of deaerators are the ___, ___, and ___.

T F **3.** Boiler feedwater can be heated above 212°F in a vented heater.

__________ **4.** The letters ppm represent ___.

__________ **5.** ___ is a condition in which the boiler metal becomes brittle because of crystalline cracking.

A. Phenolphthalein alkalinity
B. Alkaline embrittlement
C. Caustic solubility
D. Caustic embrittlement

T F **6.** Condensate will absorb oxygen and carbon dioxide as it cools.

__________ **7.** The temperature of water leaving a deaerator should be within ___°F to ___°F of the temperature of the steam entering the deaerator.

A. 3; 5
B. 5; 10
C. 10; 20
D. 30; 50

__________ **8.** The dry layup method is used for ___ boiler storage.

A. 24 hr
B. short-term
C. long-term
D. none of the above

__________ **9.** The ___ is the desired point at which an automatic controller maintains a variable condition within a process.

__________ **10.** The three basic classifications of impurities found in natural water are ___, ___, and ___.

A. dissolved solids; suspended gases; suspended solids
B. suspended gases; suspended solids; dissolved gases
C. dissolved solids; dissolved gases; suspended solids
D. none of the above

T F **11.** Scale assists the water in a boiler in removing heat from boiler tubes.

__________ **12.** Silt and algae are examples of ___ found as impurities in natural water.

A. dissolved solids
B. suspended solids
C. dissolved gases
D. suspended gases

__________ **13.** Temporary hardness of water may be reduced by ___ the water.

A. chilling
B. heating
C. aerating
D. none of the above

T F **14.** Scale is the collection of mineral deposits formed on the heating surfaces of a boiler.

________ **15.** Scale-forming impurities become concentrated in boiler water by ___.
A. solidification as water temperature increases
B. evaporation as the water boils off
C. both A and B
D. neither A nor B

T F **16.** Pretreatment of boiler water occurs inside a boiler.

________ **17.** An anion is an atom with a ___ electrical charge.
A. positive
B. neutral
C. negative
D. none of the above

________ **18.** ___ is a solution of salt and water.

________ **19.** The order in which a sodium zeolite water softener regenerates is ___, ___, and ___.
A. rinse; backwash; brine
B. rinse; brine; rinse
C. backwash; brine; rinse
D. backwash; brine; backwash

________ **20.** A ___ test is a test that determines the concentration of a specific dissolved substance.
A. reverse osmosis
B. conductivity
C. residual
D. titration

T F **21.** A hot process water softener is a pressure vessel that uses steam to heat makeup water by direct contact, and chemical injection to cause the hardness in the water to precipitate.

T F **22.** A demineralizer is a highly efficient ion-exchange process generally used for low-pressure boilers.

________ **23.** A demineralizer replaces all the cations in the inlet water with ___ ions.

T F **24.** Evaporators may operate under a vacuum, allowing less energy and a lower boiling temperature to be used.

________ **25.** On the pH scale, 0 is the most acidic and ___ is the most alkaline.

________ **26.** ___ paper turns different colors depending on the pH of the water being tested.

________ **27.** The minimum acceptable pH of boiler water is ___.

________ **28.** ___ cause(s) hardness particles in boiler water to precipitate as a heavy sludge that is removed through the bottom blowdown.

________ **29.** ___ keep(s) the hardness in boiler water in solution so that it cannot settle out on the heating surfaces.

T F **30.** Dispersants prevent solids in boiler water from sticking together to form deposits.

________ **31.** ___ are chemicals that prevent corrosion in condensate piping and steam piping.

________ **32.** A(n) ___ sample is a sample that is exactly the same as the item being tested.

T F **33.** Capacitance is the measure of the ability of an electric circuit to allow current flow.

T F **34.** Water with no impurities is a good conductor of electricity.

T F **35.** One of the primary purposes of blowdown is to remove sludge and impurities from boiler water.

36. List four factors that affect the amount and frequency of blowdown.

______________ **37.** Bottom blowdown piping for boilers should not be smaller than ___″ in diameter or larger than ___″ in diameter.

A. ½; 1 B. 1; 1½
C. 1; 2½ D. 1½; 2

T F **38.** Bottom blowdown is more effective when it is used during light-load periods.

T F **39.** Continuous blowdown is more effective when it is used during heavy steaming.

______________ **40.** The two methods of boiler layup are ___ layup and ___ layup.

A. short-term; extended B. intensive; extensive
C. wet; dry D. hot; cold

T F **41.** A deaerator is a closed heat exchanger.

______________ **42.** Hardness of water is the measurement in ppm of ___ and ___ in the water.

A. hydrogen; oxygen B. cations; anions
C. calcium; magnesium D. none of the above

______________ **43.** As water is heated, the solubility of oxygen ___.

A. increases B. decreases
C. remains constant D. increases until the boiling point is reached, then decreases

______________ **44.** Commonly used chemical oxygen scavengers include sodium sulfite and ___.

T F **45.** Permanent hardness of water may be reduced only by the use of chemicals or distillation.

T F **46.** High-pressure boilers have a greater tendency to form scale than low-pressure boilers.

T F **47.** Deaerators may operate under atmospheric pressure or a higher pressure.

______________ **48.** Natural water generally has a pH of approximately ___ to ___.

A. 3.5; 4.5 B. 4.5; 5.5
C. 5.5; 6.5 D. 6.5; 7.5

______________ **49.** The ___ pump method is the preferred method of feeding water treatment chemicals into boiler water.

T F **50.** Water conductivity is used to determine which specific impurities are contained in water.

______ **51.** Deaerators are placed ___ and ___ the feedwater pumps in a steam boiler system.
A. before; above
B. before; below
C. after; above
D. after; below

______ **52.** A(n) ___ amine is injected into a boiler during operation to raise the pH of the condensate.
A. filming
B. neutralizing
C. ecological
D. hydrazine

T F **53.** A reagent is a chemical used in a water treatment test to show the presence of a particular substance.

______ **54.** ___ is the measure of the ability of an electric circuit to oppose current flow.

______ **55.** A(n) ___ is one-millionth of a mho.

______ **56.** A feedwater storage tank beneath a deaerator is usually sized to provide boiler feedwater for a period of ___ if the water supply should be interrupted.
A. 24 hr
B. 8 hr
C. 60 min to 90 min
D. 20 min to 30 min

T F **57.** High alkalinity in boiler water contributes to the formation of CO_2.

______ **58.** The conductivity of the makeup (softened) water is 390 micromhos. The conductivity of the feedwater is 180 micromhos, and the conductivity of the condensate is 45 micromhos. What is the percentage of condensate return?
A. 1.64%
B. 60.9%
C. 76.3%
D. 80.6%

______ **59.** A desiccant is a(n) ___ agent.

T F **60.** A lever-operated gate valve is a quick-opening valve.

______ **61.** Two testing instruments used to determine the quantity of dissolved iron in a water sample are the color wheel comparator and the ___.
A. titrator
B. ion exchanger
C. spectrophotometer
D. demineralizer

62. Describe the difference between surface blowdown and continuous blowdown.

Chapter 6

Fuel Systems and Controls

The conversion of chemical energy in fuel to thermal energy in the furnace is the critical first step in the operation of the steam system. Modern environmental regulations require that the combustion process be closely controlled to prevent the formation of toxic and destructive pollutants. Ever-rising fuel costs require continual attention to combustion efficiency. Emerging combustion technologies, such as the use of waste fuels and fluidized bed combustion, present opportunities for the boiler operator to help in addressing both environmental and economic concerns.

Safety of the plant personnel and reliability of the steam generating system are paramount. Since boiler equipment and regulatory codes and standards are complex and changing, boiler operators must continually work to upgrade their skills and knowledge. In addition to preventing fuel-related accidents, this helps the boiler operator troubleshoot malfunctioning burner equipment and return the system to service with less effort.

Furnace Fundamentals

What is a combustion chamber?

A combustion chamber is the area of a boiler where the burning of fuel occurs. **See Figure 6-1.** The combustion chamber is also known as the firebox or furnace. The size and shape of the combustion chamber are designed to provide enough time for all combustible elements of a fuel to burn completely before the gases of combustion contact the tube surfaces.

When flame is visible, it is proof that combustion is still in progress. If the flame comes into contact with tube surfaces before the fuel has finished burning, the flame temperature can be lowered below the ignition temperature and the burning process at that localized point will stop. If the flame is quenched before all the fuel in the flame has been burned, the unburned carbon will deposit as soot on the heating surfaces. This is inefficient because it means that some of the fuel was not burned completely. In addition, soot acts as insulation and reduces heat transfer efficiency.

In some large boiler designs, the combustion chamber is considered a separate area from the firebox (furnace). Although the dividing line between the two is not always clear, in these cases the combustion chamber is the general area where the burning is completed before the gases enter the tube passages. In larger boiler furnaces, such as those in which solid fuel is fired, the combustion chamber is often large and high.

What basic functions are accomplished in the combustion chamber?

The correct quantity of fuel must be fed to the burner. The amount of fuel admitted to the combustion chamber is determined by the quantity of steam that must be produced by the boiler. For example, the fuel must be fed at a faster rate to make 50 Klb/hr of steam than to make 30 Klb/hr.

The correct quantity of air must be fed to the burner. Each unit of fuel (e.g., pound of coal or cubic foot of natural gas) will require a certain amount of air in order to have enough oxygen to burn all the fuel. *Primary air* is the initial volume of air that enters the furnace with the fuel for most of the combustion process. A stack (chimney) may be used to create a natural flow of air through the furnace, but in most boilers primary air is blown (forced) into the furnace by a fan. The primary air fan that supplies the air is known as the forced draft fan.

The air and the fuel must be mixed together. It is not enough to provide the correct amount of air for the quantity of fuel to be burned. The air and fuel must mix together in the correct ratio to ensure that all the fuel is burned. *Secondary air* is air mixed with the fuel to ensure that enough oxygen is available to complete the burning. Secondary air is often introduced through louvers around the outside of a gas or oil burner, or through high-velocity air jets or ports in a solid fuel-fired boiler. The high velocity of the secondary air creates turbulence in the burning gases. This helps to ensure thorough mixing of the oxygen and fuel. In solid fuel-fired boilers, the secondary air may be supplied by a separate secondary air fan.

Figure 6-1. *The combustion chamber is the area where the burning of fuel occurs.*

3 List and explain several concerns in determining boiler furnace volume.

One of the first concerns in determining the boiler furnace volume is the amount of steam demand to be satisfied. The amount of steam that must be made by a boiler determines how much fuel must be consumed.

The type of fuel used has an effect on boiler furnace volume. Types of fuel that may be burned in a furnace include coal, natural gas, fuel oil, municipal refuse, industrial wastes, and agricultural wastes. Some fuels do not lend themselves well to efficient or expedient combustion and therefore require variations in furnace design. For example, stoker coal is coal that is crushed and sized with screens to about the size of driveway gravel. Stoker coal does not lend itself well to quick and thorough mixture with combustion air, as does natural gas. Therefore, this coal will require more space and a longer time to burn.

The moisture content of fuel has an effect on boiler furnace volume. Moisture content is the percentage of water in a fuel. Fuel gases contain no moisture. Some fuel oils may have trace amounts of moisture, up to about 0.4%. Bituminous and anthracite coal have moisture content of about 1% up to about 12%. Lignite coal can have moisture content up to about 35% to 38%.

The moisture in a fuel turns to steam as the fuel burns. Since water expands as it changes to steam, fuels with a higher moisture content will require a furnace with greater volume to accommodate the steam formation as well as the other gases produced during combustion.

The Btu content of fuel has an effect on boiler furnace volume. A unit of measure of a fuel has a specified Btu content. For example, a cubic foot of a gaseous fuel may contain 1050 Btu, a gallon of a liquid fuel may contain 141,000 Btu, and a pound of a solid fuel may contain 11,500 Btu. If a fuel has a lower heating value than another fuel, then more of the first fuel will have to be burned to provide the same amount of heat as the second fuel. This requires more space in the furnace area.

The allowable flue gas flow velocity through the flue gas passages has an effect on boiler furnace volume. When burning solid fuels like coal, very high flue gas flow velocities can cause a large quatity of ash to be carried along with the flue gases. This creates abrasion in the boiler as well as high loading and wear on the pollution control equipment. In addition, very high flue gas velocities do not allow adequate time for the heat in the flue gases to be absorbed by the boiler water.

The boiler furnace volume is different when the furnace is water-cooled or refractory-lined. Much higher heat release from the fuel (Btu per cu ft per hr) is feasible when the furnace is lined with waterwalls than when the furnace is lined with refractory materials. Therefore, the furnace can be more compact.

4 What are the three T's of good combustion?

The three T's of good combustion are time, temperature, and turbulence.

Enough time must be allowed for the fuel to be fully burned before the hot gases of combustion come into contact with the boiler tubes.

Combustion must occur in a high-temperature zone to make sure that the fuel in the flue gases is sustained above the ignition temperature until fully consumed. For example, if the fuel in the flame is allowed to fall below the ignition temperature, the flame will go out. Smoke and soot will be formed and fuel will be wasted.

A turbulent area in the furnace allows the air and fuel to be fully mixed. In order for the burning to be completed, each molecule of fuel must come into contact with an adequate number of oxygen molecules. Therefore, fuels and air are mixed as completely as possible. For example, a fuel oil burner sprays fuel oil out as a mist. Each droplet can better contact oxygen and be burned.

An acronym, MATT, is also often used as a rule of thumb for efficient combustion. The letters stand for mixture, atomization, time, and temperature.

Define perfect combustion, complete combustion, and incomplete combustion.

Perfect combustion is a fire where the fuel is burned with precisely the right quantity of oxygen so that no fuel or oxygen remains and the maximum possible heat results. Perfect combustion is only attainable in a laboratory. It is not feasible to attain perfect combustion in a boiler furnace with the imperfect level of control possible with valves and dampers.

Complete combustion is a fire where the fuel is burned with a slight excess of oxygen so that the fuel is completely consumed without forming any smoke, and only a minimal amount of oxygen is left over. Complete combustion is the result desired from combustion equipment.

Incomplete combustion is a fire where the fuel is burned without the proper amount of oxygen, without enough mixing of fuel and oxygen, or at a temperature too low to allow satisfactory reaction of the fuel and oxygen. Incomplete combustion is inefficient because not all of the Btus in the fuel are released.

Which two elements in fossil fuels provide the majority of the heat?

Hydrogen and carbon and their compounds (hydrocarbons) are the desirable elements that release heat when burned with oxygen. Sulfur in the fuel is also a heat-producing element, but it is undesirable because of its adverse effect on the furnace and the atmosphere.

What is carbon dioxide?

Carbon dioxide (CO_2) is a heavy, colorless gas that results from burning carbon with the proper amount of oxygen. It is a desirable end product of burning carbon in the fuel.

What is carbon monoxide?

Carbon monoxide (CO) is a colorless, odorless, highly toxic pollutant gas. It is the result of incomplete combustion of carbon. A molecule of carbon monoxide (CO) contains half the oxygen of a molecule of carbon dioxide (CO_2). The formation of carbon monoxide during combustion is inefficient, because the combustion of carbon to carbon monoxide only produces about a third as much heat as the combustion of carbon to carbon dioxide.

What are some of the indications of incomplete combustion?

Visible smoke coming out of the stack indicates incomplete combustion. However, combustion efficiency can deteriorate to a significant degree before visible smoke appears from the stack. Soot accumulation on the boiler tubes indicates incomplete combustion. Soot is unburned fuel. Higher than normal fuel consumption to achieve the required steam production may indicate incomplete combustion.

If the facility has flue gas analysis equipment, it can be used to analyze the flue gases to determine the combustion conditions. For example, the presence of carbon monoxide in the flue gases points to incomplete combustion of the fuel.

What is meant by the flammability range of a fuel?

In order for a fuel to burn, it needs to have adequate oxygen to support the burning. The fuel will not burn if there is not enough oxygen mixed with the fuel. In other words, if the air-fuel ratio is too low, the mixture is too lean to burn. The fuel also will not burn if there is not enough fuel mixed with the oxygen. In other words, if the air-fuel ratio is too high, the mixture is too rich to burn.

The range between these two extremes is called the flammability range of the fuel, or the fuel's flammability limits. The minimum and maximum limits are different for each fuel. For example, natural gas will burn when it makes up at least 4.3% of the air-fuel mixture. This is the lower flammability limit. Natural gas will not burn when it makes up more than 15% of the air-fuel mixture. This is the upper flammability limit.

As the upper flammability limit is approached, an additional concern is that, while the fuel may continue to burn, it will produce products of incomplete combustion. For example, as the carbon in a fuel burns without adequate oxygen for complete combustion, it will produce carbon monoxide, a toxic gas.

What are the boiler operator's main duties with regard to the boiler furnace?

Safety is the boiler operator's first consideration. The boiler operator prevents explosions and fires by keeping the flame safety equipment in good working order. The boiler operator optimizes furnace efficiency by adjusting equipment to maintain the proper ratio of air and fuel in the furnace. The boiler operator monitors flue gas conditions at the stack to keep within allowable stack emissions standards. This is of particular concern when burning coal or some industrial waste fuels.

Why are baffles used on the fire side of a boiler?

Baffles direct the flow of gases of combustion along a desired path. The arrangement of the baffles causes the hot gases of combustion to come into closer contact with the heating surfaces. Baffles help the boiler claim the maximum amount of heat for making steam before the gases of combustion go out the breeching to the stack.

13 What is the path of the gases of combustion through a scotch marine package four-pass boiler and a "D" style package boiler?

For a dryback scotch marine package boiler, the burner and combustion air fan are on the front of the boiler. The flame from the burner extends into the long tubular furnace in the lower part of the shell. **See Figure 6-2.** The hot gases of combustion are directed to the lower fire tubes at the rear of the boiler by a baffle inside the rear door. The gases pass through the lower fire tubes and back to the front. At the front of the boiler, the gases are again reversed by a baffle and directed through the intermediate tubes toward the rear. At the rear, the gases reverse for the fourth pass to the front and are then discharged out to the breeching.

For a "D" style package boiler, the burner is mounted at the front of the boiler at the open portion of the D-shaped tubes. The flame from the burner passes almost to the rear of the combustion chamber. The combustion chamber is enclosed by riser tubes. The greater portion of the heat of the gases of combustion is given up through radiant heat to the riser tubes. At the rear of the combustion chamber, the gases reverse and flow back toward the front of the boiler through the convective pass of downcomer tubes. Normally, a membrane waterwall or wall tubes embedded in refractory brick separate the first and second passes. The gases leave the boiler at the front on the top or side.

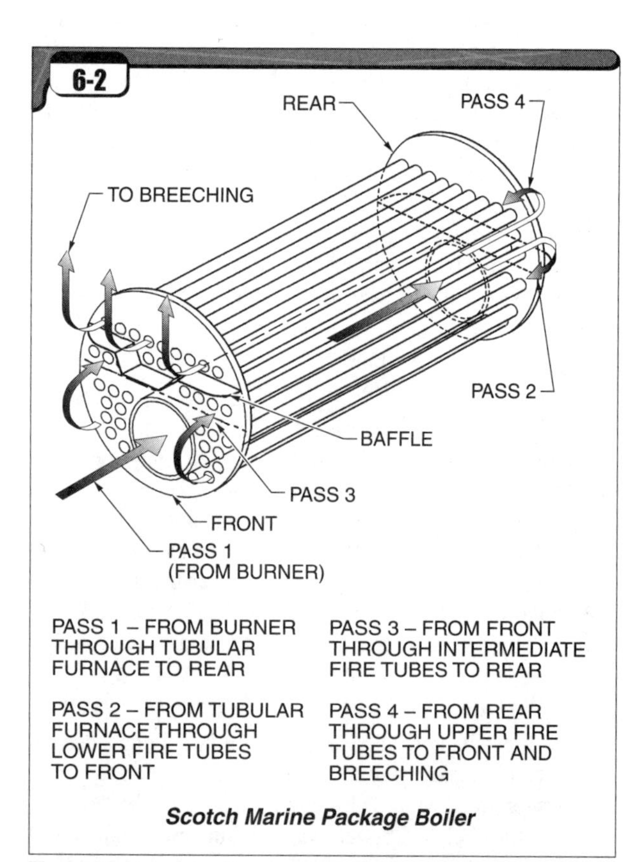

Figure 6-2. *The paths of gases of combustion in boilers are designed for efficiency.*

14 What is the function of the water seal in a large boiler furnace?

In a large boiler furnace, the ash and solids from the spent fuel are sometimes made to fall out of the furnace area through an opening in the bottom of the furnace. This opening must be sealed by some means to prevent air from leaking into the furnace.

On many large boilers, the bottom of the combustion chamber extends below the water line of a submerged conveyor. **See Figure 6-3.** A *submerged conveyor* is a heavy steel pan conveyor or apron conveyor immersed in a water trough. The water prevents air from being pulled directly into the furnace. This arrangement also causes thermal shock that shatters larger chunks of slag or clinker when the hot chunks fall into the relatively cool water.

Slag is the solid deposits that accumulate on furnace walls and boiler tubes. Slag is formed when semimolten flyash cools after it sticks to furnace walls and boiler tubes. Clinker is stoker or lump coal or other solid fuel that has fused together during combustion. Solid residue that has quickly cooled and shattered can be removed in a fairly uniform granular state.

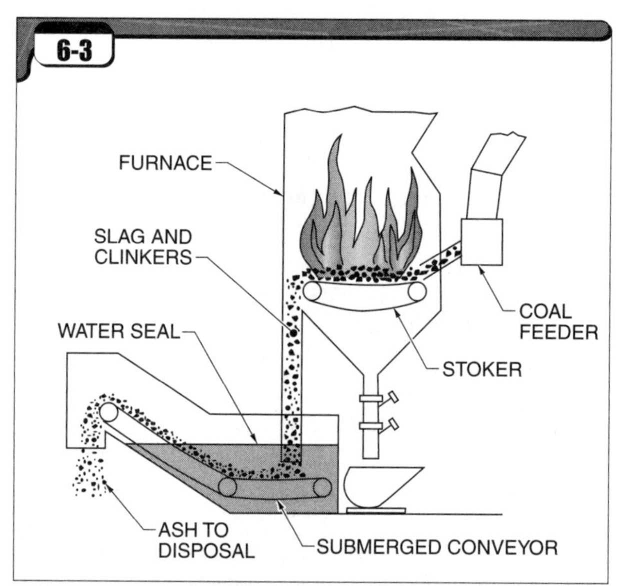

Figure 6-3. *The water seal prevents uncontrolled atmospheric air from leaking into the furnace.*

15 What forms of firebrick and furnace insulating materials are used in steam boiler installations?

The common forms of firebrick and furnace insulating materials are refractory firebrick shapes, insulating firebrick, plastic refractory, castable refractories, ceramic fiber insulation, block insulation, and board insulation.

Refractory firebrick shapes are very hard, dense, heavy firebricks. These firebricks are made of a variety of aggregate materials, but almost all include a high alumina content.

Alumina provides high temperature resistance. They are made in various shapes and curvatures and are used to construct geometric forms such as burner throats, arches, and lintels. **See Figure 6-4.**

Insulating firebrick materials are lighter, low-density firebrick aggregate materials made more for the purpose of insulation than heat resistance. Insulating firebrick materials are often layered behind refractory firebrick shapes to help reduce heat loss through the furnace walls and other similar areas of boilers.

Plastic refractory materials are those mixed with water to produce a refractory with a clay-like consistency. The term "plastic" in this context means moldable. These materials are used to fill irregularly shaped areas such as the openings around furnace view ports and coal feeder openings. They are also used to form solid baffles. Often these materials are held in place by the studs on studded waterwall tubes.

Plastic refractories include ramming and gunning materials. *Ramming materials* are plastic refractory materials that are rammed into place using heavy bars and other tools. This is needed in order to press all the voids out of the material and engage it around the waterwall studs or other anchors. *Gunning materials* are plastic refractory materials that are gunned, or sprayed, under pressure onto a surface. Gunning materials are used to coat studded waterwalls and other large, flat surfaces.

Castable refractories are firebrick aggregates that are mixed to the consistency of mortar and then trowelled, poured, or otherwise spread. Castable refractories include light castables and heavy castables. Heavy castables have higher density and better heat resistance and insulating values than light castables.

Ceramic fiber insulation materials are flexible materials used to fill furnace wall expansion joints, transition joints around furnace access doors, etc. These include common trade-named materials such as Kaowool®.

Block insulation and board insulation are parts of a family of materials used for insulation. For example, these materials are used between the furnace walls and the outer metal skin of watertube boilers to prevent heat loss from the boiler's outer surfaces. These are sometimes natural materials including diatomaceous earth and calcium silicate. Diatomaceous earth is composed of diatoms, which are tiny seashells roughly the size of grains of sand. These are mined from natural underground deposits. Block and board insulations are often composed of manufactured materials such as fiberglass and mineral wool as well.

Factoid

In order for a fuel to burn completely, all the fuel mole-cules must come into contact with an adequate number of oxygen molecules. Turbulence in the furnace helps ensure this contact. A small amount of excess air is used to ensure that there are always enough oxygen molecules.

6-4

Figure 6-4. *Refractory firebrick is used to form burner throats.*

Fuel Selection

16 What waste fuels are commonly used in boilers?

Wastes that may be used as boiler fuels are generated in a wide array of industries. The waste fuel may be solid, liquid, or gaseous. Sometimes the waste fuels are poor quality or inconsistent in nature, but their extremely low cost makes up for the inconveniences associated with handling and burning them. Common sources of waste fuels include industrial processes, municipal services, agricultural processes, and reclaimed waste.

Common industrial processes that produce waste fuels include paper mills, wood-processing operations, refineries, steel mills, chemical processing plants, municipal waste treatment, and processing of various reclaimed wastes.

Paper mills often include black liquor heat recovery boilers. Black liquor is a flammable semiliquid by-product of the processes used to break down wood fibers into pulp.

Wood-processing operations often use sawdust-fired boilers. The sawdust is obtained as a low-cost waste fuel in sawmills, cabinetmaking, furniture manufacturing, and wood building material manufacturing.

Refineries produce waste gases such as carbon monoxide in the process of refining crude oil into various products.

Steel mills that include blast furnaces produce blast furnace gas, which is predominantly carbon monoxide. **See Figure 6-5.** Steel mills also often include coke ovens. Coke ovens produce coke that is used in blast furnaces. Coke is coal that has been heated to drive off volatile vapors, leaving behind the carbon portion of the coal. These volatile vapors, called coke oven gas, may serve as boiler fuel.

Chemical processing plants produce a wide variety of waste solvents, waste vapors, and solid wastes. These may include solvents used to rinse tanks between batches of product, filter media used to strain debris from manufactured liquids, by-product vapors, and many other wastes.

A common municipal service that produces waste fuels is processing of municipal solid waste. Municipal solid waste (MSW) is trash that is frequently burned in incinerator plants. This reduces the volume of the waste that must be sent to landfills. In burning the trash, heat is generated that can be used to produce steam in boilers. The steam is generally used to operate turbine-generator equipment to generate electricity. The steam is also sold to heat buildings or operate manufacturing plants.

Municipal solid waste is burned in two main forms. Mass-burn incinerators burn the waste as received from the hauling trucks. Other incinerator plants shred the waste first in order to produce a more consistent fuel. This shredded form of municipal waste is called refuse-derived fuel (RDF). Recyclable ferrous metals and glass may be extracted before burning.

Municipal wastewater treatment plants and landfills also produce methane gas, which can be burned in boilers using gas burners of special design.

Agricultural processing operations produce a variety of wastes and by-products. These include such materials as olive and peach pits, nutshells, and bagasse. Bagasse is spent solid sugar cane fibers.

Reclaimed wastes are waste fuels refined from other forms of wastes. For example, automotive tires are shredded for use as fuel. The steel belts from the tires are recycled into other steel products.

17 What factors are weighed in selecting the fuel to be used?

The fuel must be available in ample quantity to make the quantity of steam needed. If the fuel must be transported to the site, the transportation costs must also be considered.

The initial cost of the heat derived from burning the fuel must be included in design decisions. Design engineers consider the Btus available from the fuel per dollar spent. For example, natural gas and fuel oil cost more than coal.

The initial cost of material handling equipment is also important. Some fuels require more storage and handling equipment costs than others. These costs typically include the costs associated with coal conveyors, ash silos, fuel oil tanks, and atomizing steam piping.

Solid fuels such as coal and sawdust require significant room for the fuel storage yard, inclined conveyors, and ash load-out areas.

Some fuels produce ash that must be handled and sent to landfills or recovered for other uses.

Coal and many waste fuels present obstacles in obtaining permits from the EPA and other agencies, due to air emissions and solid waste disposal regulations. In construction of a large coal-fired plant, pollution control equipment costs make up a very substantial portion of the cost of the plant. This pollution control equipment also consumes energy and this must be considered in plant operating costs.

The costs associated with equipment maintenance must be considered. These costs include such items as spare parts inventories, rigging and moving of equipment during maintenance, special tools, and contract labor.

Labor costs can be substantial for some fuels. Labor costs are associated with fuel handling, ash disposal, additional maintenance personnel, and continuous monitoring of the operation by boiler operators.

6-5

Figure 6-5. *Steel mills produce blast furnace gas that can be used as boiler fuel.*

Factoid

Many industrial, municipal, agricultural, and other processes generate waste and by-product materials that may be viable fuels for generating steam.

18 How is the rate of combustion expressed for gaseous or liquid fuels?

The rate of combustion is expressed as the number of Btu released per cubic foot of furnace volume per hour.

The rate of combustion of a gas or liquid fuel may be calculated as follows:

$$RC = \frac{H}{V_f \times t}$$

where

RC = rate of combustion (in Btu/cu ft/hr)

H = heat released (in Btu)

V_f = volume of furnace (in cu ft)

t = time (in hours)

For example, a scotch marine boiler has a furnace volume of 44.0 cu ft. If 4000 cu ft of natural gas is burned per hour and the Btu content of natural gas is 990 Btu/cu ft, what is the hourly rate of combustion? The heat released by the fuel is the amount of fuel multiplied by the Btu content of the fuel. In this case, the heat released is 3,960,000 Btu/hr (4000 × 990 = 3,960,000).

$$RC = \frac{H}{V_f \times t}$$

$$RC = \frac{3,960,000}{44.0 \times 1}$$

$$RC = \mathbf{90,000\ Btu/cu\ ft/hr}$$

Coals

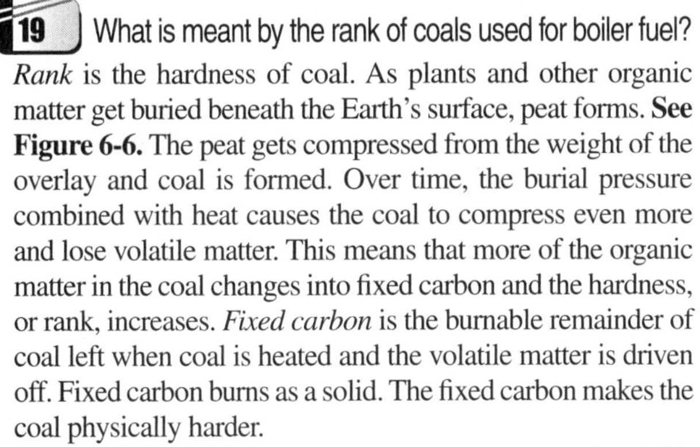

19 What is meant by the rank of coals used for boiler fuel?

Rank is the hardness of coal. As plants and other organic matter get buried beneath the Earth's surface, peat forms. **See Figure 6-6.** The peat gets compressed from the weight of the overlay and coal is formed. Over time, the burial pressure combined with heat causes the coal to compress even more and lose volatile matter. This means that more of the organic matter in the coal changes into fixed carbon and the hardness, or rank, increases. *Fixed carbon* is the burnable remainder of coal left when coal is heated and the volatile matter is driven off. Fixed carbon burns as a solid. The fixed carbon makes the coal physically harder.

Coal will also contain a certain percentage of volatile material, or volatiles. Volatiles are the constituents in coal that, when the coal is heated, distill out of the coal and burn off as gases. These gases are hydrocarbons. Hydrocarbons are any of a number of compounds composed of hydrogen and carbon atoms.

Thus, coals that contain a high percentage of fixed carbon and a low percentage of volatiles are called hard coals and have a higher rank. Coals with a low percentage of fixed carbon and a high percentage of volatiles are soft coals and have a lower rank. The two ranks of coal most commonly used in coal-fired boiler furnaces are anthracite coal and bituminous coal.

Anthracite coal is a geologically older coal that contains a high percentage of fixed carbon and a low percentage of volatiles. Anthracite is also known as hard coal.

Bituminous coal is a geologically younger coal that contains a high percentage of volatiles and a low percentage of fixed carbon. Bituminous coal is also known as soft coal. A larger furnace volume must be provided when burning bituminous coal than when burning anthracite coal. The lower rank means that a greater volume of space is required to burn the hydrocarbons.

Anthracite and bituminous coals are further broken down into subclassifications. For example, coal from a deposit in a particular location may be subbituminous or semianthracite.

A third primary rank of coal that is not widely used in boilers is lignite. *Lignite* is very young and very soft coal that has a high moisture content and low heat value. Lignite is a low-quality coal formed shortly after peat moss begins to compress into coal. Lignite produces large amounts of flyash that create slag deposits in furnaces. Lignite is also crumbly and may fall to pieces during transportation. It is very low in cost, however, and plants located close to a lignite mine may find it cost-effective.

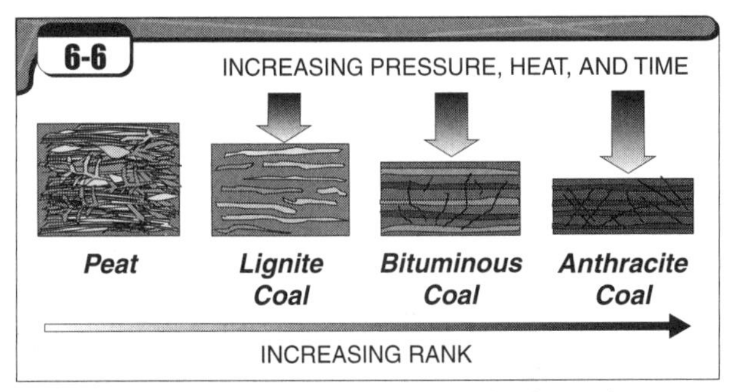

Figure 6-6. *The rank of coal refers to the coal's hardness. The primary ranks of coal are lignite, bituminous, and anthracite.*

20 What are the grades of coal?

The *grade* of coal is the size of the pieces of coal. **See Figure 6-7.** Anthracite coal is sized on round-hole screens. For example, anthracite coal of stove grade will fall through a 2 7⁄16″ diameter hole, but will pass over a 1 5⁄8″ diameter hole. There is no standard of names or screen openings by which bituminous coal is graded; however, it is generally sized on round-hole screens.

The grade of coal is a somewhat context-specific term that may also refer to the ash content of the coal, ash melting temperature, sulfur content, or Btu value. However, grade of coal primarily refers to the size of the pieces of coal.

In the past, a series of names referring to the size of the pieces of coal evolved. Depending on the rank of the coal, these terms include egg, buckwheat, rice, etc. These names are largely being abandoned and modern coal-fired plants generally refer to coal grades by the actual range of physical size desired.

BITUMINOUS		ANTHRACITE		
Grade	Size	Grade	Size	
			Over	Through
Run of mine	8″	Broken	3″	4 3⁄16″
Lump	5″	Egg	2 7⁄16″	3″
Egg	2″ TO 5″	Stove	1 5⁄8″	2 7⁄16″
Nut	1¼″ TO 2″	Chestnut	13⁄16″	1 5⁄8″
Stoker	¾″ TO 1¼″	Pea	9⁄16″	13⁄16″
Slack	UP TO ¾″	Buckwheat	5⁄16″	9⁄16″
		Rice	3⁄16″	5⁄16″
		Barley	3⁄32″	3⁄16″

Figure 6-7. *The grade of bituminous and anthracite coal is generally sized on round-hole screens.*

21 What is the proximate analysis of coal?

Proximate analysis of coal is the percentages of moisture, volatiles, fixed carbon, and ash. The proximate analysis is determined by a series of laboratory tests. Proximate analysis of coal aids the operating personnel in the choice of coal for a particular application. In a modern boiler plant, energy efficiency and pollution prevention have become more and more important. Therefore, the percentage of sulfur and the Btu content of the coal are typically included in a proximate analysis.

22 What is the ultimate analysis of coal?

The *ultimate analysis* of coal is the percentages of nitrogen, oxygen, carbon, ash, sulfur, and hydrogen (NOCASH) in the coal. The carbon and hydrogen content of coal are of particular concern to the boiler operator. For example, coal with a high carbon content will burn mostly on the grates, while coal with a high hydrogen content produces more volatile gases. Design engineers use the ultimate analysis of coal to determine combustion air requirements, furnace volume requirements, and other design factors.

Coal-Burning Equipment

23 What is the purpose of grates when firing solid fuels?

Grates support the burning fuel bed. Grates are made of cast iron and/or steel. Fixed (stationary) grates are for hand firing. Moving (traveling) grates are for automatic stoker firing. Moving grates also automatically convey the ash from the coal to a point of discharge.

Individual grate sections have holes or slots that allow air to pass through. The air is supplied to the furnace from below the grates and it flows upward through the burning fuel bed. This flow serves the dual purpose of supplying air for combustion and keeping the grates cool. Air dampers below the grates control the supply of primary air to the various sections. The ash produced by the burning of solid fuel may be removed from the grates manually or automatically.

The grates in smaller, hand-fired furnaces are usually fixed. In this case, the boiler operator shovels the coal or other solid fuel onto the grates and removes the ashes from the grates. Often the individual grate sections are tilted to allow ashes to fall through into an ash pit below. In this case, the grates are tilted much like the movement of the individual louvers in a damper. These smaller installations are becoming rare due to restrictions on emissions and due to the need for a boiler operator to be continuously on duty.

Larger grates convey the ashes from the furnace automatically. There are a number of moving grate designs. The most common types are traveling grates and chain grates. Most automatic grates are configured as a heavy continuous metal conveyor, much like the tracks on an army tank. **See Figure 6-8.** However, they move very slowly. A normal grate speed for a solid fuel-fired boiler is about 2″ to 5″ per minute. The speed at any particular time depends on the fuel-burning rate and therefore the volume of ash to be removed.

6-8

Riley Stoker

Figure 6-8. *Traveling grates are used to automatically convey ash from the furnace.*

24 How is grate surface expressed?

Grate surface is expressed in square feet (sq ft). The effective grate surface of both fixed grates and moving grates is the area, in sq ft, exposed to the furnace, or the portion that can be loaded with fuel. Some furnace design criteria are based on the fuel burned per square foot of grate surface per hour. The actual amount of fuel carried on the grate depends on the depth of the fuel on the grate.

25 How is the rate of combustion expressed for solid fuels burned on grates?

The rate of combustion is expressed as the pounds of fuel burned per square foot of grate surface per hour. Solid fuels are normally burned on a grate. The following formula may be applied to find the rate of combustion of a solid fuel:

$$RC = \frac{F}{A_g \times t}$$

where
RC = rate of combustion (in lb/sq ft/hr)
F = fuel burned (in lb)
A_g = area of grate (in sq ft)
t = time (in hr)

For example, if 2.2 tons of coal are burned per hour in a furnace with a 12′ × 16′ grate surface, what is the rate of combustion? A weight of 2.2 tons equals 4400 lb (2.2 × 2000 = 4400).

$$RC = \frac{F}{A_g \times t}$$

$$RC = \frac{4400}{12 \times 16 \times 1}$$

RC = **22.9 lb/sq ft/hr**

26 What is indicated if loose bricks or pieces of brick are on the grates?

Loose brick materials on the grates indicate wear and tear of the brickwork because of overheating or some other problem. For example, the refractory arch or baffle could have loose bricks. The main concern is that protection from the direct radiant heat from the furnace is being lost.

27 What is a bridge wall?

A *bridge wall* is a firebrick wall built across a boiler furnace. In a solid fuel-fired boiler furnace, the bridge wall supports the rear end of the grates (either fixed or moving) and encloses the underside of the grates in the rear so that the primary air is forced to pass up through the grates. The gases of combustion pass over the bridge wall. The top of the bridge wall acts as a baffle that directs the gases in the desired path and keeps the ashes and clinkers contained on the grate area.

Many hand-fired boilers and most boilers fired with retort stokers have bridge walls. Gas-fired and oil-fired boiler furnaces may also have bridge walls.

28 What is a windbox?

A *windbox* is the plenum to which the forced draft fan (primary air fan) supplies air in order to maintain enough pressure to provide proper air flow through the furnace. For example, the windbox is the enclosed area under the grates in a stoker arrangement. The primary air enters the windbox and then passes up through the grates and mixes with the fuel. On a boiler fired with a gas or liquid fuel, the windbox is the large door enclosure or plenum that surrounds the burner assembly and is pressurized with air by the forced draft fan.

29 Why should air be admitted over the fire?

Overfire air is the secondary air in a solid-fuel boiler. Overfire air creates turbulence in the flue gases above the top of the fuel bed. **See Figure 6-9.** Overfire air is admitted through rows of small nozzles so that it has a high velocity. These jets of air help the oxygen mix with combustibles in the flue gases so that greater efficiency is achieved in burning the fuel. Large solid-fuel boilers may have several rows of overfire air nozzles so that the turbulent area of the flame ranges from 1′ to 15′ above the fuel bed, depending on the size of the furnace.

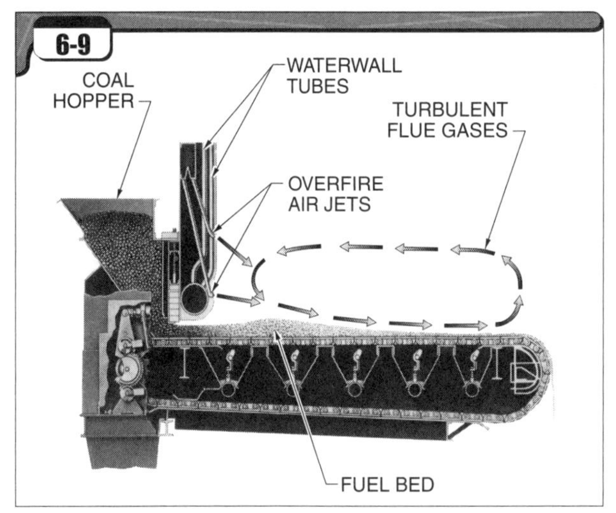

Babcock & Wilcox Co.

Figure 6-9. *Overfire air creates turbulence of the flue gases above the top of the fuel bed.*

30 What are the advantages of using a stoker?

A *stoker* is a device that automatically feeds green, unburned coal or other solid fuel to a furnace. This relieves boiler operators of this large task. The use of stokers also increases efficiency as compared to hand firing, by allowing more consistent fuel bed conditions and higher furnace temperatures.

All stokers consist of several primary parts. These include a fuel feeding mechanism, a method of supplying primary air for combustion of the fuel, a grate mechanism that supports the burning fuel mass and admits air to the fuel, overfire air nozzles to help complete the combustion process, and a system for discharging the residual ash. Stokers are classified as either underfeed or overfeed, depending on whether the fuel is fed from above or below the grates. Overfeed stokers are further classified as either mass feed or spreader feed.

Some types of stoker systems, such as retort stokers, are inseparable from the moving grates that remove the ashes from the furnace. With other stoker systems, the feeding mechanism and the grates are considered separate components.

31 What is a spreader stoker?

The two components of a spreader stoker system are the spreader stokers and the grate system. **See Figure 6-10.** Spreader stokers are used with a number of different styles of grates.

The two main parts of a spreader stoker are a spinning rotor and a coal feeding device. The spinning rotor is a cylinder about 6″ to 10″ in diameter and about 24″ long that has small paddles attached to it. This rotor is driven by a motor or a steam turbine and has an adjustable speed of approximately 750 rpm to 1000 rpm. The coal-feeding device feeds coal from a small hopper onto the rotor at a rate that may be adjusted by the boiler operator. This feed device may be a small reciprocating pusher or a small chain conveyor. The paddles on the spinning rotor strike the coal and throw it into the furnace. The coal used with spreader stokers is generally about ¼″ to 1¼″ in size.

The complete spreader stoker assemblies are about 2½′ wide. There are usually two to six spreader stokers installed across the width of the furnace. Coal is supplied to the individual spreader stokers through spouts from an overhead bunker.

In most spreader stoker systems, the coal is fed onto the rear half of a moving grate. Some smaller pieces of coal ignite and begin to burn while in suspension or before landing on the grates. The grate moves toward the spreader stokers (toward the front of the boiler furnace) and the coal burns as it is conveyed. The grate moves about 2″ to 5″ per minute, depending on the steam demand. By the time the coal reaches the point where it is dumped from the grates, the ash should be about 4″ to 6″ thick and free of red, unburned coals remaining in the ash.

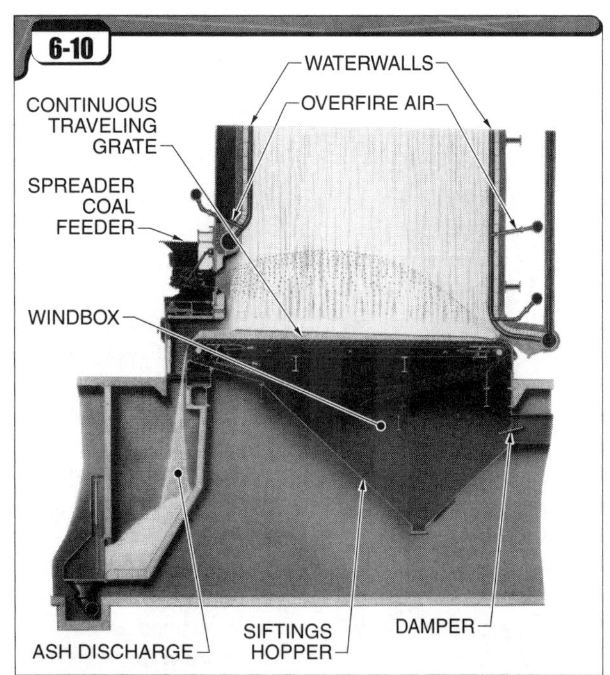

Babcock & Wilcox Co.

Figure 6-10. *Spreader stokers have a spreader and a grate.*

32 What is a chain-grate stoker?

A *chain-grate stoker* is a stoker in which the grates are composed of thousands of small staggered segments that are interlaced by support bars or rods, forming a heavy chain conveyor. **See Figure 6-11.** Coal is gravity-fed onto these grates from an overhead bunker or hopper. Coal usually enters at the front of the furnace and discharges from the rear. A fuel gate across the width of the grates controls the depth of the green coal on the grates as it enters the furnace. Combustion air from a forced draft fan flows upward through the spaces between the small grate segments. Zone dampers are used under the grates to allow the boiler operator better control in distributing the air where it is needed. For example, ash is less dense and offers less resistance to the flow of air than coal so the zone damper at the rear of the grates near the ash discharge may be throttled.

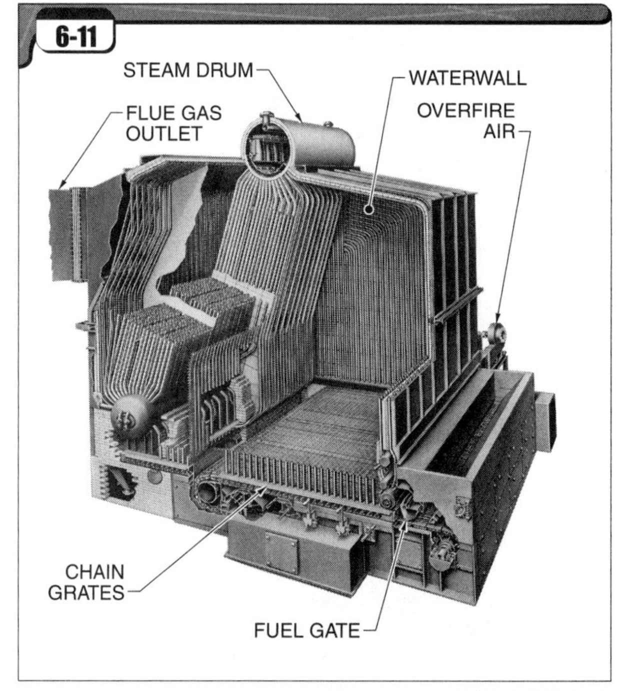

Babcock & Wilcox Co.

Figure 6-11. *Chain-grate stokers have small grate segments.*

33 What is the function of the curved ignition arch in a chain-grate stoker-fired boiler?

The *ignition arch* is the curved refractory brick arch directly above the location where green coal enters the furnace in a chain-grate stoker-fired boiler. This refractory arch becomes very hot during furnace operation. The function of the ignition arch is to radiate intense heat onto the incoming coal. This helps to ignite the coal more quickly. It also helps direct the distilled gases from the coal up and into the flame, where they are burned.

What is a vibrating-grate stoker?

A *vibrating-grate stoker* has inclined grates that vibrate, causing the fuel bed to move slowly toward the lower end. **See Figure 6-12.** The grates are typically inclined about 15 degrees from horizontal. The boiler operator may adjust the shaker to alter the period of time that the grates vibrate, how frequently the vibration periods occur, or both. When adjusted properly, the coal or other solid fuel will burn completely by the time it reaches the lowest point on the grates.

Self-contained vibrating-grate stokers are most commonly gravity-fed from an overhead hopper or bunker. However, they are also used with spreader stokers.

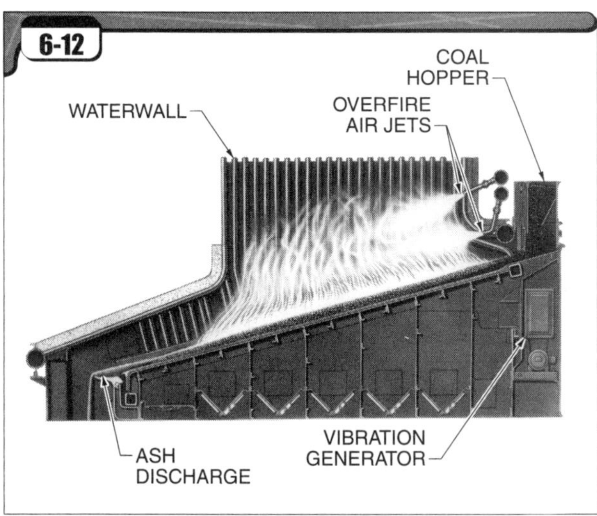

Figure 6-12. *Vibrating-grate stokers have inclined grates that vibrate, causing the ashes to fall off.*

How is the proper thickness of the fuel bed on the grates determined?

The grate maker makes specific recommendations for the proper thickness of the fuel bed on the grates. These recommendations are based on the type of equipment, type of fuel to be burned, and expected furnace temperatures.

In general, the fuel bed must be thick enough to protect the grates from the radiant heat from the furnace, which would warp the grates, but thin enough to allow ample draft from the forced draft (primary air) fan to pass through. About 4″ to 6″ of fuel and ash are normally maintained, depending on the type of equipment. The burning coal itself is normally about 1″ to 2″ thick, with the rest being ash.

How is the coal usage, in tons per hour, calculated for a stoker-fired boiler?

In order to calculate the coal usage accurately, the density (weight per cu ft) of the coal must be measured. The volume of the coal used in cubic feet per hour, times the weight per cubic foot, equals the weight of coal consumed in pounds per hour. The weight in pounds per hour is divided by 2000 to determine the weight in tons per hour.

For example, assume that a chain-grate stoker 6′ wide is burning bituminous coal. The grate speed is 4″ per minute and the coal being fed onto the grates is 4″ deep. The weight of the coal is 48 pounds per cubic foot. At what rate is coal being used, in tons per hour?

$$CR = \frac{W \times D \times S \times 60 \times D_c}{2000 \times t}$$

where

CR = consumption rate (in tons/hr)
W = width of stoker (in ft)
D = depth of coal (in ft)
S = speed of grate travel (in ft/min)
60 = minutes per hour
D_c = density of coal (in lb/cu ft)
2000 = lb per ton
t = time (in hr)

$$CR = \frac{W \times D \times S \times 60 \times D_c}{2000 \times t}$$

$$CR = \frac{6 \times 0.333 \times 0.333 \times 60 \times 48}{2000 \times 1}$$

CR = **0.96 tons/hr**

A traveling-grate stoker is 12′-6″ wide. The grate is turning 3″ per minute and the coal being fed onto the grates is an average of 4″ deep. If the coal weighs 50 lb/cu ft, at what rate is the coal being consumed in tons/hr?

$$CR = \frac{W \times D \times S \times 60 \times D_c}{2000 \times t}$$

$$CR = \frac{12.5 \times 0.333 \times 0.25 \times 60 \times 50}{2000 \times 1}$$

CR = **1.56 tons/hr**

What is a retort stoker?

A *retort stoker* is a stoker in which both the fuel and the combustion air are fed from below the combustion zone (the surface of the grates). Retort stokers are also known as underfeed stokers. Most retort stokers have only one retort and are known as single-retort units. Very large retort stokers for high-capacity boilers may have several retorts.

There are several types of single-retort stokers, but their basic operation is the same. An auger or a slowly reciprocating ram receives stoker-size coal from an overhead hopper. **See Figure 6-13.** The auger or ram plates force the coal into one end of a retort chamber. A *retort chamber* is a V-shaped trough, usually about 5′ to 12′ long, with a back plate enclosing the rear end of the retort opposite the feed ram.

After the ram pushes coal into the retort chamber, secondary ram plates carry the coal toward the rear of the retort chamber. When the retort chamber is full of coal and the ram continues to force coal into the retort, the coal presses up until it flows out of the retort chamber. The coal slowly overflows onto tuyeres and is quickly set on fire. A *tuyere* (tweer) is a special air-admitting grate designed to start combustion of the entering fuel. Tuyeres in some furnaces are a refractory shape with air holes for starting combustion.

The coal moves down from the tuyeres at a slight incline toward the sides of the furnace while passing across another row or several rows of grate bars. In many retort stokers, the grate bars have a rising and falling or lateral motion that slowly conveys the burning fuel bed and burned-out coal ashes toward the ash dump grates at each side of the stoker assembly. This movement helps to avoid clinkers by keeping the fuel bed broken up so that air may pass through. Primary air is supplied to the windbox below the grates by a forced draft fan. This air flows upward through the burning fuel bed.

The boiler operator manually operates actuators that cause the ash dump grates to open. This allows the ashes to fall into the ash pits under each side of the grates. The boiler operator removes these ashes by pulling them from the ash pits with a large metal hoe. Because a fairly large amount of unburned carbon tends to be dumped with the ash, retort stokers are less efficient than other stoker types.

The operation of multiple-retort stokers is the same, except for the direction of the retorts relative to the ash discharge. In multiple-retort stokers, the retorts are inclined, with the coal feed ram at the upper end of the retort and the ash dump at the lower end. Secondary rams help convey the burning coal and resulting ash toward the ash dump.

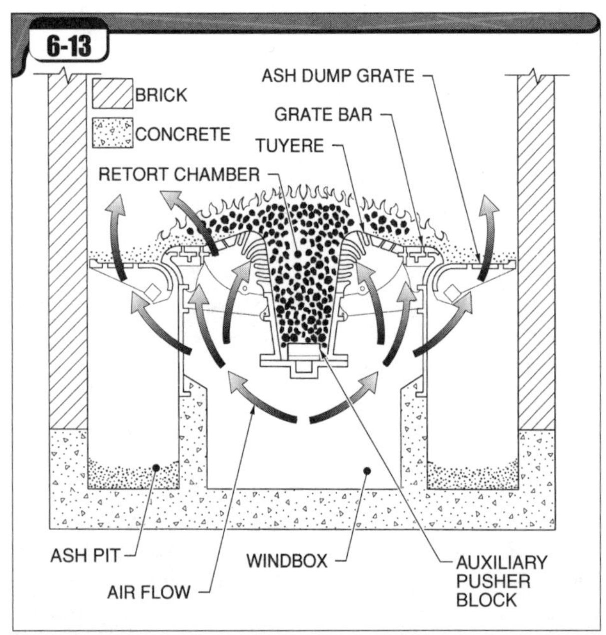

Figure 6-13. *In a retort stoker, coal is fed into the retort chamber and overflows across the tuyeres onto the grate bars.*

39 When firing coal, what is the purpose of cinder reinjection?

When firing coal using a stoker, some of the fine coal particles may be carried along with the flue gases into the boiler flue gas passages before they have fully burned. This is especially true when using spreader stokers because much of the fine coal burns in suspension. These partly burned particles are larger and heavier than flyash and tend to fall out of the flue gas stream. The particles fall into the first flyash hopper just after the boiler or below the economizer. The remaining carbon in these particles, if removed by the ash handling system, represents a loss of some of the heating value of the fuel.

The particles fall to the bottom of the flyash hopper and are dumped into a long vertical pipe called a cinder downcomer. **See Figure 6-14.** The bottom of the pipe joins with an overfire air nozzle. The overfire air thus conveys the particles back into the flame area so that the remaining Btus in the particles may be released. This results in slightly higher efficiency. Cinder reinjection is usually limited to the larger, heavier particles.

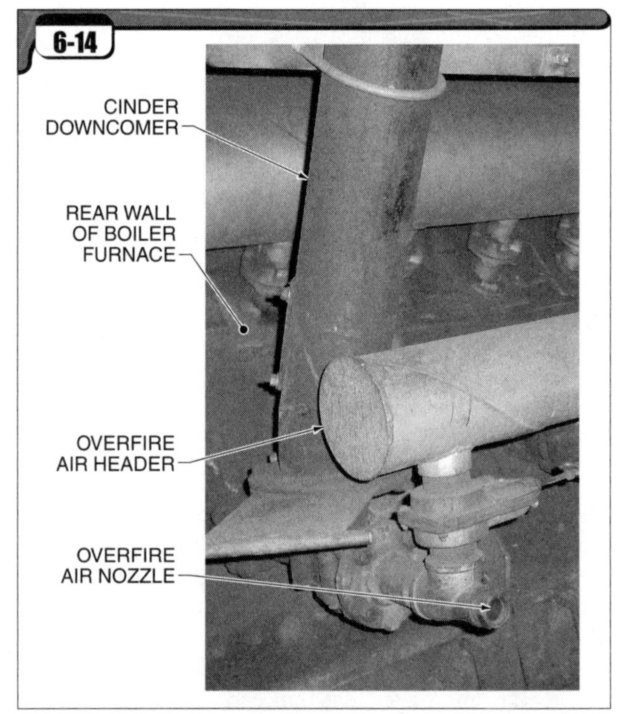

Figure 6-14. *Overfire air is used to reinject cinders into the furnace for maximum combustion efficiency.*

40 What is a shear pin? Why is it used in mechanical drive units?

A *shear pin* is a link used in a mechanical drive that is designed to break under a specific amount of shear stress. The shear stress occurs from too much torque from a drive on a shaft. For example, in a stoker drive unit, too much shear stress could be caused if the feed system jams because of a rock or other debris in the coal. In order to prevent mechanical damage or overheating of the drive, the shear pin breaks and allows the drive unit to spin freely. Once the cause of the jam is found and removed, the shear pin can be replaced and the stoker drive unit can be run.

Why is coal ground to a fine powder?

Coal is ground to a fine powder so that it can be burned in suspension using a pulverized coal burner. Coal burned in suspension is coal that is burned as a dust as it is blown into the furnace. The coal does not lie on a grate as it is burned. Burning coal in suspension is an efficient way to burn coal because the air-fuel mixture may be more closely controlled.

What is a pulverizer?

A *pulverizer* is a mill that grinds coal to a very fine powder. **See Figure 6-15.** The coal coming out of a pulverizer is roughly the consistency of flour or talcum powder. Pulverizers are made in a number of different styles.

A pulverizer is provided with a fan called an exhauster fan. The exhauster fan provides the movement of air needed to move the coal dust from the pulverizer to the pulverized coal burner at the boiler furnace.

The air is sent through a preheater before entering the pulverizer. The temperature leaving the air preheater is typically around 600°F, with some supplying air to the pulverizer as high as 825°F. Most of this heat is used in drying the coal in the pulverizers. Drying the coal also helps keep it from caking. This is important for transport of the coal to the burner and for efficient combustion. The coal dust and air mixture leaving the pulverizer is typically maintained at about 150°F to 180°F for bituminous coals and about 200°F to 210°F for anthracite coals. Bituminous coals may be burned at a slightly lower temperature because their higher content of volatile matter sustains ignition more easily. As the coal dust and air mixture enters the pulverized coal burner at the boiler furnace, it is mixed with secondary air from the forced draft fan. This creates the turbulence needed for efficient combustion and shorter flame travel.

Several types of coal pulverizers are used in modern steam plants. Common styles are attrition pulverizers, ball-and-race pulverizers, and roll-and-race pulverizers.

In an attrition pulverizer, the pieces of coal first enter a crusher-dryer section of the pulverizer. This section contains hammers that crush the coal to a fine granular size. The coal is then carried into the primary and secondary grinding sections. In these grinding sections, a rotating disc that has protruding pegs faces a stationary disc that also has protruding pegs. The pegs on the rotating disc intermesh with the pegs on the stationary disc. Pieces of coal that pass between the pegs are ground to powder. Most of the very fine grinding, however, occurs through attrition. This means that the coal particles are repeatedly dashed against the protruding pegs and against each other, causing the particles to fracture into smaller and smaller sizes.

The exhauster fans used with an attrition pulverizer are mounted on the same shaft as the rotating disc. These fans force the mixture of air and coal dust to the burner.

In a ball-and-race pulverizer, a pair of circular races face each other and large steel balls occupy the space between the races. This design resembles a ball bearing, with the races and the ring of metal balls in a horizontal direction. In these pulverizers, the top race is fixed and the bottom race rotates. The bottom race is the driving ring. The movement of the driving ring forces the balls to roll between the two races.

The pieces of coal enter through the raw coal feeder and fall into the race area. The steel balls roll over the coal, which is reduced to a fine powder. Coal is also crushed as it passes between the balls and as it is dashed against other coal. Primary air from an adjacent exhauster fan enters through the windbox area and passes through the pulverizer. The coal dust is carried to the burners by the primary air. A classifier at the outlet from the pulverizer segregates oversize coal particles and deflects them back into the race area for further size reduction.

In a roll-and-race pulverizer, roll wheels crush the coal. Heavy springs hold the roll wheels firmly against the rotating bottom race. The roll wheels in this design are vertically oriented, much like automobile wheels. In another design variation, the roll wheels are conical in shape and the coal passes between the roll wheels and the interior walls of a cone-shaped race. In this pulverizer, there is a small gap (¼″ to ⅜″) between the rollers and the race.

Primary air flowing through the pulverizer carries the coal dust through large metal tubes to the burners. The elbows and bends in these tubes are often designed to be easy to replace due to the erosive nature of the pulverized coal.

Finely ground coal may explode when mixed with air. This is especially true at high temperatures. Therefore, it is important to maintain the flame in the furnace when finely ground coal is being fed into the furnace. If finely ground coal builds up in the furnace and then ignites, an explosion can result.

How is the coal fed to the pulverizer at the proper rate?

In a pulverized coal system, the coal supply is stored in an overhead bunker. The coal feeds from the bottom of the bunker to a system used for weighing the coal and feeding it evenly to the pulverizer. One such system, called an overshot roll feeder, uses a rotating wheel that functions much like the wheel of a gristmill. Coal from the bunker flows down through a spout to fill the spaces between the blades in the rotating wheel. A leveling gate mounted over the rotating wheel scrapes off excess coal as the wheel rotates so that each space in the rotating wheel is filled level with the rim of the wheel. This results in a constant volume of coal being fed to the pulverizer with each turn of the wheel.

The total coal usage may be determined by measuring the density of the coal and multiplying by the volume moved with each revolution. The speed of the wheel is variable to allow for changes in the steam demand. This feeder is a volumetric feeder, which means that it measures the coal used by volume only. Samples of the coal must be weighed periodically to convert the volume used to estimated weight used.

Another type of feeder, known as a gravimetric belt feeder, uses a belt conveyor that moves the coal across a scale or load cell. **See Figure 6-16.** A leveling plate produces a uniform thickness of coal on the belt and the speed of the belt is adjustable to meet the steam demand. The scale or load cell weighs the amount of coal being used. This feeder is more accurate than a feeder based on the coal volume because the actual coal weight is measured by the scale or load cell and taken into account in the total coal usage.

Both designs may be referred to as mill feeders because they measure and feed the coal supply to the pulverizing mill, or pulverizer.

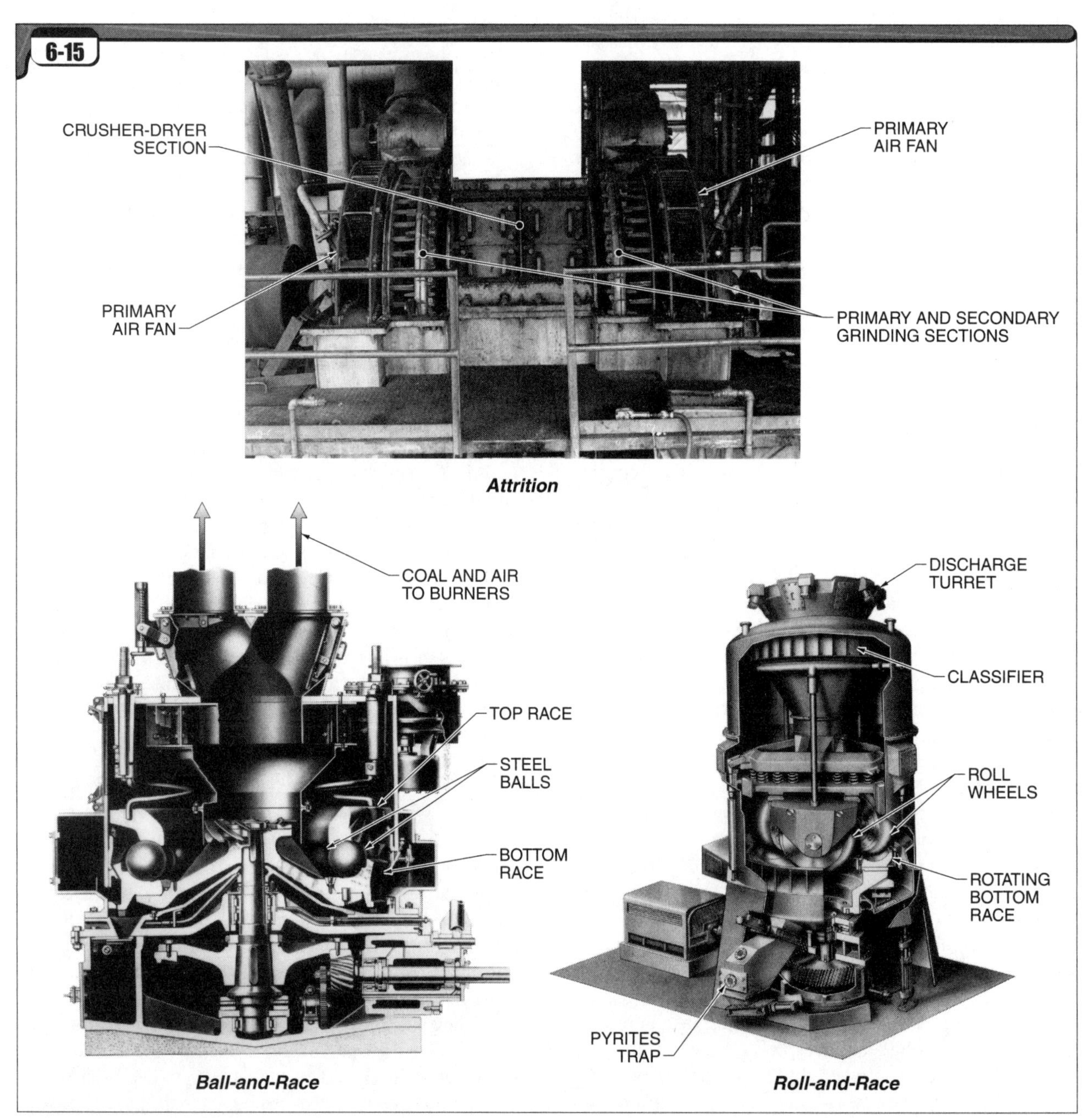

Figure 6-15. *Pulverizers grind coal to a very fine powder.*

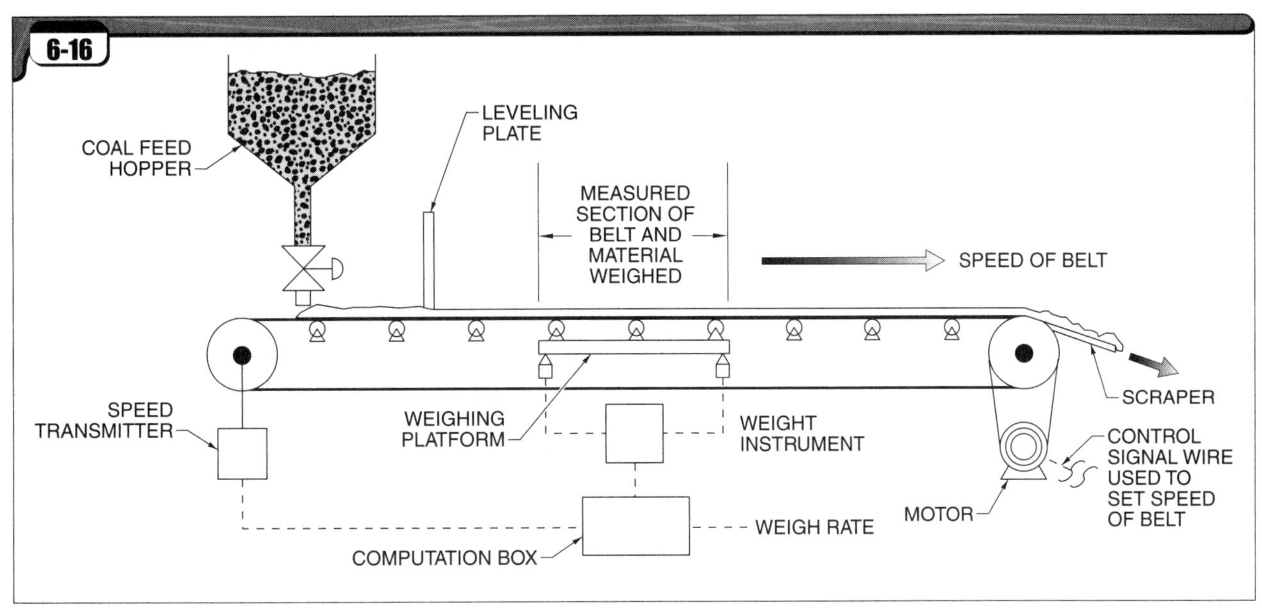

Figure 6-16. *Gravimetric coal feeders weigh the coal as it is fed to the pulverizer. This provides for accurate tracking of coal usage.*

44 Is the air blown or drawn through a pulverizer?

Both configurations are in common use. In the ball-and-race type, the air is blown through the pulverizer. These units require considerable care to keep the access ports sealed well so that coal dust does not create a housekeeping and safety issue. However, the coal does not contact the exhauster fan in this design, which reduces erosion inside the fan.

Other units, such as the conical roll-and-race design, use the exhauster to draw the air through the pulverizer. **See Figure 6-17.** This places the pulverizer under a slight vacuum, which minimizes coal dust leakage. However, the exhauster fan incurs some erosion from the coal dust so it must be reinforced with wear plates.

45 What is the purpose of a classifier on a pulverizer?

A *classifier* is a spinning set of vanes located at the coal and air outlet from the pulverizer that separates very fine coal dust from larger coal particles. It allows very fine coal dust to pass through, but throws out the larger coal particles. These particles fall back into the pulverizer for regrinding.

Attrition pulverizers are high-speed machines and often do not have classifiers. The oversized pieces of coal are thrown to the outside of the grinding area due to their own weight and are reduced in size by impact with the interior walls of the pulverizer and with each other.

46 What is a pyrites trap?

Pyrite is a common mineral consisting of iron disulfide (FeS_2). It has a pale, brass-yellow color. Pyrite is often found in veins of coal. It is often called fool's gold. Pyrite is very hard and causes wear in a pulverizer.

A *pyrites trap* is a compartment or box in the pulverizer that catches nuggets of pyrite as they are separated from coal.

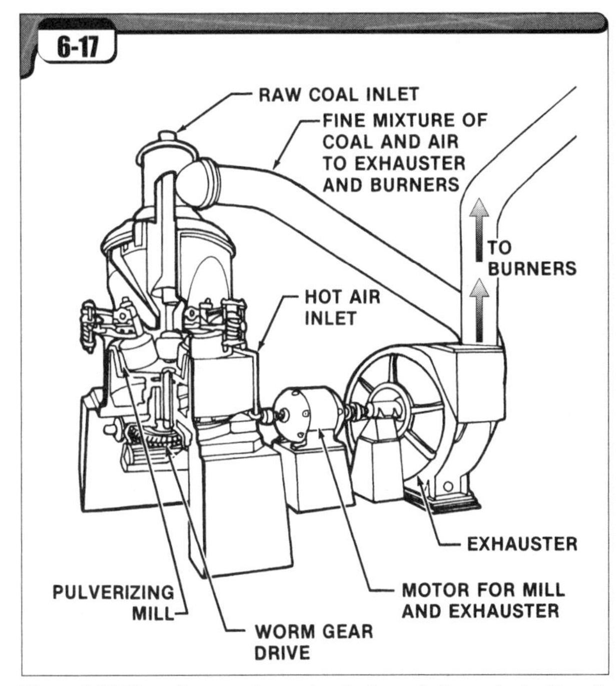

Figure 6-17. *The exhauster draws the air through the conical roll-and-race pulverizer, resulting in a slight vacuum inside.*

47 Are bituminous or anthracite coals generally better for use in a pulverizer?

Bituminous coals are softer than anthracite coals and are more suited for grinding. Bituminous coals also produce a higher percentage of volatile gases when heated. This helps the fuel ignite more quickly and maintain ignition more reliably. Therefore, bituminous coals are generally better for use in pulverizers.

48 How are metal objects separated from the coal as it enters the pulverizer?

An electromagnetic separator is installed over the top of the coal conveyors to pull ferrous items, such as wire, bolts, etc., out of the coal. Such items not only impede the grinding process, they also could cause sparks inside the pulverizer.

49 Describe a pulverized coal burner.

A pulverized coal burner consists of a tube through which the primary air and coal dust mixture enters the boiler furnace. A fixed impeller (also known as a diffuser) at the tip of the tube swirls the mixture as it flows into the furnace. **See Figure 6-18.** Secondary air from the windbox enters the furnace around the burner throat. The secondary air is swirled by air registers around the burner tube. This helps ensure complete combustion of the coal dust.

The furnace area around the pulverized coal burner becomes extremely hot (as high as 3500°F) and is subject to abrasion. Therefore, waterwall tubes adjacent to the burner throat are commonly studded tubes that are covered by a coating of castable refractory for heat dissipation and abrasion protection.

The coal dust flame may be unstable at very low startup flow rates and can cause considerable trouble with smoke until the furnace is very hot. For this reason, fuel oil or natural gas burners may be provided for initial startup. Once the furnace is heated to working temperatures, the pulverized coal burners are placed into service and the other fuel is shut down.

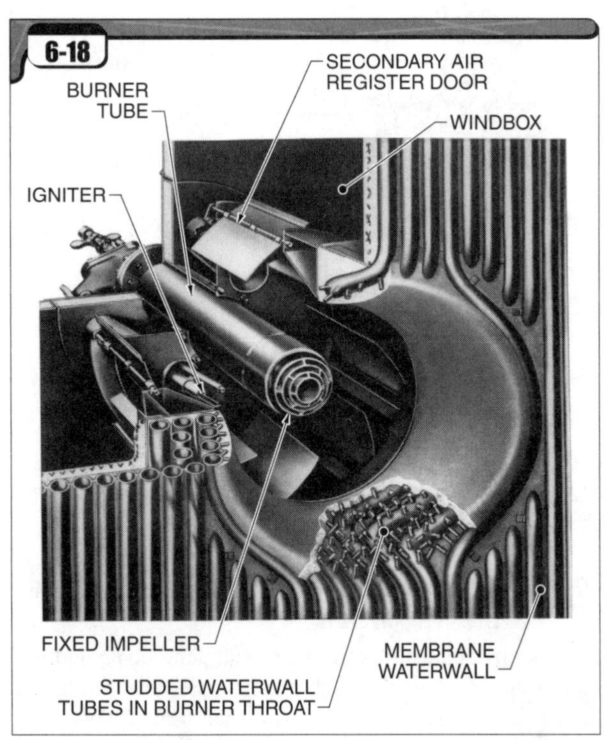

Figure 6-18. *Pulverized coal burners swirl air and coal dust as the mixture enters the furnace.*

50 What is a fluidized bed boiler?

A *fluidized bed boiler* is a boiler in which fuels are burned in a bed of inert materials such as limestone pellets or sand. **See Figure 6-19.** The bed is often supported on a surface formed by a membrane waterwall panel. The membrane panel is formed in such a way as to create a closed windbox below the bed.

Many small nozzles called bubble caps pass through the steel membrane strips and these nozzles are used to inject air into the bed material. For this reason, the part of the membrane waterwall panel that supports the bed material is known as the distributor plate. The bubble caps are small tubes, a few inches long, with the upper ends closed. Small holes are drilled in the upper end to allow the combustion air to blow in multiple directions into the bed material.

As the air flow increases, it exerts upward force on the bed material. The particles that make up the bed become semisuspended in the upward flow. As the air flow increases, air bubbles are formed in the bed material and the bed looks like a boiling liquid.

Fluidized bed boilers vary greatly in design and in the fuel used. When solid fuels such as coal are used, the fuel is delivered to the bed material from a storage bunker. The fuel may be injected into the bed at a point above or below the surface of the bed.

The two main types of fluidized bed designs are the bubbling fluidized bed and the circulating fluidized bed. In a bubbling fluidized bed, most of the bed material remains semisuspended in the bottom part of the furnace area. A relatively small amount of bed material and fuel carbon carries over into the boiler flue gas passages. In a circulating fluidized bed, a large percentage of the bed material carries over with the flue gases, falls into hoppers below the passages, and is recirculated to the furnace. There is no clear transition between the more dense bed in the bottom of the furnace and the suspended portion above.

There are several advantages presented by fluidized bed combustion. The bubbling or circulating bed material increases heat transfer rates into the boiler water. Fluidized bed combustion can also occur at lower furnace temperatures, which reduces the formation of certain pollutants. For example, the combustion zone in a fluidized bed boiler reaches about 1500°F to 1600°F, as compared to 3000°F to 3500°F in a conventional pulverized coal-burning boiler. At temperatures above about 2700°F, there is a chemical reaction between nitrogen and oxygen that forms nitric oxide, an atmospheric pollutant. Combustion at the lower temperatures in a fluidized bed boiler greatly reduces the formation of nitric oxide.

In addition, sorbent materials such as limestone may be included as part or all of the bed material. A chemical reaction known as sulfation occurs between the sulfur compounds in coal and the slightly alkaline limestone sorbent. The reaction removes much of the sulfur. This helps to prevent sulfur pollution and acid rain. The limestone is replenished in the bed material as needed and the spent bed material is tapped off or "blown down" from the bed.

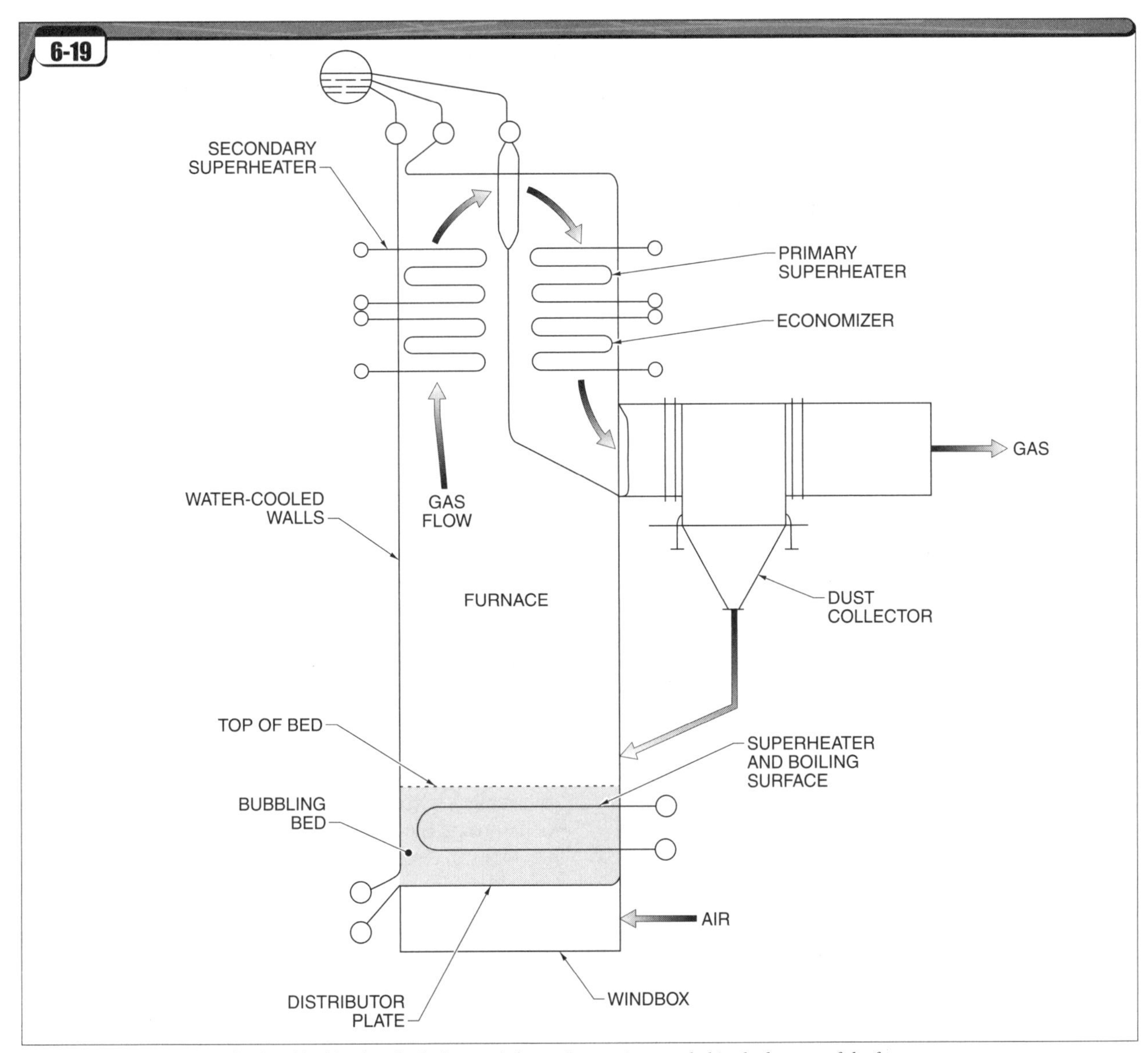

Figure 6-19. *In a bubbling fluidized bed boiler, the bed material remains semisuspended in the bottom of the furnace.*

Coal-Burning Challenges

51 What should be considered if coal is stored for more than two or three weeks?

Coal is subject to spontaneous combustion. *Spontaneous combustion* is the process where a material can self-generate heat until the ignition point is reached. The presence of moisture and oxygen in the coal causes oxidation of the coal. Oxidation generates heat. Heat in a closed area can increase until the ignition point is reached. A high percentage of sulfur in the coal can add to this problem. Coal in large coal yards is kept packed down to help keep oxygen out. If the quantity of coal kept in outdoor storage is not great, the coal can be moved around with large machinery periodically to vent off any local pockets of heat that have developed. **See Figure 6-20.**

52 What is a clinker?

A *clinker* is a mass of coal and ash that has fused together during burning. Clinkering is caused by a lack of air, an uneven fuel bed, or the use of less desirable grades of coal for the type of burning equipment used. Preventing some grades of coal from clinkering is difficult. Other solid fuels may also have problems with clinkering.

Clinkering negatively affects the operation of the boiler furnace because clinkers prevent air from passing through the fuel bed. Also, any coal fed on top of clinkers will probably not burn properly. If starved for air, the coal on the entire grate surface may become clinkered. Boiler operators should make regular observations of the boiler furnace to check for the presence of clinkers. Boiler operators use slice bars and hoes to break up and/or remove clinkers from the fuel bed.

Figure 6-20. *Coal piles are subject to spontaneous combustion.*

What is a clinker grinder?

A *clinker grinder* is a large set of steel rollers with heavy teeth that grind ash and clinkers to reduce their size before they enter the ash hoppers. **See Figure 6-21.** The ground ash and clinkers are easier to remove from the ash hoppers than unground ash and clinkers. Clinker grinders are especially helpful if the ashes are conveyed by a pneumatic conveying system to a silo.

Figure 6-21. *Clinker grinders crush clinkers and ash to reduce their physical size. This improves handling.*

What is meant by cleaning the fires?

Cleaning the fires means removing ash and clinkers from a solid fuel-fired boiler furnace.

Factoid

Operators of stoker-fired boilers must make periodic adjustments of the fuel bed to prevent the formation of clinkers.

What are combustibles in the refuse?

Combustibles in the refuse are unburned or partly burned pieces of coal in the ash coming off the grates. This ash is referred to as bottom ash. The combustible fuel is seen as red coals that fall off the grates with the ash. Combustibles in the refuse are a waste of fuel and a subsequent loss of efficiency. A boiler operator should minimize this loss by keeping the proper thickness of the fuel bed and by watching the fuel bed for proper air and coal distribution.

In most cases, the thickness of the fuel bed is controlled by regulating the speed and/or frequency of the grate movement and by adjusting the fuel gates on the feeder. The distribution of coal on the grates is controlled by adjustments made to the coal feeding equipment. For example, the paddles in spreader stokers may be adjusted so as to throw the coal in a wider or narrower arc and the speed of a spreader stoker determines how far back onto the grates the coal is thrown.

The distribution of coal on the grates may be negatively affected by unwanted segregation of coal in the coal bunkers and/or feed hoppers. Segregation is the process where very small pieces of coal (fines) may separate from the larger pieces during handling and storage. This is somewhat preventable through correct operation of the coal handling equipment. The use of nonsegregating spouts where the coal flows from an overhead bunker into the feed hoppers helps avoid size segregation as well. **See Figure 6-22.**

Zone dampers under the grates are used with some stoker-fired boilers to control the primary air flow to various sections of the grates. The zone dampers allow the boiler operator some flexibility in directing air to particular areas of the grates.

Figure 6-22. *Nonsegregating spouts help distribute coal uniformly to the coal-feeding equipment.*

Factoid

Zone dampers below the grates in a stoker-fired boiler allow the boiler operator flexibility in distributing the combustion air to the coal.

56 Describe the problems caused by holes in the fire.

A hole in the fire is an area on the grates of a solid fuel-fired furnace where the fuel bed is very thin or an area where the bare grates are exposed. Two problems are caused by holes in the fire. First, the grates may not have enough protection from the radiant heat from the furnace and may be damaged by overheating. Second, holes in the fire allow the primary air coming through the grates from below to pass through freely. The fuel on the rest of the grate may be starved for air because the thin areas are the paths of least resistance. Fuel that is remote from the thin areas may clinker.

57 How is the combustion process affected by the failure of one spreader stoker in a furnace fired with multiple spreader stokers?

The spreader stokers feed the coal in overlapping rows that extend from the front of the furnace to the back. If one spreader stoker fails, the fire dies down in that row as the remaining coal burns out. A thin area in the fuel bed from the front of the boiler to the rear results over a short time. Air begins to short-circuit through that row. Clinkering starts and smoke begins to form because the fuel on the rest of the bed is starved for air. In addition, the resulting poor burning conditions cause the steam production from the boiler to fall quickly.

58 What is abrasion of the tubes?

Abrasion (erosion) of the tubes is the damage done to the tubes by fly ash. This occurs in solid fuel-fired watertube boilers. Fly ash carried along with flue gases strikes the leading edge of the tubes, especially those closest to the furnace. The fly ash, in effect, sandblasts the tubes. A tube that is worn thin on the leading edge is more subject to rupture at that point. **See Figure 6-23.**

Some facilities use tube shields to help protect the tubes from this abrasion. Tube shields are cladding made of stainless steel alloys that are resistant to high temperatures and abrasion.

Abrasion also occurs at the rear waterwall in spreader stoker-fired watertube boilers. It results when the spreader stokers are allowed to throw the coal too far back into the furnace and the coal strikes the rear waterwall tubes. Over a period of time, this results in abrasion of the tubes.

59 What is ash fusion temperature?

Ash fusion temperature is the temperature at which ash begins to become molten. Fly ash that is carried along by the flue gases in a boiler furnace is residue from fuel that is already burned. The fly ash actually begins to melt and become sticky when it reaches a high temperature. For anthracite coals, the fly ash begins to melt at about 2800°F. For bituminous coals, the fly ash begins to melt at about 2000°F.

If these semimolten ash particles come into contact with the comparatively cool tube surfaces, the ash particles will give up their heat and harden again. This results in a slag and ash coating on the tubes. The coating insulates the tubes and reduces heat transfer. Since solid fuels produce more ash than liquid or gaseous fuels, this condition is more of a problem when burning solid fuels. The ash fusion temperature may also cause the ash in the fuel bed to clinker.

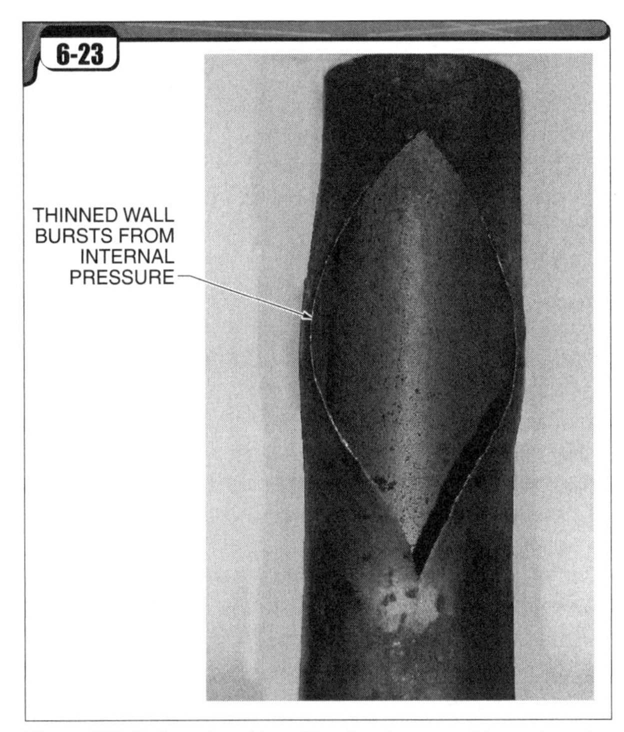

Figure 6-23. *Boiler tubes thinned by abrasion are subject to bursting from the internal pressure.*

60 What is fouling of the tubes?

Fouling of the tubes is the buildup of ash and slag on the tubes in a watertube boiler to such a degree that it restricts the flow of flue gases between the tubes. When the accumulation builds to this extent, it often requires a shutdown of the boiler to remove the fouling. If the fouling consists of hard slag (fused and melted flyash), it sometimes must be removed by mechanical chipping, shot blasting, or even dynamiting.

61 What is banking a fire?

Banking a fire is the process of greatly slowing the burning of coal or some other solid fuel. Banking consists of mounding up the red-hot coals in the furnace and covering them with a layer of green, unburned coal. Where possible, a layer of fine coal or ash is put over the top of the pile and air flow is reduced to a bare minimum. Banked coal burns very slowly while it keeps the furnace hot.

The advantage of banking a fire is that steam can be produced in a short time with small thermal strains on the boiler and furnace. All that is needed is to stir up the pile to expose the red-hot coals. The flow of air is increased and fresh coal is fed into the furnace. For example, the fire may be banked in a small boiler furnace located in a courthouse or school where the boiler is left unattended at night. The fire may also be banked in a larger boiler when the need for steam is low but expected to rise quickly within the next several hours.

Banking a fire today is not as common as it was in the past because of the difficulty in controlling smoke.

Can a furnace explosion occur under a banked boiler?

If the draft is not enough to remove volatile gases from the furnace, a buildup of ignitable smoke and gases can accumulate. This mixture can explode if adequate oxygen is introduced. In order to prevent explosive vapors from building up, the stack damper must not be closed tightly.

Why should ashes be removed frequently from the ash pit?

Ashes should be removed frequently from the ash pit because hot coals in the ashes may fuse together and form clinkers in the ash pit. Clinkers are hard to break up and remove if neglected. The process of breaking up the clinkers and removing them may expose personnel to burn hazards.

Heat from the hot ashes can also damage the grates from underneath where the grates are not protected. In addition, having the ash pit doors open for extended times allows cold air to be drawn into the furnace, which lowers efficiency.

How is the bottom ash removed from a solid fuel-fired boiler furnace?

In some cases, the ash from the grates falls into a water bath that causes the ash to cool and fracture from thermal shock. A submerged conveyor moves the ashes from the water bath and onto belt conveyors. The belt conveyors transport the ash to an outside ash pile or silo for removal.

In other plants, the bottom ash falls from the grates into the bottom ash hopper. The boiler operator opens an exterior door to access the hopper and then uses hoes to pull the bottom ash into the pneumatic conveying system. This type of system uses strong exhaust blowers to pull the ashes by vacuum through heavy steel tubes into a silo. **See Figure 6-24.**

Even though the coal is burned in suspension in pulverized coal-fired boilers, a small amount of ash falls to the bottom of the furnace. Therefore, pulverized coal-fired units often have bottom ash hoppers as described above. In some cases, the ash melts in the bottom of the furnace in a pulverized coal-fired unit. In these cases, the molten slag runs from the furnace through a funnel-shaped section and into a water bath, where it is cooled and then conveyed to an ash silo.

Figure 6-24. *Bottom ash is often held in a silo to await disposal.*

Fuel Oils

How are grades of fuel oil designated?

Grades of fuel oil are designated by number from No. 1 to No. 6. Fuel oils commonly used for fuel in steam boiler furnaces are No. 2, No. 4, No. 5, and No. 6. The heating value of fuel oil is expressed as Btu per gallon (Btu/gal) or Btu per pound (Btu/lb). Heavier, higher-numbered fuel oils produce more Btu per gallon than lighter, lower-numbered fuel oils. **See Figure 6-25.**

No. 2 fuel oil is a light, clean, distillate fuel oil with an amber color. A *distillate fuel oil* is a fuel produced by distilling crude oil. Distillate fuel oils contain the lighter, more volatile hydrocarbon elements. No. 2 fuel oil is the same oil used in residential fuel oil furnaces. Except for certain additives, No. 2 fuel is the same as diesel fuel. It is easy to burn, requiring only light filtering to remove trace impurities. It is used in most package boiler installations burning fuel oil because it requires comparatively little attention. The heating value of No. 2 fuel oil is about 141,000 Btu/gal or 19,500 Btu/lb.

No. 4 fuel oil is heavier and darker than No. 2 fuel oil. It may be a distillate, but more often it is a blend of No. 2 fuel oil and heavier fuel oils. A *blended fuel oil* is a mixture of distillate oils and residual oils that may contain some crude oil. A *residual fuel oil* is fuel oil that remains after the lighter, more volatile hydrocarbons have been distilled off. A residual fuel oil contains some of the heavy fractions from the fuel oil refining process. A fraction is one of the separate materials that come off a refinery distillation column. Heavy fractions include residual oils and tars. Lighter fractions include gasoline and kerosene. The heating value of No. 4 fuel oil is about 146,000 Btu/gal or 19,100 Btu/lb.

No. 5 fuel oil is a black residual oil that may need preheating to lower its viscosity when cold. Preheating makes the oil easier to pump. The heating value of No. 5 fuel oil is about 148,000 Btu/gal or 18,950 Btu/lb.

No. 6 fuel oil is a black residual oil, sometimes called "bunker C." A *bunker oil* is one of the heavy oils formed as crude oil is stabilized after the lighter components have been distilled off. The term bunker oil is becoming obsolete and is not often used anymore. No. 6 fuel oil often contains debris and impurities and has a very high viscosity when cold. It must be preheated to lower its viscosity so that it can be pumped and then must be further heated by a steam or electric heater for quick ignition and efficient burning when it enters the furnace. Piping carrying heavy oils is traced with steam, hot glycol/water mixture, or electric heat tracing cable to keep the oil from thickening in the piping.

Because No. 6 fuel oil contains some debris and requires more preparation before burning than other fuel oils, it is less expensive than other fuel oils. No. 6 fuel oil also has a higher heat content (Btu/gal) than other fuel oils. The lower cost and higher heat content make it attractive for a facility that is willing to tolerate the extra work, inconvenience, cost of some additional handling equipment, and cost of energy for heating. The heating value of No. 6 fuel oil is about 150,000 Btu/gal or 18,750 Btu/lb.

The estimated heating values in Btu per gallon of the various grades of fuel oil vary among refiners and suppliers. Consult the refiner or supplier for specific analyses.

	NO. 2	NO. 4	NO. 5	NO. 6
Type	light distillate	light distillate or blend	light residual	residual
Color	amber	black	black	black
°API gravity at 60°F	32	21	17	12
Specific gravity	0.8654	0.9279	0.9529	0.9861
Btu/gal.	141,000	146,000	148,000	150,000
Btu/lb	19,500	19,100	18,950	18,750

Figure 6-25. *Fuel oil commonly used for steam boilers is No. 2, No. 4, No. 5, or No. 6.*

How is the viscosity of fuel oil measured?

Viscosity is the resistance to flow of a fluid. Thick liquids have high viscosity. Thin liquids have low viscosity. For a liquid fuel, viscosity is determined by measuring the amount of liquid that will pass through a given size orifice under a controlled temperature in a given amount of time. High viscosities increase the amount of energy that must be used to pump the liquid through piping and may interfere with proper atomization.

A standard instrument used to measure viscosity is a viscometer, or viscosimeter. The measurement is often expressed in Saybolt seconds, universal (SSU). The SSU test is a standard ASTM International test that measures the time it takes for 60 cubic centimeters (cc or cm^3) of the oil sample to flow through a standard orifice at the standard test temperatures. The viscosity is normally measured at 100°F, 150°F, and 210°F. The standard amount of the oil is placed in the viscometer. The viscometer consists of a metal cup with a specific size hole in the bottom. The amount of time that it takes for the quantity of oil to drain from the cup through the orifice is measured.

For high-viscosity liquids, the measurement is expressed in Saybolt seconds furol (SSF). The SSF instrument uses a slightly larger orifice and slightly higher temperature.

Factoid

Heavy fuel oils are low in cost, but require considerable expense in handling and labor. Heavy fuels must often be heated before being pumped to the boiler. Tracing may be needed to keep the liquid pumpable.

Define the terms flash point, fire point, and pour point.

Flash point is the lowest temperature at which the vapor given off by a substance will make a flash of flame, but not continue to burn, when an open flame is passed over it. An oil or liquid fuel with a low flash point ignites more readily than one with a higher flash point. Therefore, the flash point is used as a safety consideration in transportation, handling, and storage. Heavy oils are preheated to near the flash point so that they will ignite readily when they enter the furnace.

Fire point is the lowest temperature at which the vapor given off by a substance will ignite and burn for at least 5 sec when exposed to an open flame. The fire point is also known as the ignition point.

Pour point is the lowest temperature at which a liquid will flow from one container to another. The pour point is also known as the chill point.

How is the specific gravity of fuel oil measured?

Specific gravity is the weight of a given volume of a material divided by the weight of an equal volume of water when both are measured at the same temperature. In almost all cases, the standard temperature used is 60°F. The specific gravity of fuel oil is used as a general measurement in classifying the quality and suitability of fuel oil. It is used in much the same way in which the proximate analysis of coal is used to specify the overall suitability of coal.

What is °API?

The API is the American Petroleum Institute. The term °API is used for the measurement of density for petroleum products on the API scale. The API scale is the accepted scale in the petroleum industry for measuring the specific gravity of petroleum products. The specific gravity of petroleum products may be converted to °API by applying the following formula:

$$°API = \frac{141.5}{SG} - 131.5$$

where
$°API$ = density (in degrees API)
141.5 = constant
SG = specific gravity (at 60°F)
131.5 = constant

For example, convert a specific gravity of 0.91 to °API.

$$°API = \frac{141.5}{SG} - 131.5$$

$$°API = \frac{141.5}{0.91} - 131.5$$

$°API$ = **24°**

CAUTION

Specific safety procedures must be used when entering the confined space of any fuel oil tank.

List and describe the components of a fuel oil system for a boiler using No. 2 fuel oil.

Fuel oil-fired boilers have more complex fuel delivery systems than natural gas-fired boilers. This is because the fuel oil system must condition the fuel as well as deliver the fuel safely to the burner. The fuel oil system must supply clean oil and supply it at the proper volume, pressure, and viscosity.

Note: Specific requirements for devices in a fuel oil system vary depending on insurance requirements, jurisdictional requirements, and internal corporate requirements. Not all the components described may be required in some cases. In other cases, additional equipment may be required. The design of any fuel oil system should be approved through competent engineering and safety evaluations.

A fuel oil tank is required to store the fuel oil. **See Figure 6-26.** A fill line (1) is used for filling the tank and a vent line (2) is used to vent the tank while filling. The vent line also prevents the formation of a vacuum inside the tank as the liquid level drops. A measurement well connection (3) allows the boiler operator to insert a measuring dipstick into the tank to determine the fuel oil level. Pneumercators and differential pressure level transmitters (4) are commonly used on larger fuel oil tanks. A *pneumercator* is an air-actuated liquid level measuring device.

The suction line (5) to the fuel oil pump ends a few inches above the bottom of the tank so that sludge and sediment that may accumulate in the bottom will not be drawn into the pump, lines, and burner. A second suction line with a higher inlet (6) is sometimes used as a contingency for such an event. If the fuel oil tank is installed so that the fuel oil flows to the fuel oil pump(s) under head rather than under vacuum, an antisiphon valve should be installed in the suction line(s). This prevents oil from being siphoned from the tank in the event of a leak in the pump suction line. Insurance requirements may also dictate that a spring-loaded valve with a fusible link be installed in the pump suction line in these cases. In the event of a fire, the fusible link will melt and allow the spring-loaded valve to close.

A hatch (7) is provided for cleanout of the sludge and sediment when necessary. Fuel oil tanks are sometimes buried to minimize space requirements, avoid freezing conditions, and increase safety. However, detecting leaks is more difficult, and corrosion and deterioration of buried fuel oil tanks must be considered. Buried tanks require special precautions to ensure that ground pollution does not occur. If the tanks are installed above ground, containment areas must be constructed to prevent environmental or safety hazards in the event of a spill.

Many plants use multiple fuel oil tanks for redundancy, for convenience, and to limit the size of the tank required. For example, having multiple tanks allows the boiler operators to use the older fuel oil first, or to service one tank while another is in use.

If the fuel oil tank is below the level of the fuel oil pump, a check valve (8) is used in the piping to help keep the suction piping full of oil when the pump is shut down. A manual shutoff valve is also used at the tank outlet.

A pressure gauge (9) is installed before the fuel oil strainer to allow the boiler operator to determine the pressure or vacuum in the fuel oil line before the strainer. If the tank is above the pump, pressure is created at the strainer by the head of liquid in the fuel oil tank. If the fuel oil tank is below the pump, a vacuum gauge is used instead. A thermometer (10) should also be installed at this point to allow the boiler operator to read the temperature of the fuel oil as it comes from the fuel oil tank.

Fuel oil is strained to remove impurities such as dirt and rust. These impurities must be removed or they can wear the pump, foul the burner tips, or plug the fuel oil heater. Wire mesh strainers are used to remove these impurities. For convenience, duplex strainers (11) are often used to remove impurities from fuel oil. A duplex strainer consists of two strainer baskets in separate casings that are connected with a plug valve. The plug valve switches from one strainer to the other when one side becomes plugged. This arrangement allows one side to be cleaned while the other side is in service and minimizes the number of valves and pipe fittings needed. Alternately, a pair of single-basket strainers may be installed in parallel.

If a pressure or vacuum gauge (9) is installed before and after the fuel oil strainer, the boiler operator may compare the two to measure the differential pressure across the strainer. This shows when the strainer is beginning to plug up with debris.

Fuel oil pumps (12) used for supplying fuel oil to burners are positive-displacement pumps. Gear pumps are normally used, although vane pumps, lobe pumps, or screw pumps may also be used. These rotary-type pumps may be direct drive or indirect drive. Having a steam-driven pump as a standby pump in the event of a power outage is sometimes advisable. However, steam-driven reciprocating pumps such as duplex pumps normally have a pulsation in the discharge pressure. Therefore, an air chamber (also called an accumulator) must be used to prevent surges in the fuel oil pressure to the burner. A steam turbine-driven standby pump is often a good practice.

Positive-displacement pumps used for fuel oil service are sized to provide more fuel oil than is used by the burner(s). The excess fuel oil is recirculated either to the suction side of the pump or to the storage tank by a pressure relief valve (13) installed in the discharge piping. This valve differs in use from a relief valve used for emergency pressure relief in that this valve is expected to relieve the excess fuel oil flow as a part of normal operation. Therefore, it is in reality a pressure control valve. The pressure relief valve maintains a constant fuel oil supply pressure to the burners as well as prevents pressure buildup in the piping. Such pressure could result from a downstream valve in the fuel oil line being closed while the pump is in operation. For this reason, it is critical that the pressure relief valve be placed immediately downstream of the fuel oil pump, before any manual or automatic shutoff valves.

A check valve, manual shutoff valve, and pressure gauge are installed at the discharge of the fuel oil pump. The check valve prevents backflow in cases where a primary pump and standby pump are installed in parallel. The manual shutoff valve is provided for isolation and service of the pump. The pressure gauge allows the operator to keep an eye on the discharge pressure from the fuel oil pump.

Drains (14) are used in the piping to allow the fuel oil lines to be drained for service work. Vent valves (15) should be used at high points in the piping to allow trapped air to be vented during startup of the system. Failure to vent the trapped air may lead to an erratic supply of fuel oil from the pump and inconsistent or erratic pressure delivered to the fuel oil burner.

A second strainer is often provided just before the control valves and fuel oil burner. This helps prevent any debris in the piping from lodging in the downstream control valves, safety switches, or shutoff valves. This unit may also be a duplex strainer, or a pair of single-basket strainers in parallel so that operation may be maintained while cleaning the strainer.

A pressure gauge after the second fuel oil strainer, together with the one before the strainer, allows the boiler operator to monitor the pressure drop across that strainer.

The fuel oil pressure regulating valve (16) is an adjustable pressure reducing valve that controls the fuel oil pressure delivered to the burner itself. It is initially adjusted to provide the correct pressure, and further pressure adjustments should be minor.

Low fuel pressure may cause the flame to become too lean and unstable and can allow the burner flame to move too close to the burner tip and refractory throat. If the flame moves into the burner throat, the burner and surrounding area can be damaged and carbon deposits can build up rapidly on the burner. An unstable flame may result in the flame going out and allow raw atomized fuel to enter the furnace. This could result in a buildup of explosive vapors in the furnace.

For these reasons, a low fuel oil pressure switch (17) is supplied. This safety switch causes the fuel oil to be shut off and the boiler to shut down if the fuel oil pressure is inadequate. When this switch is closed, this proves to the burner control system that the fuel oil pressure is adequate for proper burner operation. A high fuel oil pressure switch (18) is also used in many plants. If the fuel oil pressure regulating valve should fail, excessive fuel oil pressure could be delivered to the fuel oil burner. This could cause too much fuel oil to enter the burner. As a result, the air-fuel mixture in the furnace could become too rich to burn and the flame could go out. As the flow of combustion air through the furnace dilutes the mixture, however, the mixture may again become combustible. If it ignites, a furnace explosion could result.

Too much fuel oil pressure could also result in poor atomization of the fuel oil. Poor atomization causes the burning conditions to be smoky and oily soot can deposit on the boiler tubes, inside of the breeching, and in pollution control equipment.

The high fuel oil pressure switch causes the control system to shut down the boiler if too much fuel oil pressure is detected. When this switch is closed, this proves to the burner control system that the fuel oil pressure is not excessive.

A pressure gauge after the fuel oil pressure regulating valve allows the boiler operator to adjust the valve to provide the correct fuel oil pressure to the burner.

The firing rate control valve (19) controls the quantity of fuel oil delivered to the fuel oil burner in order to control the amount of steam generated. This valve is modulated based on the demand for steam. It is typically a three-way valve that diverts part of the fuel oil flow to the fuel oil burner and the remainder to a recirculation line that flows back to the fuel oil tank. The quantity of fuel oil sent to the fuel oil burner determines the quantity of steam produced. For example, on package boilers, this valve is positioned by a modulating motor, which is in turn positioned by a modulating pressure control.

A fuel oil meter (20) should be provided in the oil line to the boiler so that the boiler operator may record and track fuel usage. The meter should be provided with isolation valves and a bypass so that the meter may be serviced when necessary without removing the boiler from service.

The safety shutoff valves (SSOV) (21) are the main shutoff valves that open to admit fuel oil flow to the fuel oil burner and close to stop fuel oil flow. These valves are equipped with actuators that allow the boiler control system to open and close the valves at the proper time. On very small systems, the actuators may be solenoids. On larger burner systems, the actuators operate electrohydraulically or by motors so that they open in a slower, more controlled manner. The valves are designed to close immediately in the event that an unsafe condition is detected. Insurance requirements often call for two fuel oil SSOVs in series.

Insurers typically require that one of the SSOV actuators be equipped with a proof of closure switch. A *proof of closure (POC) switch* (22) is a sensor that detects the position of a valve to ensure that the valve closes properly. If the POC switch detects that the valve is not fully closed when it should be, this could mean that raw fuel oil is leaking through the valve and into the furnace area. This buildup of fuel could cause a furnace explosion. Therefore, if the POC switch is not confirmed closed at the proper time, the boiler control system will prevent the burner from firing.

The fuel oil burner is the last piece of equipment in the fuel oil train (23). Fuel oil is normally supplied from the SSOVs to the fuel oil burner through a flexible hose.

Most fuel oil burners are installed in the burner throat in such a way that they may be retracted and/or removed completely. For example, larger fuel oil-fired boilers often have more than one fuel oil burner so that one burner may be cleaned or serviced while another is in operation. This eliminates the need to shut the boiler down during this work. In boilers that burn more than one fuel, the fuel oil burner may be pulled into a retracted position while another fuel such as natural gas is burned. This prevents overheating damage to the idle burner.

In such cases, it is important that the fuel oil burner be restored to its correct position before it is placed in service. If the fuel oil burner is left in the retracted position, it will not be in the correct location within the burner throat and poor combustion will result. In addition, the burner and burner throat can be damaged.

To prevent such events, the fuel oil burner may be equipped with a burner position switch (24). This is also called a burner drawer switch. The burner must be slid into the correct location within the burner throat in order to satisfy this safety switch. If the burner position switch is not in the correct position, the boiler control system will not allow the burner to start.

Factoid

When the discharge valve on a pump is closed while the pump is in operation, the pump is said to be deadheaded.

What additional features and equipment must be added to the fuel oil system when heavy oils are burned?

Fuel oil heaters are required to lower the viscosity of the fuel oil in a fuel oil tank when heavy grades of fuel oil (No. 5 or No. 6 fuel oil) are used so that the fuel oil may be pumped properly. The oil should be kept in the tank at about 100°F to 120°F. Fuel oil heaters are operated by steam or electricity. **See Figure 6-27.** A fuel oil heater inside the fuel oil tank typically includes a steam heater (1) that heats the immediate area around the pump suction line. Steam is provided through a pressure reducing valve (4) and temperature-control valve (5). If steam is unavailable, an electric heater (2) may be used. This design ensures flow into the suction piping, but makes heating the entire contents of the fuel oil tank unnecessary. The thermometer in the pump suction line (6) is especially important if heavy fuel oils are used. This thermometer allows the boiler operator to confirm that the heater in the fuel oil tank is working properly.

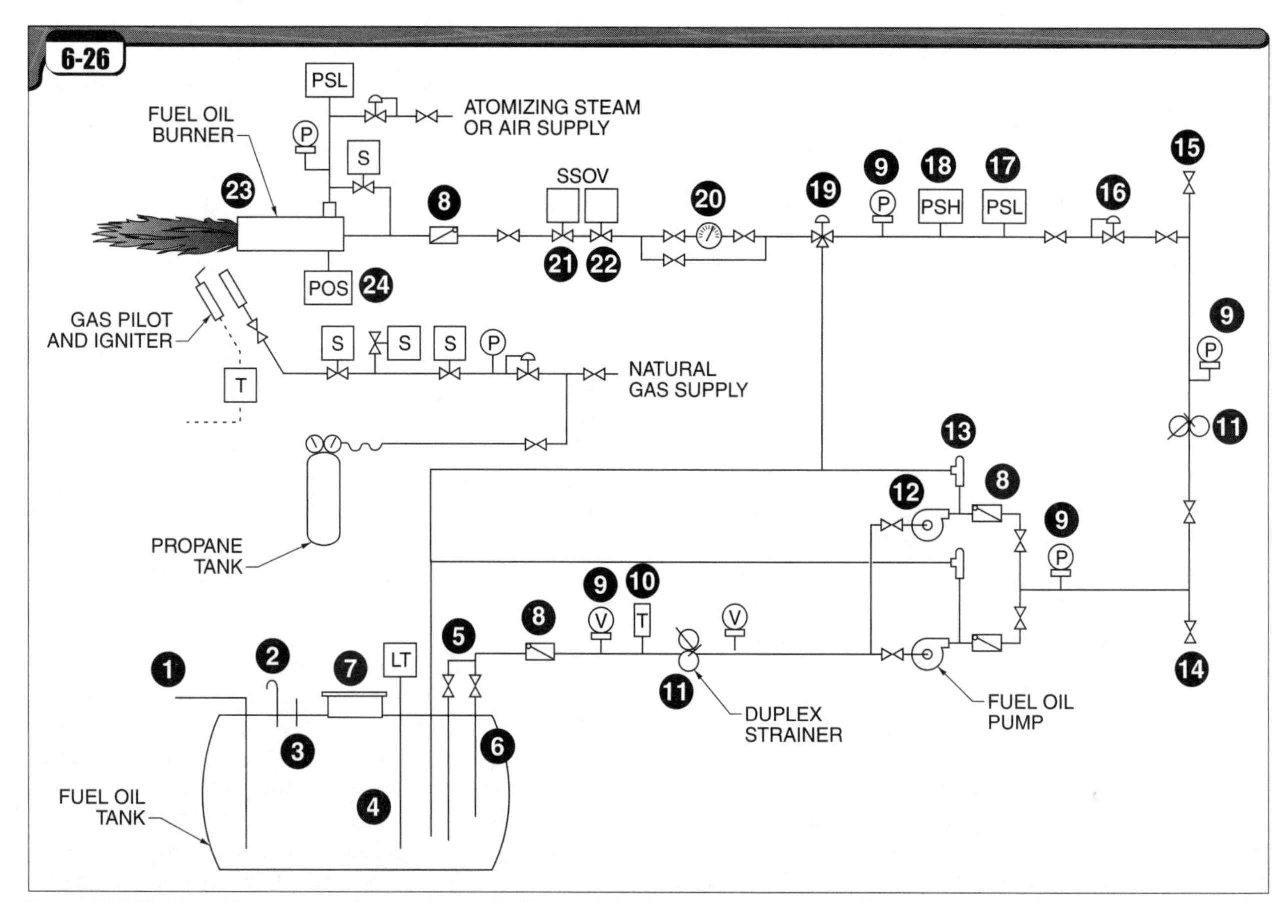

Figure 6-26. *The fuel oil system must deliver clean fuel oil to the fuel oil burner at the correct volume, pressure, and viscosity.*

Temperature switches (3) may be installed in the fuel oil tank to provide alarms if the fuel oil temperature becomes too high or too low. For example, the fuel oil temperature could become too high if the temperature-control valve on the steam line to the fuel oil heater should fail in the open position. If the valve failed in the closed position, or if the steam trap on the fuel oil heater failed in the closed position, the fuel oil temperature could fall to the point that pumping the oil becomes impossible.

When heavy fuel oils are used, the fuel oil piping should be traced (7) so that the proper fuel oil temperature is maintained in the piping. In cold climates, the entire fuel oil tank may be steam-traced to prevent the oil from thickening.

A second heater installation is supplied on the discharge side of the fuel oil pump when No. 5 or No. 6 fuel oil is used. In cold climates, No. 4 fuel oil is also sometimes preheated. This heater raises the temperature of the fuel oil to within about 10°F of the flash point so that it will vaporize and burn readily. The heater installation may consist of a combination steam and electric heater in one housing or the steam and electric heaters may be separate. The steam heater (8) is a shell-and-coil or shell-and-tube type. When No. 6 fuel oil is used, the steam normally flows through the tubes and the fuel oil flows across the outside of the tubes. This is because coke and heavy residues from No. 6 fuel oil tend to build up inside the tubes, especially if the tubes contain U-bends.

The steam supply to the fuel oil heater is controlled by a temperature-control valve (9) that throttles the steam flow based on the fuel oil outlet temperature of the heater. The electric heater (10) is an immersed resistance-type heating element that is controlled by a thermostatic device on the fuel oil outlet pipe line.

Electric fuel oil heaters are normally only used as a backup measure when a secondary boiler fuel such as natural gas is not available for making steam in the absence of fuel oil. Thus, the electric heaters would be necessary for cold plant startup when there is not any steam available.

Relief valves (12) should be provided on both the steam and electric heaters. These relief valves are necessary in the event that the temperature controls fail on the steam or electric heaters or if the isolation valves at the inlet and outlet to the heaters were accidentally left closed. In this case, expansion of the fuel oil as it is heated could cause very high pressures to develop.

An extra pressure gauge (13) is installed after the secondary fuel oil heater. The pressure gauge allows the boiler operator to determine whether the fuel oil heater is beginning to become plugged with debris. In this case, the pressure drop across the heater will increase. An extra thermometer (14) is used to determine whether the secondary fuel oil heater is working properly. For example, if the temperature-control valve that modulates the steam to the heater does not supply enough steam, the fuel oil temperature delivered to the burner will drop.

A low fuel oil temperature switch (15) detects the temperature of the fuel oil delivered to the burner. It determines whether the fuel oil heater or heaters have heated the oil to the proper temperature. If the fuel oil temperature is not high enough, the fuel will not atomize properly or burn correctly. In this case, the low fuel oil temperature switch will cause the fuel oil safety shutoff valves to be closed and the boiler will shut down. If this switch is closed, this proves to the burner control system that the fuel oil temperature is adequate for proper burner operation.

The high fuel oil temperature switch (16) causes the fuel oil safety shutoff valves to close if the fuel oil is too hot. This could occur if the fuel oil heater or heaters have failed. If the fuel oil is too hot, it will vaporize and expand too readily as it sprays from the burner. This can cause damage to the fuel oil burner and the refractory throat around the burner.

In normal operation, the part of the heated oil that is recirculated through the fuel oil tank and piping helps to maintain stable fuel oil temperature throughout the system. If the oil is heated too much, this could result in an excessive temperature in the tank and in the pump suction lines. This may cause damage to the pump(s) and cause severe pulsations in the fuel oil supply to the burner(s). Excessive fuel oil temperature can also cause coking of the fuel oil in the lines. *Coking* is the separation of heavy carbon-based fractions from the oil, resulting in precipitation of solids that may plug the piping. When the high fuel oil temperature switch is closed, this proves to the burner control system that the fuel oil temperature is below the maximum limit.

⚠ CAUTION

If a steam heater is used to heat fuel oil, the cleanliness of the condensate from the steam heater must be monitored. If a leak occurs in the steam heater, fuel oil may get into the condensate and return to the boiler. This leads to foaming and carryover in the boiler. A boiler shutdown would be required in order to clean the oil contamination from the boiler. Most plants allow the condensate from the fuel oil heaters to go to waste (11) rather than incur this risk.

What dangers are involved in heating fuel oil to high temperatures?

When fuel oil is heated to or near the flash point, substantial ignitable vapors are given off. A fire or explosion can happen if leaks or spills occur from the piping, pump seals, or other parts of the system. Fuel oil vapors can travel for great distances. If ignited, the vapors may flash back to the source of the leak.

If the tank heater fails, the oil in the tank may be heated beyond the flash point. High temperature alarms are recommended to warn boiler operators of this condition. If liquid fuel oil is trapped in a limited space with isolation valves closed, such as in the secondary fuel oil heater, the oil will expand as it is heated. This can form very high pressures inside the heater. Relief valves must be installed as appropriate to eliminate this potential hazard.

6-27

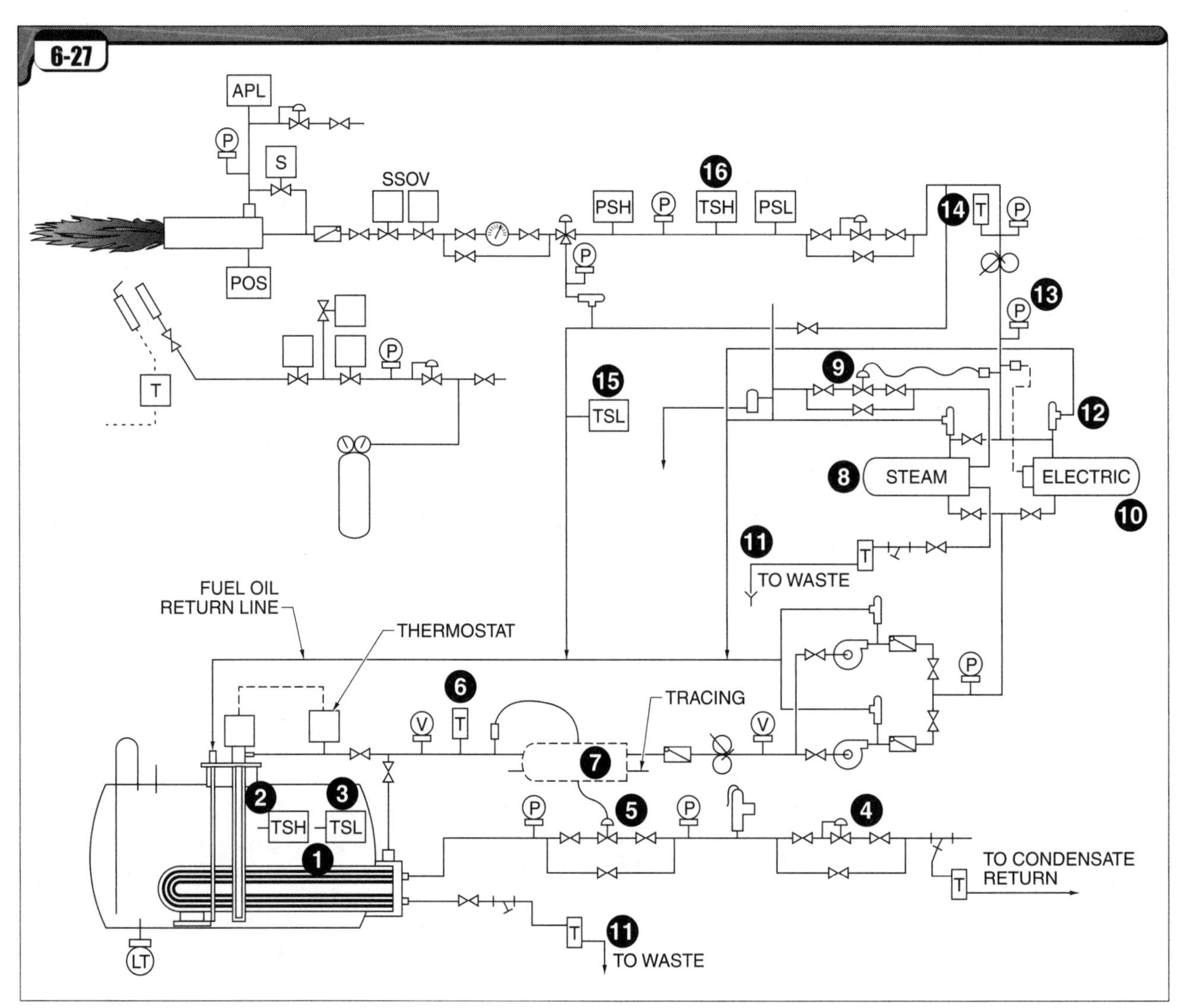

Figure 6-27. *The use of heavy fuel oils requires additional handling equipment and safety devices.*

73 How are fuel oil burners ignited?

Most commercial and industrial fuel oil burners use a natural gas or propane pilot to ignite the fuel oil. In a typical pilot system, the pilot train consists of a pilot gas shutoff cock, a pilot gas pressure regulating valve, a pilot gas pressure gauge, an automatic pilot gas shutoff valve system, an igniter, an ignition transformer, a pilot adjusting cock, and the pilot assembly itself.

The natural gas or propane supply for the pilot first passes through the pilot gas shutoff cock and then through the pilot gas pressure regulating valve. This valve reduces and automatically adjusts the gas supply pressure to the exact pressure needed by the pilot. The pilot gas pressure regulating valve is not needed if the pressure is already suitable for the pilot. The pilot gas pressure gauge indicates the pressure after the pilot gas pressure regulator. The gas then passes through the automatic pilot gas shutoff valve system. The design of this shutoff valve system may vary depending on insurance requirements. For example, the automatic pilot gas shutoff valve system may consist of a single solenoid-operated valve, a pair of solenoid-operated valves in series, or three solenoid-operated valves arranged in a double block and bleed configuration.

A *double block and bleed* is a valve configuration that consists of two automatic shutoff valves arranged in series, with a vent or bleed valve between them that vents outdoors. The solenoid actuators on these valves are wired so that when the two valves in the gas line close, the valve in the vent line opens. When the two valves in the gas line open, the valve in the vent line closes. This configuration ensures that any gas that may leak through the first valve vents to the atmosphere outside rather than into the furnace area.

The pilot is simply a tiny gas burner designed to produce a flame of the size and length needed to provide positive lighting of the main burner. The pilot-adjusting cock allows a technician to adjust the strength and length of the pilot flame. The pilot assembly receives air from the primary air (forced draft) fan and mixes it with the fuel gas in the proper proportions.

An ignition transformer and igniter are installed to provide a strong spark for lighting the pilot gas. An electrical lead from the ignition transformer goes to the igniter. The igniter is essentially a long spark plug. The igniter is typically installed in the burner assembly through a metal sleeve. An electrode tip on the end of the igniter is in close proximity to the metal sleeve and very near the tip of the gas pilot. The ignition transformer is powered at the same time as the automatic pilot gas shutoff valves. The ignition transformer is supplied with 120 V and steps up the voltage to about 6000 V to 10,000 V. The very high voltage from the ignition transformer induces a continuous spark from the electrode tip to the grounded metal sleeve. This spark provides the heat energy needed to light the pilot gas as it is introduced through the pilot assembly.

Some smaller fuel oil burners use direct spark ignition. In this case, two electrodes are provided and installed so that their tips are close together near the tip of the fuel oil burner itself. High voltage from an ignition transformer is used and a strong spark arcs from one electrode to the other. This provides the energy source to light the fuel oil burner as fuel oil is admitted.

74 What is atomization of liquid fuels?

Atomization is the process of breaking a liquid fuel stream into a mist of tiny droplets. This provides several important benefits. First, atomization maximizes the surface area of the liquid by converting a cylindrical stream of oil into countless tiny droplets. As the liquid is atomized, it also enters the hot environment of the furnace area. These two conditions serve to maximize the generation of vapors from the fuel. This is critical because the vapors produced by a liquid fuel are what actually burns. Therefore, the more free and rapid the generation of vapors, the more readily the fuel will burn. Atomization also allows the oxygen in the air to mix more quickly and efficiently with the fuel.

Viscosity is a key concern in the atomization of a liquid fuel. If the liquid has a high viscosity, it is more difficult for the burner to atomize the fuel. Therefore, heavy fuel oils are preheated to near the flash point. This ensures that the fuel will be at the best viscosity for atomization as it enters the burner assembly. The ideal temperature at which fuel oil should reach the burner for atomization and combustion varies depending on both the grade of fuel oil used and the type of burner. The burner manufacturer's recommendations for temperatures should be followed.

75 Describe the operation of a rotary cup burner.

A *rotary cup burner* is a burner that has a cone-shaped cup, usually made of brass or stainless steel, that mixes fuel oil with air. The rotating cup is mounted on the end of a motor shaft. **See Figure 6-28.** The rotating cup spins at the same speed as the motor—about 3450 rpm. The motor shaft is hollow and fuel oil is sent to the inside of the rotary cup through the hollow shaft. As fuel oil exits the shaft, it is thrown to the wall of the rotating cup by centrifugal force and migrates to the rim of the cup. A fan rotor is also attached to the motor shaft and air from the fan mixes with the fuel oil as it sprays off the rim of the cup. Fixed louvers inside the fan shroud cause the flow of combustion air from the fan to rotate in a direction opposite that of the spinning cup. The opposing flows of air and fuel oil ensure thorough mixture of the two.

Rotary cup burners are less common than other fuel oil burners and are primarily limited to smaller commercial and industrial boiler systems.

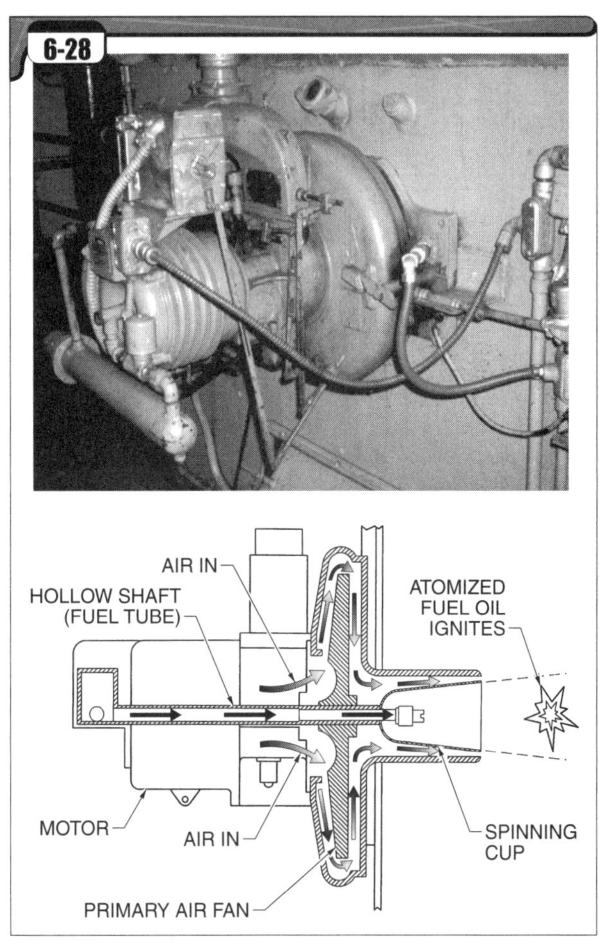

Figure 6-28. *Fuel flows through the motor shaft in a rotary cup oil burner and is thrown from the rim of a spinning cup.*

76 List and describe three classes of fuel oil burners with regard to their methods of atomization.

The three classes of fuel oil burners are steam-atomizing, air-atomizing, and pressure-atomizing fuel oil burners. **See Figure 6-29.**

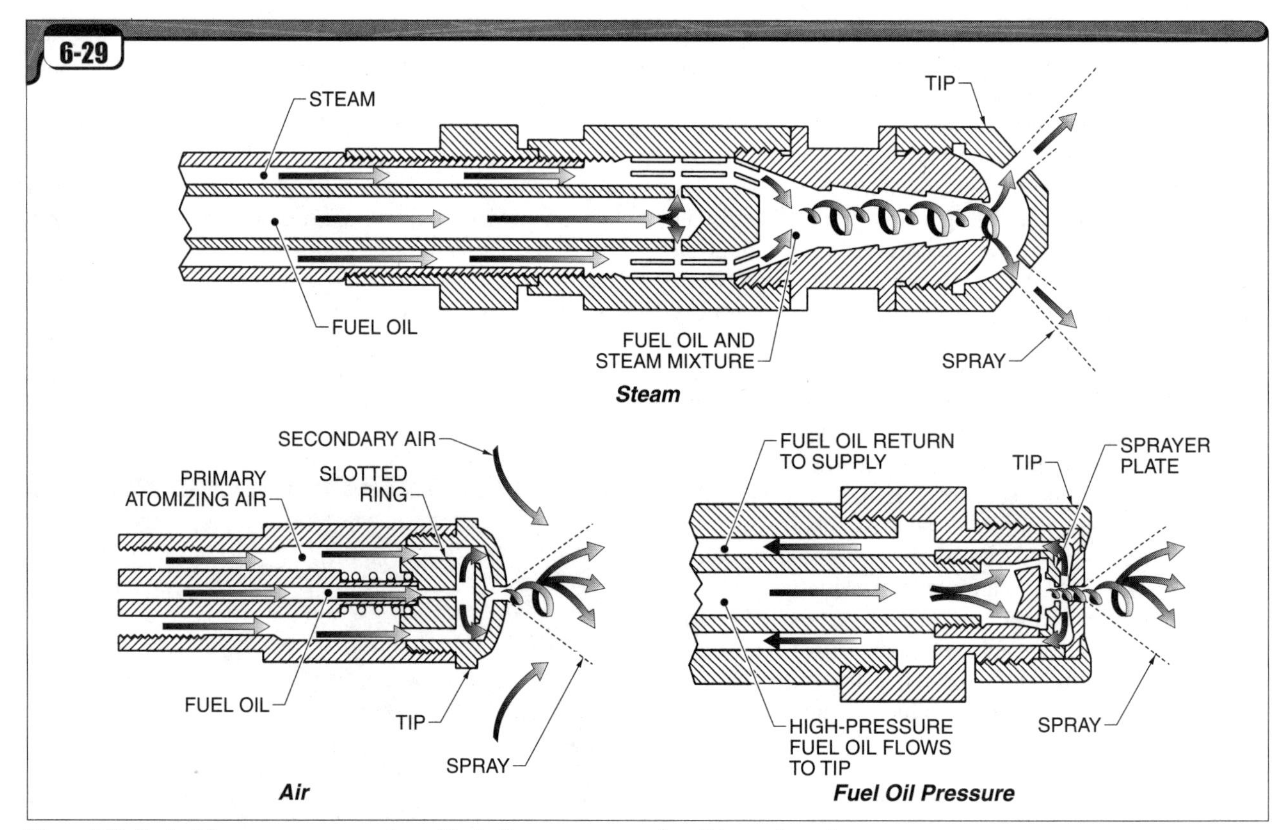

Figure 6-29. *Fuel oil burners use steam, air, and fuel oil pressure to atomize oil into a fine mist.*

Steam-atomizing fuel oil burners are typically 3′ to 8′ long. This depends on the size and construction of the burner windbox. They have a double-walled tube that consists of a tube inside a tube. The fuel oil enters the burner through the inner tube. Steam enters through the outer tube. The steam passes through a ring of tangential slots at the tip of the burner, converging with the stream of fuel oil. The mixture blasts from holes in the burner tip as a long, swirling spray.

Primary air is admitted through a diffuser around the fuel oil burner. The diffuser causes the air to swirl, which creates the turbulence needed to mix the fuel and air. Depending on the burner size and grade of oil used, secondary air is often admitted through louvers around the diffuser. Steam atomization has the benefit of preheating heavy fuel oils and lowering their viscosity. This causes the fuel oil to release vapors more readily.

The quantity of steam used for atomization is typically about 0.1 lb to 0.2 lb of steam per pound of oil. The steam pressure is generally maintained at about 20 psig higher than the fuel oil pressure at the burner.

Air-atomizing fuel oil burners are usually of the same basic design as steam-atomizing oil burners. In fact, the same fuel oil burner is often used for either steam or air atomization. However, the primary air used with air-atomizing fuel oil burners is also the atomizing medium. Atomizing air passes through a ring of tangentially oriented slots in the spray tip. The air flow converges across the fuel oil stream, which breaks up the stream.

Secondary air is brought in through a diffuser or louvers around the outside of the burner tube for complete burning. These burners are often used on package firetube and watertube boilers burning No. 2 and/or No. 4 fuel oil. An air compressor or air pump must be used to produce the atomizing air. Air atomization is not often used on large fuel oil burners due to the cost of generating the compressed air.

The atomizing air pressure is kept about 20 psig higher than the fuel oil pressure at the burner. This differential pressure (atomizing medium vs. fuel pressure), for both steam and air atomization, is the main factor in controlling the size of the oil droplets. Smaller droplets generally vaporize more readily, but the cost of the atomizing medium is significant. Design engineers therefore try to strike a balance between optimum droplet size and minimum cost.

Pressure-atomizing fuel oil burners create atomization by causing the fuel oil to flow under high pressure through a sprayer plate or plug in the tip of the burner and then through a small orifice, into the furnace. Fuel oil entering the burner flows through the center tube. The fuel oil flows through the tangential slots in the sprayer plate or plug and converges into the middle with a large amount of energy, because of the high pressure from the fuel oil pump. This high pressure creates a spinning shaft of fuel oil in the burner tip.

When a part of this fuel oil passes through the orifice at the tip of the burner, it breaks up and forms a cyclone spray that becomes a hollow, cone-shaped flame. The part of the fuel oil that does not go through the orifice and into the furnace flows back to the supply through passages in the burner tube.

In pressure-atomizing burners used in very small commercial boilers, the fuel oil pressure may be as low as 100 psig. In large industrial and utility boilers, however, pressure-atomizing fuel oil burners use fuel oil pressures as high as 600 psig to 1000 psig. Pressure-atomizing burners generally do not modulate well; therefore, they are limited to ON-OFF applications and multiple burner applications where more or fewer burners may be placed in service.

77 Why is the fuel oil burner's position in the burner throat important?

The burner throat formed by refractory bricks is designed to control the flame shape and to ensure that the burner flame is directed into the furnace properly. If a fuel oil burner is inserted too far into the burner throat, the flame may whip around the diffuser that surrounds the burner. If the burner is retracted too far, the oil spray may strike the burner throat. This can cause a buildup of carbon and cause heat damage to the burner throat. The flame will tend to flutter or pulsate.

78 What prevents a fuel oil burner from being operated if the required atomizing medium is not available?

An atomizing-pressure proving switch is used to detect the pressure of the atomizing steam or air. When the atomizing steam or air pressure is high enough to close this switch, this proves to the control system that the atomizing medium is available. If the atomizing steam or air pressure drops to an unsafe point during burner operation, the atomizing-pressure proving switch opens. This causes the control system to stop the power to the fuel oil SSOVs and the burner shuts down.

79 How may the fuel oil burner be cleaned without shutting down steam production in the boiler?

An auxiliary burner is provided if it is necessary to clean the fuel oil burner without shutting the boiler down. This allows the boiler operator to clean and service one burner while the other burner is in operation. The auxiliary burner is started and stabilized and the primary burner is shut down for service.

80 What care should be given to burner nozzle tips?

Burner nozzle tips must be kept clean of carbon and ash deposits. The presence of carbon and ash deposits can cause the atomization to deteriorate. The carbon and ash deposits eventually foul the holes in the tip. This often results in oil dripping from the burner. The dripping oil can cause carbon and ash to form on the end of the tip. To prevent buildup, the holes in the burner tip should be cleaned out on a regular basis. The holes in the burner tip may also become oversized due to the erosive properties of the steam. The burner manufacturer provides a small gauge known as a "go/no go" tool. One end is a pin the same size as the holes in a new burner tip. It should fit snugly into the holes. The other end is very slightly larger in diameter and should not fit into the holes. When the larger end of the tool fits into the tip holes, this indicates too much erosion of the holes and the tip should be replaced.

When assembling the nozzle, the sprayer plate or plug must be kept tight against the cap on the tip, or the atomization will not have the swirling energy that it should have. **See Figure 6-30.** The size of the tangential atomization slots should also be checked periodically for damage due to erosion.

When burning heavy oils (No. 5 or No. 6), the fuel oil burners should be inspected and cleaned if necessary on each shift. For lighter fuel oils, the burner will not need to be serviced as frequently.

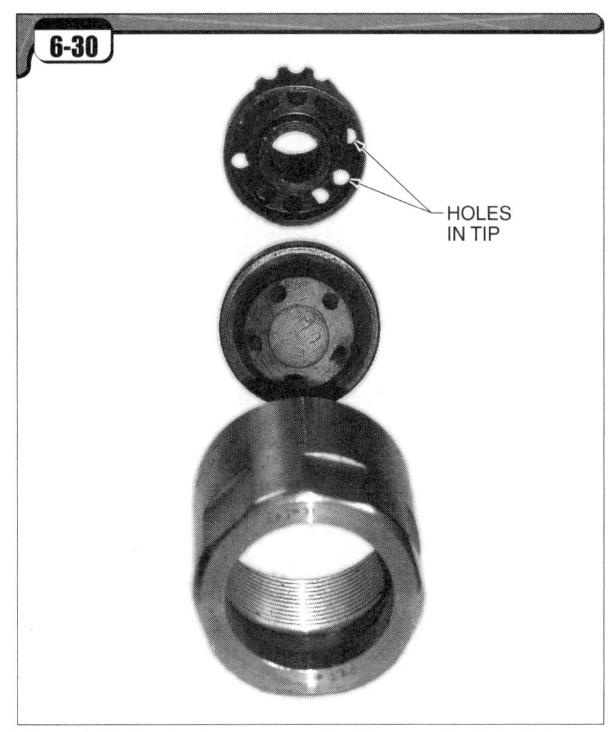

Figure 6-30. *The boiler operator should clean the fuel oil burner frequently to prevent carbon and ash from fouling the holes in the tip assembly.*

81 What would happen if the burner nozzle tip came off the fuel oil burner?

If the burner nozzle tip came off the fuel oil burner, the atomizing effect would be largely lost and a stream of fuel oil, instead of a mist, would be pumped into the furnace. Combustion quality would drop off immediately, forming smoke, carbon monoxide in the flue gas, and potentially dangerous buildups of vapors in the furnace. To help avoid this trouble, the interior rifling or tangential slots of the burner tip that impart the swirling action to the fuel oil mist are oriented so that the reactive force tends to tighten, rather than loosen, the tip of the fuel oil burner.

Fuel Gases

82 What is a combination burner?

A combination burner is a burner assembly that can use more than one type of fuel, either separately or, in rare cases, at the same time. For example, a typical combination burner may burn either No. 2 fuel oil or a gas. **See Figure 6-31.**

Combination burners are in wide use today due to the volatile nature of fuel prices and availability. This allows a steam-using facility to use the fuel that is cheaper at any given time. In facilities such as hospitals where steam must always be available, combination burners also provide the benefit of a backup fuel supply if the main supply is interrupted for some reason.

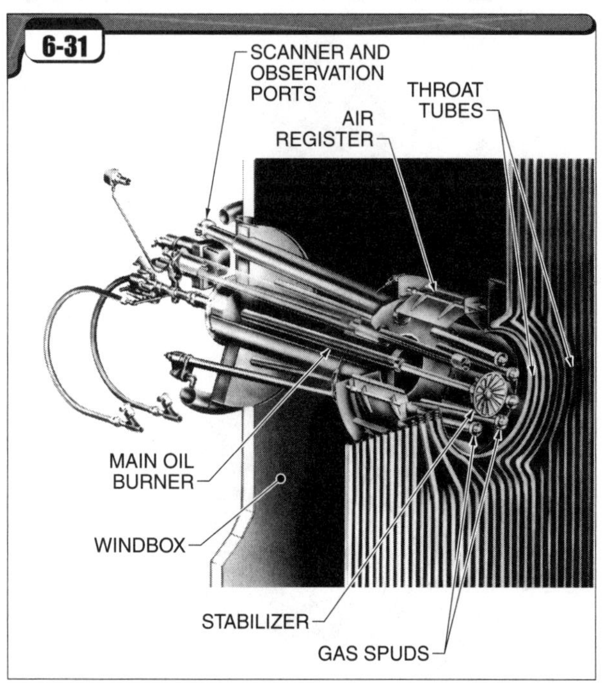

Figure 6-31. *Combination burners can burn more than one type of fuel.*

83 List several types of gas that can be used as boiler fuel.

Natural gas is obtained from natural sources in the earth, as opposed to manufactured (by-product) gas. Natural gas consists primarily of pure methane with small amounts of ethane and sometimes propane. Natural gas contains about 950 Btu/cu ft to 1150 Btu/cu ft.

Butane is a by-product gas derived from the refining of crude oil into gasoline. Butane also occurs naturally and is found in varying amounts at the casing head of oil wells. Butane is dense and liquefies under slight pressure. It is sold as a liquid. Because of its commercial value, it is only rarely used as boiler fuel. Butane contains about 3200 Btu/cu ft to 3260 Btu/cu ft.

Coke oven gas consists mainly of hydrogen and methane, but also contains sulfur and other elements. It is produced when coke is made from coal in coke ovens. Coke is used as fuel in blast furnaces in the manufacture of steel. Coke oven gas contains about 500 Btu/cu ft to 600 Btu/cu ft.

Blast furnace gas is a by-product gas recovered from blast furnaces. Blast furnaces are used to smelt iron ore in steel mills. Blast furnace gas is mainly composed of carbon monoxide and contains about 90 Btu/cu ft to 100 Btu/cu ft.

Propane is a liquefied hydrocarbon gas derived from the petroleum refining process. Propane is also known as liquid petroleum gas (LPG). Propane liquefies under pressure and is sold as a liquid. Propane contains about 2500 Btu/cu ft.

Methane gas is a bacterial by-product gas formed in wastewater treatment plants, landfills, and in other areas. This type of methane gas contains a mixture of pure methane and carbon dioxide. Depending on the source, methane gas contains about 500 Btu/cu ft to 750 Btu/cu ft.

84 What is a therm of natural gas?

A *therm* is a unit of measure indicating 100,000 Btu. The amount of natural gas required to comprise one therm may vary, depending on the heating value of the natural gas. Natural gas used in industrial plants is usually metered and billed in decatherms. A *decatherm* is 10 therms, or 1,000,000 Btu.

85 What equipment is provided in a typical industrial gas valve train for control and regulation of the gas supply to a boiler?

A typical industrial gas valve train consists of a main gas valve train and a pilot gas valve train.

The main components of the main gas valve train are the manual shutoff valves, a gas flowmeter or flow transmitter, a main gas pressure regulating valve, a low gas pressure switch, safety shutoff valves, and a high gas pressure switch.

A manual shutoff valve (1) is the first and most important piece of equipment in the gas valve train. **See Figure 6-32.** It allows the boiler operator to shut off the gas supply to the boiler in the event of an emergency. Note that the use of slow-opening valves such as gate or globe valves in a natural gas line is not recommended because they cannot be closed quickly.

Gas valve trains generally use lubricated plug valves for manual shutoff and isolation. A plug valve is similar to a ball valve, except that the interior plug is conical in shape rather than spherical and the passageway is rectangular rather than round. The 90° quick-shutoff action of a plug valve or ball valve is an important feature of gas valves for emergency shutoff. A lubricated plug valve uses a special sealant, similar in texture to heavy grease, to provide a seal between the conical plug and the valve body. The sealant makes the valve less susceptible to leak-through due to foreign materials in the gas line. Such materials are more likely to score the tight-fitting seating surfaces of other valve designs. Ball valves are used in smaller gas lines, but are limited to about 3″.

It is desirable to have a separate flowmeter or flow transmitter (2) for each boiler's fuel system. A flowmeter displays the quantity of fuel used. A flow transmitter transmits this information to a computer terminal or panelboard controller for display in a control room. These devices allow the boiler operator and plant engineering personnel to track the efficiency of each individual boiler.

The main gas pressure regulating valve (pressure regulator) (3) is a pressure reducing valve that lowers the gas pressure to the pressure required by the burner. Almost all main gas pressure regulating valves contain a flexible diaphragm in the actuator portion. The pressure downstream of the main gas pressure regulating valve is applied to the diaphragm and helps position the internal valve to maintain the proper downstream pressure. The movement of the diaphragm is balanced by compression of a spring inside the regulating valve.

Because the diaphragm in main gas pressure regulating valves can fail, a vent line to the atmosphere is required above the diaphragm. For example, if the diaphragm should develop a hole or tear, natural gas could leak into the boiler room and create a hazardous condition. The vent line above the diaphragm ensures that any gas leakage will be carried safely outside to the atmosphere.

The low gas pressure switch (4) is used to shut down the gas burner if the gas supply pressure should become too low. For example, this can occur if the main gas pressure regulating valve fails or if the gas supplier does not provide sufficient gas pressure. A hazard of low gas pressure is that the velocity of the incoming air-fuel mixture may fall to the point where the flame is allowed to migrate upstream into the burner assembly. This can allow the burner to overheat.

Low gas pressure may also result in an unstable flame. If the flame falters and goes out, raw fuel can be admitted to the furnace. This can result in a furnace explosion if it is not detected by other safety controls. As long as the gas pressure after the main gas pressure regulating valve is satisfactory, the low gas pressure switch remains closed. If the gas supply pressure drops to an unsafe level, the switch opens and interrupts the power supply to the main safety shutoff valves. This causes the valves to fail in the closed position and shuts the burner off. The low gas pressure switch is also known as a vaporstat.

The safety shutoff valves (SSOVs) (5) admit the fuel to the gas burner when open and stop the gas flow to the burner when closed. Insurance requirements dictate various designs of these valves. Smaller commercial and industrial boilers have two safety shutoff valves in series. On larger gas burners, a double block and bleed configuration is required. In this case, two main SSOVs are provided in series, with a solenoid-operated vent valve to the atmosphere between them. In these installations, the SSOV after the vent line is known as the blocking valve.

Some larger automatic gas valves open against the direction of flow and close with the direction of flow. For this reason, when the SSOVs close and the valve in the vent line opens, the differential pressure across the first SSOV is increased. This helps the valve to close more tightly.

The type of valve actuators used on SSOVs varies as well. On the smallest commercial and industrial boiler burners, two solenoid-operated SSOVs may be provided. Slightly larger burners may be provided with one solenoid-operated SSOV and one electrohydraulically operated SSOV.

Large burners normally require two electrohydraulically operated SSOVs. These valve actuators contain a small hydraulic pump. The actuator, when energized, pumps up hydraulic pressure to open the main gas valve. As long as the control circuit remains energized, the electrohydraulic actuator will hold the valve open. If the circuit is opened by an interlock, such as low water in the boiler, the power to the electrohydraulic actuator is lost and a strong spring snaps the valve shut. These gas valve actuators are hydraulically dampened to limit the speed at which the valve can open. For example, typical opening times of these actuators are 13 or 26 seconds. For safety reasons, the closing speed is never delayed.

As in the case of fuel oil SSOVs, insurance requirements often require proof of closure (POC) switches (6) on SSOV actuators for natural gas. If the POC switch cannot prove that the SSOV is in the closed position, this can mean that raw fuel is leaking into the boiler furnace. In this case, the boiler control system must not allow the burner to light.

The high gas pressure switch (7) is used to shut down the burner if the gas pressure becomes too high. This can occur if the main gas pressure regulating valve fails in the open position. Excessive gas pressure can cause an uncontrollable burner flame. This can result in damage to the boiler from excessive heat input. If enough gas is admitted to the furnace, the air-fuel mixture can become too rich to burn. This can cause the flame to be pushed away from the burner and the flame can go out. As the air flow dilutes the mixture, however, it may again become combustible. If it ignites, a furnace explosion could result. The high gas pressure switch prevents this from happening.

Another manual shutoff valve (8) should be provided just before the burner. This manual shutoff valve is provided mostly for testing and troubleshooting. For example, closing the manual shutoff valve allows the second (blocking) SSOV to be tested for leak-through. The manual shutoff valve also allows for functional testing of all the main gas valve train components without the potential for leaking fuel into the furnace.

The gas flow control valve (9) controls the rate of fuel flow to the burner. It is modulated by an actuator. During periods of low steam demand, the valve is throttled to admit a low fuel flow rate and during heavy steam demand it is modulated to admit a high fuel flow rate.

On smaller boilers, the gas flow control valve is often a butterfly valve. A modulating motor that also actuates the linkage to an air damper is often used to actuate the butterfly valve. The modulating pressure control positions the modulating motor as necessary to adjust the burner firing rate to maintain a constant steam pressure. Adjustable linkages and shaped cams are provided to allow burner tuning technicians to fine tune the air-fuel mixture throughout the firing rate range of the burner. On larger boilers, the gas flow control valve is usually either a globe valve design or a ball valve with a V-shaped port. These designs provide more precise flow control than butterfly valves.

The equipment provided for a gas pilot on a gas-fired boiler is in most cases identical to that provided for a fuel oil-fired boiler. The gas line for the pilot branches from the main gas line after the first manual shutoff valve in the main gas valve train because the manual shutoff valve in the main gas line should also shut off the pilot gas. The main components of the pilot gas valve train are the pilot gas manual shutoff valve, pilot gas pressure regulating valve, automatic pilot gas shutoff valve system, pilot adjustment cock, gas pilot, and the ignition transformer and igniter.

A separate pilot gas manual shutoff valve is provided (10) where the pilot gas line branches from the main gas line so that the pilot gas line may be shut down separately if desired.

A separate gas pressure regulating valve (11) must be supplied if the pilot gas pressure is to be lower than the pressure of the main gas supply to the burner.

The automatic pilot gas shutoff valve system consists of ON/OFF controls. The gas supply to the pilot is rarely modulated. It is either ON or OFF. For this reason, solenoid-operated valves are used in the gas line to the gas pilot. Depending on insurance requirements and the size of the pilot, the required number and configuration of these valves in the pilot gas line may vary. For example, a given boiler may require one solenoid-operated valve, two solenoid-operated valves in series, or three solenoid-operated valves configured in a double block and bleed configuration (12).

A small pilot adjustment cock, consisting of a small plug valve (13), is provided to allow the burner tuning technician to adjust the gas flow to the gas pilot to obtain the correct flame length.

The gas pilot (14) is a small, simple burner designed to produce a gas flame of adequate length and strength to light the main burner reliably.

An ignition transformer (15) is energized at the proper time to provide high voltage to an igniter (16). The igniter in turn provides a strong spark to light the gas pilot.

Describe three types of gas pilots with regard to their method of control.

A *standing pilot* is a gas pilot that is always lit. Standing pilots are used in many small commercial boilers as well as in direct-fired water heaters and other commercial and residential appliances with small gas burners. Although very simple and low in cost, standing pilots consume energy even when it is unnecessary.

An *intermittent pilot* is lit at the appropriate time to light the main burner and then stays on during the entire period that the burner is on. Intermittent pilots are also generally used in smaller boilers, industrial paint curing ovens, rooftop air-heating units, etc.

An *interrupted pilot* is lit at the appropriate time to light the main burner and then extinguished as soon as the main burner is lit. For example, interrupted pilots on commercial and industrial boilers typically stay on for about 10 sec after the main gas safety shutoff valves open. Then the solenoid-operated valves in the pilot gas line close and the pilot is extinguished.

Trade Tip

***The last manual shutoff valve in the gas train allows for testing of all the components in the train without the possibility of leaking raw fuel into the furnace. The boiler operator's rule of thumb for this is** "No Fuel, No Boom!"*

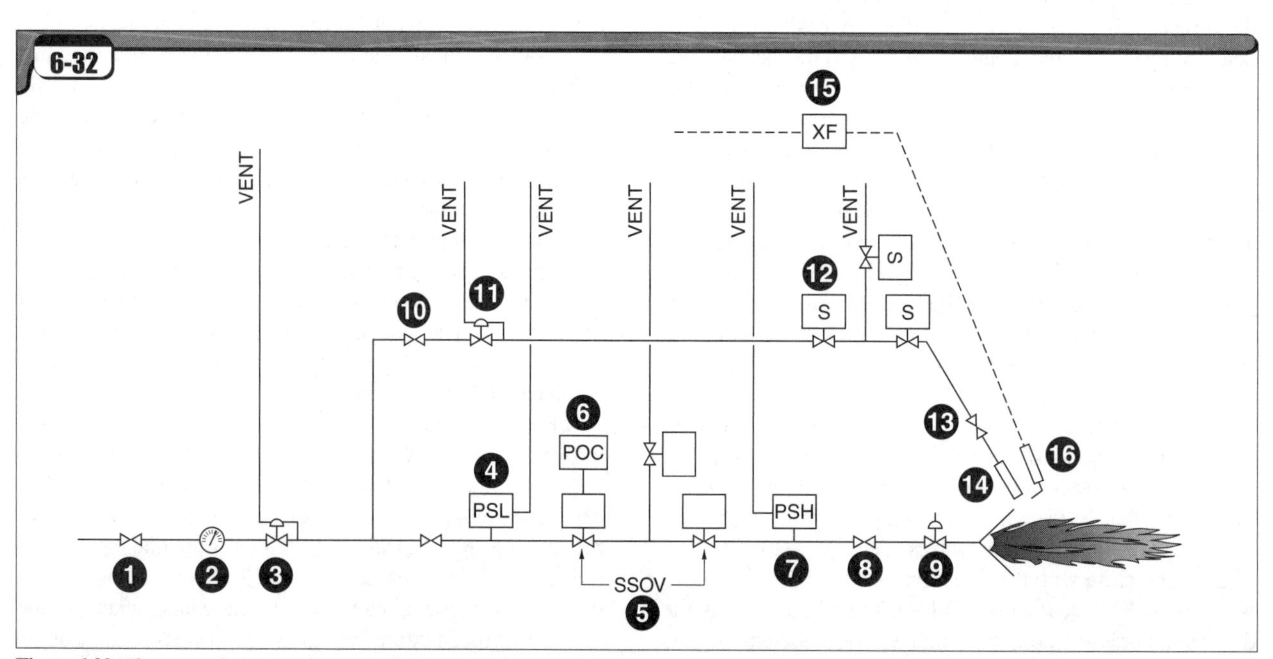

Figure 6-32. *The natural gas supply to a boiler burner consists of the main gas valve train and the pilot gas valve train.*

Combustion Safety Systems

87 What is a flame safeguard system?

A *flame safeguard system* is the collection of automatic control devices that ensure safe operation of the combustion equipment. **See Figure 6-33.** Flame safeguard systems are used on boilers that use gaseous fuels, liquid fuels, or solid fuels fired in suspension. A flame safeguard system is also known as a burner management system, or BMS.

The heart of a flame safeguard system is the control logic. This is the electronic circuitry or computer software that directs the burner-related control devices to operate in such a manner and sequence that safe furnace conditions are always maintained. The control logic resides in either a stand-alone control device (known as a flame safeguard control) or in software that serves as the burner's "brain." Note that a flame safeguard control is a single part of the larger flame safeguard system.

In most commercial and industrial boilers, a stand-alone flame safeguard control serves as the heart of the flame safeguard system. Flame safeguard controls consist of hardware components housed in a control panel. These components receive and process information from sensing devices like low- and high-gas pressure switches. The software-based flame safeguard systems consist of software logic and graphic displays as well as hardware components. The software is usually programmed as part of a larger overall boiler plant computer control system.

The two fundamental classes of flame safeguard controls are based on their level of complexity. These classes are primary controls and programming controls. A *primary control* is a flame safeguard control that consists of the relays and electronics required to safely start, run, and stop the burner under orders from an external control device such as a pressure switch. Primary controls do not employ a sequence of operation. They are mainly used with small burners that are either ON or OFF. However, some accept plug-in modules that allow for a furnace purging period before the burner lights.

A *programming control,* or *programmer,* is a flame safeguard control that consists of all the components needed to safely perform a desired sequence of operations for a larger commercial or industrial burner. For example, a programming control steps the burner though a furnace purging period, ignition of the pilot, establishment of the main burner flame, automatic modulation of the burner, and purging of the burner assembly after burner shutdown.

Older programming controls use a motor-driven shaft that turns very slowly. Cams of different shapes on the shaft contact electrical switches at the proper time, which causes sequenced actions in the firing cycle. For example, one cam will start the forced draft fan and another will open the gas valve to the pilot. The shape of the cam determines how long the various switches are held closed.

Modern programming controls are electronic devices. These control systems cause the same actions in the firing sequence as the cam type, but do not depend on mechanical movement. These modern systems have a number of advantages. One major advantage is the ability to aid the boiler operator in troubleshooting when the firing sequence does not proceed properly. These systems have self-diagnostic functions and can display messages that indicate the cause of a problem. Controllers also store information related to hours of operation, number of burner cycles, and fault histories. In addition, they may be used to communicate information to remote computers or pagers.

Whether the flame safeguard system is based on a primary control, programming control, or software, all flame safeguard systems also serve as a central command center for shutting down the burner equipment if an unsafe condition occurs. For example, the low water fuel cutoff control, low and high gas pressure switches, high steam pressure limit switch, and other sensors are all monitored by the flame safeguard control. If any of these or other unsafe conditions is detected, the flame safeguard control interrupts the power supply to the fuel valves and the fuel valves fail closed.

6-33

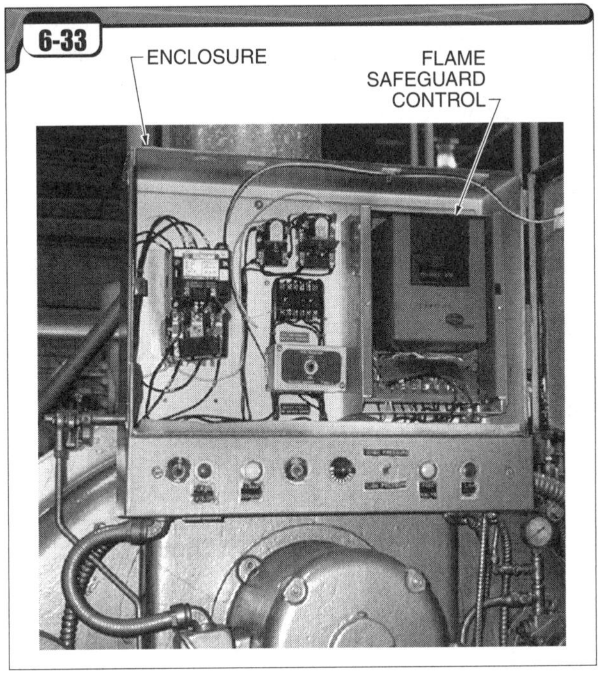

Figure 6-33. *Flame safeguard controls on modern boilers are electronic devices that monitor all conditions related to safe burner operation.*

Trade Tip

The boiler operator should exercise the manual gas shutoff valves periodically to ensure that they will operate freely in an emergency. The lever handles should never be removed from these valves because it is possible to place the handles in the wrong position when reinstalling.

How is the flame safeguard control or burner management software informed that required actions for safe operation of the boiler burner equipment have taken place?

The flame safeguard control or software cannot look at pressure gauges, gauge glasses, thermometers, or other devices that require sight. Therefore, field devices must be installed that inform the flame safeguard control or software of the status of various measured conditions. This is called proving the condition. **See Figure 6-34.**

When the low water fuel cutoff control float is lifted by the water level in the boiler and the float-operated switch is closed, this proves to the flame safeguard control or software that a safe water level is present in the boiler. When the low gas pressure switch receives adequate gas pressure from the main gas pressure regulating valve, the diaphragm in the switch is deformed upward and the switch is closed. This proves to the flame safeguard control or software that the gas pressure is adequate for safe operation. Similarly, a pressure switch installed on the fuel oil line after the fuel oil pump must be closed to prove that the fuel oil pump is running and providing adequate fuel oil pressure for correct operation of the fuel oil burner. All conditions that are necessary for safe operation must be proven to the flame safeguard control or the control will not allow the fuel valves to remain open.

This is a vital fact for the boiler operator to keep in mind when trying to troubleshoot problems with the boiler. It does not matter whether the boiler operator can see that there is water in the gauge glass or whether the fuel pressure indicated on a pressure gauge is correct. These and other conditions must be proven to the flame safeguard control or software or it will not allow the burner system to operate.

Factoid

Atomization of fuel oil breaks a stream of fuel oil into countless tiny droplets. This maximizes the surface area of the fuel oil and allows for more efficient mixing of the fuel vapors with oxygen.

What must be done before lighting any gas-, fuel oil-, or pulverized coal-fired boiler?

Before lighting any gas-, fuel oil-, or pulverized coal-fired boiler, the furnace must be purged of any buildup of ignitable vapors or dust to prevent an explosion. **See Figure 6-35.** This step is done by running the combustion air fan(s) for a prescribed period of time with the dampers wide open. The fan(s) must move a set number of exchanges of air through the fire side of a boiler before the burner is lit. This means that the furnace volume and the volume of the flue gas passages in cubic feet are calculated and the fan(s) must run several times that amount of air through the furnace before fuel ignition is allowed to occur.

Four exchanges of air is the minimum per most nationally accepted codes. The National Fire Protection Association (NFPA) requires a minimum of eight air exchanges before the burner is lit. Which standard must be adhered to depends on the plant's insurance requirements and on local governmental agencies having jurisdiction.

With most smaller boilers, the purge period before ignition (prepurge) is a timed function. The required purge period is preprogrammed into the flame safeguard control. The flame safeguard control confirms that the furnace has been purged for the prescribed period of time before allowing the pilot to light. This purge period is typically 30 sec to 60 sec. In larger boilers, the purge air flow is measured. In this case, the required number of cubic feet of air, as measured by a flowmeter, must run through the furnace before the flame safeguard control allows the pilot to light.

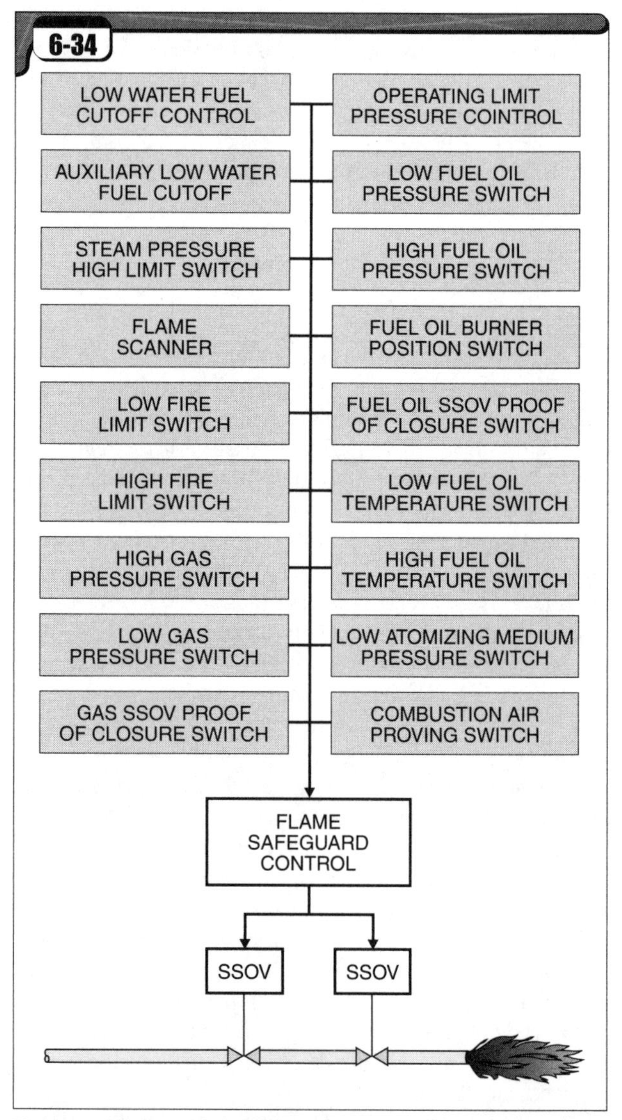

Figure 6-34. *Safe conditions must be proven by field devices for the flame safeguard control to allow the burner to operate.*

Figure 6-35. *Failure to purge the furnace of fuel vapors or coal dust can lead to an explosion.*

Factoid

Duplex strainers or two strainers piped in parallel are used on commercial and industrial fuel oil burner systems so that the boiler operator may clean the strainer without shutting down the burner.

How is the flow of combustion air and purge air proven?

The flow of combustion air and purge air is proven with an air pressure switch, air flow switch, or air flowmeter. **See Figure 6-36.** For example, air pressure produced by the fan may be used to cause the air pressure switch to activate. Air pressure switches are common on package boilers. In some cases, an air flow switch is used to prove a minimum air flow during the purge cycle. An air flow switch uses a small paddle, or sail, that extends into an air duct. The flow of air pushes on the sail, causing it to deflect. This closes the switch. Once the switch is activated, it permits the purging period timer in the programming-type flame safeguard control or BMS software to begin timing the purge. These devices, when used on boilers, are collectively known as combustion air proving switches.

On larger boilers, an air flowmeter counts, or totalizes, the cubic feet of air that have passed through the furnace area. The pilot is lit only after the required number of cubic feet of air have passed through the furnace. A minimum flow rate is also often required. This ensures that the flow of purge air is brisk enough to carry out any fuel vapors that may be present due to leaking fuel valves.

Factoid

Grates in a solid fuel-fired boiler support the fuel bed. Grates may be fixed or moving.

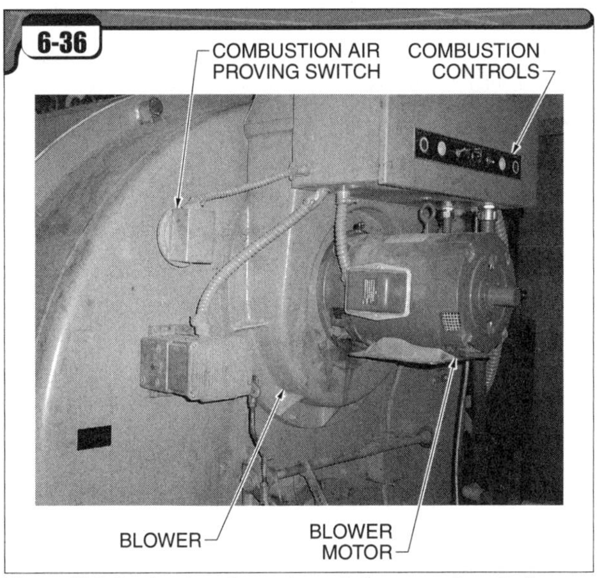

Figure 6-36. *Combustion air proving switches on package boilers are usually pressure switches.*

Why might the flame safeguard control require proof that the burner's forced draft fan motor is energized?

Because combustion air proving switches must detect very slight pressures in the air ducts, these devices are somewhat delicate. Therefore, combustion air proving switches tend to be more problematic than other safety devices. For this reason, some insurance concerns require additional levels of safety protection during the purge period.

It is common practice to have a set of auxiliary contacts in the blower motor control circuit that close and provide proof to the flame safeguard control that the blower motor is energized. This also provides a way to confirm the correct operation of the combustion air proving switch. For example, if the combustion air proving switch indicates suitable purge air flow but the forced draft fan motor is not energized, the combustion air proving switch is most likely faulty and the flame safeguard control will not allow the firing sequence to continue.

Describe the function of a high fire limit switch.

A high fire limit switch proves that the forced draft fan (combustion air blower) damper is in the fully open position when starting the purge period during the light-off of the burner. **See Figure 6-37.** This switch, along with the combustion air proving switch, proves to the programming flame safeguard control that an adequate purge air flow is available.

On package boilers, the high fire limit switch is often an auxiliary switch inside the modulating motor enclosure. It is a microswitch that is activated by a cam on the modulating motor shaft. In this case, it is often referred to as an end-of-travel switch, or end switch. When this switch is closed, it indicates that the modulating motor, and therefore the combustion air damper, is in its maximum position.

While having the high fire limit switch inside the modulating motor enclosure is convenient, this design is potentially subject to errors. In this case, the switch is not actually verifying that the damper is fully open. It is verifying that the damper motor is trying to open the damper. For example, the damper linkage could fall apart, the damper actuating arm could slip on the damper shaft, or other mechanical problems could occur. In these cases, proof could be supplied to the programming control that the damper is open when in fact it is closed or partially closed. To avoid this potential, especially on larger boilers, the high fire limit switch may be located on the damper assembly itself.

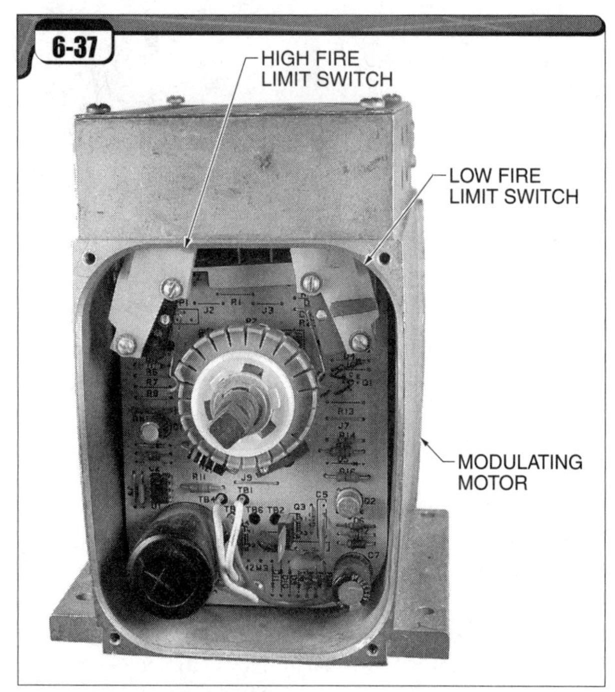

Figure 6-37. *The high fire and low fire limit switches on package boilers are generally found inside the modulating motor enclosure.*

93 What is the function of a low fire limit switch?

A low fire limit switch proves that the forced draft fan (combustion air blower) damper is in the minimum position before allowing the gas pilot to try to light. If the damper is not in the minimum position, the air flow may be too great and the pilot could be blown out.

The low fire limit switch also helps ensure that minimum electrical current will be drawn by the forced draft fan when the fan is started. With the damper in the minimum position, the fan simply freewheels during startup because it does not have to move any significant volume of air. This is similar in concept to pedaling a bicycle while the rear tire is lifted off the ground.

If the fan is started while the damper is open, the motor must perform the work of moving air at the same time that it is performing the work of bringing the fan rotor up to operating speed, and the current draw is consequently very high. This results in high wear on the motor, the fan and motor shafts, the shaft coupling, the bearings, etc.

Like the high fire limit switch, the low fire limit switch is usually found inside the modulating motor on package boilers. On larger boilers, it is often found on the fan damper shaft.

⚠ WARNING

Safety devices should never be bypassed or defeated except during testing under controlled conditions by thoroughly trained and competent service technicians.

94 What is the function of explosion doors on a furnace?

Explosion doors are designed to blow open and relieve excess pressure in the furnace, rather than having damage occur to furnace walls. **See Figure 6-38.** Explosion doors sometimes have a weighted latch handle that holds the door closed under normal conditions. A high furnace pressure on the inside of the door causes the weighted latch to release.

Some explosion doors are held closed by springs that compress and allow the door to open if the furnace pressurizes. Other explosion doors are held in place by their own weight.

Figure 6-38. *Explosion doors blow open to relieve excess pressure on the furnace walls.*

95 What is the purpose of a flame scanner?

A *flame scanner* is a device that proves that the pilot and main burner flames have been established and remain in service. A flame scanner is used to make a distinction between the presence of actual flame and the presence of just a fuel and air mixture. The flame scanner prevents explosions in the furnace by informing the flame safeguard control to stop the firing sequence if the fuel and air mixture does not ignite immediately or if the flame goes out during operation.

If the flame scanner detects failure of the flame during ignition or during normal operation, the flame safeguard control interrupts the power to the burner control circuit. This causes the automatic fuel shutoff valves (SSOVs) to fail closed.

It is also important to note that the flame safeguard system uses the flame scanner to prove that there is not a flame in the furnace when there is not supposed to be a flame. For example, if the flame scanner detects a flame in the furnace during the purge period, this indicates an unsafe condition. In this case, the flame safeguard control stops the light-off sequence. Safe conditions then have to be restored and the flame safeguard control has to be manually reset in order to attempt to start the burner again.

A flame scanner is often called a Fireye®. Fireye Co. is one manufacturer of flame scanners and related equipment. Other names for these devices include flame sensors or flame detectors. However, Fireye is widely used by boiler operators to refer to all flame-detecting devices.

List four types of automatic flame-detecting devices used for firing boilers.

The two basic groups of modern flame-detecting devices are optical and ionization sensors. Optical flame-detecting devices sense the various forms of light radiation emitted by flame. These include infrared, visible, and ultraviolet scanners. Ionization flame-detecting devices use the fact that the presence of flame causes static charges to be applied to tiny particles in the air. A flame rod is an example of an ionization flame-detecting device. **See Figure 6-39.**

Infrared flame scanners sense light radiation frequencies lower than those visible to the eye. Modern infrared flame scanners use a lead sulfide cell that is sensitive to infrared light radiation. Infrared light radiation from the flame causes the resistance across the cell to drop and allows current flow through the cell. If the flame goes out, the resistance increases and the flame safeguard control senses that the flame is no longer present. The flame safeguard control interrupts the fuel valve circuit and closes the fuel valves.

Lead sulfide infrared flame sensors can be used on gas and fuel oil flames and in some pulverized coal applications because infrared light radiation is present in these flames.

Ultraviolet flame scanners are optical devices that sense light radiation frequencies higher than those visible to the eye. The sensing tube in an ultraviolet flame sensor contains two electrodes (cathode and anode) that are surrounded by a gas.

When the sensing tube senses ultraviolet radiation from the flame, the cathode emits electrons that ionize the gas. The gas becomes conductive and current flows between the two electrodes. When current flows, the voltage between the two electrodes drops sharply and stops the current flow. If ultraviolet light radiation remains, the cycle will repeat indefinitely. This alternate energizing and de-energizing is known as firing and quenching the tube.

The flame safeguard control monitors the resulting pulsating electrical signal. If the flame goes out, the sensing tube loses its conductive quality, the pulsating signal stops, and the flame safeguard control causes the fuel valves to fail closed. Ultraviolet flame sensors may be used on gas or fuel oil flames.

Photocell flame scanners are optical devices that sense visible light from the burner. Modern photocells for flame sensors contain cesium oxide that becomes conductive when exposed to visible light. Additionally, the cesium oxide coating on the cathode side of the photocell element makes the element a true rectifier. A *rectifier* is an electrical device that converts alternating current (AC) into direct current (DC).

Since the anode side of the photocell element has no coating of cesium oxide, it does not produce electrons when exposed to visible light. This principle makes the photocell produce direct current only. If the flame is lost, the direct current signal will be lost and the flame safeguard control will cause the fuel valves to fail closed. Cesium oxide photocells will not work properly with gas flames since these flames do not produce enough visible light. These sensors are usually used with fuel oil but may be used with pulverized coal.

Flame rod flame sensors are not optical in nature. They work on the principle of flame ionization. A tiny current can actually be conducted by and through a flame. The flame rod sensor uses two electrodes. Alternating current is supplied to one electrode. The other electrode is at least four times as large and is grounded. The larger size of the grounded electrode makes it easy for electrons to flow toward the grounded electrode, but difficult for electrons to flow the other way.

The resulting effect is that the flame rectifies the electrical current passing through it from AC to pulsating DC. The control circuit is designed to monitor this DC signal. If the flame is lost, the DC signal is lost and the flame safeguard control causes the fuel valves to close. Flame rod flame sensors are used with gas, but seldom with fuel oil because the very high temperatures encountered in a fuel oil flame can damage the device.

Industrial fuel burners introduce a very large amount of potentially explosive air-fuel mixture to the furnace in a short time. Therefore, all types of flame sensors used with commercial, institutional, and industrial boilers must respond quickly to a loss of flame. Modern flame sensors must de-energize the burner circuit and cause fuel valves to fail closed within 4 sec. Common thermal-type flame detecting devices, such as the thermocouples used in residential water heaters, do not respond quickly enough to the loss of flame and are therefore not used in commercial and industrial burners.

Infrared light radiation may be produced by hot refractory brick. Under certain conditions, an infrared flame scanner cannot distinguish between the infrared light radiation produced by a flame and that produced by hot refractory brick. Never change the type or model of flame scanner used for a particular application without consulting the burner equipment manufacturer first.

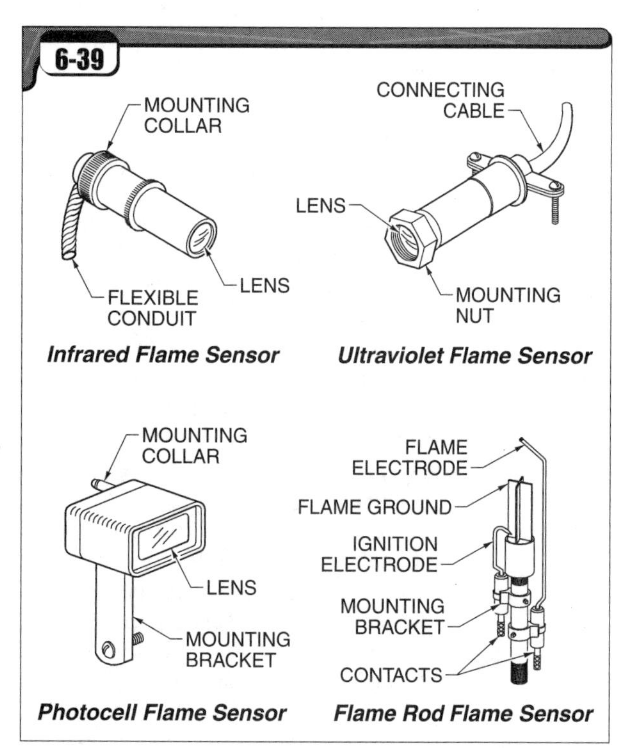

Figure 6-39. *Automatic flame sensors sense the presence of a flame in the burner.*

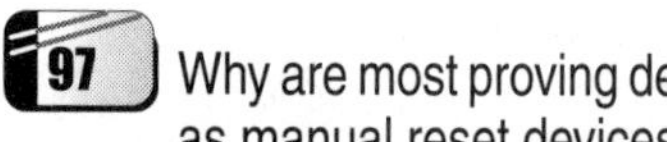

97 Why are most proving devices on boilers configured as manual reset devices?

A *manual reset device* is a safety device that must be manually reset by the boiler operator before the flame safeguard control will attempt to restart the burner equipment. **See Figure 6-40.** If a safety device detects an unsafe condition, such as high gas pressure or loss of flame, it is desirable to have the boiler operator made aware of the condition. The use of manual reset detection devices serves this goal. The fact that the controls must be manually reset requires intervention by the boiler operator. It is very important that the boiler operator investigate the cause of the automatic shutdown rather than simply resetting the controls.

When a manual reset is tripped on a specific safety device, it usually indicates that another device is not working properly. For example, if the manual reset button is tripped on a high gas pressure switch, it most likely indicates that the main gas pressure regulating valve is leaking through or otherwise not working properly. A manual reset button tripped on a low fuel oil temperature switch most likely indicates that the fuel oil heater is not working properly.

Sometimes a single manual reset button is provided in the boiler main control panel. In these cases, field devices must be reset using this central reset button, as opposed to having individual reset buttons on several field devices.

If a safety device causes shutdowns of the boiler, the boiler operator should assume that the device is detecting an unsafe condition rather than assuming that the safety device itself is faulty.

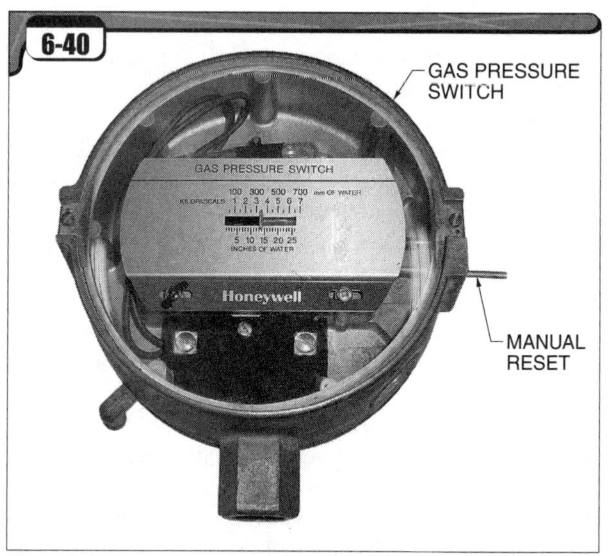

Figure 6-40. *Manual reset devices make the boiler operator aware of an unsafe condition that caused the device to trip.*

98 What is the turndown ratio of a modulating burner?

The *turndown ratio* of a modulating burner is the ratio of the maximum firing rate of the burner to the minimum firing rate. For example, if the minimum firing rate of a burner is one-fourth of its maximum firing rate, the burner has a turndown ratio of 4:1. If the minimum firing rate of the burner is one-tenth of its maximum firing rate, the burner has a 10:1 turndown ratio.

Due to the limitations of controlling fluid flows (specifically fuel and air) with valves and dampers, it is not feasible to turn down a large industrial burner until the flame is the size of a candle flame. There is a minimum flame size at which the flame is stable and free of operational problems.

A factor that affects the turndown ratio of a burner is the flame propagation rate. The *flame propagation rate* is the rate at which the flame can ignite the incoming air-fuel mixture, in feet per second. This is similar to salmon swimming upstream in a rapidly flowing river. If the river slows down, the salmon travel upstream faster. If the river speeds up, the salmon have difficulty making any headway.

Flashback is the condition where a flame travels upwind and into the burner assembly. The air-fuel mixture must be traveling fast enough so that the air flow across the burner assembly keeps it cool enough to prevent damage by overheating. Therefore, if the burner is turned down too far, the flame can migrate into the burner and damage the burner. For this reason, the turndown ratio of a burner is limited by the burner design.

Trade Tip

The industry term "light-off" refers to the processes that occur during the starting sequence of a burner.

Describe the automatic firing sequence of a gas- or light oil-fired package boiler.

When the main pressure control (pressurestat or other steam pressure control device) senses a drop in steam pressure, it provides a call for burner start to the flame safeguard control (programmer). The programmer first checks its internal circuitry. This is a built-in, nearly instantaneous feature.

The programmer then polls the field devices to determine that all required prestart requirements are met. For example, the gas pressure or oil pressure proving switches must indicate that the fuel pressure is satisfactory, the low water cutoff must indicate that a safe water level exists in the boiler, the proof of closure switch on the SSOV must indicate that the fuel valve is closed, the combustion air damper must be in the minimum position, etc. This series of checks is also nearly instantaneous.

If all the field devices confirm that conditions are safe, the programmer starts the forced draft fan with the fan damper closed. The auxiliary contacts in the fan's motor control circuit provide feedback to the programmer that the fan is energized. The programmer then energizes the modulating motor that controls the fan damper and fuel flow control valve. The modulating motor drives the fan damper wide open, closing the high fire limit switch. No fuel is admitted at this time because the SSOVs are still closed.

Within a few seconds after starting the forced draft fan, the combustion air proving switch should detect the development of air pressure or air flow in the combustion air ductwork. The switch provides proof to the programmer that the combustion air fan is performing properly. This allows the prepurge timer in the programmer to begin timing the purge period.

A time delay of a few seconds is designed into the programmer's internal logic. This delay is to allow the fan to come up to speed and pressurize the combustion air ductwork before the programmer requires proof of combustion air. This is necessary because the fan cannot pressurize the ductwork instantly when it is energized.

The damper stays fully open during the predefined purge period. This allows the maximum flow of air through the furnace in order to completely purge any flammable gases inside. After the purge period times out, the programmer signals the modulating motor to reverse, driving the fan damper to the minimum position. The low fire limit switch in the modulating motor must prove the minimum position of the damper.

At this point, the igniter is energized with about 6000 V to 10,000 V from the ignition transformer to produce a strong spark to light the pilot gas. The programmer opens the automatic pilot gas shutoff valves, allowing the pilot to light.

The *pilot trial for ignition,* or *PTFI,* is a period of about 5 sec to 10 sec for the flame scanner to sense the presence of flame from the pilot. The exact amount of time required depends on the size of the burner and insurance requirements. If the flame scanner does not confirm the presence of light from the gas pilot during this period, the programmer shuts the system down and a flame failure alarm and manual reset condition occur. If the flame scanner senses the pilot flame, then the main fuel SSOVs (gas or fuel oil) are opened by the programmer.

If the boiler is being fired on fuel oil, the atomizing medium (compressed air or steam) is also started. This allows fuel (and atomizing medium) to flow to the burner. The *main trial for ignition,* or *MTFI,* is a period of about 5 sec to 10 sec for the flame scanner to sense the presence of flame from the main flame. The programmer then closes the automatic pilot gas shutoff valves and de-energizes the ignition transformer.

If the flame scanner does not sense flame after the pilot is shut off and the programmer shuts the system down, a flame failure alarm and manual reset condition occur. This most likely means that the main flame did not light. It could also be a failure of the flame scanner.

If the main flame is proven by the flame scanner, the programmer allows the modulating pressure control to take control of the modulating motor. The programmer continues to monitor the information provided by all of the field safety devices.

Because low steam pressure is what caused the burner to start initially, the modulating pressure control now causes the modulating motor to drive the combustion air (forced draft) damper and fuel flow control valve to their high fire positions. This causes the boiler to generate steam at a relatively fast rate.

As steam pressure rises in the boiler, the modulating pressure control adjusts the modulating motor to lower the firing rate. The modulating pressure control and modulating motor then continually adjust the firing rate in an effort to maintain constant steam pressure.

Due to the fixed turndown ratio of the burner, a point may eventually be reached where the boiler produces steam faster than the steam system needs the steam, even with the burner at the lowest firing position. The upper setting on the main pressure switch is reached and the SSOVs are closed.

If the boiler is fuel oil-fired, the programmer causes a solenoid-actuated valve to admit the atomizing medium (steam or compressed air) into the fuel line to cause the remaining fuel oil in the burner assembly to be purged out. This is necessary so that the fuel does not leave residue in the burner. This residual fuel blows into the furnace and burns for a few additional seconds.

The programmer then times the post-purge period. After the burner is shut down, the forced draft fan continues to run for a specified time, typically about 15 sec to 25 sec. This postpurge is used primarily to remove any remaining air-fuel mixture from the burner assembly. This ensures that the flame cannot travel upstream into the burner assembly when the forced draft fan is shut down.

When the boiler pressure again drops to the lower setting on the main pressurestat or other pressure control, this cycle repeats.

Trade Tip

When a manual reset device causes a burner shutdown, it is an indication that another device did not function.

Can a boiler furnace be fired with more than one gas burner?

Many large boilers have more than one gas burner. The controls for multiple burners can be set so that if one burner cannot generate enough heat to satisfy the steam demand, another burner is started. If additional steam is needed, the process continues until the steam demand is met or until the boiler is at maximum capacity.

Is it good practice to light a burner with a burner that is already lit?

It is not good practice to light a burner with another burner. A substantial amount of fuel could be introduced into the furnace from the unlit burner before it finally is ignited by the lit burner. This is especially true if the burners are spaced more than a short distance apart.

Factoid

The quantity of combustion air used should always be limited to only the amount needed for efficient combustion. Any additional air is counterproductive, as it results in heat being carried up the stack, cooling of the furnace, and other problems.

Combustion Air Requirements

Why must the boiler room have an unobstructed air supply?

Most burners on commercial and industrial boilers draw their combustion air supply from the boiler room. Air must therefore be able to enter the room from outside, or the room will soon be under a vacuum. This can impair the performance of the boilers because inadequate combustion air may be available to mix with the fuel.

Boiler rooms are equipped with large louvered openings to allow air to flow in from the outside. **See Figure 6-41.** In cold climates, the openings may be equipped with steam heating coils to warm the incoming air. The sizes of these openings are engineered to ensure an adequate flow of combustion air. Therefore, they should never be covered over.

Why is the quantity of air used for combustion important?

Each combustible element in the fuel must combine with adequate oxygen in the combustion air in order to burn completely. Thus, combustion is a predictable chemical reaction. This is demonstrated in the following examples:

Carbon	$C + O_2$	$\rightarrow CO_2$ + Heat
Hydrogen	$H_2 + ½O_2$	$\rightarrow H_2O$ + Heat
Sulfur	$S + O_2$	$\rightarrow SO_2$ + Heat
Methane	$CH_4 + 2O_2$	$\rightarrow CO_2 + 2H_2O$ + Heat
Propane	$C_3H_8 + 5O_2$	$\rightarrow 3CO_2 + 4H_2O$ + Heat

Each of the elemental constituents in the fuel (hydrogen, sulfur, carbon, nitrogen, etc.) has a known molecular weight and the oxygen also has a known molecular weight. Therefore, the weight of the combustion products (e.g., CO_2, H_2O) equals the sum of the weights of the elements and the oxygen. When the ultimate analysis of the fuel is known, design engineers can determine the amount of air required to burn a unit of the fuel (e.g., 1 lb of coal). This is an important factor in properly sizing the equipment. For example, if a given quantity of fuel is to be burned, it is important to have a combustion air fan that can deliver the proper quantity of air and deliver it as efficiently as possible.

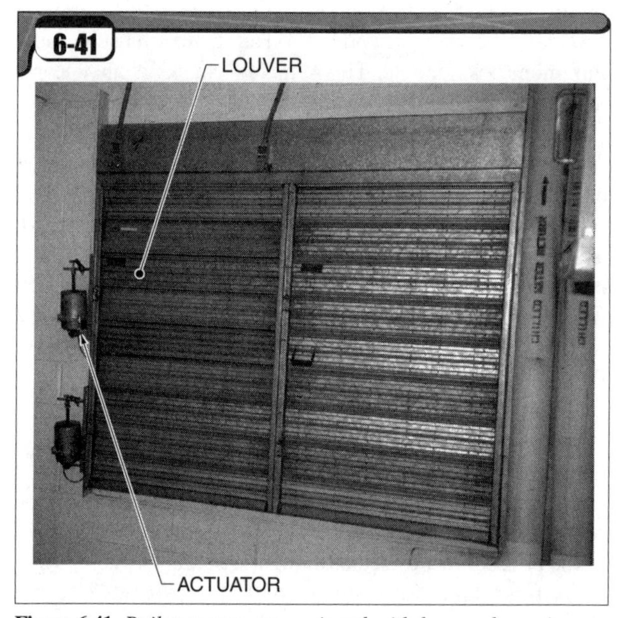

Figure 6-41. *Boiler rooms are equipped with louvered openings to ensure an adequate combustion air supply for burner equipment.*

What is excess air?

Excess air is the amount of air added to the combustion process over and above that which is theoretically necessary. The word "excess" in this case does not necessarily mean too much.

Stoichiometric combustion is the process of burning a fuel with precisely the amount of air required so that no unburned fuel or unused oxygen remains. Stoichiometric combustion is not feasible in a boiler furnace because of the imperfect degree of control afforded by valves and dampers.

Additionally, the fuel and oxygen cannot be mixed completely enough for perfect combustion. For example, as the combustion process proceeds, the free oxygen molecules become fewer and fewer. In addition, there are molecules of nitrogen, CO_2, SO_x (sulfur oxides), and H_2O present that make it harder for the remaining molecules of fuel to find and react with the oxygen molecules.

In order to achieve the maximum efficiency from the fuel, slightly more air is supplied than the theoretical amount of air needed, to ensure that sufficient oxygen is present to burn all the fuel. This can be indirectly checked by measuring the oxygen content of the flue gases going up the stack. The flue gases leaving the boiler should contain about 3% to 7% oxygen, depending on the fuel, the firing rate, and the type of firing equipment used.

The amount of excess air used should be kept to the minimum required to ensure complete combustion. This is because the oxygen is the only portion of atmospheric air that contributes to the combustion process. The remainder of the air (predominantly nitrogen) passes through the furnace unchanged. However, heat from the combustion process is used in heating this remainder and this volume of air carries some of that heat up the stack to the atmosphere. The use of too much excess air also results in lower flame temperatures, which reduces heat transfer efficiencies. In addition, it creates higher flue gas velocities and subsequent increased erosion of tube surfaces, and it increases overall air pollutant emissions.

What amount of excess air should be used for good combustion?

The amount of excess air used depends on the fuel and the performance of the combustion equipment. Efficient combustion is always affected by the quality of atomization, turbulence in the furnace, and other mechanical factors. As a general rule, natural gas requires 10% to 15% excess air, fuel oil requires 10% to 30% excess air, coal burned on grates requires 50% to 75% excess air, and pulverized coal requires 20% to 30% excess air.

The carbon monoxide content of the flue gases is another valuable reference in adjusting excess air. Assuming that atomization, fuel oil preheating, and other mechanical factors are in order, the appearance of carbon monoxide in the flue gas indicates incomplete combustion and the requirement for more air. It is desirable to operate with the lowest possible percentage of excess air without developing carbon monoxide.

It is usually necessary to use more excess air at lower firing rates than at high firing rates. This is because the lower velocity of the air-fuel mixture at lower firing rates results in less turbulence in the furnace and the fuel molecules will subsequently come into contact with fewer oxygen molecules. Increasing excess air at lower firing rates thus provides more oxygen to solve this problem.

Why should the measurement of oxygen in the flue gases exiting the boiler be made as close to the boiler flue gas outlet as possible?

Because most larger boiler furnaces operate under a very slight vacuum, any air that leaks into the furnace area dilutes the flue gases and thus drives the oxygen content of the flue gases higher. Taking the oxygen measurement as near the boiler's flue gas outlet as possible minimizes the likelihood of this dilution and thus produces more accurate measurement results.

What is an Oxygen (O_2) trim system?

An *oxygen (O_2) trim system* is an automatic control system that makes fine adjustments in the amount of combustion air used in order to minimize excess air. This system measures the oxygen content of the flue gases exiting the boiler and compares the measurement to a predetermined setpoint. The control then makes fine adjustments in the combustion air damper to add or reduce combustion air. By maintaining excess air at optimum levels, the combustion efficiency of the boiler is increased.

What is NO_x?

The abbreviation NO_x is used to represent a group of gases that are produced as a result of the chemical reaction between nitrogen and oxygen at very high temperatures (greater than about 2700°F). These gases are oxides of nitrogen. The most common nitrogen oxides are nitric oxide (NO) and nitrogen dioxide (NO_2). These gases are environmental pollutants. They combine with water vapor in the upper atmosphere to form nitric acid (HNO_3). This results in acid rain.

In addition, nitrogen dioxide is a light brown photochemical oxidant gas with a pungent, irritating odor, and is a principal component of smog. It can cause various forms of respiratory system ailments.

Minimizing the use of excess air, and thus the quantity of nitrogen available for this reaction, is important in preventing these pollutants from forming. Fuels that contain nitrogen (called fuel-bound nitrogen or chemically-bound nitrogen) result in higher emissions of NO_x. Some fuels such as natural gas result in lower emissions of these gases.

The use of low-NO_x burners is becoming widespread as industry is making efforts to comply with new emissions standards governing these pollutants. *Staged combustion* is the process of mixing fuel and air in a way that reduces the flame temperature and NO_x generation in a burner. Low-NO_x burners are designed to use staged combustion.

Staged combustion works by limiting the availability of oxygen in the primary stages of combustion, through the use of flue gas recirculation (FGR), or both. In FGR systems, a portion of the spent flue gases being discharged to the stack is conducted back to the combustion air inlet. This slightly dilutes the quantity of oxygen in the overall mixture and reduces the temperature of the mixture. This causes the combustion process to happen at a slower rate.

Combustion Efficiency

What are higher heating value and lower heating value of a fuel?

Higher heating value (HHV), also known as gross heating value, is the total heat obtained from the combustion of a specified amount of the fuel under perfect (stoichiometric) combustion conditions. To determine the higher heating value correctly, both the fuel and the combustion air are at 60°F when the combustion starts and the resulting products of combustion are cooled to 60°F before the heat release is measured.

Not all of this heat content of the fuel is available when the fuel contains some amount of hydrogen, however. This is because when hydrogen burns, it reacts with oxygen to form H_2O and heat is used in evaporating this water.

Lower heating value (LHV), also known as net heating value, is the quantity of heat remaining after subtracting the latent heat used in evaporating the water formed in the combustion of the hydrogen. The latent heat used in evaporating the water is not available for making steam in a boiler.

110 What is the greatest loss in a boiler plant?

The greatest loss in a boiler plant is heat up the stack. It can be minimized, but not eliminated, through the use of heat recovery equipment such as economizers and air preheaters. Heat loss can also be minimized through preventive efforts such as maintaining the combustion equipment in top condition, keeping the heating surfaces clean on both fire and water sides, and reducing excessive combustion air.

It is necessary to allow enough heat to remain in the flue gases to keep the water vapor formed in the combustion of hydrogen in the form of steam. If too much heat is extracted from the flue gases, the steam will begin to condense and can be very corrosive to the stack, breeching ductwork, and pollution control equipment.

111 Why are automatic combustion controls used in modern boiler operations?

Automatic combustion controls relieve the boiler operator of the tedious duty of manually monitoring and regulating combustion conditions. Automatic combustion controls also result in a more efficient operation, although they should not be trusted exclusively.

Automatic combustion controls differ from combustion safety controls. Automatic combustion controls regulate the quantities of fuel and combustion air delivered to the furnace as determined by the plant load. Some automatic combustion controls continually make fine adjustments of the combustion conditions to maximize combustion efficiency. Combustion safety controls prevent the combustion equipment from being operated until safe furnace conditions are established and stop the fuel supply if unsafe furnace conditions are detected during operation.

112 What is an Orsat analyzer?

An *Orsat analyzer* is a flue gas analyzer that measures the percentages of carbon dioxide, carbon monoxide, and oxygen in flue gases. A measured sample of the flue gas is pulled into the analyzer by a squeeze bulb or air pump. **See Figure 6-42.** The analyzer has three separate chambers, each used for quantifying one of these gases. The relative quantities of each are used in determining what adjustments need to be made to the fuel-burning equipment to achieve higher combustion efficiency.

Various forms of flue gas analyzing apparatus are available that are similar in function and operation to the Orsat analyzer. For example, single-analysis devices of this type measure one variable in the flue gas only, such as carbon dioxide or oxygen.

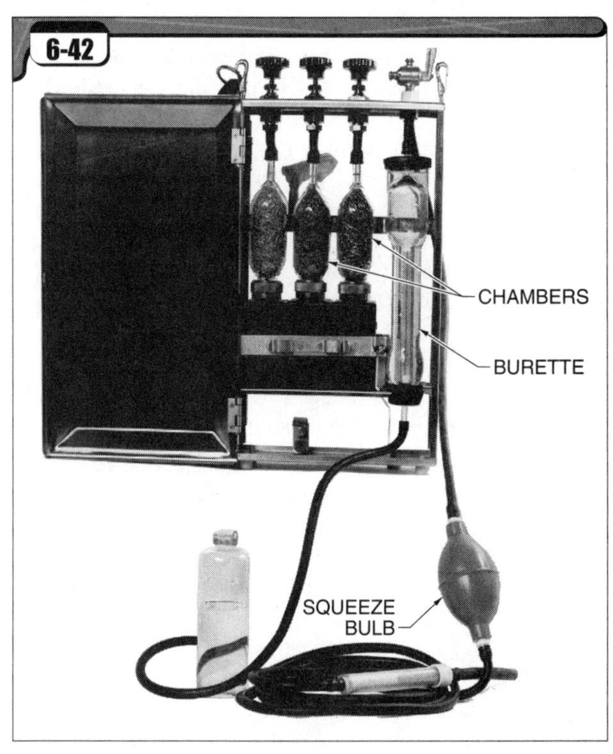

Figure 6-42. *Orsat analyzers measure the percentages of carbon dioxide, carbon monoxide, and oxygen in flue gases.*

113 What additional functions are provided by electronic flue gas analysis equipment?

Electronic flue gas analyzers are available that are capable of performing the same measurements as the Orsat analyzer. However, electronic flue gas analyzers are also capable of additional measurements. In addition, these instruments can perform automatic calculations, store data, create trends and histories, etc.

In addition to measuring carbon dioxide, carbon monoxide, and oxygen, electronic flue gas analyzers can be configured to measure flue gas temperature and smoke, as well as furnace draft. From these measurements, the equipment can compute the percentage of excess air being used and the percentage of combustion efficiency. If so equipped, these analyzers may also be used to measure the emissions of pollutants such as NO_x.

In large industrial, institutional, and utility boilers, environmental permits often dictate the use of continuous emissions monitors, or CEMs. As the name implies, CEMs are used 24 hours a day to document the emissions of regulated pollutants from these sources. In addition to direct measurement of specific pollutant gases and opacity, CEMs in coal-burning plants typically include calculated emissions of oxides of sulfur, or SO_x. For example, automatic calculations are often used to determine sulfur emissions in pounds per million Btu worth of fuel burned.

114 What percentage of carbon dioxide in the flue gases is considered desirable?

The percentage of carbon dioxide considered desirable in the flue gases depends on the fuel and the type of combustion equipment used. It is desirable to convert all of the carbon in the fuel to carbon dioxide, but none to carbon monoxide. About 10% carbon dioxide is desirable for a gas-fired boiler, 12% to 13% for light fuel oil-fired boilers, 13% to 14% for heavy fuel oil-fired boilers, and 14% to 15% for coal-fired boilers.

The ratio of carbon to hydrogen in the fuel affects the maximum amount of carbon dioxide possible. For example, more carbon dioxide is developed when burning coal than when burning natural gas because coal contains more carbon.

115 With excessive draft, will the carbon dioxide content of flue gases be higher or lower?

With excessive draft, the carbon dioxide content of flue gases will be lower because a given quantity of fuel (a pound of coal, for example) contains a fixed amount of carbon, and therefore can only produce a fixed amount of carbon dioxide. If more air is added after all possible carbon dioxide is formed, the air will simply dilute the carbon dioxide. This will lower the percentage of carbon dioxide in the flue gases.

Factoid

Flame color is not an adequate indication of combustion efficiency. Combustion equipment should be tuned based on the analysis of the flue gases.

116 If the percentage of carbon dioxide in the flue gases is low, what should be done to increase it?

If an analysis of the flue gas shows the percentage of carbon dioxide is low and there is some carbon monoxide, then there is either a lack of oxygen or the oxygen and carbon are not mixing well enough for efficient combustion. If there is not enough oxygen, as indicated by the oxygen measurement, the primary air flow should be increased to provide more oxygen.

If the flue gas analysis shows both carbon monoxide and residual oxygen, the reaction between the carbon and oxygen did not occur completely and carbon monoxide was formed rather than carbon dioxide. This likely means that the oxygen and carbon are not mixing well enough for efficient combustion. More turbulence should be created by increasing the secondary air flow. This will result in better mixing of the carbon and oxygen. If fuel oil is used, the atomization may not be adequate. This could occur due to low pressure of the atomizing medium or partial plugging of the holes in the burner tip. Additionally, the furnace temperature may be too low for efficient combustion. This is often an issue during boiler startup.

If there is no carbon monoxide evident in the flue gas, too much combustion air (excess air) is being used, since the absence of carbon monoxide in the flue gas analysis shows that all the carbon was burned with adequate oxygen.

117 How does altitude affect fuel combustion in a boiler furnace?

The density of air decreases as altitude increases. If the air is less dense, a larger volume of air is required to complete the combustion process. Fuel burners and combustion air blowers sold for high-altitude installations are normally selected to handle the greater air and gas volumes associated with the lower air densities. Even so, a competent burner tuning technician should adjust the burner settings when the new boiler is commissioned.

118 What is an opacimeter?

Opaque means impervious to light. Opacity (the quality of being opaque) of flue gases is expressed as a percentage, with 0% being invisible and 100% being opaque smoke and/or ash particles. An *opacimeter* is an automatic indicator that measures the amount of light blocked by the smoke and ash going up the stack. The opacimeter consists of a light source and receiver, a recorder for generating permanent records, and an indicating panel. **See Figure 6-43.** The indicating panel often includes a transmitter function to transmit the opacity reading to a remote boiler plant control room.

The light source shines a beam of light to the receiver across the inside of the breeching. In some cases, both the light source and receiver are on the same side of the breeching and a mirror is located on the opposing side. Any smoke or ash present restricts the amount of light detected by the receiver. This is converted by the opacimeter to an opacity reading of the smoke and ash density as a percentage.

Opacity of flue gases is sometimes a problem when burning fuel oils as well as when burning coal or other solid fuels. When burning oil, the visual appearance of the smoke can be an aid in determining the appropriate adjustments. For example, if the smoke is light brown, the oxygen level in the combustion zone is too low. If the smoke is white, too much oxygen is being introduced, causing the fuel oil to be atomized to a near-aerosol state. Black smoke indicates incomplete combustion.

There should not be any smoke when burning fuel oil under complete combustion conditions. With heavy oils there may be a slight particulate haze. *Particulate* is fine ash particles from a burner that ultimately settle back to earth. The particulate matter is gritty and often contains trace amounts of sulfur. The particulates can cause environmental damage and deterioration of rivers and streams, buildings, automobile paint, etc. The EPA regulates the acceptable level of particulate matter issuing from the stack.

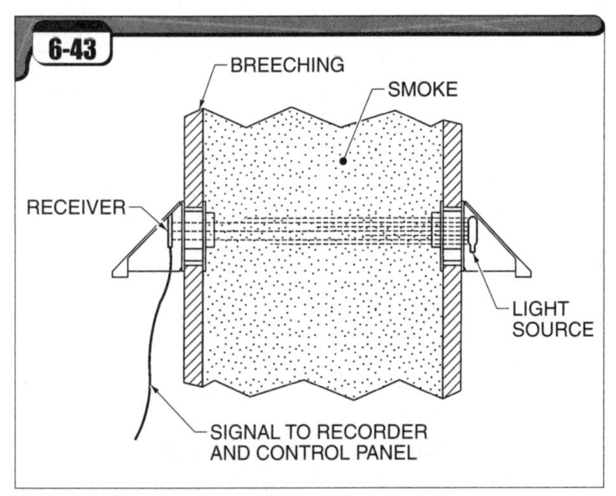

Figure 6-43. *Opacimeters measure smoke density.*

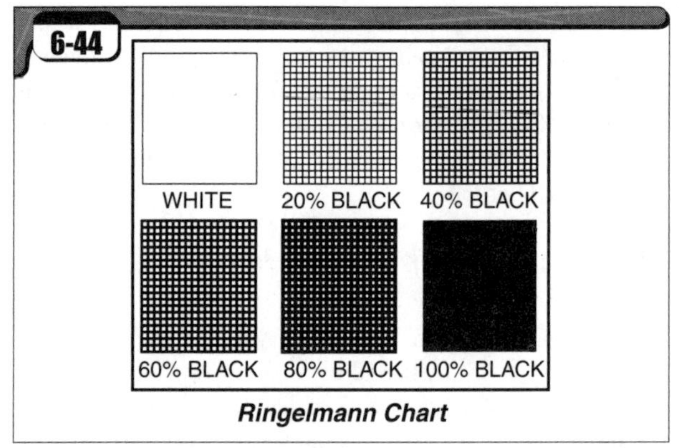

Figure 6-44. *The Ringelmann chart measures smoke opacity.*

119 What is a Ringelmann chart?

A *Ringelmann chart* is a comparison chart used to measure opacity. Sections of the chart are shaded in 20% increments from white to black, with 0% being white and 100% being black. **See Figure 6-44.** The shade of smoke coming out of the stack is compared with the shaded sections of the chart to determine opacity. Modern environmental permit requirements require corrective action for stacks operating above 20% opacity.

To use the Ringelmann chart, the observer should be standing 100′ to 1300′ from the smoke. The observer's line of sight must be perpendicular to the smoke. The chart is placed 50′ in front of the observer near the line of sight. The observer then compares the smoke density with the chart. The use of Ringelmann charts and other manual comparison methods of determining opacity has been largely replaced by opacimeters. In jurisdictions where opacity readings are made by boiler operators, documented training and testing is required.

120 What is a pyrometer?

A *pyrometer* is an instrument that measures temperatures above the temperature range of mercury thermometers. In a boiler furnace, a pyrometer measures refractory temperatures by sensing the infrared rays produced, and flue gas temperatures by direct contact thermocouple measurement.

121 What is a thermocouple?

A *thermocouple* is a device used to measure temperature consisting of two dissimilar metals joined together. These metals produce a reaction when heated and a small electrical voltage (potential) is created. Depending on the types of metals used in the wires and the voltage produced (in millivolts), the temperature can be extracted using known formulas. Example pairs of joined metals that produce this reaction are copper and constantan or Chromel® and Alumel®.

Heat Transfer Efficiency

122 List two reasons why the formation of soot on the heating surfaces should be avoided.

Soot acts as insulation. Soot is an effective insulator on boiler tubes and other heating surfaces. Therefore, more fuel must be burned to generate the same amount of steam. **See Figure 6-45.** This can result in a significant drop in the efficiency of the boiler. For example, a $\frac{1}{16}$″ soot coating can increase fuel consumption over 5%. When the soot thickness is increased to $\frac{1}{8}$″, the fuel consumption can increase over 9%.

Soot can cause metal deterioration. Sulfur combines with moisture in the flue gases and forms sulfuric and/or sulfurous acids that cause rapid deterioration of the metal surfaces.

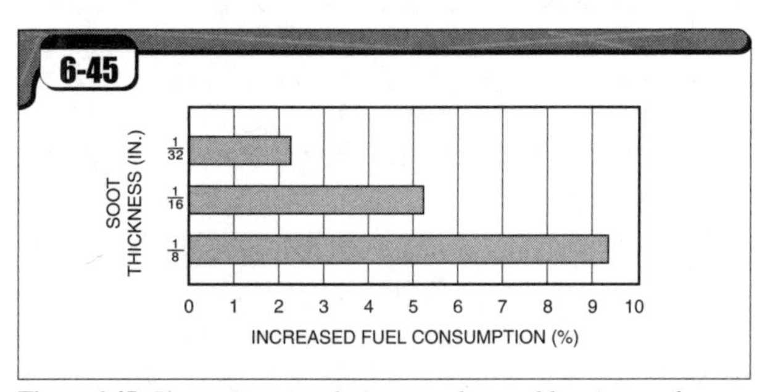

Figure 6-45. *Soot acts as insulation on tubes and heating surfaces.*

123 Describe two methods by which soot may be removed from the boiler tubes.

The two common methods by which soot may be removed from boiler tubes are soot blowers and flue brushes.

Soot blowers are steam or air pipes that are operated at specified intervals to blast soot and flyash accumulations from tube surfaces in watertube boilers. An automatic or manually operated valve is connected to a pipe that distributes the steam or compressed air across the tube banks. **See Figure 6-46.**

If the area of the flue gas passages where the soot blower is installed is not too hot, a stationary corrosion-resistant pipe with nozzles along its length is permanently installed in the passage. As the soot blower is operated, the pipe slowly rotates and the steam or compressed air from the nozzles blows between the rows of tubes. The steam should not blow directly against the tubes because this can cause thin spots in the tubes and these thin spots could burst.

A retractable soot blower is used if the device must operate in a location in the furnace where the gases are hot enough to damage the soot blower when it is idle. This soot blower also has a rotating pipe, but has a T-shaped nozzle on the end of the pipe. This nozzle points at right angles to the pipe. The movement of the soot blower is designed to blow between the rows of tubes as the nozzle comes around on each side. This unit automatically extends fully into the furnace and then backs out, blowing steam the entire time. Steam must continue to flow through the pipe to prevent overheating and warping.

A much shorter version of the retractable soot blower is used for deslagging waterwalls. This type normally has a 90° nozzle as opposed to a T-shaped nozzle. It extends through the furnace walls for a few inches and blows steam as it rotates, removing ash and slag from the waterwall tubes.

Soot blowers in coal-fired watertube boilers are normally operated on each shift. This frequency may be varied, however, based on plant experience. Soot blowers in specific sections of the boiler system may be operated less frequently. For example, soot blowers in the economizer section may only need to be operated once per day.

Long bottle-type flue brushes are used to remove soot from the inside of the tubes in a firetube boiler. The boiler must be shut down to do this. These flue brushes are often used in conjunction with a heavy-duty vacuum cleaner that draws in the soot as it is brushed loose. This helps minimize the mess that would otherwise result.

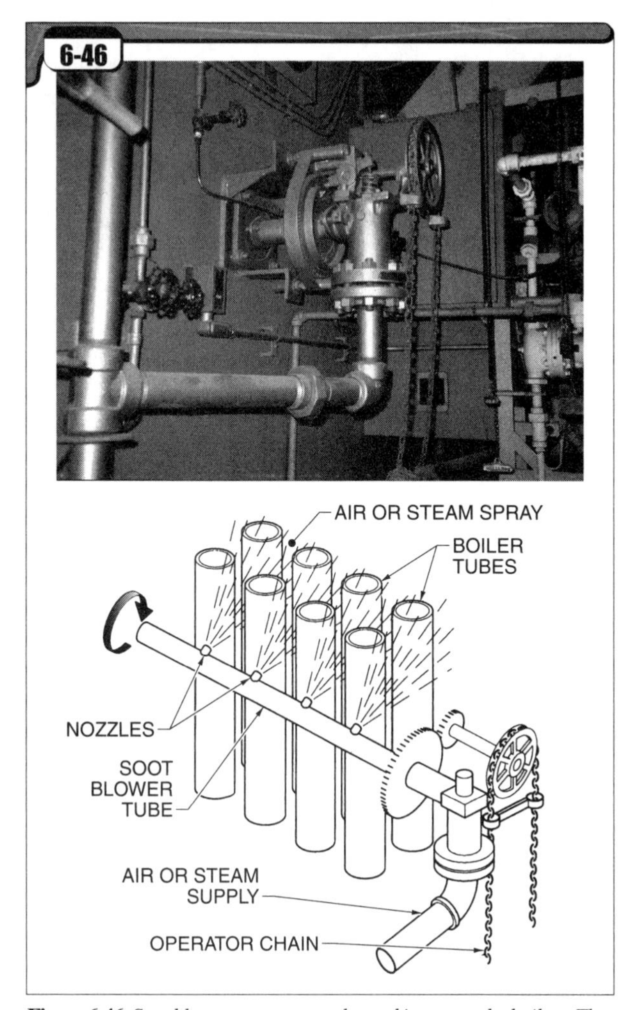

Figure 6-46. *Soot blowers are commonly used in watertube boilers. They may be stationary or retractable.*

Factoid

Although the boiler operator should understand the concepts of efficient combustion, burner tuning is potentially dangerous and should be left to experienced, factory-trained technicians equipped with proper flue gas analysis equipment.

124 What is the purpose of turbulators in a firetube boiler?

Turbulators, or *retarders,* are devices that swirl the hot gases of combustion as the gases pass through the center of the tube so that the gases come into more efficient contact with tube walls where heat transfer occurs. **See Figure 6-47.** Turbulators also increase the velocity of the flow of flue gases slightly. The increased velocity scrubs cooled flue gases away from the tube walls and replaces them with hot gases of combustion to increase efficiency. Turbulators are most often found in older firetube boilers that have 3″ and 4″ tubes. Smaller tubes used in modern firetube boilers increase the flue gas velocity, resulting in more efficient heat transfer.

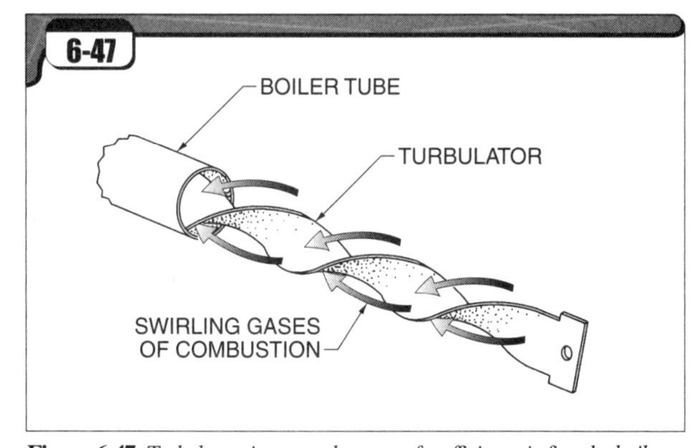

Figure 6-47. *Turbulators increases heat transfer efficiency in firetube boilers.*

125 What is fuel-to-steam efficiency?

Fuel-to-steam efficiency is the percentage of heat content of the fuel that is transferred into the boiler water. For example, a fuel-to-steam efficiency of 85% indicates that 85% of the

Btus in the fuel supplied to the furnace are present in the steam for use where needed. The majority of the remaining heat goes up the stack. **See Figure 6-48.** Some heat is also lost as radiant heat from the shell, furnace walls and ductwork, external appliances, and other parts of the boiler. Additionally, fuel-to-steam efficiency drops as scale accumulates in the boiler and as soot or ash accumulates on the fire side of the tubes. The fuel-to-steam efficiency is a good overall efficiency analysis of the boiler because it compares the Btu input in the fuel to the Btu output available in the steam.

A measurement that is similar to the fuel-to-steam efficiency is the steam rate of the boiler. The steam rate of a boiler is the quantity of steam produced per unit of fuel. The steam rate can be a valuable tool in planning boiler usage to minimize plant costs. For example, a boiler that generates an average of 11.2 pounds of steam per pound of coal will be significantly less expensive to operate over the long term than a boiler that averages 10.7 pounds of steam per pound of coal. Boiler operators should track the steam rate on individual boilers as an indication of when boiler maintenance is called for.

Other efficiency ratings include combustion efficiency and thermal efficiency. Combustion efficiency refers to the effectiveness of the combustion equipment in converting the available chemical energy in the fuel into Btus, using the minimum amount of air. Thermal efficiency of a boiler refers only to its efficiency as a heat exchanger in transferring heat from one side of the tubes to the other side. Thermal efficiency does not take into account the radiant, convective, or other losses.

The steam load in a plant averages 100 Klb/hr. The plant has two coal-fired boilers. The steam rate of Boiler #1 averages 11.3 lb of steam per lb of coal and Boiler #2 averages 10.5 lb of steam per lb of coal. If the coal is $40/ton and the plant operates 24 hr/day, 365 days/yr, how much may be saved annually by operating Boiler #1 rather than Boiler #2?

$$FS = \frac{\left(\frac{SL}{SR_2} - \frac{SL}{SR_1}\right)}{2000} \times (24 \times 365) \times P$$

where
FS = fuel savings (in \$)
SL = average steam load (in lb/hr)
SR_1 = steam rate Boiler #1 (in lb steam/lb coal)
SR_2 = steam rate Boiler #2 (in lb steam/lb coal)
2000 = conversion constant (in lb/ton)
24 = conversion constant (in hr/day)
365 = conversion constant (in days/year)
P = coal price (in \$/ton)

$$FS = \frac{\left(\frac{100{,}000}{10.5} - \frac{100{,}000}{11.3}\right)}{2000} \times (24 \times 365) \times 40$$

$$FS = \frac{(9523.81 - 8849.56)}{2000} \times 8760 \times 40$$

$$FS = \frac{674.25}{2000} \times 350{,}400$$

FS = **\$118,000/yr**

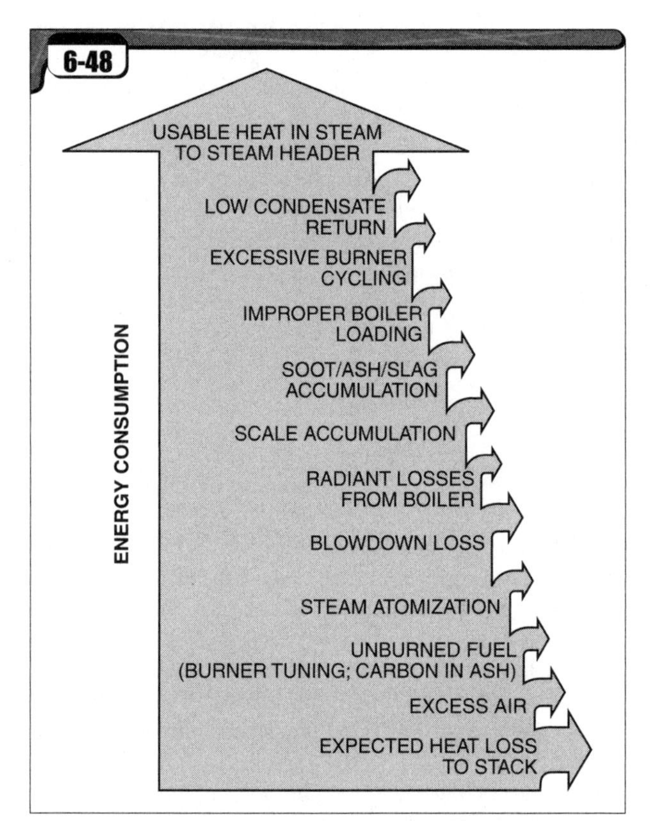

Figure 6-48. *Fuel-to-steam efficiency is affected by energy losses.*

A fuel oil-fired boiler's steam rate averages 114 lb of steam/gal. of fuel oil. The boiler operates 24 hr/day, 350 days/yr. The fuel oil costs $0.58/gal. and the load is 50 Klb/hr. What is the annual fuel cost?

$$FC = \frac{SL}{SR} \times 24 \times 350 \times P$$

where
FC = fuel cost (in \$)
SL = average steam load (in lb/hr)
SR = steam rate (in lb of steam/gal. of fuel oil)
24 = conversion constant (in hr/day)
350 = conversion constant (in days/yr)
P = price (in \$/gal.)

$$FC = \frac{50{,}000}{114} \times 24 \times 350 \times 0.58$$

$FC = 438.59 \times 24 \times 350 \times 0.58$
FC = **\$2,137,000/yr**

128 How does the acid dewpoint limit the amount of heat recovery that may be attained in a boiler?

The acid dewpoint is a function of the relative humidity of the flue gases and air. *Relative humidity* is the ratio of the actual amount of water vapor in the air to the greatest amount possible at the same temperature. At any given temperature, air can hold a predictable, fixed amount of moisture. Cold air cannot hold as much moisture as warm air.

Because water vapor is formed in the combustion of fuels, the flue gases contain some moisture. Moisture may also be added by fuel oil burners that use steam for atomization. Solid fuels also contain some moisture. If the flue gases in the furnace area are 2000°F, they can hold more moisture than when they cool down. If the flue gases are cooled too much before they pass out of the breeching and stack, the moisture may begin to condense in these areas. Since the fuel normally contains at least a small amount of sulfur, the water vapor and sulfur-containing compounds combine to form sulfuric and sulfurous acid vapors.

The *acid dewpoint* is the temperature at which acidic vapors begin to condense out of the flue gases. Acidic vapors in the flue gases begin to condense at a higher temperature than water vapor. These acids cause corrosion of the economizer, ductwork and hoppers, fans, stack, and other components. Some heat is necessarily carried out through the stack because flue gases cannot feasibly be cooled below the acid dewpoint.

Some steps may be taken to minimize acid dewpoint corrosion. The corrosion occurs readily on duct and hopper walls because the flue gases are cooled there as heat is radiated from the surface. Insulating exposed ductwork and hopper walls helps prevent this heat loss and thus keeps the surface hotter. Bypassing heat recovery equipment during periods of very low load on the boiler also helps prevent acid dewpoint corrosion.

129 What constitutes an acceptable heat loss up the stack?

Boiler operators should frequently check the temperature of the flue gases going up the stack. As a general rule for natural gas-fired boilers, the flue gas temperature should not be allowed to drop to lower than 50°F above the saturated steam temperature, or to be greater than 150°F above the saturated steam temperature. For example, if the steam temperature is 338°F (100 psig), the exiting flue gas temperature from the boiler proper should be between approximately 388°F and 488°F at rated boiler load. These temperatures may be somewhat lower if the ductwork and downstream equipment are insulated and/or constructed of stainless steel. However, these temperatures will need to be somewhat higher if a fuel containing more sulfur is used or if steam atomization is used. Any evidence of moisture seeping or dripping from breeching ductwork is cause for immediate investigation.

If the flue gases exiting the boiler run hotter than 150°F above the saturated steam temperature, the boiler operator should investigate whether the heating surfaces are fouled with soot or ash on the fire side or scale on the water side.

Furnace Maintenance

130 What is flame impingement?

Flame impingement is a condition where flame from burning fuel continually strikes the boiler surfaces or refractory brick. This contact causes localized boiler hot spots and potential overheating. It also results in soot deposits as the flame is quenched by contact with the surface and cooled below the ignition temperature. Burner orientation and flame shape should be adjusted to avoid flame impingement.

131 List some causes of extra furnace maintenance.

Excessive furnace temperature can damage boiler metal, warp the grates, and cause spalling of the refractory brick in the furnace and burner throats. Spalling is damage to the refractory materials by chipping, hairline cracking, and flaking. **See Figure 6-49.** This may be caused by overfiring the boiler or by operating with inadequate combustion air.

Excessive thermal cycling is repeated fluctuation of furnace temperatures from very hot to considerably cooler within short time periods. This often occurs in automatically fired boilers that are oversized for the duty they perform. Each time the burner starts and stops, cool purge air is followed by high temperature from the burner, followed by cool postpurge air. Over a period of time, this can cause refractory spalling and metal fatigue. In firetube boilers, it also tends to cause the tube ends to leak due to expansion and contraction. Flame impingement causes soot deposits and local damage to boiler tube surfaces.

Damage from slag deposits (stalactites) may occur when burning solid fuels with a low ash fusion temperature results in the formation of slag deposits on the furnace walls. These deposits break away and fall when they become large and heavy. They often pull away pieces of the refractory when they fall.

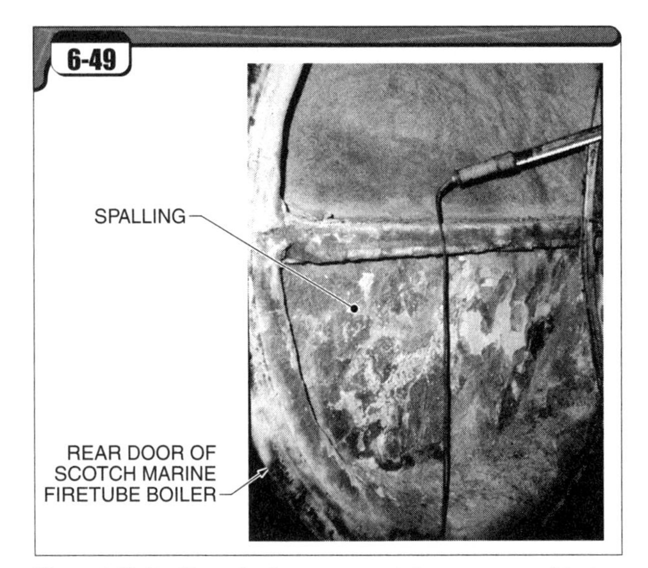

Figure 6-49. *Spalling of refractory materials appears as chipping, hairline cracking, and flaking.*

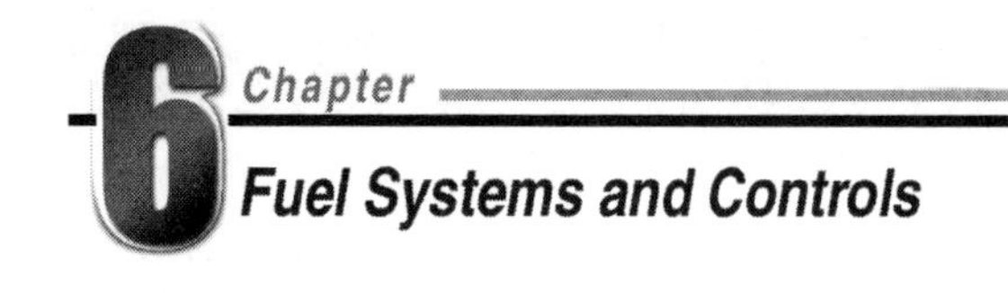

Name: ____________________ Date: ____________________

______ **1.** The ___ switch confirms that the main fuel valves are closed at designated times.
A. POC B. SSOV
C. FGR D. MTFI

______ **2.** The combustion of fuel in a boiler occurs in the combustion ___.

______ **3.** ___ is the solid, hard mass formed when coal or another solid fuel burns under poor furnace conditions.
A. Pyrite B. Lignite
C. Clinker D. Slack

______ **4.** ___ air is the initial volume of air that enters the furnace with the fuel for most of the combustion process.

T F **5.** Baffles direct the flow of hot gases of combustion along a desired path.

______ **6.** ___ support the burning fuel bed in a solid fuel-fired boiler.

______ **7.** A(n) ___ is a firebrick wall built across the boiler furnace to support the rear end of the grates.

______ **8.** ___ is the hardness of coal.
A. Grade B. Rank
C. Slag D. Fly ash

T F **9.** Bituminous coal is softer coal than anthracite coal.

T F **10.** Anthracite coal contains a low percentage of fixed carbon and a high percentage of volatiles.

T F **11.** Grate surface is expressed in square feet.

______ **12.** ___ is very soft coal that has a high moisture content and low heat value.
A. Anthracite B. Bituminous
C. Lignite D. Stoker

______ **13.** ___ is the size of coal.
A. Grade B. Rank
C. Clinker D. none of the above

______ **14.** If the flame is quenched before all the combustible elements in the flame have been burned, the unburned carbon will deposit as ___ on the heating surfaces.
A. flyash B. soot
C. volatiles D. slag

T F **15.** NO_x is an abbreviation for nitrous oxide.

______________ **16.** The ultimate analysis of coal is the percentages of ___ in the coal.
A. sulfur and nitrogen
B. hydrogen and carbon
C. oxygen and ash
D. all of the above

______________ **17.** The two classes of fluidized bed boilers are the ___ fluidized bed and the ___ fluidized bed.
A. volumetric; gravimetric
B. proximate; ultimate
C. attrition; classifying
D. bubbling; circulating

T F **18.** Clinkering can be caused by a lack of combustion air.

T F **19.** Combustibles in the refuse are burned fuel in the ash.

T F **20.** Banked coal burns very slowly while it keeps the furnace hot.

______________ **21.** A(n)___ stoker is a stoker in which the grates are comprised of thousands of small staggered segments that are interlaced by support bars or rods, forming a heavy chain conveyor.
A. retort
B. vibrating-grate
C. chain-grate
D. underfeed

______________ **22.** Coal burned in ___ is burned as a dust as it is blown into the furnace.

______________ **23.** A(n) ___ is a grinding mill that grinds coal to a very fine powder.

T F **24.** Thick liquids have low viscosity.

______________ **25.** Grades of fuel oil are designated by number, from No. ___ through No. ___.
A. 1; 4
B. 1; 6
C. 2; 4
D. 2; 6

______________ **26.** ___ fuel oil is fuel oil that remains after the lighter, more volatile hydrocarbons have been distilled off.

______________ **27.** Specific ___ is the weight of a given volume of a material divided by the weight of an equal volume of water.
A. weight
B. volume
C. gravity
D. none of the above

T F **28.** Fuel oil heaters are required in a fuel oil tank to lower the viscosity of heavier grades of fuel oil.

T F **29.** A fuel oil pump supplies more fuel oil than is used by the burner(s).

______________ **30.** Of the several types of gas that can be used as boiler fuel, ___ contains the most Btu per cubic foot.

______________ **31.** A ___ of natural gas contains 100,000 Btu.
A. gallon
B. pound
C. cubic foot
D. therm

______________ **32.** The NFPA requires a minimum of ___ air changes in the furnace before a burner is lit.
A. two
B. four
C. eight
D. no minimum per the NFPA

T F **33.** A flame scanner confirms that the pilot and main burner flames are OFF at the proper time, as well as confirming that they are ON at the proper time.

______________ **34.** Four types of automatic flame sensors used for firing boilers are ___.
A. infrared, ultraviolet, LED, and flame rod
B. infrared, photocell, LED, and flame rod
C. infrared, ultraviolet, photocell, and flame rod
D. infrared, photocell, LED, and ultraviolet

T F 35. Solid coal is normally burned on a grate.

______________ **36.** When the burner flame ignites the incoming air-fuel mixture so quickly that the flame travels upstream into the burner assembly, this is known as ___.
A. flareback
B. flashback
C. fluidization
D. impingement

______________ **37.** When using flue brushes to remove soot from the inside of the tubes in a firetube boiler, the boiler must be ___.
A. ON and under full pressure
B. ON and under low pressure
C. OFF
D. all of the above

______________ **38.** ___ combustion is another name for perfect combustion.
A. Stoichiometric
B. Pristine
C. Spontaneous
D. Ultimate

T F 39. The greatest loss in a boiler plant is heat up the stack.

______________ **40.** Turbulators are also known as ___.

______________ **41.** ___ are the two elements in fuel that provide the majority of the heat.
A. Hydrogen and carbon
B. Carbon and carbon monoxide
C. Hydrogen and carbon dioxide
D. none of the above

______________ **42.** The highest temperature on a summer day reaches 97°F with 67% relative humidity. When the sun goes down in the evening, the temperature drops and the ___.
A. relative humidity drops because cooler air can hold more moisture
B. relative humidity increases because cooler air cannot hold as much moisture
C. relative humidity remains constant
D. temperature drop has no effect on humidity

______________ **43.** ___ is moisture that precipitates out of air as the temperature drops.

T F 44. Moisture vapor is formed in the combustion of fuels.

______________ **45.** The three T's of good combustion are ___.
A. time, temperature, and toxicity
B. turbulence, toxicity, and time
C. time, temperature, and turbulence
D. none of the above

______________ **46.** ___ air is the amount of air added to the combustion process over and above that which is theoretically necessary.

______________ **47.** ___ is the name given to the visible ash haze that leaves the stack and ultimately settles back to earth.
A. Residual
B. Particulate
C. Pyrite
D. Sorbent

T F **48.** The carbon dioxide content of flue gases is higher with excessive draft.

T F **49.** A Ringelmann chart is a comparison chart used to measure opacity of smoke.

______________ **50.** A(n) ___ measures temperatures above the range of mercury thermometers.

______________ **51.** A(n) ___ provides the very high voltage needed to induce the formation of a spark at the tip of the igniter.

A. flame rod
B. ignition transformer
C. ionization rod
D. pneumercator

T F **52.** The difference between the higher heating value and the lower heating value of a fuel is the portion of the heating value used in evaporating water formed in the combustion of the carbon in the fuel.

______________ **53.** What is the specific gravity of a brine solution that weighs 9.5 lb/gal. at 60°F?

______________ **54.** Convert a specific gravity of 0.95 to °API.

______________ **55.** If 3 tons of coal are burned per hour in a furnace with a $12' \times 15'$ grate, what is the rate of combustion in lb/sq ft/hr?

T F **56.** Oxidation generates heat.

______________ **57.** ___ air is the secondary air in a unit fired by solid fuel.

T F **58.** No. 4 fuel oil is darker than No. 2 fuel oil.

T F **59.** Natural gas contains more Btu per cubic foot than propane.

______________ **60.** What does the acronym MATT stand for, with regard to combustion?

61. Why should combustion safety devices such as low gas pressure switches or low fuel oil temperature switches require the boiler operator to manually reset the device if it causes the burner to shut down?

62. Explain the difference between an intermittent pilot and an interrupted pilot.

______________ **63.** A chain-grate stoker is 14′ wide and is traveling 2.5″ per minute. The coal is fed onto the grates in a layer 5″ thick. The coal weighs 48 lb/cu ft. At what rate is the coal being consumed, in ton/hr?

______________ **64.** The ___ ratio of a modulating burner is the ratio of the maximum firing rate of the burner to the minimum firing rate.

A. propagation
B. modulation
C. turndown
D. ionization

T F 65. Cinder reinjection in stoker-fired boilers typically results in a slight efficiency increase.

______________ **66.** A watertube boiler burns No. 2 fuel oil and a steam meter reading shows that the boiler generated 1,620,000 lb of steam over a 24-hour period. During the same period, the boiler consumed 14,465 gal. of fuel oil. What was the steam rate of the boiler during this period?

T F 67. The breeching ductwork should be insulated to help avoid acid dewpoint corrosion.

______________ **68.** Excessive thermal cycling should be avoided because ___.

A. it can lead to leaks at tube ends due to excessive expansion and contraction
B. it leads to spalling of the refractory
C. it can lead to metal fatigue in the boiler
D. all of the above

T F 69. When flame is visible, it is evidence that combustion is still in progress.

T F 70. The measurement of oxygen in the flue gases should be taken at the boiler breeching, as near the boiler outlet as possible.

Chapter 7

Draft and Flue Gas Systems and Controls

The draft system equipment plays an important role in optimizing boiler system efficiency, but is often neglected or misunderstood. In operating this equipment, a balance is sought between providing enough oxygen and turbulence for efficient combustion of the fuel and too much loss of heat up the stack. In order to reach this balance, the boiler operator needs to know how to measure and control the operation of a draft system.

The draft system also includes the equipment needed to control air pollution that results from burning fossil fuels. These pollutants cause a variety of problems. The problems may include negative impacts on human health, forests and vegetation, fish and wildlife, and building materials. The boiler operator is usually responsible for monitoring and responding to the indications of the pollution-control equipment performance.

Natural Draft

1 What is draft?

Draft is the movement of air and/or gases of combustion from a point of higher pressure to a point of lower pressure. Because air and combustion gases are fluids, there must be differential pressure in order to have flow.

The draft system in a boiler serves several functions. First, it provides the oxygen needed to burn the fuel. The draft system then causes the hot combustion gases to pass across the heating surfaces of the boiler where the gases can transfer heat to the boiler water. The draft system also carries the spent gases out of the furnace area, making room for new combustion air and fuel to burn in the furnace. The final part of the draft system, the chimney or stack, dissipates the combustion gases well up into the air and away from the people below.

2 Describe the principle of natural draft.

Natural draft is draft that occurs without mechanical aid. Natural draft occurs because air, like all fluids, flows from a point of higher pressure to a point of lower pressure. As air is warmed, its density decreases and it becomes lighter. The lighter air rises and leaves an area of low pressure under it. Cool air moves in to fill the lower pressure area and replaces the warm air. As this cooler air warms up, the cycle repeats itself. In boiler plants, the combustion of fuel creates a continual source of heat. Therefore the natural draft is constant.

3 How is natural draft harnessed for doing work?

The upward flow of heated air and gases may be harnessed by passing the updraft through a stack or chimney. If a constant source of heat exists at the bottom of the stack, a constant natural draft is produced within the stack. The slight vacuum produced toward the bottom of the stack is used to draw the spent gases from the boiler furnace and help induce new combustion air to flow into the furnace.

The terms "stack" and "chimney" are used synonymously in most cases. However, in certain contexts, a chimney is considered to be of masonry construction while a stack is considered to be of steel construction.

A considerable percentage of smaller natural gas-fired commercial, institutional, and industrial boilers use natural draft. The burners in these smaller boilers are called atmospheric burners, because no fan or blower is provided. The combustion air is simply drawn in from the surrounding room. Atmospheric burners are generally ON/OFF or two-stage designs. They normally do not modulate. Solid fuel-fired boilers using natural draft are quickly becoming rare due to both efficiency and environmental issues.

4 What determines the amount of draft created by a stack?

The differential pressure that may be created by the stack is determined by the height of the stack and the difference between the inside temperature and outside temperature of the stack. The volume of gases that may pass through the stack is determined by the inside diameter (and therefore the cross-sectional area) of the stack and the temperature of the gases. The temperature of the gases is a factor because the volume of the flue gas changes with temperature changes.

The amount of available draft that is actually used at any given time can be controlled by using a damper. **See Figure 7-1.** A *damper* is a device that partially or totally inhibits the flow of air or combustion gases through ductwork. For example, a damper installed in the breeching between the boiler and the stack allows the boiler operator to determine how much of the available draft is used at any given time. Dampers may be controlled manually or automatically.

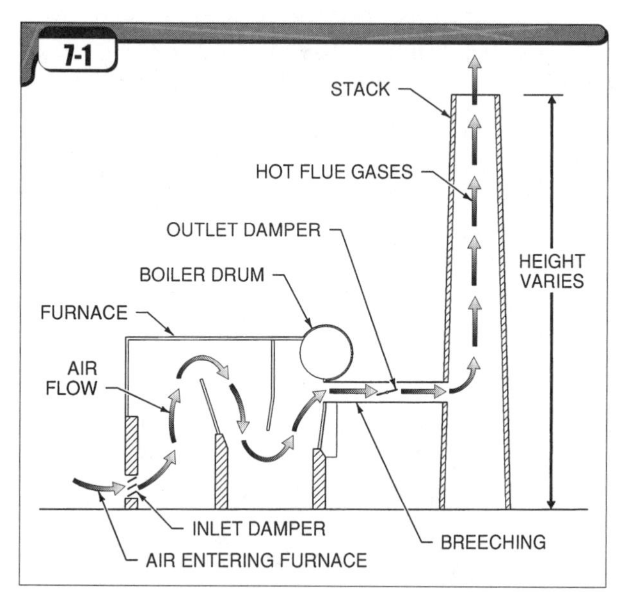

Figure 7-1. *The amount of draft is determined by the height of the stack and the differential temperature between the inside of the stack and the atmosphere.*

5 What is a barometric damper?

A *barometric damper* is a free-swinging adjustable balanced damper used on smaller boilers to automatically limit the amount of air pulled through the combustion chamber. **See Figure 7-2.** This helps avoid excessive draft, which could adversely affect the burner operation. For example, excessive draft can occur if the outside air is very cold or if it is very windy outside. Such conditions can cause air to be drawn through the furnace too briskly, especially in atmospheric burner installations. This causes the flame to be pulled away from the burner and makes the flame unstable. This also results in too much excess air being used and inadequate flame residence time in the furnace. These conditions result in reduced efficiency.

A barometric damper is installed in the breeching of a small boiler. If the draft becomes excessive and the vacuum in the breeching thus becomes too strong, the barometric damper opens in proportion to the excessive draft. This allows air from thc room to be drawn directly into the breeching, so that the draft in the furnace area remains constant.

A similar component known as a draft hood performs the same function as a barometric damper. A draft hood looks like an inverted funnel in the vertical breeching ductwork just above a small boiler. The flue gases exiting the boiler are directed upward through a short section of ductwork. This ductwork terminates inside the inverted funnel, with an open space around the short duct section. This arrangement allows air from the room to be pulled through the open space if the draft becomes excessive. This works much like a vacuum breaker fitting.

As air is drawn directly into the breeching through a barometric damper or draft hood, the flue gases rising through the stack are cooled somewhat. The air being drawn in and the resulting lower temperature of the flue gases help reduce the excessive draft in the furnace area. Although these arrangements work well, they can create problems in some cases. For example, too much cool air infiltration can cause water vapor in the flue gases to fall below the dew point and condense in the breeching and stack. This leads to corrosion.

Figure 7-2. *Barometric dampers are used on smaller boilers to limit the amount of air pulled through the combustion chamber.*

Mechanical Draft

6 What is draft loss?

Draft loss is the loss of available draft due to friction and other pressure losses as the flue gases flow through the combustion gas passageways. For example, this friction occurs as the flue gases pass through the tubes of firetube boilers or across the tubes of watertube boilers.

Until the early 1900s, all stationary boilers used natural draft. As the demand for steam in industrial and institutional facilities increased, larger and larger chimneys were needed to provide the draft needed to burn the additional fuel. This problem increased with demands for more efficient boiler systems. For example, the need for higher boiler efficiency resulted in smaller tubes being used in firetube boilers in order to increase heat transfer. The smaller tubes create friction, or draft loss, as the flue gases flow through the smaller passageways.

The use of economizers, pollution control equipment, and air preheaters in large boilers also contributes to draft loss. A limit was reached beyond which the chimney height requirements made the chimney too expensive or too difficult to construct. A solution to this limitation was found in the use of fans (blowers) to provide the needed movement of combustion air and flue gases.

What is mechanical draft? When does mechanical draft become necessary?

Mechanical draft is draft created by fans and blowers. Mechanical draft becomes necessary when a stack relying on natural draft would be too expensive, would not allow adequate flexibility in control of the draft, or would otherwise be impractical. The great majority of industrial boilers now use mechanical draft in one form or another. These fans may be driven by electric motors or by steam turbines.

Centrifugal fans move the air or flue gases in a radial direction, much like a centrifugal pump. Small impeller-type centrifugal fans are often mounted directly on package boilers to provide combustion air to the burner equipment. **See Figure 7-3.** Larger stand-alone fans are used in plants that require higher combustion air flow rates or flue gas flow rates. These fans are often squirrel-cage centrifugal fans. Vane-axial fans are often used in very large boiler plants. A vane-axial fan looks much like an airplane propeller.

What is forced draft?

Forced draft is the discharge of combustion air from a fan into a furnace to combine with the fuel for combustion. A forced draft (FD) fan provides air to the windbox beneath the grates in a solid fuel-fired furnace or to the windbox around the burner in a gas- or liquid fuel-fired furnace.

An FD fan is also known as the primary air fan. In pulverized coal-fired boilers, however, an FD fan provides secondary air, while the pulverizer exhauster fan provides the primary air.

What is induced draft?

Induced draft is the use of a fan to simulate the effect of a stack by drawing the combustion gases from the furnace and through the flue gas passages. An induced draft fan is commonly known as an ID fan. While a stack is still needed to dissipate the gases into the atmosphere, the stack can be shorter than in a natural draft installation. An induced draft fan creates a slightly negative (lower than atmospheric) pressure in the furnace. The negative pressure keeps smoke and hot gases of combustion from leaking out.

The induced draft fan also discharges the spent gases toward the stack so that they can be removed. Induced draft fans are almost always used in conjunction with forced draft fans. It is uncommon in modern boiler installations for an induced draft fan to be provided without a forced draft fan also being provided.

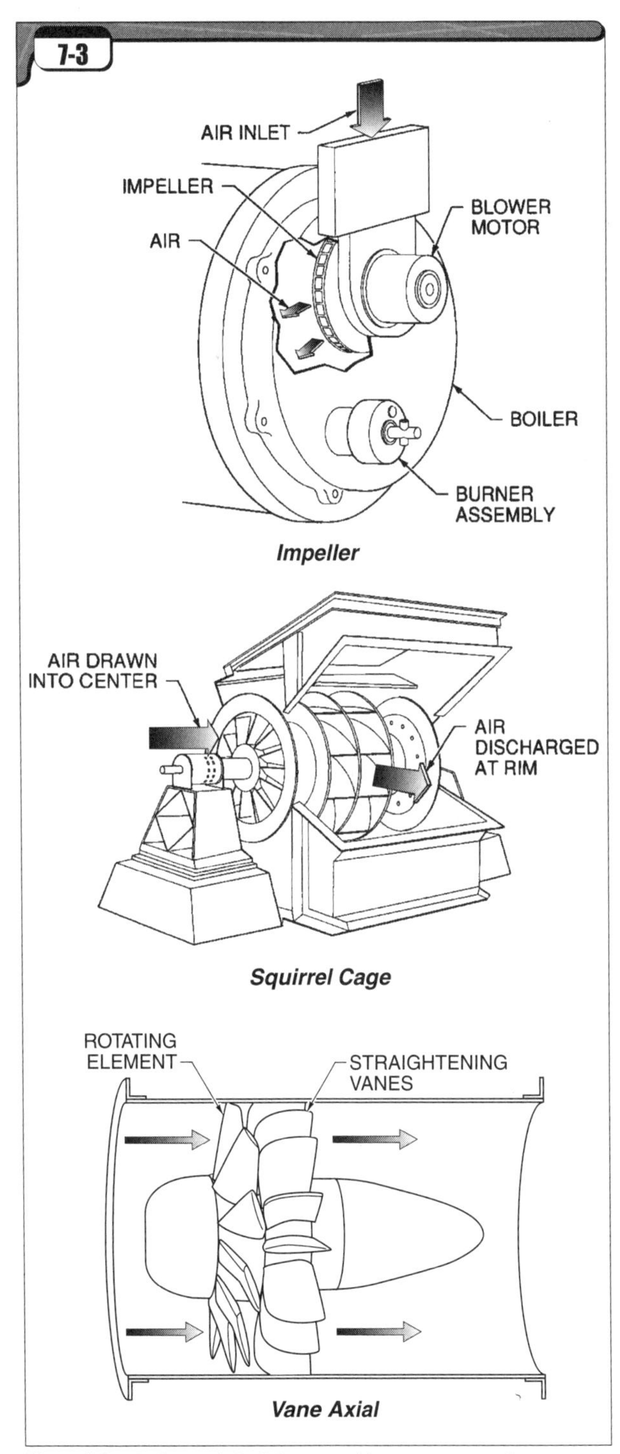

Figure 7-3. *Mechanical draft is created by one or more fans.*

Where are the forced draft fan and induced draft fan located in the boiler setting?

A forced draft fan is located so that it discharges air under a positive pressure to the furnace for combustion. **See Figure 7-4.** It can be located anywhere in relation to the boiler furnace. On smaller package boilers it is almost always mounted on the front of the boiler. On large industrial and utility boilers, it may be many feet away from the furnace and connected to the furnace by ductwork. For example, in some installations the FD fan is located on the top floor of the building housing the boiler, while in others the FD fan is located in a basement area.

An induced draft fan is located in the breeching so that it draws the spent flue gases through the outlet ductwork from the furnace and exhausts them to the stack. Like the FD fan, this may involve many feet of breeching ductwork.

On large boilers, the induced draft fan is commonly located at the base of the stack such that it discharges directly into the stack. However, in some installations the discharge of the ID fan forces the flue gases through pollution control equipment before the gases enter the stack. In cases where the ID fan must handle flyash-laden flue gases, the fan blades are reinforced with wear plates to help resist erosion.

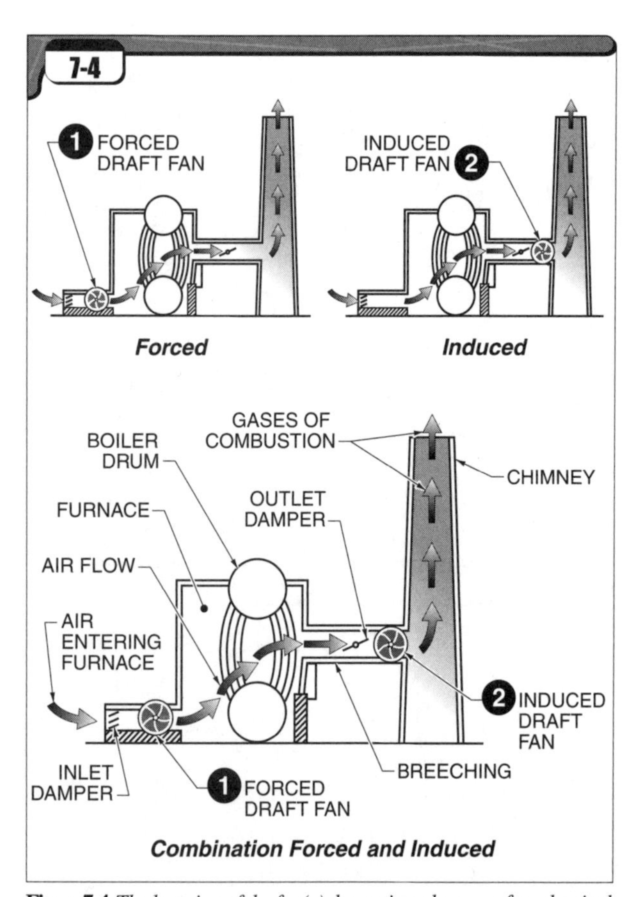

Figure 7-4. *The location of the fan(s) determines the type of mechanical draft system.*

What is a pressurized furnace installation?

A pressurized furnace boiler installation has a furnace pressure greater than atmospheric pressure. Most arrangements of this type use a comparatively short stack and rely mainly on the forced draft fan to push the gases through the unit.

Pressurized furnaces are primarily associated with package firetube boilers and small to medium-sized package watertube boilers. The furnace areas and breeching ductwork of these package boilers are sealed with gaskets to prevent leakage of the pressurized gases. However, a few very large power plant boilers use pressurized furnaces as well. These require considerable attention to keep all furnace leaks sealed so that flyash, noxious gases, and smoke do not leak out.

Not all boilers using forced draft fans have pressurized furnaces, however. For example, in a great many installations an ID fan provides adequate draft to remove the combustion gases and maintain a slight negative pressure in the furnace even though forced draft is used.

What is a balanced draft installation?

A balanced draft installation uses both a forced draft fan and an induced draft fan. The two fans work in unison to move the combustion air and the resulting flue gases through the flue gas passages, while maintaining a furnace pressure just below atmospheric pressure.

It is desirable to have a very slight negative pressure in the furnace in most larger boiler designs. This is because it is difficult to eliminate all the small leaks around furnace doors, inspection hatches, soot blower penetrations, etc. By maintaining a very slight negative furnace pressure, any leakage will be small amounts of air being drawn into the furnace rather than noxious gases, flame, and ash blowing out. For these reasons, balanced draft furnace configurations are the most common type in larger industrial and utility boilers. Normally the furnace pressure is maintained between approximately –0.05″ WC and –0.2″ WC.

Draft Measurement and Control

How are draft pressures measured?

Draft pressures are measured in inches of water column, using manometers or panel-mounted draft gauges. **See Figure 7-5.** In general, manometers are used within a few feet of the point of draft pressure measurement, while panel-mounted draft gauges are used for more remote locations such as control rooms.

A *manometer* is an instrument that measures draft by comparing pressure at two locations. In its basic form, a manometer is a clear glass or plastic U-shaped tube that contains water. A scale in the middle of the U is graduated in inches, beginning with zero in the middle. If pressure is applied to one side of the tube, the water will lower on that side of the tube and rise on the other side of the tube. If vacuum is applied to one side of the tube, the water will rise on the vacuum side of the tube and fall on the other side. The sum of the two measurements is the draft pressure (above or below atmospheric) in inches of water column.

Inches of water column is abbreviated ″WC or in. WC, and may be expressed as positive or negative, depending on whether the reading is above or below atmospheric pressure. For example, if the draft as read on a manometer is 2 inches of water column below atmospheric pressure, this would be abbreviated as –2″ WC. If the draft is 0.15 inches of water column above atmospheric pressure, this would be abbreviated as +0.15″ WC.

Inclined manometers work on the same principle as U-tube manometers, but the inclined tube provides a more precise reading. This is because having the tube full of water on an incline allows finer increments to be read.

Panel-mounted draft gauges read in the same units as manometers (inches of water column), but use a flexible diaphragm or bellows and linkages to cause movement of a pointer. Although instruments of this type do not contain water, they are often referred to as manometric instruments.

Panel-mounted digital controllers also read in units of inches of water column. Controllers contain the electronics and programming that can automatically modulate a damper to maintain the draft at a desired setpoint.

Factoid

Although panel-mounted draft gauges do not contain water, the term "manometric" is often used to refer to all draft-measurement devices.

14 What draft pressures are typical in a boiler installation using natural draft?

The strongest negative draft pressure will be found at the base of the stack. A negative pressure of –1″ WC is typical at the stack damper, depending on the height of the stack and atmospheric conditions. The pressure in the furnace area is usually about –0.1″ WC, depending on the draft loss across the boiler and the breeching ductwork.

15 Is the forced or induced draft fan larger in capacity?

The induced draft fan must be considerably larger in capacity than the forced draft fan. The induced draft fan must remove the gases of combustion that are produced by the burning fuel. These gases of combustion are formed when the mixture of combustion air and fuel burns, and when the mixture burns it expands dramatically in volume.

The induced draft fan must also remove all air that may leak into the furnace through openings or cracks, as well as any steam used for atomization of liquid fuel. *Inleakage air* or *tramp air* is air that leaks into a furnace. In addition, the ID fan must remove the steam produced by any moisture that evaporates from solid fuel and the steam that results from the combustion of hydrogen in the fuel.

The driving unit on the induced draft fan is typically an electric motor or steam turbine. The driving unit must provide the input power necessary to help the ID fan accomplish these objectives as well as additional power required to overcome friction created by obstructions installed in the flue gas stream. For example, the boiler's steam-generating tubes, economizers, flue gas cleaning equipment, and air preheaters are all obstructions to flue gas flow.

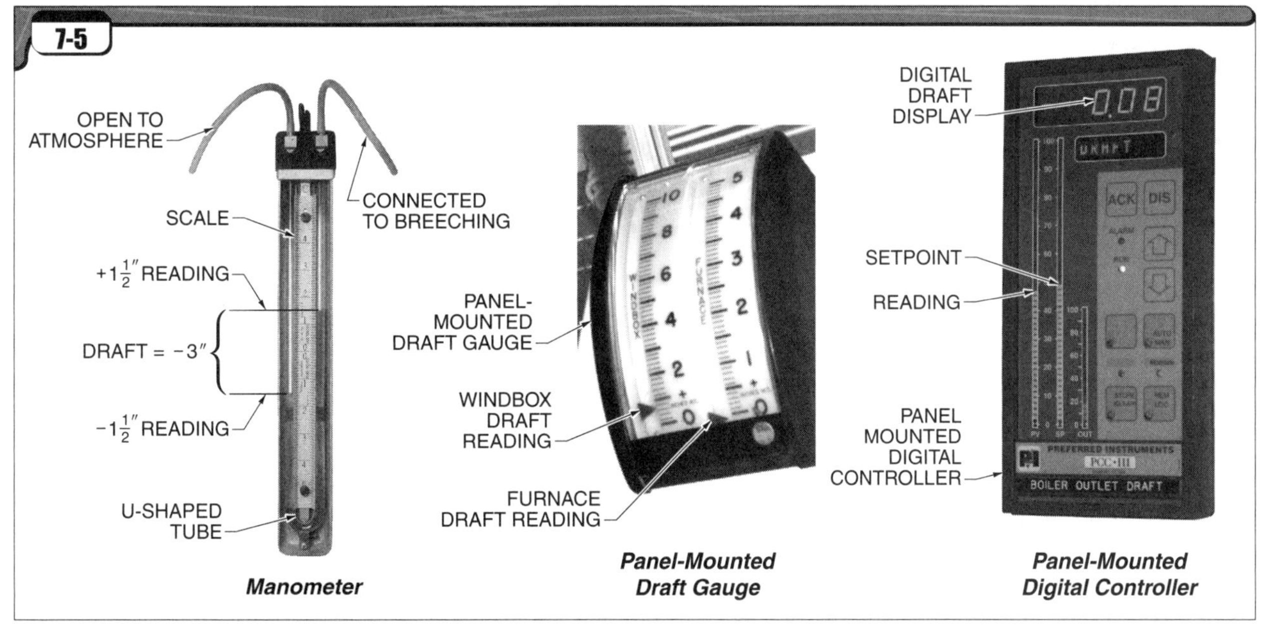

Figure 7-5. *A digital draft gauge or a manometer may be used to measure draft.*

Trade Tip

Most pneumatic fan damper actuators have a provision for removing the pneumatic signal and manually locking the damper in one position when necessary.

When using an induced draft fan, where is the strongest negative draft found?

The strongest negative draft is found in the breeching. The breeching area is closer to the induced draft fan, which creates the negative draft pressure. **See Figure 7-6.**

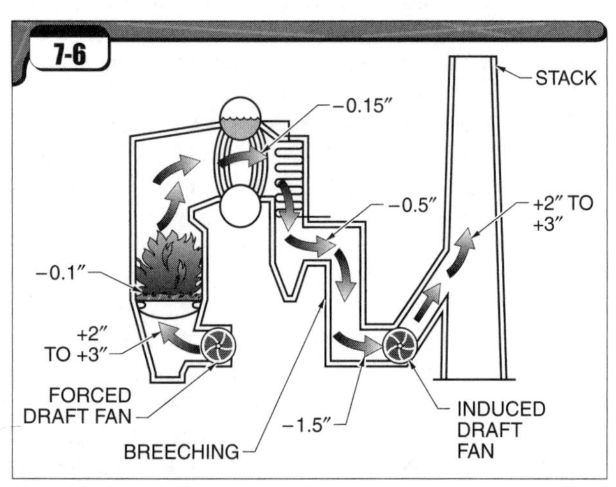

Figure 7-6. *The draft pressure is the most negative near the induced draft fan inlet.*

What is the unit of measure of flue gas flow?

Flue gas flow is expressed in standard cubic feet per minute (SCFM) or in actual cubic feet per minute (ACFM). Because the volume of flue gases changes depending on temperature, atmospheric pressure, and relative humidity, it is difficult to directly compare the flows of flue gases at different temperatures. The SCFM calculation is used to convert a flue gas flow at any given temperature, pressure, and relative humidity to an equivalent flow at 68°F, with atmospheric pressure at sea level and 50% relative humidity. By using these conversions, all flue gas flows can be compared on an equal basis. Automatic instrumentation is used to convert these figures. The ACFM is the measurement of an air or flue gas flow without these conversions, and is of limited use in day-to-day boiler system operation.

How is the damper on a large fan controlled from a remote location?

A damper is controlled remotely with a damper actuator. **See Figure 7-7.** A damper actuator is very similar to a valve actuator. The control signal from the control panel or computer console is sent to the actuator, which converts the signal into a proportional mechanical movement. This movement is transmitted to the damper through a linkage. The majority of damper actuators are pneumatically controlled devices but some are electrically operated.

If the control signal is electric (4 mA to 20 mA), a converter (transducer) is required to change the electric signal to a proportional pneumatic signal (3 psig to 15 psig) that can be used by the damper actuator.

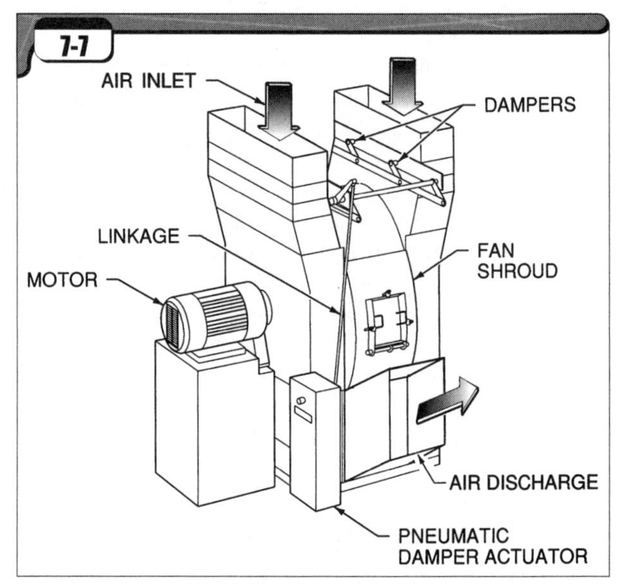

Figure 7-7. *Dampers are controlled with damper actuators.*

What is the difference between primary and secondary air?

Primary air is intended to provide the great majority of the oxygen needed for combustion of fuel. Secondary air is intended to provide the turbulence needed for ensuring complete combustion of the fuel.

In solid fuel-fired boilers, secondary air is generally admitted through small ports or nozzles, which give a high velocity to the air. These jets of air cause turbulence of the flame and mix the combustibles and oxygen together so that the highest possible percentage of combustibles is burned.

In burners for liquid or gaseous fuels, secondary air may be admitted through a ring of dampers, also known as registers or louvers, around the burner. This arrangement swirls the air as it enters the furnace. A diffuser, sometimes known as an impeller, in the burner throat also helps swirl the fuel and air mixture; often in the opposite direction. These actions maximize turbulence of the flame.

In some modern burners of special design, air may be injected in or around the burner at very specific points. *Tertiary air* is combustion air added to a burner in addition to the secondary air.

How does secondary air differ from overfire air?

Overfire air is secondary air introduced over the fire in a solid fuel-fired furnace. This air creates turbulence in the flame over the fuel bed and causes close contact of the combustibles with oxygen at a high temperature. In a solid fuel-fired boiler, overfire air is particularly important in preventing smoke.

What limits the amount of forced draft that may be used?

Forced draft is limited by the ability of the induced draft fan or stack to remove all of the flue gases. The furnace would become pressurized if the flue gases were not removed. Excessive forced draft (high excess air) also robs heat from the furnace, which reduces efficiency.

Why should motor-driven fans be started against a closed damper?

When the damper is closed, the fan rotor may be spun with almost no resistance. This greatly reduces the electrical current required to bring the fan from a stop to operating speed. This is comparable to the effort required to begin pedaling a bicycle with the rear wheel lifted off the ground. Once the fan motor has reached operating speed, the boiler operator may begin to modulate the damper open as needed.

Many larger motor-driven fans are equipped with limit switches on the damper shaft. The limit switches are part of the fan motor's circuitry, and the damper must be in the closed or minimum position in order for the limit switches to allow the motor to be energized. This ensures that minimum starting current will be required to start the motor, and increases the motor life.

Trade Tip

In addition to testing steam traps, ultrasonic listening devices are helpful in detecting air and flue gas leaks. They are also helpful in detecting early failure of bearings.

Draft-Related Problems

Why is positive draft undesirable in a large boiler furnace?

Positive draft is the condition wherein the pressure inside the boiler furnace becomes greater than the pressure outside the furnace. Positive draft is undesirable in a large boiler furnace because hot gases, flame, ash, and smoke can blow out of any openings or cracks in the furnace. The hot gases, flames, and smoke create safety and health hazards to personnel in the area. The ash creates housekeeping problems. Due to the potential for furnace pressurization, a boiler operator should not stand directly in front of an open furnace view port or other furnace opening. Always wear appropriate personal protective equipment (PPE) when working in the vicinity of a furnace opening.

How is the furnace draft affected by the operation of soot blowers?

Because soot blowers blow a substantial amount of steam or compressed air into the furnace area, it is possible that the furnace may become pressurized when the soot blowers are operated. To counteract this, the boiler operator may adjust the ID fan damper setting to provide a slightly more negative furnace pressure prior to starting the soot blower operation. This will help ensure that the furnace pressure is maintained at –0.1″ to –0.15″ WC, even with the soot blower operating sequence being executed.

How should the induced draft fan be interlocked with the other fans?

The induced draft fan should be electrically interlocked so that if the ID fan stops, the other fans will also stop. Additionally, the primary air fan and the secondary air fan should be wired to start and run only when the induced draft fan is running.

Interlocking prevents the furnace from becoming pressurized if the induced draft fan fails. Pressurization would cause hot gases, flames, ash, and smoke to blow out of any openings in the furnace.

In addition, some boiler control systems include a similar interlock that causes the burner(s) and fans to shut down if the furnace becomes pressurized, even if the ID fan is still running. This interlock is intended to cover contingencies other than failure of the ID fan. For example, the furnace may become severely pressurized if the ID fan damper should fail in the closed position. In such a case, the ID fan could continue to run but not be able to remove the flue gases from the furnace.

What are the disadvantages of having cracks or gaps in the furnace walls?

Cracks or gaps in the furnace walls allow air to leak into the furnace, which decreases combustion efficiency. Air that leaks into a furnace absorbs heat from the gases of combustion. This air also takes away some degree of control over the furnace conditions and may overload the induced draft fan. Localized acid dewpoint corrosion can occur because moisture in the flue gases can condense at points of cool air inleakage.

If the furnace pressure rises even slightly above atmospheric pressure, smoke, flame, ash, and hot gases of combustion can blow out through these openings. This creates hazards to personnel and poor housekeeping conditions.

Factoid

Like centrifugal pumps, motor-driven fans should be started with the damper closed.

How can a boiler setting be tested for leaks?

A boiler setting can be tested for leaks by holding a smoke tube or smoldering wick around the joints of the furnace area and watching the smoke. If the smoke is drawn in, air is leaking into the furnace at that point. Such tubes or wicks are available from HVAC (heating, ventilation, and air conditioning) supply firms. They are generally used in the installation of new ventilation ductwork to detect leaks.

In areas where using this kind of ignition source would not be acceptable, an alternative method would be the use of an ultrasonic listening device to detect the noise of air leaking into or out of the opening. This ultrasonic device may be the same device used for testing steam traps, but using a different sensing attachment for this purpose.

What effect does a heavy accumulation of soot and slag on the boiler tubes have on the furnace draft?

In watertube boilers, a heavy accumulation of soot and slag constricts the openings between the tubes, causing increased resistance to the flow of flue gases through the tube banks. As the boiler is fired at higher ratings, this resistance results in difficulty in maintaining the correct negative pressure in the furnace. Soot blowers help to remove the ash accumulation and minimize draft loss.

In firetube boilers, heavy soot accumulations also create resistance to the flow of combustion air and flue gases. This results in pressurization of the furnace area, and noxious gases can be emitted from any small leaks. This condition also results in loss of combustion efficiency as the flow of combustion air to the fuel and mixture of oxygen with the flue gases is affected.

What could result from a damaged flue gas baffle?

A damaged flue gas baffle can result in localized hot spots and water circulation problems as hot flue gases short-circuit through the baffle. It also results in a loss of overall boiler efficiency, because very hot flue gases are allowed to short-circuit past boiler heating surfaces and exit the stack.

At the moment when a flue gas baffle incurs damage, the temperature of the flue gases leaving the boiler will likely increase significantly. This occurs as the flue gases short-circuit past heating surfaces without giving up heat to the boiler water. Thus, a sudden increase in the stack temperature (the temperature of the gases leaving the boiler) is an indication to the boiler operator of possible baffle damage.

A gradual increase in stack temperature, on the other hand, is more likely to indicate gradual accumulation of foreign material on the heating surfaces. For example, scale may slowly accumulate on the water side of the tube surfaces, and soot, ash, and slag may slowly accumulate on the flue gas side of the tubes. For these reasons, boiler operators should monitor and log the stack temperature frequently.

List three reasons, other than a damaged baffle, for excessive flue gas temperatures developing at the stack.

- Excessive soot deposits on the fire side of the tubes.
- Scale deposits on the water side of the tubes.
- Overfiring and forcing the boiler to produce more steam than it is designed for.

Why might a boiler installation suffer from insufficient draft, even if the fans and ductwork are sized properly?

If the boiler is equipped with an FD fan and the FD fan is inside the building, it is possible that the combustion air inlet vents to the building are partially or completely blocked. When combustion air is drawn from inside the building, combustion air must be able to enter the building from the outside.

A common mistake made by engineers and architects is failure to provide the boiler room with adequately sized combustion air inlet vents. Boiler operators who may not understand the importance of these vents frequently block them or partially block them in an effort to maintain warmer boiler room conditions during cold weather. These conditions may starve the combustion equipment for oxygen.

One method of avoiding this problem is to install the FD fan outside and connect it to the furnace with ductwork. This is very frequently done in large boiler installations. If the FD fan is to be installed inside, a good compromise between cold boiler room conditions and poor combustion conditions is to duct outside air directly to or near the FD fan inlet. **See Figure 7-8.** However, the ductwork must be large enough in size to avoid presenting a restriction to the air flow.

A boiler that has historically had adequate draft may suffer from inadequate draft conditions for other reasons as well. For example, accumulation of ash, soot, and slag on boiler tubes may restrict the openings between the tubes, increasing the draft loss. Air infiltration through significant leaks in the furnace walls or flue gas passages may rob capacity from the draft system.

It is good practice to install manometric draft gauges to be able to detect the draft pressure on both sides of each significant piece of equipment in the flue gas stream. The differential pressure between various points may then be used to reveal the location of restrictions. For example, fouled passages between the boiler's steam-generating tubes will result in high differential pressure of the flue gases across the boiler.

Flue gas analysis is another diagnostic tool that is helpful in determining the cause of draft problems. For example, if the oxygen reading in the flue gas rises and the carbon dioxide reading drops between the furnace and the stack, tramp air is most likely leaking into the furnace at some point. Note, however, that information of this nature is only useful if the boiler operator knows what these readings should be.

The appearance of the fire and the stack plume are not precise enough indicators to use to tune the combustion equipment, but the boiler operator should observe them to become familiar with the normal appearance of both. An abnormal appearance (a dark fire or a hazy stack) is indication of combustion- and/or draft-related problems.

Various mechanical problems may contribute to this problem as well. For example, FD or ID fan damper linkages may have slipped. This could allow the damper(s) to remain partially closed even though the damper actuator is in the wide-open position. Flow-measuring equipment that is out of calibration may indicate lower-than-actual combustion air or flue gas flows. Flow indicators and draft gauges, like most control devices, should be calibrated often enough to maintain their credibility. Damaged baffles may partially restrict the flue gas flow as well.

7-8

Figure 7-8. *Outside air for combustion may be ducted to the forced draft fan inlets.*

Draft System Optimization

32 What is a combustion air preheater?

A *combustion air preheater* is a piece of equipment provided to preheat the combustion air to some degree before the combustion air enters the furnace. Combustion air preheaters may use waste heat from flue gases, heat from live steam, or heat from process sources. They are often referred to simply as air preheaters. The design of these air preheaters varies widely. Common air preheater designs include plate, regenerative, tubular, and steam-air coil designs.

As fuel and air enter the furnace, the mixture of the two must be heated to the ignition temperature before ignition will take place. For efficient combustion, the fuel-air mixture must then be maintained above the ignition temperature until the combustion process is complete. The heat necessary for accomplishing these two objectives is normally extracted from the fuel-air mixture already burning in the furnace. Thus, the portion of the heat used for establishing and maintaining stable combustion conditions is no longer available for generating steam in the boiler. The concept of the combustion air preheater, then, is to preheat the combustion air so that more heat from the combustion process remains available for generating steam.

In order to realize a cost savings, the source of heat used in the air preheater must be less expensive than the heat derived directly from the fuel. For this reason, most air preheaters extract a portion of the heat remaining in the flue gases being discharged to the stack and use this heat to preheat the combustion air. Alternately, waste heat from a plant process may be used. If steam is used, as in a steam-air coil preheater, the steam is usually derived from an inexpensive or waste fuel, such as sawdust, blast furnace gas, or waste refinery gases.

The preheated combustion air is usually limited to about 350°F in boiler furnaces using stokers, because the combustion air flowing across the grates helps protect them from overheating. In units firing pulverized coal, the preheated air temperature may be as hot as 825°F.

Air preheaters that extract heat from the flue gases are practical where fairly constant flows of hot flue gases and combustion air are maintained and the combustion air flow and fuel flow are separately controlled. Power generating plants and paper mill operations are examples of good applications for air preheaters. Air preheaters are much less practical for package boilers that may cycle on and off, or have fuel valves and air dampers controlled by common linkages. This is due to the fact that the combustion air changes substantially in volume as it heats and cools.

When the air is heated and expands, more air volume must be provided in order to provide the same amount of oxygen. When cooler air is used, less air volume is required. When the fuel and air are controlled by common linkages, it is not feasible to vary the air flow adequately to compensate for the change in air volume.

Some plants using package boilers take advantage of combustion air preheat to a lesser degree by installing ductwork to draw in the combustion air from a point high in the building. For example, if the boilers are installed inside a building with high ceilings, the combustion air may be drawn from a point near the ceiling, where hotter conditions generally exist. This also cools these higher areas and makes them less oppressive to work in.

In cases where heat recovery from the flue gases is called for but an air preheater is impractical, an economizer is generally used.

33 What is a tubular air preheater?

A *tubular air preheater* consists of tubes enclosed in a shell where flue gases heat up incoming combustion air. **See Figure 7-9.** A tubular air preheater is similar to a large shell-and-tube heat exchanger. The hot flue gases flowing toward the stack pass through the tubes of the preheater. The cold combustion air passes across the tubes and picks up the heat. Baffles increase the heat transfer efficiency by causing the cold air to come into maximum contact with the hot tubes.

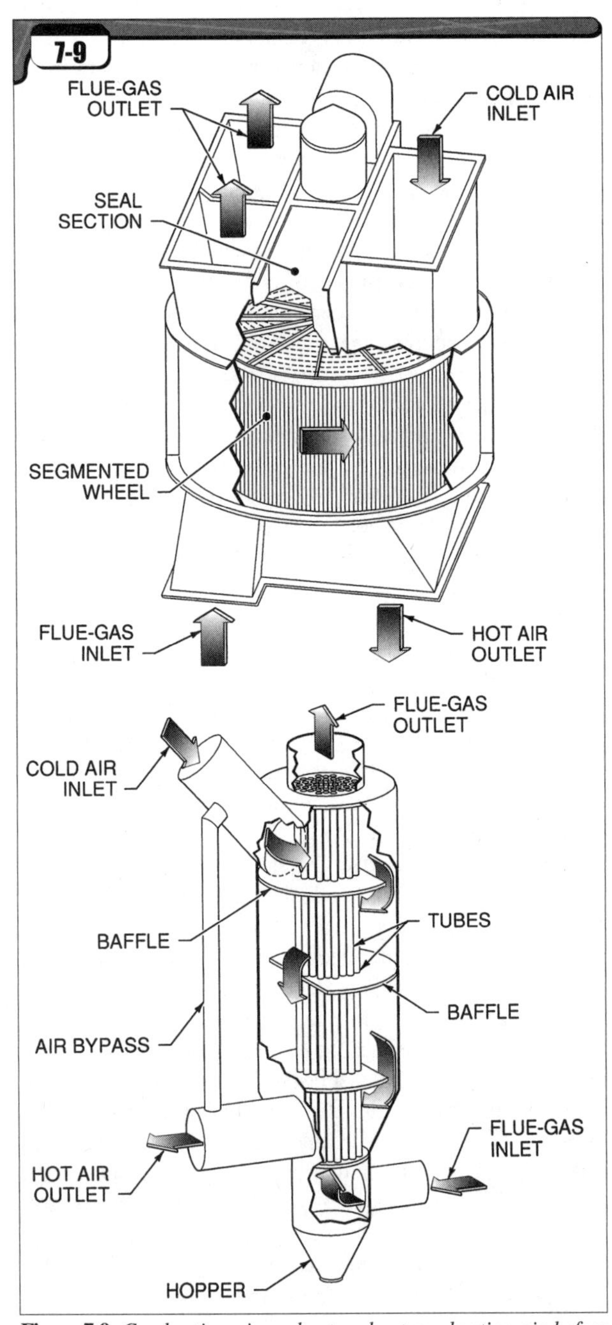

Figure 7-9. *Combustion air preheaters heat combustion air before the air enters the furnace.*

Factoid

Preheating the incoming combustion air with a source of low-cost or free heat helps maximize boiler system efficiency.

34 Describe a regenerative air preheater.

A *regenerative air preheater* consists of a rotating segmented wheel that is 8′ to 20′ in diameter and about 3′ to 6′ high. The size of the wheel depends on the size of the boiler. Corrugated metal is placed in each segment with the corrugations running semi-parallel to the flow of air and flue gas through the unit.

The regenerative air preheater turns at 1 rpm to 2 rpm. It is divided into two halves with a seal section between the two halves. The spent hot flue gases flowing toward the stack pass through one side of the unit as it rotates and heat the corrugated metal plates. As the heated portion of the preheater passes through the sealed section and into the other side, primary air flowing through the other side of the preheater on its way to the furnace picks up the heat from the corrugated plates. Thus, combustion air is preheated before it enters the furnace.

Regenerative air preheaters are constructed in both horizontal and vertical configurations. In either configuration, bearings support the rotating wheel and maintain its alignment. These bearings and the drive unit for the preheater operate in severe conditions due to the flyash and heat normally found in these areas.

35 What is a plate-type air preheater?

A plate-type air preheater is similar in operation to a tubular air preheater, except that the heat transfer surfaces are flat plates rather than tubes. The plates are configured so that the hot flue gases pass across one side of the plates while the combustion air passes across the other side.

36 What is a steam-air coil air preheater?

A steam-air coil air preheater is very similar in design and function to an automobile radiator. Steam passes through a grid of tubes, and the tubes pass through metal fins. As air passes across the metal fins, the fins dissipate heat from the steam into the air stream. The steam-air coil air preheater is configured as a slab and is installed between a pair of flanges in the primary air ductwork.

The flow of steam to the steam-air coil is modulated to maintain a constant air outlet temperature. As the steam transfers heat to the combustion air, condensate is formed inside the tubes and runs to the bottom of the coil where it is drained by steam traps.

37 Why are combustion air preheaters often supplied with a bypass duct and bypass damper?

A bypass may be employed with any air preheater to bypass a portion of the combustion air or flue gases around the heater. The bypass may be used for more accurate control of the exit air temperature. It is also used to limit the heat transfer in the preheater when reclamation of heat from the flue gases could result in precipitation (condensation) of acidic gases from the flue gas stream. This situation could occur when the boiler is operating at very low firing rates.

Because of the potential for acid dewpoint corrosion, combustion air preheaters should be inspected closely during boiler outages.

Why are soot blowers provided in the regenerative air preheaters used in solid fuel-fired boilers?

Due to the intricate stamped configuration of the metal heat transfer surfaces in a regenerative air preheater, these units are more susceptible to fouling by flyash accumulation than other designs. **See Figure 7-10.** For this reason, a soot blower is normally provided for these units. This soot blower may be cycled on a scheduled basis, or it may be operated as dictated by the differential pressure across the air preheater.

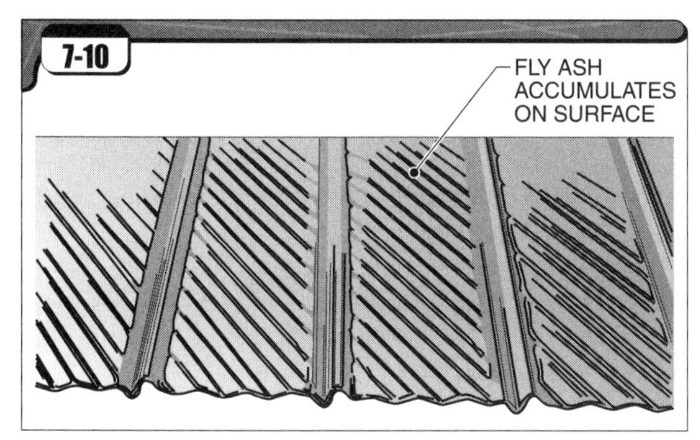

Figure 7-10. *Soot blowers are often necessary for removing ash accumulation from the stamped plates in a regenerative air preheater.*

How is the expansion and contraction of metal ductwork provided for?

Where possible, the ductwork is suspended or supported in such a way that it can freely expand and contract with temperature changes. If this is not feasible, flexible expansion joints are installed in the ductwork to provide for expansion and contraction. These expansion joints may be made of corrugated metal or of fabrics reinforced and coated to withstand high temperatures.

If the expansion joint is located where the potential exists for condensation of acidic flue gases, the material of construction must be corrosion-resistant. It is good practice to insulate expansion joints as well, to minimize the condensation of acidic gases.

Stack Emissions and Environmental Issues

What is an air pollutant?

An *air pollutant* is a substance that produces an adverse effect on humans, animals, vegetation, or materials. These adverse effects may include health hazards, destruction of forests, or acidic deterioration of buildings and surfaces. Air pollutants may be classified as primary pollutants, secondary pollutants, or hazardous pollutants.

A primary pollutant is one that is emitted directly from the emission source. Primary pollutants include particulate matter, sulfur dioxide, and others.

A secondary pollutant is a pollutant formed by the reaction of primary pollutants with other reactive materials in the atmosphere. For example, nitrogen oxides (NO_x) may react with certain hydrocarbons from automotive exhausts or factory emissions to form photochemical oxidants, or smog.

A hazardous pollutant is material that poses a substantial risk of serious illness by itself. Common hazardous pollutants are heavy metals and organic compounds. Metals of this nature include lead and mercury. Mercury often enters the environment from the combustion of coal. Hazardous organic compounds include benzene and vinyl chloride.

What are the major air pollutants of concern in the operation of boilers?

A number of pollutants are emitted when fossil fuels are burned in commercial, industrial, and utility boilers. **See Figure 7-11.** These are known as fuel-dependent air pollutants, and their formation may be due to incomplete combustion, the type of fuel burned, and the design and operation of the combustion equipment.

The major air pollutants associated with the combustion of fuels in boilers are particulate matter, sulfur oxides, nitrogen oxides, hydrocarbons, and carbon monoxide. Ozone, a secondary pollutant, is formed by the chemical reaction between hydrocarbons and nitrogen oxides.

POLLUTANT	PROBLEM CAUSED
Particulate matter	Deposits in respiratory passages; increases exposure to toxic substances
Sulfur dioxide	Causes bronchoconstriction and odor; results in acid deposition (acid rain)
Nitrogen oxides	Results in atmospheric discoloration (brown haze); promotes formation of photochemical oxidant (smog); results in acid deposition (acid rain)
Hydrocarbons	Along with nitrogen oxides, contributes to photochemical oxidant (smog) and ozone
Carbon monoxide	Impairs oxygen transport in blood; impacts central nervous system

Figure 7-11. *Pollutants resulting from combustion of fossil fuels can have adverse effects on human health and the environment.*

What is an emission factor, as related to air pollutants from boilers?

An *emission factor* is an expression of the rate of pollutant production per unit of fuel input. For example, if 4 lb of sulfur dioxide are produced per million Btu of coal burned, this emission factor would be expressed as 4 lb/MMBtu.

(When "M" is used to represent one thousand, then "MM" is one thousand thousand, or one million.) Emission factors are used in establishing some plant environmental permits. For example, a power plant's sulfur dioxide (SO_2) emissions may be restricted to 1.2 lb/MMBtu worth of coal burned.

43 What are the major laws relating to air pollutants from boilers in the U.S.?

The U.S. Congress passed the Clean Air Act in 1963. Amendments were made to the Clean Air Act in 1970, 1977, and 1990. This major legislation called for federal agencies to establish standards for air quality to protect the public health and welfare. The 1970 revision empowered the Environmental Protection Agency (EPA) to establish the National Ambient Air Quality Standards (NAAQS), which are the basis for most of the environmental operating permit restrictions faced by commercial, industrial, and utility boiler plants.

The later amendments to the Clean Air Act gave the EPA the authority to establish New Source Performance Standards (NSPS). The basis for these regulations was that older boilers, industrial furnaces, and other sources of air pollutant emissions would be phased out over a period of time due to age and obsolescence. Newer boilers and other air emissions sources would be of more modern design and more capable of achieving higher standards for air quality. The NSPS thus creates higher performance standards for boilers and other air emissions sources constructed after August 1971.

The NSPS also contains requirements for the installation of continuous monitoring equipment to measure concentrations of stack emissions such as sulfur dioxide, nitrogen oxides, and either oxygen or carbon dioxide. This standard allows for reasonable exceptions. For example, natural gas contains only tiny traces of sulfur; therefore, boilers burning natural gas are not required to have a continuous monitor for SO_2.

In general, the federal Environmental Protection Agency (USEPA) sets the standards for ambient air quality. State and local jurisdictions must then develop regulations and enforcement actions to assure compliance with the federal standards.

Flue Gas Cleaning

44 List three ways that particulates are removed from flue gases.

Particulates are flyash particles that are carried along with flue gases. Particulates are removed from flue gases with mechanical dust collectors, electrostatic precipitators, and filter baghouses.

45 What are mechanical dust collectors?

Mechanical dust collectors use centrifugal force to separate particulates from flue gases. The flue gases flow through multitube units. Each tubular collector (precipitating tube) consists of an outer tube and a concentric inner tube. **See Figure 7-12.** The inlet to the space between the two tubes is covered by stationary vanes that swirl the flue gases. The flue gases flow downward through the outer tube and must reverse to pass upward through the inner tube. The swirling action in the outer tube and the reversal of the gases when they enter the inner tube create centrifugal force that separates flyash from the flue gases. The flyash falls into a hopper below the tube.

Mechanical dust collectors are efficient in removing large dust particles, because these heavier particles are the most affected by centrifugal force. Very small and light flyash particles tend to pass through mechanical dust collectors. These dust collectors should be inspected during each boiler outage because the particulate matter causes erosion of the metal tubes. This results in holes in the tubes and reduced dust removal efficiency.

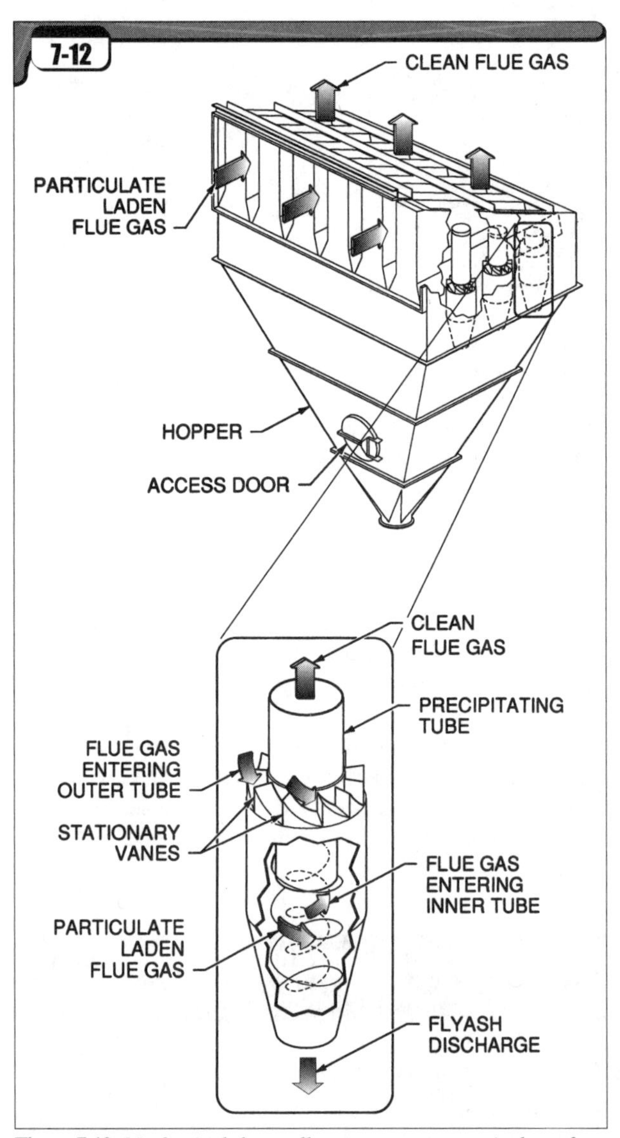

Figure 7-12. *Mechanical dust collectors separate particulates from flue gases by centrifugal force.*

Trade Tip

The swirling flyash in mechanical dust collectors is erosive. Therefore, mechanical dust collectors should be inspected frequently for holes and subsequent short-circuiting.

46 What is an electrostatic precipitator?

An *electrostatic precipitator* is a device used to separate flyash particles from the flue gas stream before it goes out the stack. **See Figure 7-13.** Electrostatic precipitators give the particles an electrical charge that causes the particles to be attracted to grounded collector plates. Electrostatic precipitators are generally used in conjunction with mechanical dust collectors so that the mechanical dust collectors capture the heavier flyash particles and the electrostatic precipitators capture the lighter, powdery flyash.

An electrostatic precipitator contains a set of vertically hung steel plates called collector plates. The collector plates are approximately 1′ apart and are oriented so that the flue gases flow between them. A group of electrodes is suspended between each of the plates. The electrodes may be wires hanging with a weight on the bottom to keep them straight, a grid of thin-wall tubing, or flat plates. The grid or plate designs often have small protruding pins that allow the electrical energy to more readily dissipate from the electrode surface.

A perforated baffle is installed at the inlet to the electrostatic precipitator. The flue gases are directed through this baffle. The perforated baffle is similar to a screen, with many holes cut into it in a grid design. These holes are usually about 2″ to 3″ in diameter and spaced within 2″ to 3″ of each other. The perforated baffle causes the flue gas flow to slow down and spread out evenly. This increases the efficiency of the electrostatic precipitator by ensuring that the flyash is distributed more evenly and does not flow through in one heavy central stream.

In operation, electrical transformers step up the voltage supplied to the electrostatic precipitator to about 40,000 V to 50,000 V. Rectifiers convert the alternating current to direct current. The electrode grids (or wires) are energized with a high-voltage, low-amperage direct current electrical charge. The high voltage on the electrodes is continuously maintained at very slightly less than that required to discharge across the gap to the adjacent grounded collector plates. This creates a strong electrical field, similar to that created when lightning is about to strike.

The particulates passing between the plates become electrically charged by the electrical field and are then attracted to the grounded collector plates. The dust accumulates on the collector plates in a fairly even layer. A set of mechanical rappers taps on the collector plates, causing the accumulated flyash to fall from the plates and into the hoppers below. From the hoppers, the flyash is conveyed to the ash removal system.

Some electrostatic precipitators, called wet precipitators, use water sprays instead of mechanical rapping to remove the dust from the collector plates. In these units, a portion of the precipitator is deenergized and isolated from the remaining sections. Water sprays wash the accumulated dust from the collector plates, and the dust slurry is removed by a wet ash conveying system.

Electrostatic precipitators are highly efficient in flyash dust removal. A dust removal efficiency of 99% is common in modern units, depending on the flyash particle size. Facilities that use electrostatic precipitators can usually maintain opacity levels at very low values.

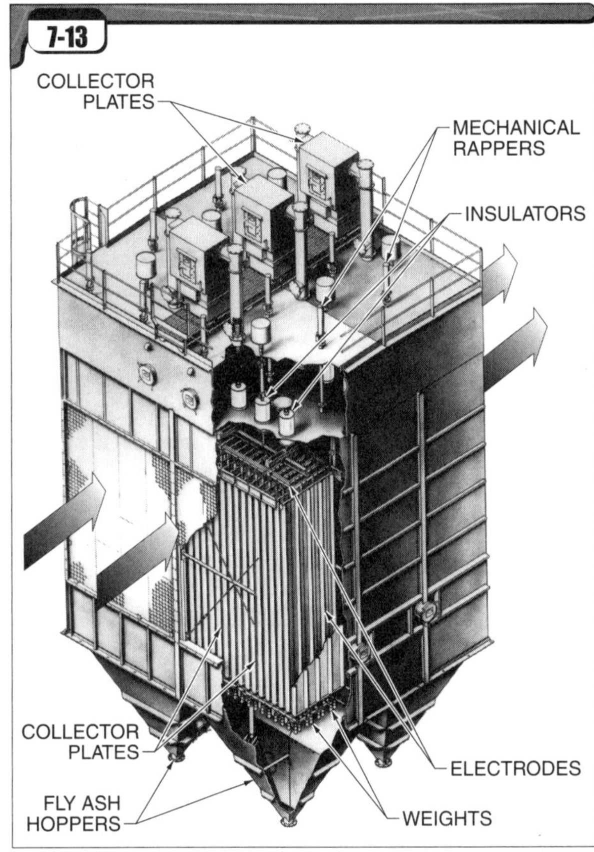

Figure 7-13. *Electrostatic precipitators separate electrically charged particulates from flue gases by attracting them to grounded collector plates.*

47 Describe a filter baghouse.

A filter baghouse separates particulates from flue gases by trapping flyash on fabric surfaces, operating in much the same way as a common household vacuum cleaner. **See Figure 7-14.** Filter baghouses are also known as fabric filters. A typical filter baghouse contains two or more compartments, each containing many vertically oriented fabric tubes. The fabric tubes are suspended from an upper tube sheet and are supported internally by wire cages. Flue gases enter the filter baghouse compartments through a plenum connected to the breeching. The flue gases most commonly flow through the fabric tubes from the outside and exit through the center of the tubes. Thus, the flyash collects on the outside of the tubes. The flue gases exit the tubes at the upper ends and are directed to an outlet plenum. This plenum is connected to another section of breeching leading to the stack. In other baghouses, the dirty flue gas enters the interior of the bags and exits the exterior.

The particle size of the flyash is very often smaller than the pore size of the fabric tubes, so the tubes are often preconditioned prior to placing the baghouse in service. To precondition the fabric tubes or bags, the boiler operator introduces a dust material supplied by the filter baghouse manufacturer or other vendor into the flue gas stream prior to startup of the boiler. This dust material forms a dust cake on the outside of the bags and effectively reduces the pore size to less than the size of the dust particles.

As a layer of flyash thickens on the fabric tube surfaces, the tubes become more efficient in filtering, but the differential pressure across the baghouse increases. When the differential pressure reaches a predetermined point, the bags must be cleaned. In many filter baghouses, one section of the tubes is removed from service by isolating that section with isolation dampers. This section of the tubes is automatically cleaned while the others remain in service.

Several bag cleaning methods are used. In baghouses where the dirty flue gas enters the bag from the interior, the reverse air principle is used. In this case, a section of the bags is isolated from the rest and air is introduced from the exterior of the bags. This causes the bags to collapse somewhat, cracking the flyash accumulation from the interior surface. The flyash falls into a collection hopper below.

In designs where the dirty flue gas enters the fabric tubes from the outside, the tubes are either shaken by a timed shaking system or cleaned with compressed air using a pulse jet system. In the shaker design, a mechanical system shakes the bags in a circular motion, causing a downward ripple effect through the bag. This dislodges the flyash from the exterior.

The pulse jet cleaning system is the most common bag-cleaning system. In many cases, the pulse jet system allows the bags to be cleaned while the system is operating so the bags do not have to be isolated from service. In this system, a large compressed air receiver is provided to store compressed air. A series of compressed air manifolds is configured so that each manifold extends across a row of filter bag ends. The manifolds are arranged so that nozzles spaced along the manifold will blow air into the open end of the filter bag in the opposite direction from the normal flue gas flow. A large solenoid valve is provided at the inlet to each air manifold. An automatic control system sequentially snaps each solenoid valve open and closed. A pulse or puff of air inflates the bags slightly and dislodges the accumulation of dust from the outside of the bag. The dust falls from the outside surfaces of the bags into ash hoppers below. From there, the dust is conveyed to the ash removal system. In most baghouses using either the timed shaking or pulse system, half of the bags are isolated for cleaning while the other half remain in service.

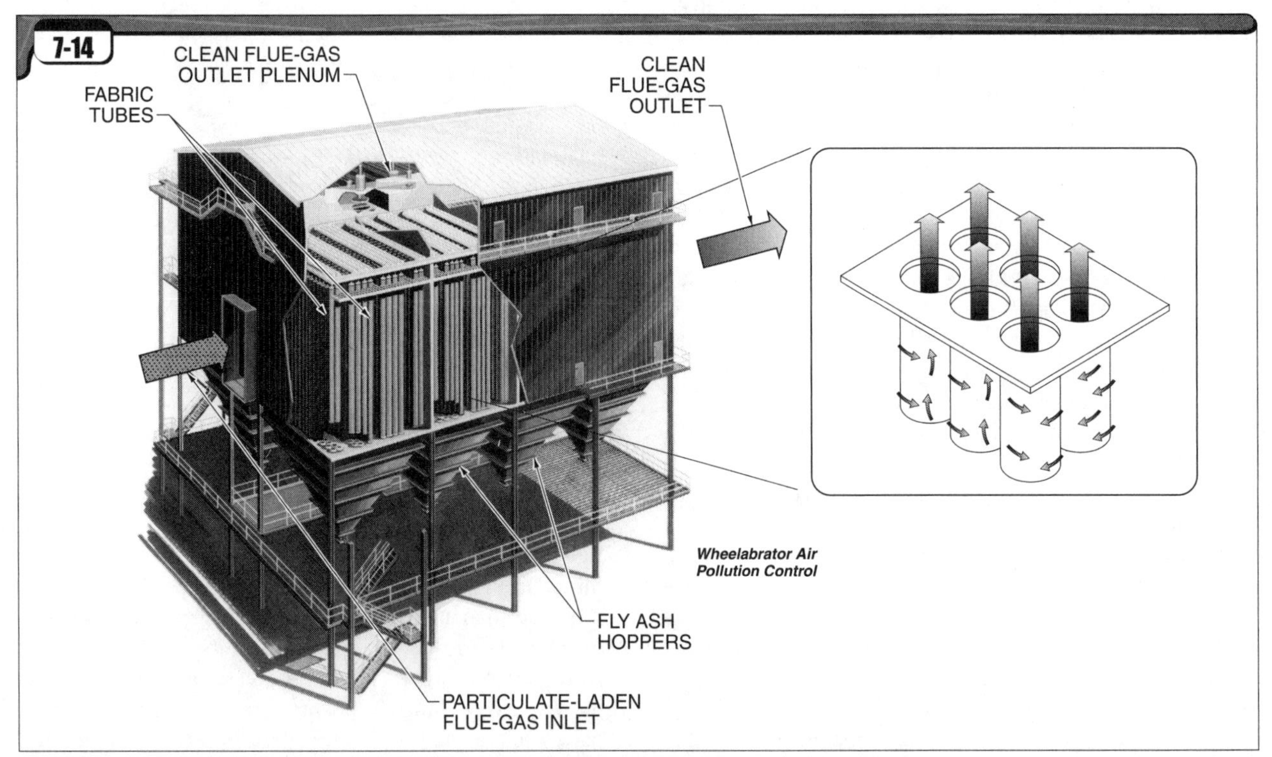

Figure 7-14. *Filter baghouses separate particulates from flue gases by trapping the flyash on fabric tubes.*

48 What is a flue gas scrubber?

A flue gas scrubber removes acidic constituents from the flue gas stream leaving the boiler. Flue gas scrubbers are most commonly used with coal-fired boilers, which discharge acidic sulfur dioxide in the flue gases. Flue gas scrubbers are also known as flue gas desulfurization (FGD) systems.

In most flue gas scrubbers, an alkaline solution of lime (calcium oxide) and water is mixed with the flue gases to neutralize the acids. **See Figure 7-15.** A few flue gas scrubbers for special applications use caustic soda (also alkaline) instead of lime, but these cases are rare due to the expense of the caustic soda.

Caustic soda, if used, is usually delivered to the plant in a liquid form. It may be added directly to the scrubber system. If lime is used, a solution of lime and water must be made before it is added to the scrubber system because lime is a solid. In order to make the solution, the lime is conveyed through a slaker. A *slaker* is a conveyor in which lime is mixed with water to make a soluble paste. The lime paste is then mixed with additional water and this mixture is pumped into the flue gas scrubber. The alkaline solution is mixed with the flue gases by one of several methods.

In a dry scrubber, spray nozzles spray the alkaline solution downward into the flue gas stream. Special nozzles are used to enhance the spray dispersion. The objective is to finely atomize the lime solution into a mist that is readily mixed with the flue gases. In some scrubbers, the flue gas flow through the scrubber is also oriented downward at this point. In other scrubbers, the flue gas flow enters near the bottom of the scrubber and flows upward as the lime solution is sprayed downward. The chemical reaction between the droplets of alkaline lime solution and acidic gases results in the formation of nonsoluble particles of calcium sulfate salts. Much of the water used is evaporated into steam that is released to the atmosphere. The salts that are formed are entrained in the flue gases and removed by filter baghouses.

In a wet scrubber, the alkaline solution cascades down through a packed bed of corrosionproof material such as plastic or stainless steel alloys. This fill material (also known as packing) may take many forms. A widely used fill material resembles hollow plastic golf balls with many holes or slots. The design of the fill material is intended to maximize its surface area and allow maximum contact of the flue gases with the circulating alkaline solution. The flue gases generally pass upward through the packed bed as the alkaline solution cascades downward. The scrubber and stack are made of corrosion-resistant materials. These materials may include certain castable refractory materials, alloys of stainless steel, or fiberglass-reinforced plastic (FRP) to withstand the harsh conditions. When FRP is used, a quench section protects the FRP from melting by cooling the flue gases before they enter the scrubber. The quench uses evaporative cooling, in which heat from the flue gases is used to evaporate part of the water spray in the quench section. This cools the flue gases.

The neutralizing reaction forms solid salts. In a wet scrubber, these salts remain in the circulating liquid (known as scrubber liquor). The concentration of the salts is controlled by blowing down a portion of the liquor and replacing it with fresh water/lime solution. The blowdown must be treated in an appropriate water treatment process before being discharged from the plant.

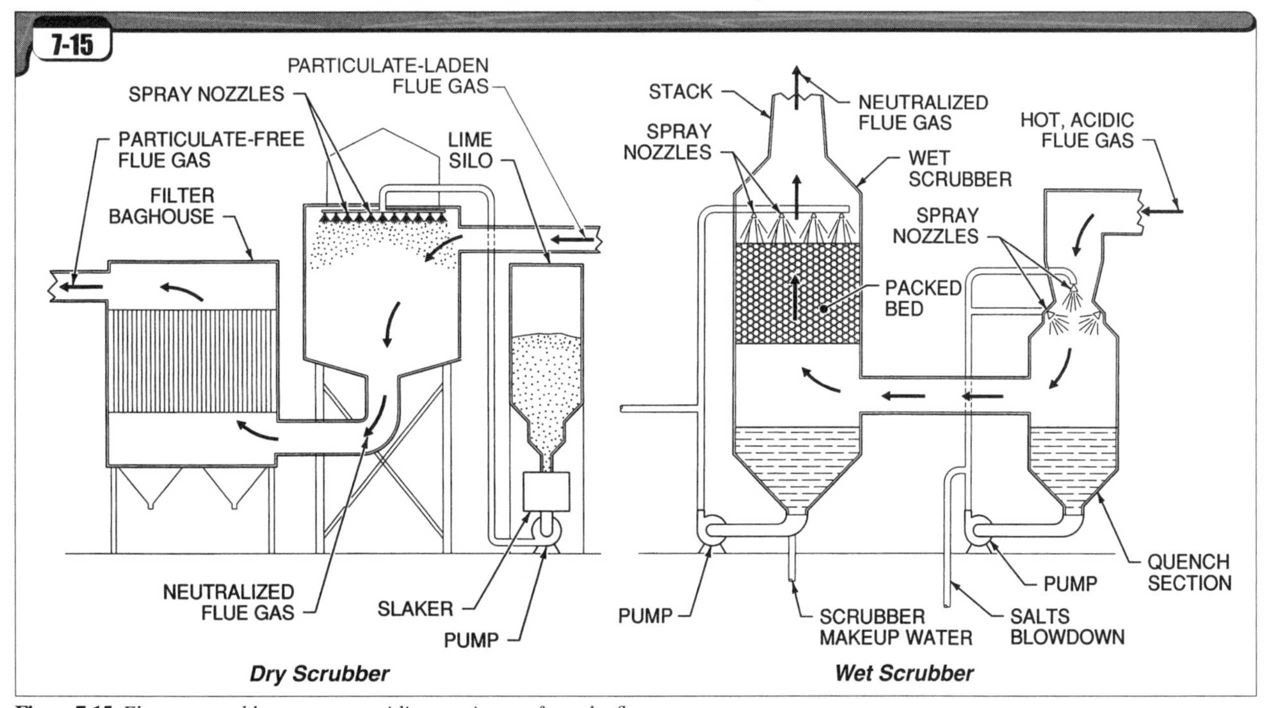

Figure 7-15. *Flue gas scrubbers remove acidic constituents from the flue gas stream.*

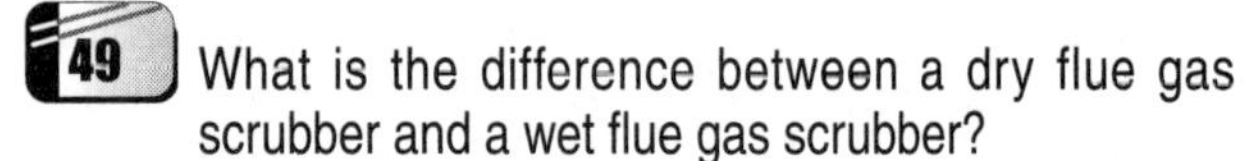

49 What is the difference between a dry flue gas scrubber and a wet flue gas scrubber?

Dry and wet in this context refer to the condition of the waste salts as the material is removed from the scrubber system. Dry flue gas scrubbers typically produce waste in a damp cake form. The waste material from wet flue gas scrubbers is discharged in the form of brine. The brine is often run through a dewatering process to minimize the volume of waste before disposal.

Flue Gas Analysis

50 What is a continuous emissions monitoring system?

A continuous emissions monitoring system (CEMS) is the set of instrumentation used to measure the concentration of a selected gaseous component or the opacity of the flue gas stream. **See Figure 7-16.** There are two basic groups of CEMSs. An *in situ CEMS* is a monitoring system that directly measures the concentration of a specific constituent in the stack, without conditioning, and provides a readout of that concentration for the boiler operator. An *extractive CEMS* is a monitoring system that withdraws a sample of the flue gas stream, conditions the sample, analyzes the conditioned sample, and then provides a readout of the flue gas condition. Extractive systems are considerably more common than in situ systems, due to the severe conditions of dirt, moisture, and corrosion encountered by in situ systems. In an extractive system, the conditioning equipment removes the particulates and moisture and cools the sample.

CEMSs are used for detecting the concentrations of oxygen (O_2), carbon dioxide (CO_2), carbon monoxide (CO), oxides of nitrogen (NO_x), sulfur dioxide (SO_2), and opacity.

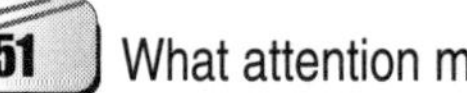

51 What attention must be given to a CEMS?

Continuous emissions monitoring system equipment requires some basic care in order to assure accuracy of the measurements. This care is typically provided by a plant instrument technician, often with the assistance of the boiler operator. The equipment is usually calibrated on a daily basis and inspections are made of the components. For example, the probe through which the sample is drawn is checked for blockage and the sample tubing leading to the conditioning system and analyzer are checked for leakage. Filters that are used to remove particulate must be cleaned or replaced on a routine basis, and the sample pump must be inspected for proper operation.

52 What is the overall path of air and combustion gases through a large boiler and the flue gas system?

In a typical situation, a forced draft fan blows air through a burner or stoker. For a coal-fired boiler, there is often an overfire air fan that blows air into the furnace just above the fuel bed. The gases of combustion flow past a superheater, through the boiler, and through an economizer before passing through the pollution control equipment. **See Figure 7-17.**

The pollution control equipment includes a mechanical dust collector and an electrostatic precipitator or filter baghouse, the CEMS, and the exhaust stack. An induced draft fan maintains the draft necessary to move the air through the equipment and up the stack.

Factoid

Sulfur dioxide emissions result in sulfuric acid being formed as the flue gases condense in the atmosphere.

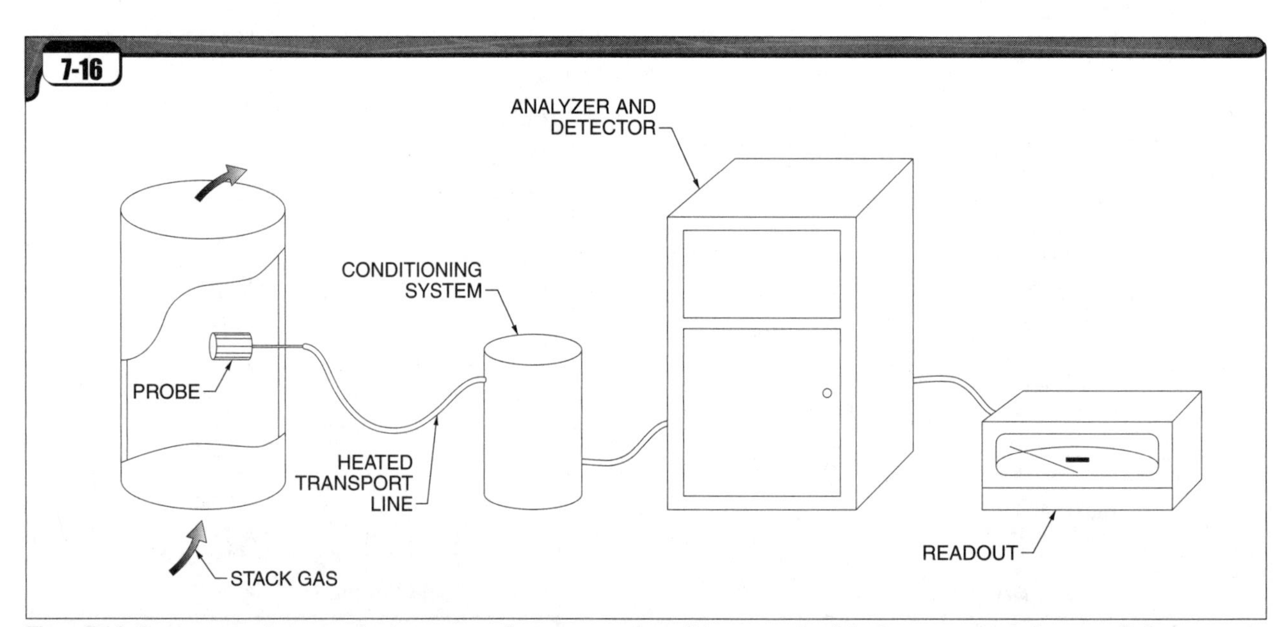

Figure 7-16. *Continuous emission monitoring systems often extract and condition a sample of the flue gas stream prior to analysis.*

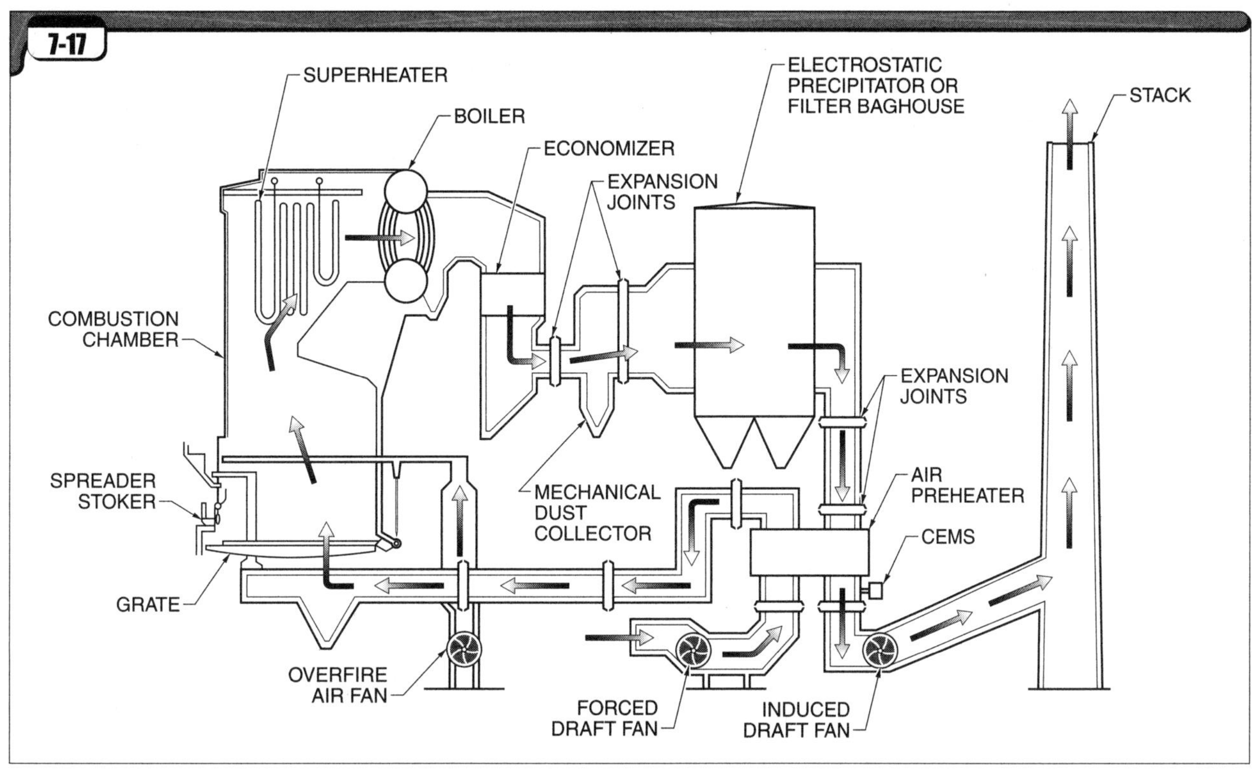

Figure 7-17. *Air and combustion gases move through a complex system that provides the right quantity of air required for complete combustion. The flue gases are also cleaned before being discharged to the atmosphere.*

Chapter 7

Draft and Flue Gas Systems and Controls

Trade Test

Name: ______________________ Date: ______________________

T F **1.** Natural draft occurs without mechanical aid.

T F **2.** By using a fan, a shorter stack can be used.

______________ **3.** ___ is an air pollutant emitted by boilers, and is the primary cause of acid rain.
A. Particulate matter
B. Sulfur dioxide
C. Nitric oxide
D. Carbon monoxide

______________ **4.** The amount of draft in a natural draft boiler is determined by the ___ of the stack and the difference in ___ between the inside and outside of the stack.
A. diameter; inches
B. circumference; temperature
C. weight; inches
D. height; temperature

T F **5.** A fairly constant source of heat at the bottom of a stack produces an irregular natural draft.

______________ **6.** ___ draft is the discharge of combustion air from a fan into the furnace to combine with the fuel for combustion.

______________ **7.** The induced draft fan is commonly known as the ___ fan.

______________ **8.** Draft pressures are expressed in ___.
A. psi
B. inches of mercury
C. inches of water
D. none of the above

______________ **9.** A(n) ___ is an instrument that measures draft pressures.

T F **10.** A pressurized furnace has a furnace pressure lower than atmospheric pressure.

______________ **11.** The strongest negative draft is found ___ when using an induced draft fan.
A. over the fuel bed
B. in the breeching
C. at the top of the stack
D. none of the above

T F **12.** The induced draft fan in a balanced draft system is larger in capacity than the forced draft fan.

______________ **13.** ___ air is secondary air introduced over the fire in a solid fuel-fired furnace.

______________ **14.** ___ prevents the furnace from being pressurized if the induced draft fan fails.

______________ **15.** The flow of flue gases is measured in ___.
A. standard cubic feet per minute
B. actual cubic feet per hour
C. pounds per minute
D. pounds per hour

______________ **16.** ___ air provides the great majority of the oxygen needed for combustion of the fuel.
A. Primary
B. Secondary
C. Makeup
D. Superheated

T F **17.** Positive draft is desirable in a large boiler furnace.

T F **18.** Cracks or gaps in the furnace walls allow air to leak into the furnace, which increases combustion efficiency.

T F **19.** A heavy accumulation of soot and slag on boiler tubes can cause increased resistance to the flow of flue gases through tube banks.

______________ **20.** A damaged flue gas baffle can produce a sudden ___ in the stack temperature because the hot flue gases could short-circuit through the boiler.

T F **21.** A bypass may be used with a combustion air preheater for more accurate control of the exit air temperature.

T F **22.** Expansion and contraction must be provided for in ductwork because of temperature changes.

______________ **23.** ___ are flyash particles that are carried along by the flue gases.

______________ **24.** A natural gas burner that has no blower or fan is called a(n) ___ burner.

______________ **25.** A(n) ___ separates particulates from flue gases by trapping the flyash on fabric tubes.
A. mechanical dust collector
B. electrostatic precipitator
C. filter baghouse
D. none of the above

______________ **26.** A(n) ___ separates particulates from flue gases by centrifugal force.
A. mechanical dust collector
B. electrostatic precipitator
C. filter baghouse
D. none of the above

______________ **27.** A(n) ___ converts alternating current to direct current.

T F **28.** A wet flue gas scrubber produces brine waste.

T F **29.** Combustion air preheaters are of barometric, atmospheric, or electrostatic construction.

T F **30.** The boiler operator should wear appropriate PPE when working in the vicinity of an open furnace opening, and stand to the side of the opening.

______________ **31.** ___ draft is draft created by one or more fans.

T F **32.** Induced draft creates a negative pressure in the furnace.

______________ **33.** A(n) ___ draft installation uses a forced draft fan and an induced draft fan to maintain furnace pressure just below atmospheric pressure.

______________ **34.** A dust-removal efficiency of ___% is common in modern electrostatic precipitators.

______________ **35.** A(n) ___ is a conveyor in which lime is mixed with water to produce a soluble paste.

T F **36.** A drop in the percentage of CO_2 and a rise in the percentage of O_2 in the flue gases between the furnace and the stack indicates inleakage of air somewhere in the breeching.

______________ **37.** A(n) ___ damper is a free-swinging adjustable balanced damper used on smaller boilers to automatically limit the amount of draft pulled through the combustion chamber.

38. Describe the four functions of a boiler's draft system.

_______________ **39.** ___ is the movement of air and/or gases of combustion from a point of high pressure to a point of lower pressure.

_______________ **40.** Air that leaks into a furnace is known as ___.

T **F** **41.** Warm air rises and cool air descends.

_______________ **42.** A forced draft fan discharges air under a(n) ___ pressure to the furnace.

_______________ **43.** The electric control signal for a damper actuator is typically ___.

A. 2 mA to 10 mA
B. 4 mA to 10 mA
C. 4 mA to 20 mA
D. none of the above

T **F** **44.** Overfire air creates turbulence in the flame.

_______________ **45.** A regenerative air preheater turns at a speed of 1 rpm to ___ rpm.

T **F** **46.** Particulates may be removed from flue gases by electrostatic precipitators.

T **F** **47.** Flue gas scrubbers remove alkaline constituents from the flue gas stream.

_______________ **48.** A motor-driven fan should be started with the damper ___.

A. wide open
B. halfway open
C. closed
D. either B or C, depending on design

49. What are the major air pollutants associated with combustion of fuel in boilers?

_______________ **50.** A natural gas burner that uses a natural draft fan is known as a(n) ___ burner.

T **F** **51.** The forced draft fan is commonly known as the primary air fan.

_______________ **52.** The induced draft fan is commonly located at the base of the ___.

_______________ **53.** When ___ is applied to one side of the tube of a manometer, the water will rise in the other side of the tube.

A. pressure
B. vacuum
C. mercury
D. none of the above

T **F** **54.** Excessive forced draft absorbs heat from the furnace and reduces efficiency.

T **F** **55.** Scale deposits on the water side of the tubes can cause excessive flue gas temperatures.

T **F** **56.** If furnace pressure goes slightly above atmospheric pressure, smoke, dust, and hot gases of combustion can puff out of the cracks and gaps in a furnace wall.

T **F** **57.** An electrostatic precipitator separates electrically charged particulates from flue gases by attracting them to grounded collector plates.

T **F** **58.** Excessive flue gas temperatures can result from damaged baffles.

T **F** **59.** Mechanical dust collectors are most efficient in removal of larger dust particles.

T **F** **60.** An inclined manometer is capable of reading in smaller increments than a U-tube manometer.

Chapter 8

Instrumentation and Control Systems

The ability to control the operation of a boiler and utility plant depends on the availability and correct functionality of instruments. Basic indicators and gauges provide boiler operators with information about the condition of the operation at specific points. Pneumatic and electronic control systems are able to automatically perform many of the tedious and routine tasks that would otherwise have to be performed by the boiler operators. These tasks include making changes to fuel and combustion air flows, maintaining safe water levels in boilers and other vessels, and many other routine tasks.

In addition to providing for automatic operation, control systems provide information needed to optimize safety, maximize efficiency, and ensure compliance with environmental regulations. Boiler operators need to understand the data provided by the control system in order to achieve these goals. Raw data is of no value unless it is used. Boiler operators should also learn the function of the various components of the control systems and be able to tell whether each is doing what it is supposed to do. This job is made much easier through the use of piping and instrumentation diagrams, or P&IDs.

Process Measurement and Instrumentation

What are the four basic variable conditions that are measured by boiler plant instrumentation?

The basic variable conditions measured by boiler plant instrumentation are pressure, temperature, level, and flow. The great majority of equipment and process conditions that the boiler operator is responsible for are measured as one of these variable conditions.

Pressure conditions that are measured include boiler and steam header pressure, feedwater pressure, and fuel pressure delivered to burners. In addition, the boiler operator also monitors or controls manometric pressures in ductwork, differential pressures across strainers, and the vacuum in vacuum-type condensate return systems.

Temperature conditions include steam temperature, feedwater temperature, fuel oil temperature, preheated combustion air temperature, and air temperature in comfort heating systems. One of the simplest temperature indicators, the sense of touch, is used in monitoring bearing temperatures and the temperature of tubing carrying cooling water.

A boiler operator typically monitors the water level in boilers, deaerators, and water towers supplying makeup water. Chemical feed tank levels must be measured in order to continuously maintain water quality conditions. In addition, bearing sumps on pumps, fans, and other rotating equipment must be kept filled to prescribed levels with lubricating oils to prevent equipment damage.

Of the four fundamental conditions above, flow is generally the most important to a boiler operator. This is because pressures, temperatures, and levels are often controlled by changing a flow. For example, the steam header pressure is controlled by balancing the steam flows entering the header from the boilers with the flows leaving the header that go to the steam-using equipment. The level of water in a large boiler is controlled by changing the flow of incoming feedwater to match the sum of the rates of steam flow and boiler blowdown flow. The rate of steam generation is controlled through the flow rates of the fuel and combustion air.

For this reason, most things that are continuously controlled in a steam plant involve the manipulation of either control valves or dampers. Of course, flow, level, temperature, or pressure may be controlled in certain instances by instantaneous control as well. Instantaneous control normally consists of starting a pump motor or opening a solenoid valve.

There are a number of other conditions that do not fit into these four basic categories of pressure, temperature, level, or flow. These include conditions of liquids, such as conductivity, pH, or specific gravity. They may also include the speed of fans in rpm, flue gas opacity or oxygen content in percent, and others.

What roles do direct-reading gauges and instruments serve in the operation of a steam plant?

Direct-reading gauges and instruments are those gauges that are located at the point where the measurement is being made. The individual gauges and instruments provide information to the boiler operator about the condition of the pressures, temperatures, levels, flows, and other variable conditions within the boiler system. Accurate measurement of such conditions is the first and most important part of being able to control the system. In addition to providing a direct readout, many direct-reading gauges are part of a system that provides data to an automatic control system.

Factoid

The great majority of the process conditions that the boiler operator monitors are pressures, temperatures, levels, or flows. Pressures, temperatures, and levels are usually controlled indirectly by controlling a fluid flow.

What roles do automatic controls serve in modern plant operation?

From the time boilers were first developed, steam plants were controlled manually. The boiler operator had to monitor and respond appropriately to continual changes in the steam header pressure, boiler water level, and many other plant conditions. The requirement for unwavering concentration made the work tedious. Many explosions and other disasters resulted from lapses of attention. Boiler system efficiencies were low.

The demand for safer operation, rising fuel and labor costs, and the demand for larger and more complex steam plants drove the development of automatic controls that improve efficiency and safety. As a result, boiler plant operation became safer, more precise and efficient, and much less mentally exhausting for the boiler operators. At the same time, great strides were made in boiler construction materials, water chemistry, and other areas. Automatic controls have been continuously improved over time from pneumatic controls to modern digital controls. **See Figure 8-1.**

Automatic controls were developed to improve boiler plant safety, efficiency, and emissions control. For example, many instruments are continually monitoring the variable conditions within the boiler system. These measurements are communicated to an automatic burner management system (BMS) that will quickly stop the fuel feed if an unsafe furnace condition is detected. The level of water in the boiler is also automatically controlled to ensure safe operation.

In order to optimize efficiency, automatic combustion controls continually adjust combustion conditions in order to help maximize fuel-to-steam efficiency and avoid emissions of atmospheric pollutants. In addition, such controls serve to automatically maintain documentation of operating conditions.

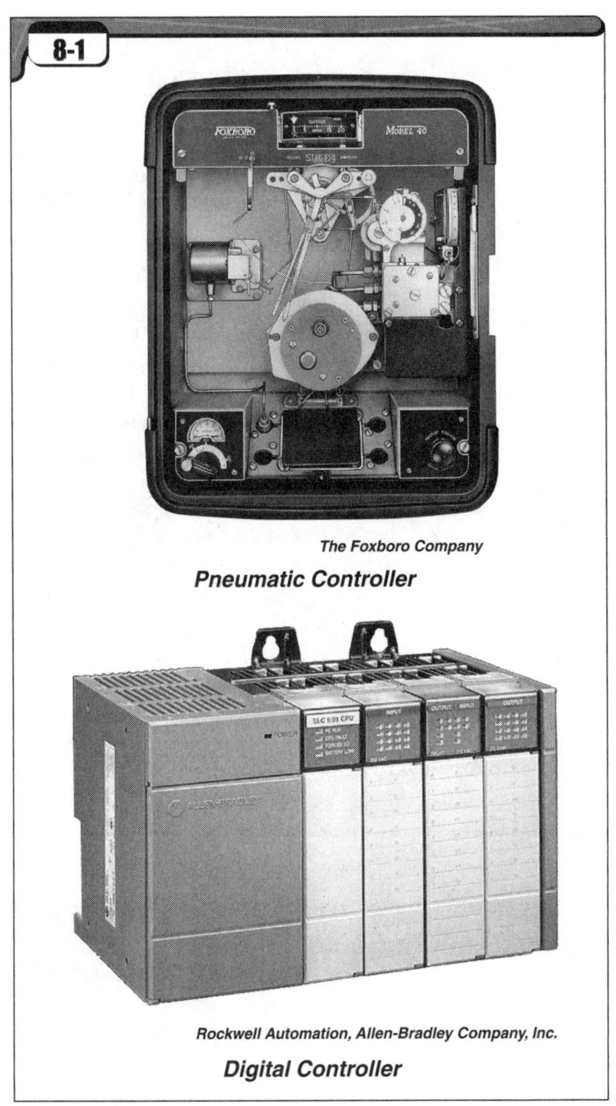

Figure 8-1. *The evolution of modern automatic controls has increased boiler plant efficiency and safety, and reduced emissions.*

Describe several stand-alone instruments used to measure pressure.

The most common device used to measure pressure is the Bourdon tube mechanical pressure gauge. A Bourdon tube is a hollow tube compressed into an oval and curled into a question-mark shape. When a pressure is applied to the inside of the tube, the tube expands and extends the tip. The tip of the tube moves a pointer that registers as a pressure on a gauge. Bourdon tube mechanical pressure gauges are available in many ranges. Variations of the basic pressure gauge include vacuum gauges and compound gauges.

Manometers and panel-mounted draft gauges measure very slight pressures in inches of water column. A manometer is a U-tube filled with liquid. When the ends of the tube are connected to a pressure or a vacuum, the liquid in the tube moves. The amount of movement represents the applied pressure.

Differential pressure gauges have two pressure-sensing connections and display the difference between the pressures at two measured points. Note that differential pressure gauges typically read in pounds per square inch of differential, or psid. **See Figure 8-2.**

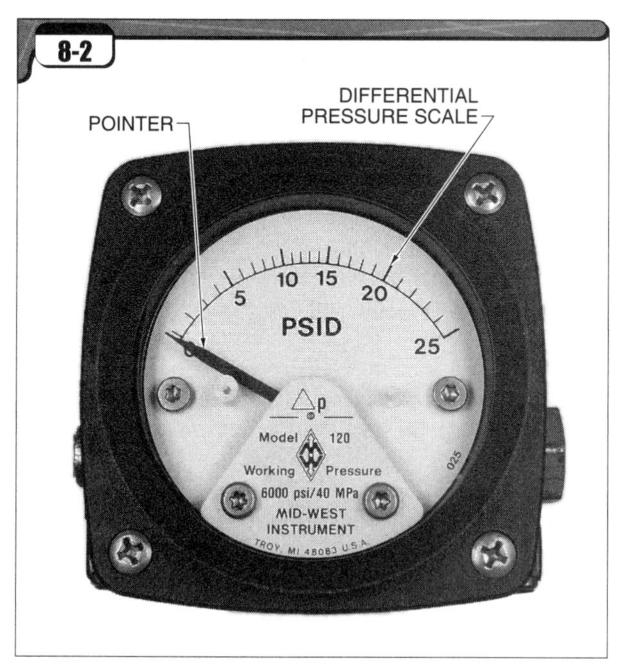

Figure 8-2. *Differential pressure gauges display the difference between pressures at two locations.*

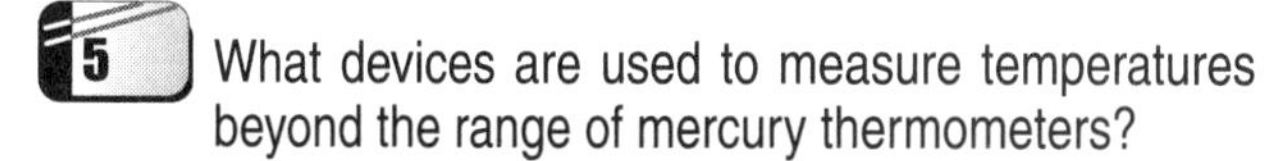

5 What devices are used to measure temperatures beyond the range of mercury thermometers?

Mercury boils at about 675°F. Therefore, mercury thermometers cannot be used above this temperature. In fact, mercury thermometers develop significant inaccuracies above about 575°F. Thermocouples, RTDs, and infrared thermometers are used for measurement of temperatures beyond the capability of mercury-filled thermometers. **See Figure 8-3.** Thermocouples produce a tiny measurable voltage in proportion to the temperature sensed. *RTD* is an acronym for resistance temperature device. The resistance of an RTD changes with changes in temperature. An electric circuit can be used to measure the change in resistance that occurs within an RTD.

Infrared thermometers are also used for a great many applications. An infrared thermometer detects and measures the infrared heat being emitted from the surface of an object and displays the measurement on a readout. Infrared thermometers do not require contact with the object in order to measure the temperature. In fact, these devices often measure accurately from many feet away. This is a major advantage when measuring heat from electrical switchgear or from piping and equipment that is high off the floor, and in many other applications where the instrument needs to be kept away from the heat of the process.

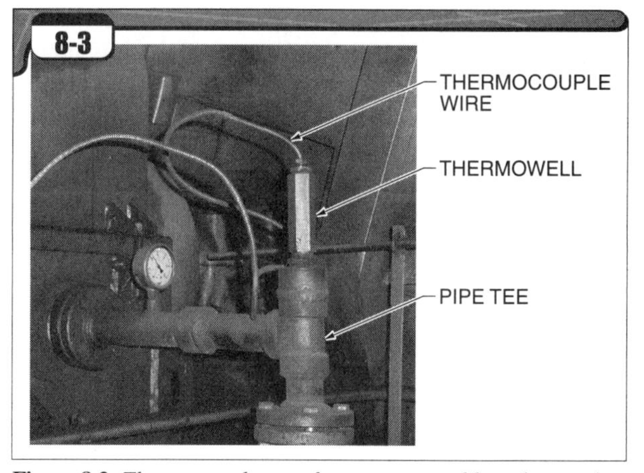

Figure 8-3. *Thermocouples produce a measurable voltage when exposed to heat. A thermowell protects the thermocouple and allows the thermocouple to be removed for maintenance.*

6 What is a thermowell?

A *thermowell* is a receptacle into which a temperature-sensing instrument is inserted. Most commonly, thermowells are used with thermocouples. The two main reasons to use a thermowell are to protect a thermocouple from damage caused by corrosive conditions, and to allow a thermocouple to be inserted into a pipe or process equipment containing a fluid that would otherwise leak out when the thermocouple is removed. This allows the thermocouple to be serviced or replaced without draining and/or depressurizing the process.

If the thermocouple does not firmly contact the wall of the thermowell, a thermally conductive paste is used to fill the space between the thermocouple and the interior wall of the thermowell. This is necessary because the air is a poor conductor of heat and will create an error between the temperature sensed by the thermocouple and the actual process temperature.

7 What is a bubbler control, as used for measuring liquid level?

A *bubbler control* is a set of components that use a small flow of compressed air or another gas to detect the level of liquid in a vessel such as a storage tank. In operation, a tiny metering valve is used to admit a small amount of compressed air through a tube that extends vertically to very near the bottom of the vessel. **See Figure 8-4.** The pressure near the bottom of the vessel depends on the depth of the liquid. A greater depth of liquid in the vessel will result in greater pressure at the bottom. For example, there will be 0.433 psig per foot of water. When the air pressure in the tube rises very slightly above the pressure exerted by the elevation of the liquid, the air will leave the tube and bubble up to the surface of the liquid.

The instrument air line is connected to a device that converts the instrument air pressure required to overcome the head of the water into a usable output. The output may be in the form of a visible indication of the liquid level, or it may be a transmitter used to send the level measurement to a remote location for display there. Direct-reading indicators used with bubbler controls often consist of a liquid-filled column with graduation marks or a panel-mounted gauge. When an indication for a remote location is needed, a transmitter is used.

The transmitter contains a bellows or diaphragm that converts the instrument air pressure into a proportional movement of a linkage. The linkage adjusts the 4 mA to 20 mA electrical output from the transmitter to represent the depth of the liquid. For example, when the level in the vessel is at the lowest point measurable by the transmitter, the transmitter will produce a 4 mA signal. When the level in the vessel is at the highest point measurable by the transmitter, the transmitter will produce a 20 mA signal. An electronic display is then used to convert the electrical signal into a visible readout.

Bubbler controls are only able to measure the depth of the liquid, not the volume. For example, if the vessel containing the liquid is a vertical cylindrical tank, then each inch of the tank represents the same volume. However, if the cylindrical tank is horizontal, an inch of liquid near the center of the tank will be much more volume than an inch of liquid from near the bottom or top of the tank. For this reason, displays for bubbler controls sometimes must be engineered and calibrated for the specific application. In addition, the accuracy of bubbler controls is affected by the specific gravity of the liquid.

Bubbler controls are often referred to as pneumercators. Pneumercator is the name of a company that manufactures bubbler controls, but this name is often used to refer to all similar controls.

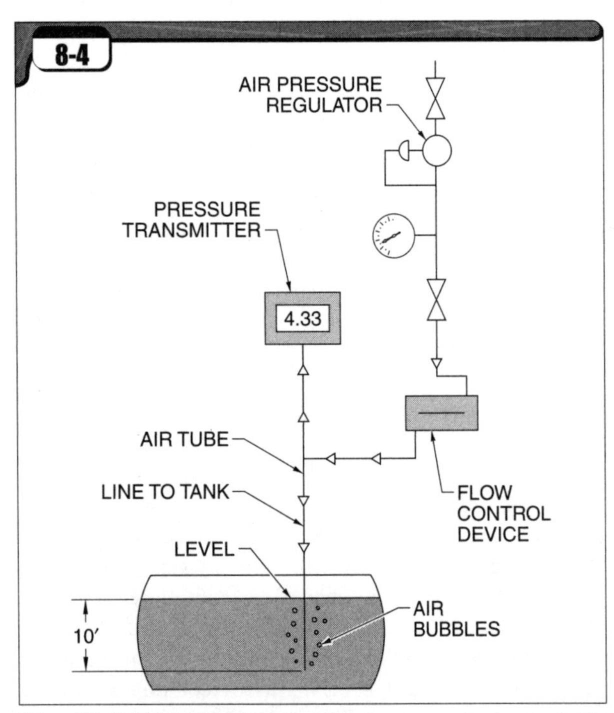

Figure 8-4. *Bubbler controls convert the pressure required to overcome a head of liquid into a level measurement.*

Describe several self-contained devices that are typically used to measure the flow of liquids.

There are many varieties of stand-alone flowmeters. The most common are positive-displacement meters and rotameters. **See Figure 8-5.** Positive-displacement meters repeatedly fill a chamber with liquid to measure the liquid flow, much like repeatedly filling a bucket before dumping its contents. For a bucket of a given size, the number of bucketfuls per unit of time represents the flow rate. In most positive-displacement flowmeters, the rotating parts have tight tolerances so that very little liquid slips between them. Examples of positive-displacement flowmeters include turbine flowmeters, lobe-type flowmeters, and wobble-plate flowmeters.

Turbine flowmeters operate in a fashion similar to a revolving door or turnstile. Liquid fills the spaces between the vanes of the turbine wheel. The greater the flow rate, the faster the turbine wheel turns.

Lobe-type flowmeters work like a positive-displacement lobe pump except that the liquid movement makes the lobes turn rather than the other way around.

Wobble-plate (nutating disc) flowmeters are very commonly used for water, light fuel oils, and other liquids. The meter contains a pillbox-shaped measuring chamber, which in turn contains a disc that surrounds a spherical hub. As liquid flows through the flowmeter, the disc is made to wobble, or nutate, in a motion similar to a coin slowly spinning on a tabletop. The liquid alternately enters above and then below the disc, forcing its way through the flowmeter by virtue of the differential pressure. The liquid does not cause the disc to rotate but to wobble in a continuous rippling motion. On each cycle of the disc, one compartment is being filled while the other compartment is discharging fluid. The continuous movement of the disc provides a steady, nonpulsating flow.

A spindle shaft protrudes from the top of the spherical hub. Since the disc is always diagonal as it wobbles, the spindle shaft moves in a cone-shaped path with the wide end of the cone at the outer end of the spindle shaft. A mechanical linkage or magnetic pickup converts the movement of the spindle shaft to drive a meter that indicates the flow rate. The meter in turn often drives a counter to indicate the volume of liquid that has passed through the flowmeter over a period of time.

A *rotameter* is a variable-area instrument used for measuring rate of flow. It consists of a tapered vertical tube having a cone-shaped interior. A float, or plunger, is forced by the flowing fluid to move in a vertical path. As the float rises, the area around the float increases because of the taper of the tube. This creates more area for the flow to occur in. If the flow rate is constant, the plunger reaches a point in the taper of the tube where the upward flow rate causes it to settle at a constant level. The tapered tube has graduation marks on the side, and the position of the plunger indicates the flow rate passing through the tube.

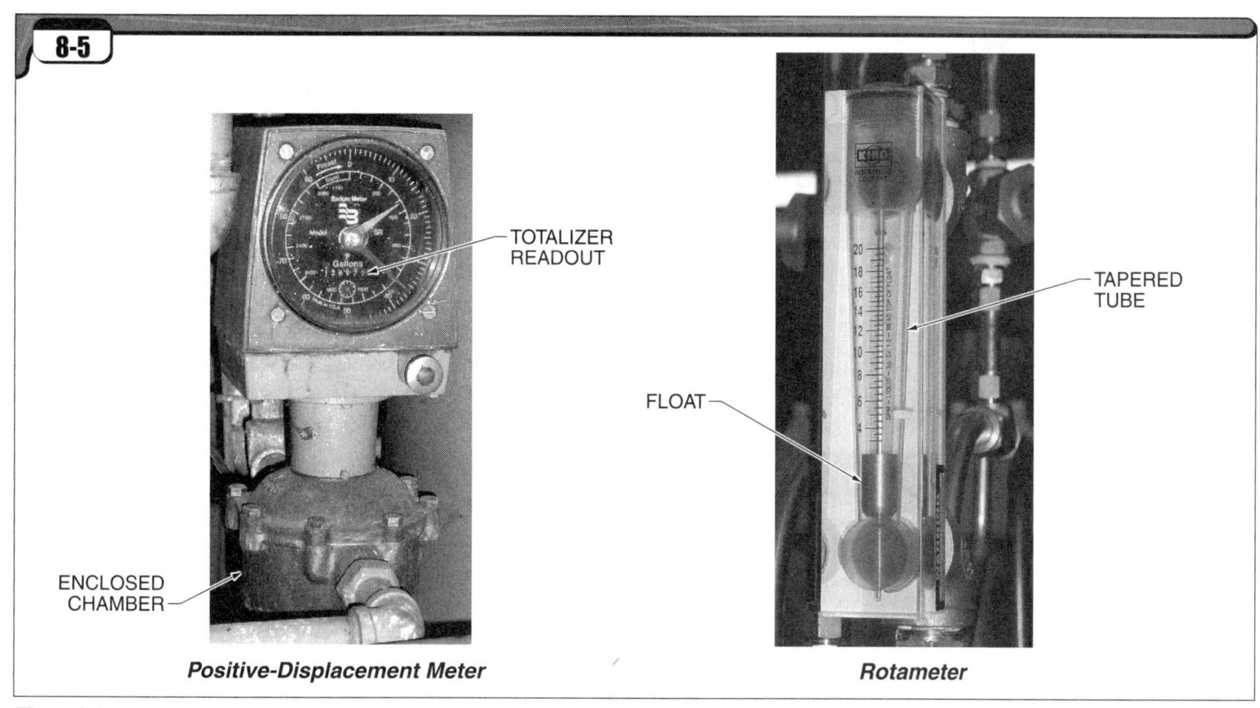

Figure 8-5. *Positive-displacement flowmeters and rotameters are widely used stand-alone flow meters.*

Automatic Control Terminology

9 What is local control? What is remote control?

A *local control* is a control device that is installed directly on or very near the equipment on which it is being used. **See Figure 8-6.** For example, a heat exchanger can have a temperature sensor and local controller to modulate the flow of fluids through the exchanger to control the temperature of the fluid exiting the exchanger. A mechanical float-operated valve that controls the water level in a deaerator is a local control. Another example of local control is a conductivity controller that is mounted on a column next to a boiler and controls an automatic continuous blowdown valve.

A *remote control* is a control device that is installed a considerable distance from the equipment on which it is being used. Electric or electronic remote control devices use wire to communicate the measured condition to the control device, or commands from the control device to a valve or damper actuator. Pneumatic remote controls use tubing and instrument air pressure instead of wire. Remote controls are most commonly used to allow monitoring and control of many sub-processes from one location, such as a control room or central control panel.

10 What is a process, in the context of control systems?

A *process* is the collection of equipment and actions required to accomplish a desired objective. A process in a manufacturing plant often involves control of a large array of machines, conveyors, and utilities to produce a product. In the instrumentation and control field, however, a process may consist of controlling only one variable condition, such as the liquid level in a tank or the temperature of fluid in a heating system.

Figure 8-6. *A local control is installed at the location of the controlled equipment.*

11 What is a process variable?

A *process variable* is the condition that is being controlled. It is the condition that may vary within the process. For example, the process variable may be the water level in a boiler or the steam header pressure. In most cases, it is desirable to keep the process variable at a constant value. The control system exists for the purpose of keeping the condition constant.

12 What is a setpoint?

A *setpoint* is the desired point at which an automatic controller maintains a variable condition within a process. A common example is the setting on a residential thermostat. If the setpoint on the thermostat is 70°F, the thermostat will operate the furnace or air conditioning equipment as necessary in order to maintain the room temperature at or about 70°F.

13 What is a control loop?

A *control loop* is the collection of control devices and other components necessary for automatic control of a process or subprocess. Generally, it is desirable for the control loop to control the process variable with little human intervention.

The term "control loop" is derived from the fact that automatic control of a process variable is a repeating, never-ending sequence of events. As soon as the control sequence has completed, it starts again. Thus, we call this a control "loop."

14 What are signals, in instrumentation terminology?

A *signal* is the language that the control devices use to communicate with each other. For example, the controller must use a signal to communicate instructions to an actuator. The signal is usually a small electrical current or air pressure.

An *input signal* is the flow of control information provided to a control device. An *output signal* is the flow of control information leaving a control device and traveling to another device. The control device may be a local or remote controller, or it may be a local "smart" valve.

15 What is an analog signal? What is a digital signal?

An *analog signal* is a continuous signal used in the ongoing control of a continuous process. An analog signal can take on any value within the limits of normal operation. Typical examples of analog signals are a 3 psig to 15 psig instrument air signal or a 4 mA to 20 mA electrical signal. A milliamp, or mA, is one one-thousandth of an ampere (A). The normal electrical circuit in a residence is 15 A to 20 A. For example, analog signals are used in the continuous control of the water levels in boilers and other vessels or the temperature of fuel oil being supplied to a burner.

A *digital signal* is an instantaneous signal used to implement a one-time action. For example, digital signals are used to start motors, shift the position of solenoid valves, or trigger alarms. In addition, newer control systems use digital signals to communicate measurements taken on a process and also communicate the status of the control device.

16 What is a permissive?

A *permissive* is a process condition that must be met before a certain action may be taken. For example, the low water fuel cutoff switch must be satisfied that there is an adequate water level in the boiler before the boiler control system will permit the burner to be started. It may therefore be said that the low water fuel cutoff switch is a burner start permissive.

It is very common for boilers and other mechanical equipment to have several permissives that must be met before the equipment may be started. There may also be several layers of permissives. As an example, during the light-off sequence of a large burner, there are permissives for each step in the sequence.

Control Loop Components

17 What basic events or steps occur in the operation of a control loop?

The basic steps that occur in the operation of a control loop are measurement, communication, comparison, computation, adjustment, and repetition. **See Figure 8-7.**

- A measurement device must measure the condition of the process variable.
- The measurement device must communicate the condition of the process variable to a local or remote controller.
- The controller must compare the process variable to the setpoint so that the difference between the two may be determined.
- The controller computes the amount of change required in the process to match the process variable to the setpoint.
- The controller provides an output signal that becomes the input signal to an actuator. The actuator converts the signal into a proportional movement of a control valve or damper.
- The process is measured by the measurement device, and the loop repeats.

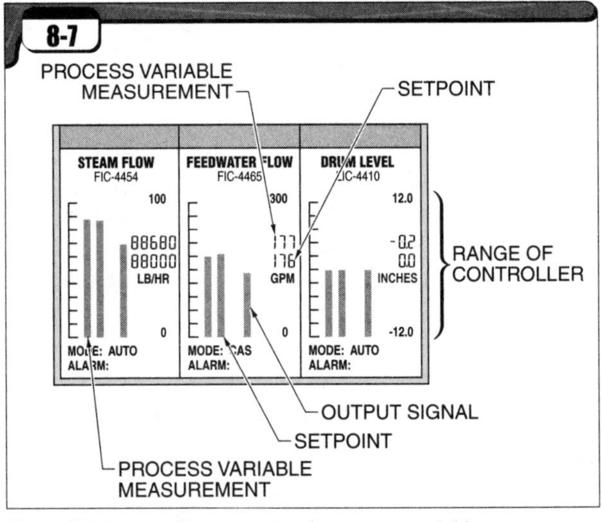

Figure 8-7. *A controller compares the process variable measurement to the setpoint in order to determine the appropriate output signal.*

Factoid

A boiler operator should understand the function of each component in a control loop to the point of being able to ascertain whether each is working properly. This is a great aid to instrumentation technicians and helps them minimize the time required to return malfunctioning instruments to normal service.

Describe the component parts of a typical modern control loop.

Steam from a boiler is used in many industrial processes. For example, the steam may be used to heat up a chemical reactor used to produce an industrial chemical. A typical control loop is temperature control of the reactor. **See Figure 8-8.**

A *primary element,* or *sensor,* is a device that measures the process variable and produces a usable output in the form of a mechanical movement, electrical output, or instrument air pressure output. This output is used by other components in the control loop. The primary element does not make any change to the process, it only measures the process variable. A temperature sensor is a primary element.

A *transmitter* is a device that conditions a low-energy signal from the primary element and produces a suitable signal for transmission to other components and devices. For example, the transmitter may send the signal to an indicating device or to a controller that will use the signal to maintain the process at the setpoint. The transmitter does not make any change to the process, it only communicates information.

The transmitter must be designed to measure the range of the process variable that will be encountered in the process. In addition, the transmitter and other control components must be calibrated for the process once they are installed. The primary element and transmitter are often combined into one component. For example, a modern pressure transmitter may contain a flexible diaphragm that is actually the primary element. Changes in the process pressure result in changes in the deformation of the diaphragm. The movement of the diaphragm is converted into a proportional electric or pneumatic output signal from the transmitter.

A *controller* is a device that takes an input signal, compares the signal to a setpoint, performs a computation, and sends an output signal to a final control element. A controller is the brain of the control loop. It serves several functions. The four main functions are as follows:

- A controller receives the process variable measurement signal from the primary element/transmitter.
- A controller compares that signal to the setpoint that is set at the controller.
- A controller computes the amount of control action needed to bring the process variable into compliance with the setpoint.
- A controller sends its output signal to a final control element, generally a valve actuator or damper actuator, to initiate the change.

A *transducer,* or *converter,* is a device that converts one type of control signal into another type of control signal. The purpose of using a transducer is to allow all the components to "speak the same language." For example, if an electric or electronic controller is used with a pneumatically operated valve actuator, these two devices cannot communicate directly. The electric or electronic controller only understands electrical signals and the valve actuator only understands pneumatic signals. The transducer functions like an interpreter so that the two can understand each other.

As an example, a typical transducer might change 4 mA to 20 mA electric signals from the controller to a 3 psig to 15 psig pneumatic signal that the valve actuator can use. Other signal ranges may be used as well.

A transducer may also be used to change one type of signal to another on the input side of the controller. For example, a primary element for a pressure loop may convert the pressure reading into a proportional mechanical movement. The transducer then changes this movement into a proportional electric signal that can be sent to the controller.

Many modern digital devices have integrated transducers. For example, some transmitters convert the signal from one form to another, such as mV from a thermocouple to a milliamp or digital signal, or ohms from an RTD to a milliamp or digital signal. Controllers may also convert signals from one form to another. They can have modules added to receive digital or analog mV inputs and to send in the other form.

A *final element* is the device that actually causes the change in the process. The great majority of final elements are valves or dampers. However, final control elements may also be variable frequency drives used to speed up or slow down electric motors or other devices.

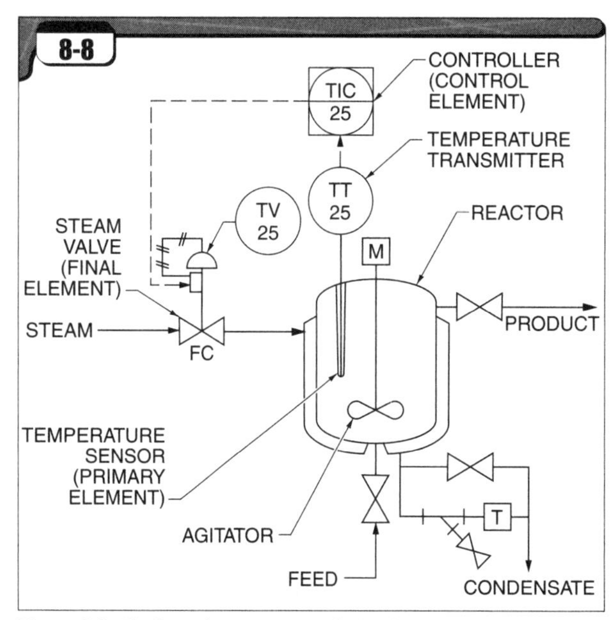

Figure 8-8. *The flow of steam is controlled with a control loop consisting of a primary element, a final element, and a control element.*

Factoid

Oxygen trim systems can have a significant effect on overall boiler efficiency and are quite common on large industrial and utility boilers.

What types of primary elements are commonly used in boiler plant control systems?

Many primary elements for pressure control convert the sensed pressure into a proportional mechanical movement that can be used by the transmitter or controller. For example, a diaphragm capsule contains a flexible diaphragm that has a plunger attached at its center. Deformation of the diaphragm by the pressure produces a linear movement of the plunger. If an electronic controller is used, the diaphragm capsule is typically built into a transmitter that is connected by piping to the process. **See Figure 8-9.** The movement of the diaphragm is converted into a 4 mA to 20 mA output signal that is proportional to the process pressure. The movement of the diaphragm may also be used to make adjustments inside a pneumatic controller.

When measuring temperature, the primary element is usually a sensing bulb, thermocouple, or RTD. Sensing bulbs are very commonly used when the controller is locally mounted. The temperature-sensing bulb is connected to a capillary tube. A *capillary tube* is a long, small-diameter tube, usually of stainless steel or a copper alloy. Capillary tubes are filled with specific liquid or gaseous materials and are used to convert changes in temperature at the sensing bulb to movement of another device. The movement happens when the contents of the bulb expand or contract with changes in temperature.

The capillary tube may be connected to a Bourdon tube found inside a pneumatic controller. Changes in the process temperature cause the volume of the material inside the capillary tube to change and produce a proportional movement of the Bourdon tube. The movement of the Bourdon tube is used by the controller to send a proportional output signal to a valve actuator or damper actuator.

Thermocouples and RTDs are commonly used as primary elements with electronic controls. The small signal produced by the thermocouple or RTD is converted inside a transmitter to a 4 mA to 20 mA output signal to the control system.

Liquid levels in deaerators, process tanks, and other vessels are very often locally controlled using a displacer control. A *displacer* is a long, fairly heavy cylindrical float of small diameter used to measure liquid level in a tank or vessel. Displacers work on the principle that an object's weight is affected by immersion in liquid. For example, a stone, when lifted under water, does not seem to weigh as much as when lifted on shore.

While the displacer is technically a float since it is hollow inside, it is heavy and therefore only slightly buoyant. For this reason, the difference between its vertical position when not immersed and when fully immersed is small (about ¾″). The displacer is connected through a mechanical linkage to an instrument air valve assembly inside the attached pneumatic controller. The small vertical movement of the displacer is converted into a proportional instrument air signal used to position a control valve actuator.

When using electronic controls, pressure transmitters are usually used to measure the level of liquid in vessels. If the tank is only under atmospheric pressure, a basic transmitter may be used. The transmitter is installed on the outside of the tank at the bottom. The weight of the column of liquid pressing down on the transmitter is converted into a reading of the level in the tank. The transmitter must be calibrated to take into account the specific gravity of the liquid being measured.

If the tank is under pressure, a differential pressure transmitter is used. A differential pressure transmitter compares two pressure inputs and produces an output signal based on the difference between the two. One of the sensing lines for the transmitter is connected to the top of the vessel and reads only the pressure inside the vessel. The other sensing line is connected to the bottom of the vessel and reads the pressure inside plus the weight of the column of liquid. This is necessary so that the output from the transmitter will represent only the difference between the two readings. This method gives the height of the column of liquid and compensates for the pressure in the vessel.

Capacitance probes are also widely used to measure the liquid levels in tanks. The electrical capacitance of the probe corresponds to the level of liquid in contact with the probe. A transmitter is used to convert this measurement into a usable electrical signal.

The most common primary element for measuring the flow of pressurized fluids (steam, water, and natural gas) is the orifice plate. An orifice plate is a solid plate with a circular hole of a precise size machined through the center. The edges of the metal at the rim of the hole are precisely machined as well. An orifice plate is installed between two flanges in a pipe. Because it creates a measured and precise obstruction in the pipe, there is a slight difference in pressure between the upstream side and the downstream side of the orifice.

The greater the flow rate of the fluid, the greater the differential pressure will be across the orifice. A differential pressure transmitter measures the pressure drop across the orifice and automatically calculates the flow rate based upon this differential pressure. Fluids that contain debris or are sticky are generally not well suited to measurement with an orifice plate. For these applications, other flow measurement equipment is selected.

Trade Tip

During boiler plant upset conditions, a boiler operator often faces several successive alarms within a few seconds. DCS systems have the ability to identify which alarm occurred first. This can be a great help in determining the root cause of the upset.

Figure 8-9. *Many types of primary elements are used, depending on whether they are measuring flow, pressure, level, temperature, or another process variable.*

What type of final control elements are commonly used in boiler plant control applications?

Most valve and damper actuators are pneumatically controlled. Diaphragm-and-spring and piston actuators are the most common types. Some electric actuators are also used.

Diaphragm-and-spring actuators use a flexible rubber diaphragm located in a diaphragm chamber. **See Figure 8-10.** The diaphragm is reinforced by a backing plate. The actuator stem, which operates the valve stem or damper arm, is attached to the backing plate. When instrument air is applied to the actuator, the instrument air deforms the diaphragm, pushing on the backing plate and causing the actuator stem to move. The movement of the diaphragm and actuator stem is opposed by a spring. The spring's primary purpose is to return the actuator to the desired position when the instrument air signal is removed from the diaphragm. Diaphragm-and-spring actuators are calibrated to use the available instrument air signal. For example, they may be configured to use a 3 psig to 15 psig signal or a 5 psig to 35 psig signal.

Actuators are matched with appropriate valves to create a great variety of configurations to meet specific requirements.

For example, the diaphragm actuator and valve may be configured so that application of the instrument air signal opens the valve, or so that the air signal closes the valve. In addition, the actuators can be configured for the desired fail-safe condition. "Fail-safe" refers to the position to which the actuator should fail if the instrument air supply is lost. The instrument air can be lost if the air compressor fails. Actuators may be configured to fail so that the valve or damper is wide open, so that the valve or damper is fully closed, or so that the valve or damper locks into the last controlled position.

Piston actuators are used to control rotary valves, such as ball valves or butterfly valves. A piston inside the actuator is forced by the instrument air signal to travel lengthwise through the actuator cylinder. The piston rod is machined as a rack gear and the actuator stem is machined as a pinion gear. Thus, when the piston strokes through the cylinder, the actuator stem rotates the valve.

Like diaphragm-and-spring actuators, piston actuators may be equipped with a strong spring that forces the valve to the desired position when the instrument air signal is removed. Alternately, piston actuators may be configured with dual pistons. Dual pistons require the removal of the instrument air signal from one end of the actuator and application of the instrument air signal to the other end in order to shift the actuator's position.

Electric actuators contain reversible electric motors and gearing to control the motor speed to a desired output shaft speed. The control signal received from a controller is used to energize the motor in one direction or the other and cause movement of a valve or damper.

21 What is a feedback transmitter?

A *feedback transmitter* is a transmitter used to provide confirmation that a valve or damper actuator has made a change as commanded. The feedback transmitter detects the position of the valve or damper and provides an indication of this position to a remote control station. For example, the position of the valve or damper typically is shown in percent on a computer screen. This allows the boiler operator to visually confirm that the valve or damper actually responds as instructed by a controller.

22 What is characterization, with regard to valves?

Characterization is the relationship between the amount a valve is open and the amount of flow through the valve. For example, one might assume that if a plain gate valve were 25% open, then 25% of the maximum available flow would result. However, fluid flow through valves is not so simple. As a gate valve is opened, the geometrical area of the passageway increases more quickly than the rise of the stem would suggest. In other words, though the stem may rise from 10% to 20% open, the area of the opening through the valve may increase by 50%. This type of flow characteristic is called quick-opening. **See Figure 8-11.**

In addition, the differential pressure across the opening changes with the valve position. Thus, the flow through the valve will often increase sharply with only a small adjustment of the valve position. The same is true for plain dampers, such as the simple plate damper found in the ductwork leading from a wood stove to a chimney.

The types of valves and dampers typically used in industrial applications often need to provide a much more uniform adjustment of the flow rate. This is called a linear flow characteristic. To illustrate this, assume that the performance of a linear valve or damper was graphed with the horizontal line of the graph representing valve or damper position from 0% to 100% open and the vertical line representing flow from 0% to 100% of available flow.

When a linear valve or damper opens 30%, the flow rate through the valve is 30% of the available maximum flow. When the same valve opens 50%, the flow rate is 50% of the available maximum flow. The resulting chart shows a straight line, or linear flow characteristic, for this valve. This is a good scenario because it is usually easier to control flow through a linear valve. In reality, this rarely occurs and adjustments must be made to compensate for nonlinearities. For example, when a standard gate valve is opened only 20%, the flow rate jumps to about 60% of the available maximum flow. This type of valve would be a poor choice if accurate control of the flow were desired.

Some valves are designed so that when the valve opens a certain percentage, the flow increases by the same percentage. For example, when the valve opening doubles, the flow doubles. This is called an equal-percentage flow characteristic.

Some valve designs with otherwise poor control characteristics may be modified to produce more linear performance. For example, if the passageway through a ball valve is formed in the shape of a V rather than a circular hole—this is known as a V-port ball valve—the performance of the valve becomes much more linear. Valves or dampers that have been modified to provide more linear flow characteristics are referred to as linearly characterized.

23 Why is instrument air cleanliness important in process control systems?

Control devices that work on compressed instrument air, such as control valve actuators and damper actuators, need clean, dry compressed air. Moisture buildup in actuator tubing can freeze during cold weather, blocking the sensing lines and rendering the actuator inoperative. Moisture and compressor oils in the compressed air system can also transport dirt and impurities that will accumulate in system components such as air pressure regulators. Dirt trapped in regulators can cause erratic operation of the regulators.

Extremely clean, dry compressed air is particularly critical in pneumatic controllers. These controllers contain tiny orifices, called nozzles, that are easily blocked by dirt. This can cause total failure of the controller.

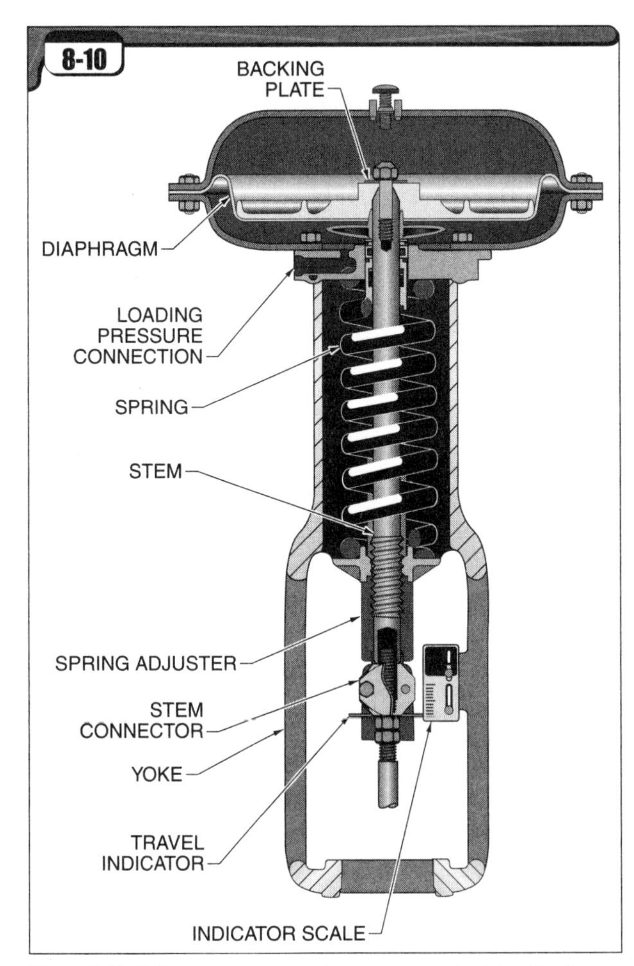

Figure 8-10. *Valve actuators are often constructed with a diaphragm and spring in order to precisely control the flow of a fluid.*

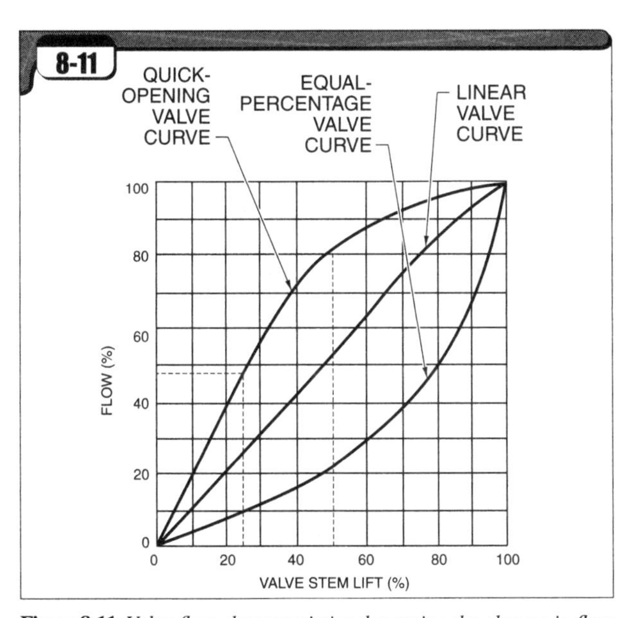

Figure 8-11. *Valve flow characteristics determine the change in flow for a given change in opening.*

Process Control Strategies

24 What is single-loop control?

Single-loop control is the use of a controller to control a single process variable without any influence from any other controller or process variable. Single-loop control is adequate for many simple control applications, such as controlling the pressure in a heat exchanger or the level in a makeup water tank.

25 What is the difference between automatic and manual modes on a controller?

When the controller is in automatic mode, it automatically acts to correct the deviation between the process variable measurement and the setpoint. It does so by determining the amount of change necessary in the process and adjusting the final control element (valve or damper actuator) to cause that amount of change.

When the controller is in manual mode, the boiler operator adjusts the final control element manually. This is done by varying the controller's output signal to the valve or damper actuator. Manual mode is normally used only during startup or shutdown operations until the process stabilizes. Then the controller is switched to automatic or cascade mode. It is occasionally necessary to switch a controller to manual mode for troubleshooting purposes or during process upsets as well.

Before switching from manual mode to automatic mode, it is good practice to bring the setpoint to very near the process variable. If the two are considerably apart, the final control element (the valve or damper actuator) may suddenly "lurch" as the automatic control system responds to the large deviation between the two. This is hard on the equipment and may cause a process upset as other control systems respond to the sudden change. The practice of having the process variable and setpoint at the same value before switching is known as "bumpless transfer."

26 What is cascading control?

Cascading control is a control scheme in which one process variable is measured and used to set the setpoint of another controller. For example, the steam flow measurement from a boiler is often used to influence the feedwater flow rate, as a means of controlling the boiler drum level. One controller uses the steam flow measurement to adjust the setpoint of another controller that manages the feedwater flow rate. When controllers are set to control one from another in this fashion, this is known as having the controller set on cascade mode.

Trade Tip

Dry instrument air is critical when actuators are exposed to freezing conditions. Ice blockage of instrument air tubing can cause loss of equipment control.

27 What is a programmable logic controller?

A *programmable logic controller (PLC)* is a small computer that may be configured, or programmed, to control a wide variety of processes. **See Figure 8-12.** For boiler system applications, PLCs are most often used to control batch or sequential processes. These include such applications as sequential starting of a series of several conveyor belts and bucket elevators in a coal delivery system, or regeneration of several water softeners or demineralizers.

PLCs have simplified such processes considerably through the use of standard components and many self-diagnostic features. The software logic is easily changed, as opposed to the complicated wiring changes that would be necessary with earlier control systems.

The software configured in the PLC receives input signals from field devices, calculates appropriate actions based on preprogrammed setpoints and logic, and initiates output signals in response. The input signals are usually presence or absence of electrical current from devices such as pressure switches, level switches, limit switches, or pushbutton stations. PLCs also make the use of auxiliary components and functions such as timers, cycle counters, and interlocks much easier than do hard-wired electrical relays.

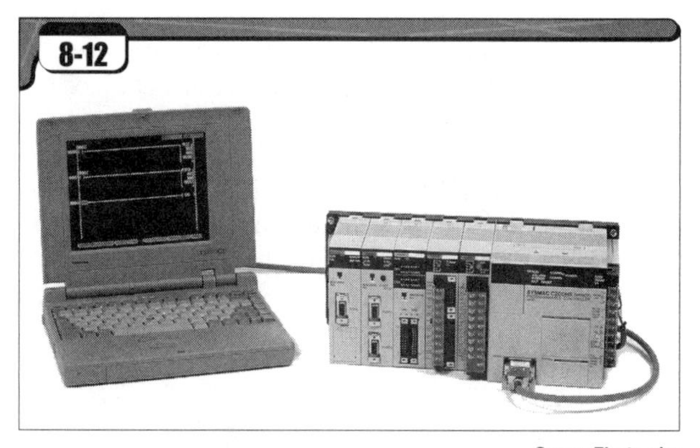

Omron Electronics

Figure 8-12. *A PLC is a solid-state control device that is programmed to automatically control a process or machine.*

28 What is a distributed control system (DCS)?

A distributed control system (DCS) is a microprocessor-based process control system that allows the integration of many microprocessor control units so that all of them function as one system. The DCS is now the standard for plant process control. Distributed control means that the electronic hardware necessary for control of the plant equipment is distributed to multiple physical locations in the plant. This reduces the wiring and conduit necessary, compared to having all wiring and conduit run to a central control room. From the field locations, large volumes of control information are routed to a central processing unit (CPU) through a data highway cable system. **See Figure 8-13.**

Making changes to control systems using panelboard-mounted single-loop electronic or pneumatic controllers is also a cumbersome and time-consuming process. Distributed control systems allow controls engineers to configure and change control loops easily and quickly from a computer terminal.

DCS technology also provides boiler operators with a clearer visual representation of the various plant processes. Graphic displays are used to provide a visual representation of the process or a portion of it. These displays use color coding of the various process flows, standard equipment symbols, and on-screen indications of equipment status. For example, a pump symbol may be configured to change from one color to another when the pump is started. Equipment symbols may also be made to flash or otherwise show upset status when there is a problem. The operator interface for many modern DCS installations consists of a standard computer keyboard, monitor, and mouse.

In order to make the DCS more intuitive for veteran boiler operators used to working with older control systems, controls engineers very often create faceplate screens. On these displays, the various individual controllers appear as boxes that look similar to older panel-mounted controllers. Colored bars represent the controller setpoint, the process variable, and the output signal from the controller.

One of the most useful aspects of many DCS installations is the ability to identify which was the first of several alarm conditions during a process upset. This is known as first-out capability. For example, when an interlock condition causes a piece of equipment to shut down, several alarms may result as other equipment and processes are in turn upset. If the DCS is configured for first-out capability, it is able to identify which upset occurred first. This helps the boiler operator quickly determine which piece of equipment was the cause of the upset and which alarms resulted from effects of the cause.

Modern DCS installations are reliable, but, like any electronic device, they may fail without warning. Therefore, backup (redundant) DCS terminals are usually provided. This usually consists of multiple operator terminals that look identical, but which are separately powered and receive information from multiple microprocessors that communicate but are separate.

Figure 8-13. *A distributed control system provides many benefits over older panelboard-mounted controls.*

29 What is PID control?

PID stands for proportional integral derivative control. This is a type of control that is designed to initiate a control action based on the present, past, and anticipated future condition of a process variable. Like any controller, a PID controller receives a signal from a primary element, or sensor. The controller then compares the process variable measurement to the setpoint in order to determine the amount of measurement error. *Error* is the amount of deviation of a measurement from the setpoint. Based on this error, the controller produces a control action in the final control element that is influenced by the proportional, integral, and derivative functions built into the controller.

After receiving the signal from the primary element, a controller calculates the error. The proportional function of the controller produces a command to drive the process variable toward the setpoint. The command is proportional to the size of the error and is based on the present measurement.

The integral function of the controller calculates how far apart the process variable and setpoint have been over a period of time. The result of this calculation is used in determining the amount of adjustment required to continually minimize the error. This uses the mathematical function of integration and is based on the past measurements.

The derivative function of the controller calculates the rate of change of the process variable with respect to time and this information is used to influence the magnitude of the controller's response in order to bring about the desired change at the desired rate. This uses the mathematical function of taking a derivative and is based on the anticipated future measurements.

What types of recorders are used in process control systems?

Recorders of various kinds are used for automatically documenting many conditions in the boiler plant. These conditions include pressures, temperatures, levels, flows, pH, conductivity, electrical or compressed air consumption, opacity, and many more. Recorders help boiler operators and engineering personnel track plant efficiencies and costs, and are highly valuable in troubleshooting or predicting usage of plant equipment. For example, the data from recorders helps determine the exact time when a problem occurred and the order in which events take place in response. They also help identify and trend peak equipment load times during the day. Recorders in use include chart recorders, digital chartless recorders, and dataloggers.

The two primary types of chart recorders are clock chart recorders and strip chart recorders. Clock chart recorders use circular paper charts that are segmented into hours. Circular graduation lines on the chart (in a fashion similar to the rings in a tree trunk) represent degrees of magnitude. For example, each ring on a specific chart may represent a change of 10 psi.

Ink pens of various colors move inward or outward on the chart with changes in the measured variable. This creates a jagged line that documents the condition of the measured variable at specific times of the day. In many cases, several colored pens document conditions on a single clock chart. For example, a single chart may show the steam production from each of four boilers over a 24-hour period.

Strip chart recorders are similar in function to clock chart recorders except that the paper chart itself takes the form of a long strip of paper rather than a circular disc of paper.

Digital chartless recorders have an appearance that is very similar to chart recorders, but the "chart" is an electronic display that has the appearance of a paper chart. **See Figure 8-14.** These recorders eliminate the need for replacing the paper charts because the historical data is stored in computer memory.

Dataloggers are computer printers that are connected to a DCS computer. Dataloggers produce a chronological record of each alarm, each change made to the system by the boiler operator, and any other relevant information. The primary benefit of the datalogger is in clarifying the sequence in which events take place during normal operation and during upset conditions.

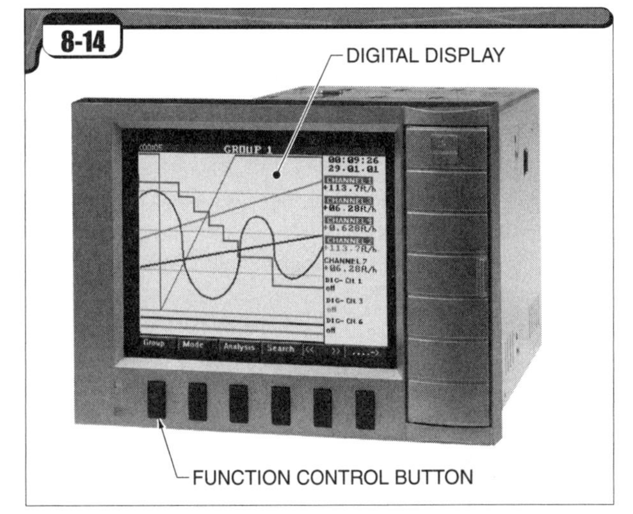

Endress+Hauser

Figure 8-14. *Chartless digital recorders are quickly replacing older chart recorders.*

What is a totalizing flowmeter? How are these important to boiler plant efficiency?

A *totalizing flowmeter* is a flow measuring device that not only measures on an instantaneous basis (flow rate), but also measures total flow over time. For example, a residential water meter or natural gas meter totalizes the flow through the meter over time.

Totalizing flowmeters incorporate a component (or, in the case of computerized meters, a software function) known as an integrator. An *integrator* is a calculating device that combines multiplication and addition. It continually multiplies the instantaneous flow rate by time and adds the product of this multiplication to the previous total.

This allows plant personnel to easily extract data used to track boiler system efficiency. For example, the total fuel consumption over a 24-hour period can be compared to the total steam production over the same period. This reveals how efficiently a boiler plant performed on a given day as compared to historical performance. It also helps plant maintenance and engineering personnel predict when a boiler should be removed from service for maintenance and tuning work. If the amount of fuel used per thousand pounds of steam begins to steadily increase, for example, it may indicate the need for cleaning of the tube surfaces, burner adjustments, investigation of furnace air leaks, or other service work.

DCS installations are especially suited to the use of totalizing flowmeters because many necessary plant calculations may be performed automatically by the computer equipment. For example, automatic reports can be produced to help track stack emissions and wastewater discharge flows against environmental permit limitations. In addition, utilities usage, condensate return, and steam consumption may be easily tracked by a plant department. By internally charging each department for utilities consumed, efficient operation is encouraged.

32 What is a plant master?

A *plant master* is the master controller that calculates and distributes the steam production requirements to two or more boilers when they are used to maintain the pressure in a common steam header, such as when the boilers are installed in battery. **See Figure 8-15.** A pressure transmitter senses the steam header pressure and passes this information to the plant master controller. Two or more pressure transmitters are often provided in case one pressure transmitter fails. Based on the amount of difference (error) between the actual steam header pressure and the plant master's setpoint, the plant master calculates and distributes the steam production requirements for each of the boilers. It then sends an output signal, which is called the firing rate demand (FRD) signal, to each boiler.

Combustion-Related Controls

33 What is a boiler master?

A *boiler master* is a controller that determines the appropriate quantities of fuel and air that need to be provided to a boiler's combustion equipment in order to produce the required quantity of steam. When multiple boilers feed steam to the same steam header, as is usually the case, the boiler master receives a firing rate demand signal from a plant master. The boiler master then sends output signals to the combustion air damper actuators and the fuel valve actuators in order to increase the fuel and air feeds in the correct proportions.

34 What is biasing?

Biasing is a control function that allows a particular boiler to be set up to respond more quickly or more slowly than the other boilers and carry a larger or smaller share of the steam production load. The bias function of a controller is usually something the boiler operator can adjust as needed through software commands or with a dial or other adjustment on a panel-mounted controller.

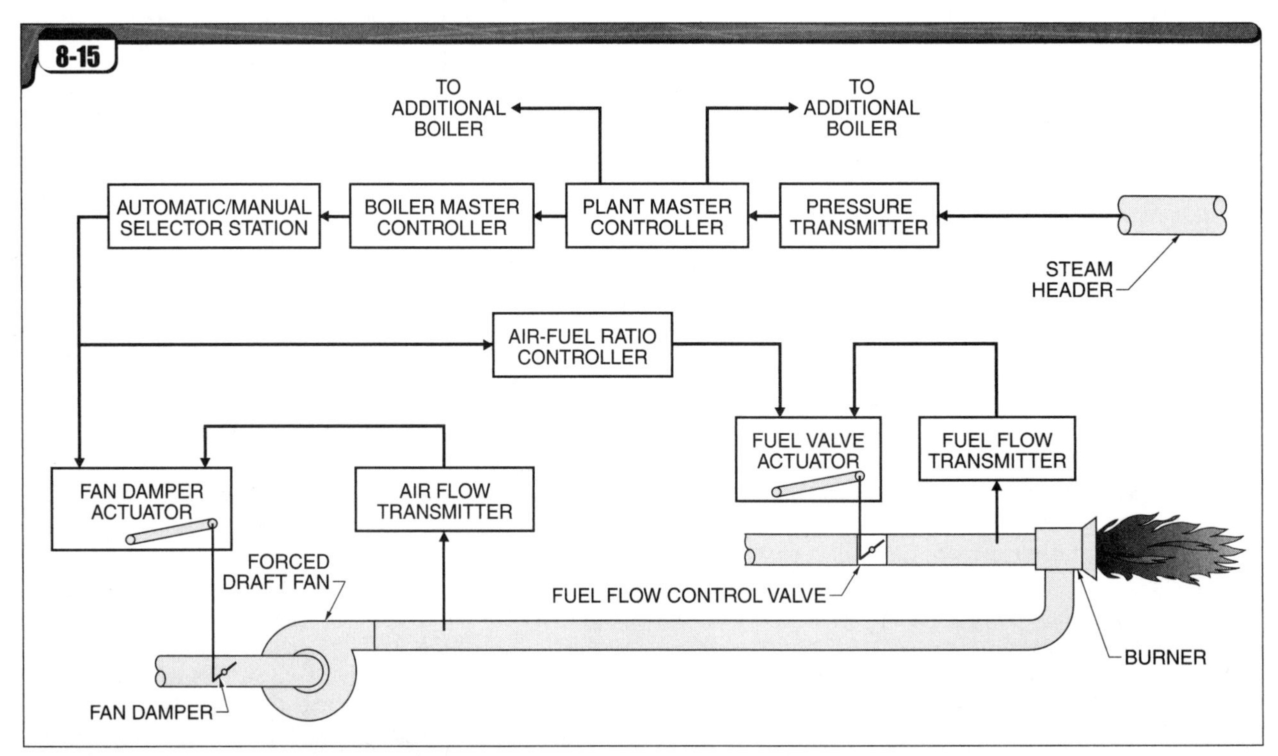

Figure 8-15. *A plant master controller distributes steam production requirements to two or more boilers.*

The total steam demand (load) equals the sum of the loads applied to the individual boilers, and each boiler satisfies a percentage of the total steam demand. For example, if two boilers are in service and equally loaded, each satisfies 50% of the overall demand for steam. If three boilers are in service and equally loaded, each satisfies 33 ⅓% of the overall demand for steam.

Usually, the boiler master controls for each boiler may be set either to manual mode or to automatic mode. When in manual mode, the boiler master does not respond to directions from the plant master. The boiler operator changes the firing rate on that boiler manually. If the boiler master is set to automatic mode, the boiler master responds to the commands from the plant master.

When in automatic mode, if any one boiler is adjusted (biased) to carry a higher percentage of the total steam load than the other boilers, the other boilers will automatically respond to carry less of the total steam load. For example, assume that four boilers are in service and providing steam to a common steam header. A plant master controller senses the total steam demand and allocates a portion of the total steam load to each of the four boilers. If the boiler operator biases one boiler to carry 15% more of the steam load than the other boilers, the other three boilers will respond by automatically carrying 5% less of the load each. Remember, the total steam production equals 100% of the total steam demand.

Similarly, if one boiler is set to manual mode so that the firing rate will stay constant, the other boilers will automatically adjust their firing rates continually so that the total steam production from all the boilers still equals 100% of the steam demand from the plant processes. In this case, veteran boiler operators would refer to the boiler in manual as being "base loaded" and the others as "carrying the swing." The "swing" refers to the changes in the steam demand from the plant processes.

In reality, the total steam production from all the boilers may amount to somewhat more or less than 100% of the steam demand from the plant processes at any given time. This may occur as the boilers increase or decrease their firing rates to maintain a constant steam header pressure.

What is an oxygen (O_2) trim system?

An *oxygen (O_2) trim system* is an automatic control system that makes fine adjustments in the amount of combustion air used in order to minimize excess air. **See Figure 8-16.** This system's purpose is to optimize combustion efficiency by optimizing the quantity of excess air used in the combustion process.

If too much air is provided, an excessive amount of the heat from the combustion process is used in heating the extra air. In addition, the velocity of the resulting diluted flue gases is excessive as it passes across the boiler heating surfaces. The excessive velocity of the flue gases results in the boiler not having adequate time to absorb an appropriate amount of heat from these gases, and the heat is instead carried up the stack. By limiting the quantity of air to only that amount needed to ensure complete combustion of the fuel, an oxygen trim system minimizes this loss.

Oxygen trim systems may be used in cases where independent control of the fuel and air is provided. When oxygen trim systems are used on package boilers on which the fuel and air adjustments are rigidly linked together, an extra linkage must be provided to allow the combustion air to be adjusted independently of the fuel.

The system consists of an oxygen sensor, a transmitter, and an oxygen controller. The oxygen sensor measures the percentage of oxygen present in the flue gases going to the stack. This measurement is relayed by the transmitter to the oxygen controller. The oxygen controller influences the position of the forced draft fan damper and/or secondary air fan damper to cause a desired change in the oxygen content of the flue gases.

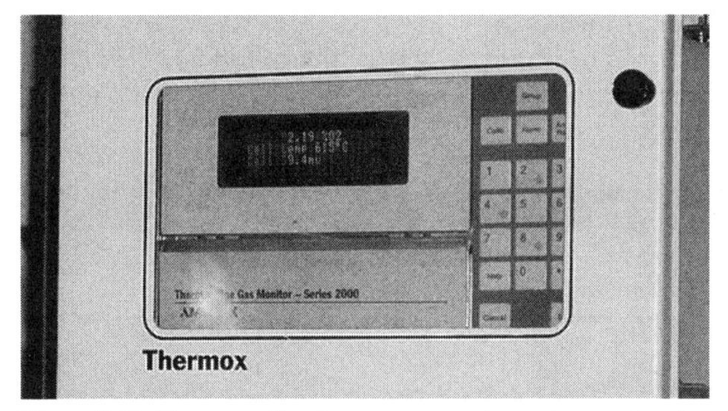

Figure 8-16. *The controller in an oxygen trim system makes fine adjustments to the combustion air dampers depending on the oxygen content of the flue gases.*

What is cross-limiting, with regard to boiler combustion control systems?

Cross-limiting is the configuration of boiler combustion controls such that, with an increase or decrease in the firing rate, the combustion conditions never become oxygen-deficient. If at any time there is an inadequate amount of oxygen available to burn all the fuel, some of the fuel will remain unburned or partially burned. Unburned fuel results in smoke, carbon monoxide emissions, and lowered efficiency. This also results in the risk of a furnace explosion if a significant volume of unburned fuel and air mixture is allowed to accumulate. Cross-limiting is used in gas-fired, liquid fuel-fired, and pulverized coal-fired boilers. It is not a concern in stoker-fired boilers.

The possibility of an oxygen-deficient condition is created when there is a change in the firing rate. This is due to the compressible nature of the air being delivered by the combustion air fans and the slow response of pneumatic damper actuators. When variable-speed fans are used, there is also a lag time involved in making a change in the fan speed due to the mass of the fan rotor. For these reasons, when the fuel valve and combustion air dampers open farther to admit more fuel and air to the furnace, the fuel will tend to arrive at the furnace first. The extra flow of combustion air must "catch up" with the extra flow of fuel. This can result in a fuel-rich condition for a short time.

To counteract this, fuel/air cross-limiting control uses control functions known as high selectors and low selectors. For example, a controller may be programmed to choose the higher of two signals as the one it responds to.

The fuel demand is the amount of fuel required to generate the amount of steam required. In a typical control scheme, the fuel demand is calculated. This required fuel demand calculation is compared to the actual fuel flow measurement. The controller chooses the higher of these two and uses that as the setpoint for the air flow. This is the high selector function. The combustion air damper is then adjusted to admit the required air flow to the furnace.

At the same time, the low selector function in the controller compares the calculated fuel demand to the calculated fuel flow that the current air flow would be able to handle. The lower of these two is used as the setpoint for the fuel flow. The effect is that the air flow and fuel flow are continually compared in order to assure that a fuel-rich condition is avoided.

Factoid

Distributed control systems (DCSs) are the modern standard for process control systems.

Control Diagrams

37 What are piping and instrumentation diagrams (P&IDs)?

Piping and instrumentation diagrams are concise drawings that show the various flows and instrumentation associated with a controlled process. **See Figure 8-17.**

P&IDs generally use standard symbols recognized by the Instrumentation, Systems, and Automation Society (ISA). Standard symbols are provided for control devices such as control valves and transmitters and for various kinds of piping and wiring. In addition, standard symbols are used to convey instrument location information, such as whether the devices are at local or remote locations, and whether a signal is carried by wire, by computer software, or by pneumatic tubing.

38 How are P&IDs an aid to the boiler operator?

P&IDs are extremely helpful to the boiler operator in learning a new process quickly and in troubleshooting scenarios. Using P&IDs, a boiler operator can learn how a process is controlled without having to spend time tracing out piping and wiring. By becoming familiar with the symbols and numbering conventions, the boiler operator can quickly identify control devices in the control room and in the field and be able to tell whether they are working properly or not.

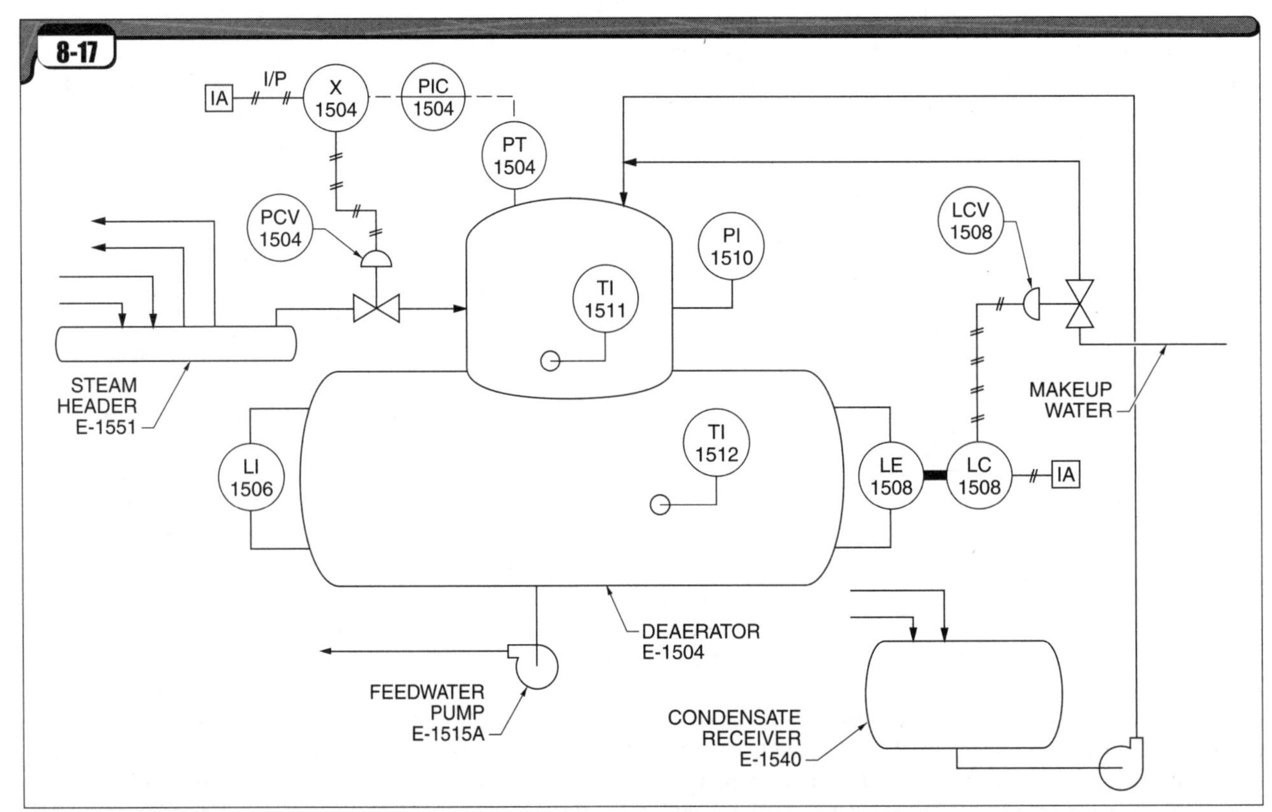

Figure 8-17. *Piping and instrumentation diagrams (P&IDs) provide boiler operators with concise information about the operation of the various pieces of equipment in the system.*

Chapter

8 Instrumentation and Control Systems

Trade Test

Name: ____________________ Date: ____________________

T F 1. A local control is a control device that is installed directly on or very near the equipment on which it is being used.

______________ 2. RTD stands for ___.
- A. reference temperature detector
- B. resistance temperature device
- C. represented temperature derivative
- D. resistance temperature differential

T F 3. Infrared thermometers do not require contact with an object in order to measure its temperature.

______________ 4. A ___ control is a set of components that uses a small flow of compressed air or another gas to detect the level of liquid in a vessel such as a storage tank.
- A. rotameter
- B. PID
- C. programmable
- D. bubbler

______________ 5. The four basic variable conditions measured using boiler plant instrumentation are ___.

T F 6. A boiler master receives a steam demand signal from the plant master and determines the appropriate quantities of fuel and air that should be provided to a boiler's combustion equipment.

______________ 7. ___ combustion controls ensure that an oxygen-deficient condition is not encountered, especially during load changes.
- A. Nitrogen-trimming
- B. Cascading
- C. Single-loop
- D. Cross-limiting

T F 8. A setpoint is the desired point at which an automatic controller maintains a variable condition within a process.

______________ 9. A(n) ___ signal is a continual signal used in the ongoing control of a continuous process and a(n) ___ signal is an instantaneous signal used to implement a one-time action.
- A. digital; analog
- B. reference; response
- C. analog; digital
- D. integral; derivative

______________ 10. A(n) ___ is a receptacle into which a thermocouple is inserted.
- A. orifice
- B. thermowell
- C. RTD
- D. nutator

______________ 11. The most common primary element used for measuring the flow of pressurized fluids is a(n) ___.
- A. capillary tube
- B. diaphragm capsule
- C. capacitance probe
- D. orifice plate

______________ **12.** Most valve and damper actuators are ___ controlled.

13. What is the difference between a process variable signal and an output signal?

T F **14.** Programmable logic controllers are often used to control batch or sequential processes.

______________ **15.** ___ is now the standard for plant process control.
- A. Distributed control system (DCS)
- B. Programmable logic controller (PLC)
- C. Proportional Integral Derivative (PID)
- D. Burner management system (BMS)

16. Describe two reasons why instrument air must be dry and free of debris.

______________ **17.** ___ is a control scheme in which one process variable is measured and that measurement is used to set the setpoint of another controller.

T F **18.** Differential pressure transmitters are very often used to measure the level of liquid in pressurized vessels such as deaerators and boilers.

______________ **19.** A process condition that must be met before a certain action may be taken is known as a(n) ___.
- A. interlock
- B. process variable
- C. permissive
- D. final control element

T F **20.** Piping and instrumentation diagrams (P&IDs) help the boiler operator quickly learn how the various control schemes in the boiler plant work.

______________ **21.** A ___ is a master controller that calculates and distributes the steam production requirements to two or more boilers in order to maintain a constant steam header pressure.
- A. plant master
- B. boiler master
- C. programmable controller
- D. cascading control

______________ **22.** A(n) ___ system is a control scheme that has the ability to adjust the amount of combustion air provided to a boiler's combustion equipment depending on the oxygen content of the flue gases being discharged to the stack.
- A. cascading control
- B. oxygen trim
- C. cross-limiting
- D. distributed control

T F **23.** Valves or dampers that have been modified to provide more linear flow characteristics are referred to as linearly characterized.

Chapter 9 Boiler Operation and Maintenance

By their very nature, boiler plants contain many potential hazards, such as confined spaces, hot surfaces, rotating shafts, and electrically energized equipment. A boiler operator must prevent accidents and injuries that can result from these hazards by following safe practices and procedures. Understanding and following emergency procedures also ensures that injuries are prevented and equipment damage is kept to an absolute minimum during upset conditions.

A boiler system and its equipment represent a very substantial investment for the facility using the equipment. Boiler operators help ensure long equipment life, operating efficiency, and equipment availability by following accepted procedures. During planned outages, solid inspection techniques help boiler operators detect and correct problems at an early stage and avoid more costly problems later. Sometimes the equipment malfunctions but boiler operators who possess good troubleshooting skills are able to return the equipment to full functionality in minimum time. Especially in manufacturing and utility plants, this maximizes plant revenue and profitability and minimizes maintenance costs.

Safe Practices

What is OSHA and what is its significance for boiler operation?

OSHA is the Occupational Safety and Health Administration. OSHA's mandate is to ensure that all employers provide their employees with a workplace that is free from recognized hazards that could cause harm. Individual states have agencies that administer occupational safety and health regulations, and such agencies must comply with the minimum federal standards.

Every employer is required to implement workplace rules, install signs and labels, install protective barriers, and take other steps necessary to ensure compliance with safety standards. Adequate training must be provided to employees to ensure that they are knowledgeable and competent in the practices necessary to ensure their safety and health.

OSHA publishes its requirements in the Code of Federal Regulations (CFR). A *code* is a regulation or law. The portion of the Code of Federal Regulations that applies to workplaces contains standards for known safety and health-related hazards. A *standard* is an accepted reference, practice, or method. Individual standards are published to address significant hazards to life or health. A number of these standards apply to safety issues encountered by boiler operators. The primary OSHA standards that apply to boiler systems are as follows:

- Permit-Required Confined Spaces
- Control of Hazardous Energy (lockout/tagout)
- Hazard Communication (Right to Know)
- Fall Protection
- Personal Protective Equipment (PPE)
- Respiratory Protection

What other types of agencies or organizations publish information related to boiler safety?

Governmental laws and regulations often defer to recognized standards created by nongovernmental bodies. For example, the ASME *Boiler and Pressure Vessel Code* is recognized as the authoritative standard for fabrication and repair of pressure vessels. Codes and standards are created and maintained by technical societies, private organizations, and industry trade associations.

A *technical society* is an organization made up of personnel having expertise in a particular subject and a common professional interest. For example, The National Board of Boiler and Pressure Vessel Inspectors, commonly referred to as the National Board or NB, promotes safety to life and property through uniformity in the construction, installation, repair, maintenance, and inspection of boilers and pressure vessels.

A *private organization* is an organization that develops standards from an accumulation of knowledge and experience with materials, methods, and practices. For example, the National Fire Protection Association (NFPA) and Underwriters Laboratories, Inc. (UL) are private organizations that serve to ensure safety of the equipment and systems used in commercial, institutional, industrial, and utility facilities. The American National Standards Institute (ANSI®) is a prominent private organization that helps identify needs for national standards. ANSI serves as a national clearinghouse and consensus-building organization for many standards applicable to boiler operation. For example, the ANSI standards for personal protective equipment (PPE) are recognized nationwide.

Insurance companies providing coverage of the facilities encompassing the boiler system often have additional requirements as well. For example, an insurance company may require that their insured facilities install specific controls or devices and that they be installed in a particular fashion.

A *trade association* is an organization that represents the producers of specific products. For example, the American Boiler Manufacturers Association (ABMA) is a trade association that represents manufacturers of commercial, industrial, institutional, and utility steam-generating equipment and auxiliary equipment. Such trade associations promote uniformity, safety, and environmentally friendly applications of the products and services of their members.

3 What is a confined space?

A *confined space* is a space that has limited or restricted means for entry and exit, has unfavorable natural ventilation such that a dangerous atmosphere could exist inside that does not naturally vent out, and is not intended for continuous employee occupancy. In addition, a confined space falls under the OSHA Permit-Required Confined Spaces standard when it may contain a hazardous atmosphere, when it may contain an engulfment hazard where a person could drown or be suffocated by material within the space, when it has an internal configuration which could cause entrapment or asphyxiation by inwardly converging walls or a floor that slopes downward and tapers into a smaller cross section, or when it may contain any other serious safety or health hazard. **See Figure 9-1.**

Boilers, deaerators, fuel oil tanks, coal bunkers, and a number of other spaces found in boiler systems meet the definition of confined space and therefore fall under the OSHA standard. The hazards of confined spaces that threaten boiler operators are most commonly those of hazardous atmospheres and the entry of hazardous fluids into the space.

Hazardous atmospheres that may be encountered include an oxygen-deficient atmosphere, combustible gases, and toxic vapors. An oxygen-deficient atmosphere may occur due to displacement of oxygen by leaking gases or vapors or through a combustion process in which oxygen is consumed and products of combustion are produced. The interior of a boiler is very often an oxygen-deficient atmosphere due to the use of nitrogen during boiler layup. Flue gas passages are often oxygen-deficient due to the oxygen being consumed during combustion. Oxygen-deficient air can lead to suffocation, causing death or severe injury. Hazardous atmospheres may also be created as a result of activities such as welding, painting, or cleaning with solvents within the confined space.

Hazardous atmospheres are eliminated by removing the source of any hazardous gases or vapors and thoroughly ventilating the space. Continuous ventilation may be appropriate in many cases. Confined spaces such as boilers also pose hazards in the form of contact with scalding steam, hot water, or other fluids that could enter the boiler while a person is inside.

To eliminate the hazards associated with entry into a confined space, a confined space entry permit procedure must be established and rigidly followed. The permit procedure is spelled out in the OSHA Permit-Required Confined Spaces standard. Substantial training is required for all boiler operators or other personnel who must perform work in confined spaces.

Figure 9-1. *Boiler room work involves entry into permit-required confined spaces such as boiler drums, deaerators, and breeching ductwork.*

4 What is the Control of Hazardous Energy standard?

The OSHA Control of Hazardous Energy standard is often referred to as the lockout/tagout standard. It is the standard related to preventing the release of hazardous forms of energy when such releases could harm personnel in the workplace. *Lockout* is the use of locks, chains, and other physical restraints to prevent the operation of specific equipment. *Tagout* is the use of a danger tag at the source of the hazardous energy to indicate to other personnel that the device is not to be operated until personnel working on the equipment have removed their lockout devices and the equipment is safe to operate. Stored hazardous energy may be present in the form of captive hydraulic pressure, suspended equipment that could fall by gravity, spring-loaded devices, etc.

What is the procedure for locking out a valve?

A valve may be disabled with a lock and chain, or other special lockout devices may be used. When locking the valve with a lock and chain, the valve should be firmly closed. A piece of chain is then run through the handwheel of the valve and around the pipe or around the valve yoke. **See Figure 9-2.** The chain is locked with a padlock belonging to the individual who will be performing the work on the system. A lockout tag is attached that states the date and the reason the equipment is out of service. The lockout tag system should identify the keyholders for all locks.

Special lockout devices prevent valves from being operated by disabling the operating means. For example, a special bracket may be locked onto a ball valve in such a way as to prevent the lever handle from being turned. Other special enclosures are available to prevent the handwheel of a gate or globe valve from being rotated. Even when a valve is locked in the closed position, a boiler operator should never assume that it holds tightly and allows no fluid to leak through. Piping connections after the valve should be opened with extreme caution.

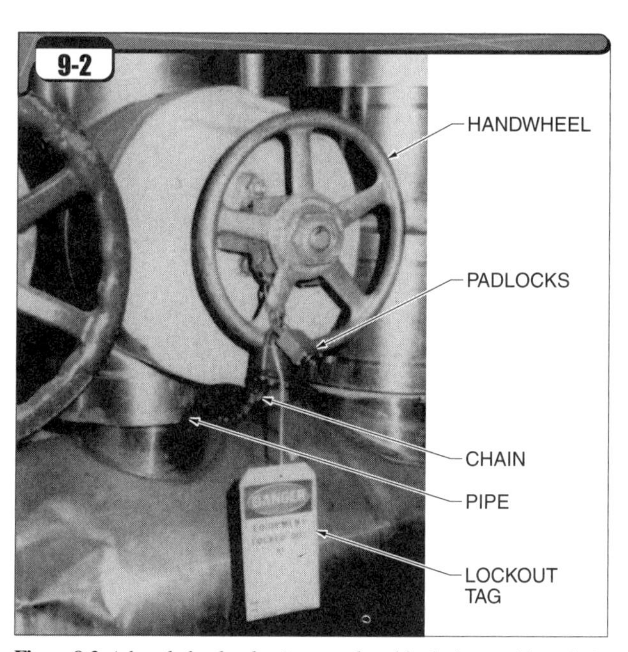

Figure 9-2. *A handwheel valve is tagged and locked out with a chain and padlocks.*

What is the procedure for locking out an electrically powered piece of equipment?

The disconnecting means (disconnect) for the overcurrent protection, such as fuses or circuit breakers, of a piece of equipment should be padlocked in the open (OFF) position. **See Figure 9-3.** Some facilities also require that the main fuses be pulled.

Personal safety locks must be installed by every person who will be performing work on the equipment. Tags stating the nature of the work being performed, the date, and the name of the keyholder of each lock should be placed on the breaker. Finally, the boiler operator should try to start the equipment to confirm that the power is OFF. Portable power tools and other devices that are plugged into electrical receptacles may be disabled using a lockbox.

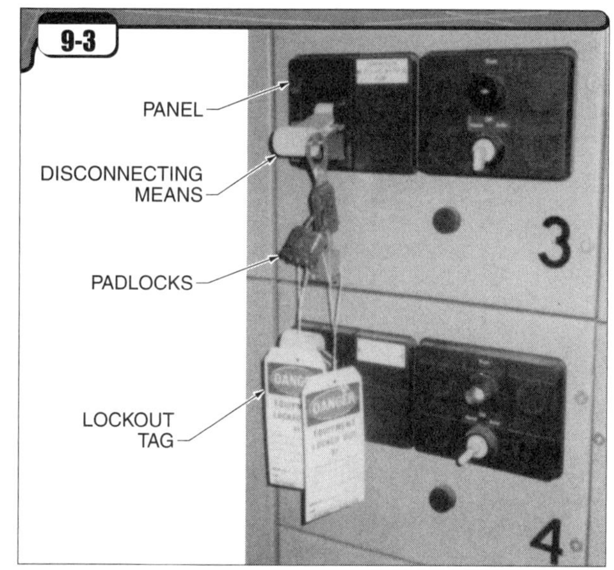

Figure 9-3. *Electrical equipment is tagged and locked out with padlocks and special devices.*

What is the Hazard Communication standard?

The OSHA Hazard Communication standard is the standard applying to the communication of important safety information to employees in the workplace. The Hazard Communication standard is also often referred to as the Right to Know (RTK) standard. A major focus area of the Hazard Communication standard is container labeling. Containers of any kind that contain hazardous materials must be labeled or marked with the identity of the material and appropriate warnings regarding the nature of the hazards it presents. Although the appearance of the labeling may vary between manufacturers, certain information is mandatory. This includes information related to health hazards, fire hazards, and toxicity. **See Figure 9-4.**

The NFPA Hazard Signal System is helpful in identifying the hazards associated with a given material. It uses a four-color diamond-shaped sign or placard to quickly communicate the degree of hazard severity for each of four categories. The four categories are health (blue), flammability (red), reactivity (yellow), and any special hazard (white). The relative severity of each of these hazards is also identified by a number from zero to four, with four being the most severe hazard.

The health hazard rating represents the likelihood of a material to cause temporary or permanent injury or incapacitation as a result of contact, inhalation, or ingestion. The flammability hazard rating indicates the degree of likelihood of the material burning. The reactivity hazard rating indicates the degree of likelihood of the material exploding or releasing energy by itself or by exposure to other substances or conditions. The special hazard rating identifies specific special properties and hazards of a material. For example, a material may react violently with water, it may be unstable or explosive when heated, or it may possess oxidizing properties. The special hazard rating is particularly valuable to emergency responders.

These hazard ratings are important in helping boiler operators determine what special work conditions may be necessary with specific materials, and what personal protective equipment (PPE) may be appropriate when working with or handling such materials.

Material safety data sheets (MSDSs) also provide boiler operators with important information about potentially hazardous materials encountered in the workplace. An MSDS is a set of printed materials used to communicate chemical hazard information from the manufacturer, importer, or distributor of a material to the employer. All chemical products used in a facility must be inventoried and have an MSDS.

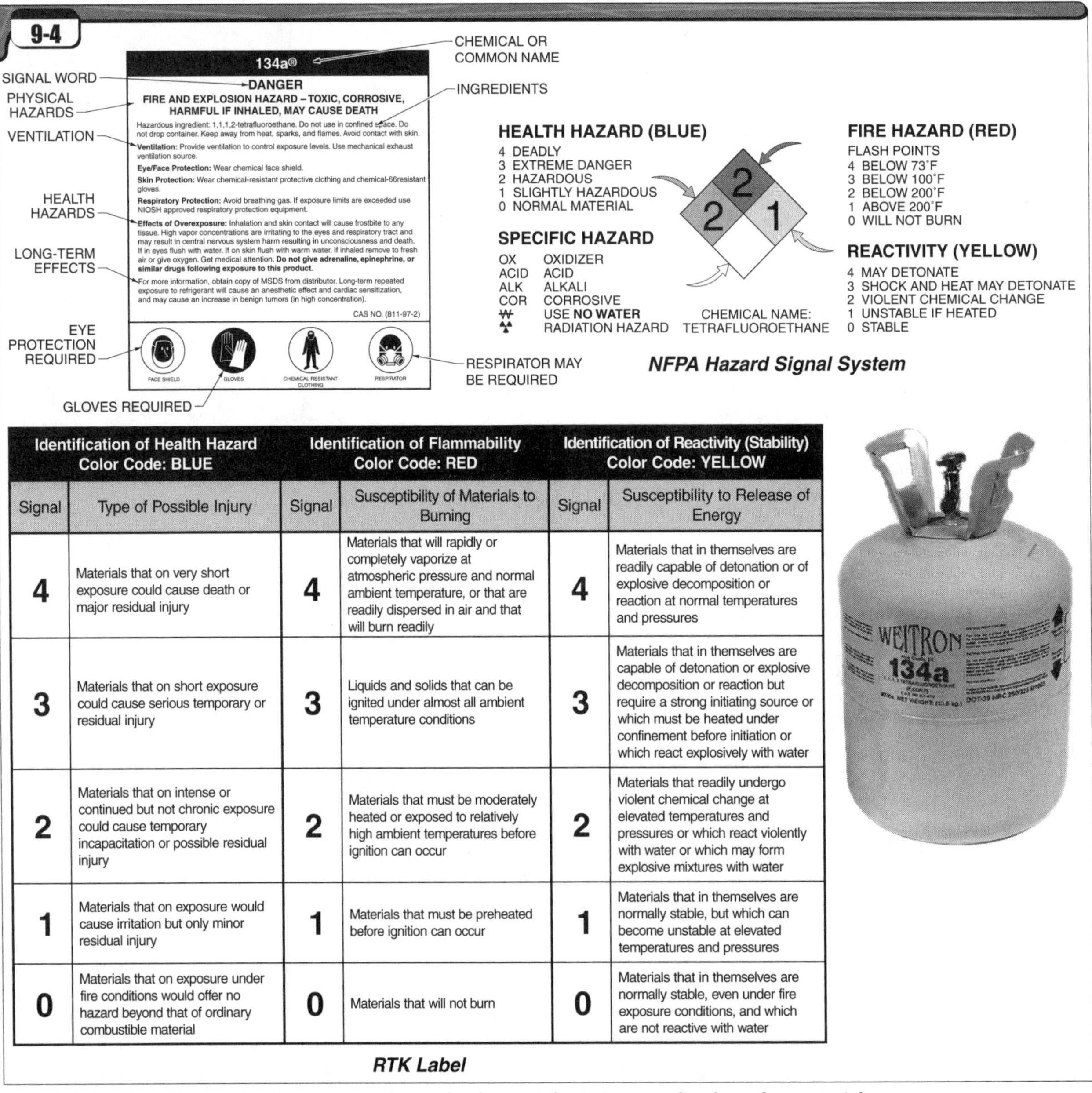

Identification of Health Hazard Color Code: BLUE		Identification of Flammability Color Code: RED		Identification of Reactivity (Stability) Color Code: YELLOW	
Signal	Type of Possible Injury	Signal	Susceptibility of Materials to Burning	Signal	Susceptibility to Release of Energy
4	Materials that on very short exposure could cause death or major residual injury	4	Materials that will rapidly or completely vaporize at atmospheric pressure and normal ambient temperature, or that are readily dispersed in air and that will burn readily	4	Materials that in themselves are readily capable of detonation or of explosive decomposition or reaction at normal temperatures and pressures
3	Materials that on short exposure could cause serious temporary or residual injury	3	Liquids and solids that can be ignited under almost all ambient temperature conditions	3	Materials that in themselves are capable of detonation or explosive decomposition or reaction but require a strong initiating source or which must be heated under confinement before initiation or which react explosively with water
2	Materials that on intense or continued but not chronic exposure could cause temporary incapacitation or possible residual injury	2	Materials that must be moderately heated or exposed to relatively high ambient temperatures before ignition can occur	2	Materials that readily undergo violent chemical change at elevated temperatures and pressures or which react violently with water or which may form explosive mixtures with water
1	Materials that on exposure would cause irritation but only minor residual injury	1	Materials that must be preheated before ignition can occur	1	Materials that in themselves are normally stable, but which can become unstable at elevated temperatures and pressures
0	Materials that on exposure under fire conditions would offer no hazard beyond that of ordinary combustible material	0	Materials that will not burn	0	Materials that in themselves are normally stable, even under fire exposure conditions, and which are not reactive with water

Figure 9-4. *The NFPA Hazard Signal System provides quick-reference information regarding hazardous materials.*

8 What is the Fall Protection standard?

The OSHA Fall Protection standard is the standard applying to the prevention of falls from areas such as ladders, scaffolds, catwalks, and elevated platforms. The standard applies to both fall prevention and fall arrest. Fall prevention is made possible through the use of railings, ladder enclosures, barriers, and other restraints. **See Figure 9-5.** Fall arrest is made possible through the use of full-body harnesses and lanyards that decelerate the fall and distribute the forces of the fall arrest throughout the body to prevent further injury. Waist belts are not acceptable as fall arrest devices because of their potential to cause serious spinal and organ damage if a fall occurs. Waist belts are only acceptable as a means of fall prevention.

When using fall arrest measures, the fall arrest equipment must be attached to an attachment point that is capable of arresting the forces created by the person's weight and the force caused by acceleration. Acceleration forces can be many times the person's weight. The OSHA Fall Protection standard stipulates that such attachment points be capable of supporting a weight of 5000 lb. The Fall Protection standard requires that appropriate fall protection and/or fall arrest measures be taken when working at an elevation of 6 ft or more. It also applies to potential falls from ground level to a belowground elevation, such as a pit.

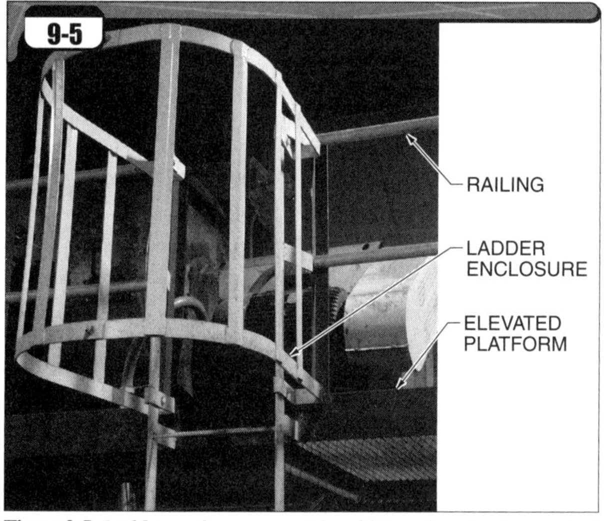

Figure 9-5. *Ladder enclosures provide additional protection against falls from elevated platforms and walkways.*

What is the Personal Protective Equipment (PPE) standard?

The OSHA Personal Protective Equipment (PPE) standard is the set of requirements governing the use of protective devices and equipment worn by a worker to prevent personal injury. All PPE must meet the requirements of the OSHA standard as well as applicable ANSI standards.

Protective headgear, such as hard hats or bump caps, must be worn where overhead hazards are present. When working around boiler-related equipment, overhead obstructions such as piping and conduits often create the potential for head injury. Safety shoes or boots with toe caps made of steel or other approved protective material are used to prevent foot injuries from falling objects.

The potential for eye injury is presented by flying objects, splashing of liquid materials, welding and grinding work, and many other situations. For basic protection against flying debris, for example, approved safety glasses are worn. However, safety glasses offer only very limited protection against splash hazards or blowing dusts. For these hazards, other eye protection such as a full face shield or chemical splash goggles must be worn. **See Figure 9-6.**

Heavy cotton or polyester gloves are suitable for keeping hands clean while touring the plant. However, gloves made of polyester or other synthetic materials should not be worn when the potential exists for the gloves to be exposed to extremely hot surfaces because the material may melt and stick to the skin. For protection from normal cuts, abrasions, splinters, etc., leather gloves should be worn. Neither fabric nor leather gloves protect the skin from hazardous liquids because these gloves absorb the liquids and hold them against the skin. Therefore, when handling water treatment chemicals or other materials that may be hazardous to the skin, appropriate chemical protective gloves should be worn. A number of specialty gloves may be needed for specific plant operations. These include hot work gloves suitable for handling hot thermocouples and other high-temperature objects, insulating gloves suitable for electrical work, and others.

Hearing protection is important in most boiler plants. Pumps, fans, steam turbines, air compressors, and other equipment emit noise that, over time, may cause permanent hearing loss. Hearing protection may take a number of different forms, including foam earplugs, molded earplugs, and ear muffs.

Respiratory protection may take many forms, depending on the respiratory hazard encountered. Respiratory protective equipment may be negative-pressure or positive-pressure. A negative-pressure respirator is one that requires the wearer to draw in the supply of air through any of a number of filters or cartridges that remove specific forms of contaminants from the air. Employees must be medically examined and approved to use a negative-pressure respirator because the respiratory system must work harder to overcome the resistance posed by the respirator. In addition, the specific respirator to be used must be fit-tested to the employee to ensure that the respirator is able to make a tight seal to the face and prevent inleakage of contaminated air. No facial hair that may interfere with the seal area of the respirator is permissible. Therefore employees must be clean-shaven in the area where the seal occurs.

A positive-pressure respiratory protective device is one that delivers a supply of purified air to the employee. Such devices include self-contained breathing apparatus (SCBA), powered air-purifying respirators (PAPR), supplied-air respirators, and loose-fitting supplied air hoods. Due to the many variables and regulatory requirements that apply to respiratory protection, an industrial hygienist or safety professional should oversee the plant's respiratory protection policy.

Protective clothing may be needed to protect the boiler operator from contact with heated materials, from chemical contact, from irritants to the skin, or from other hazards. For basic protection from dusts and dirt, disposable protective garments of paper-like materials or Tyvek® are used. For protection from skin irritants or chemical contact, specific materials of construction are necessary. For protection from hot surfaces or hot materials, Nomex® or other heat-resistant fabrics are appropriate. Aluminized insulated fabric pull-on sleeves are convenient when handling hot fuel oil guns or when working around steam piping.

PPE should be the last line of defense. Hazards should be removed through the use of engineering controls wherever it is reasonable to do so. For example, the first way a hazardous atmosphere should be managed is through changes that prevent the hazardous atmosphere from being created. The next way to manage a hazardous atmosphere is through the use of ventilation. When these methods are not feasible, the last method to protect people from a hazardous atmosphere is through the use of respirators. Handrails should be used to protect a location that presents a fall hazard. The least desirable way of protecting employees is having employees wear harnesses and lanyards.

Figure 9-6. *A face shield, apron, and gloves provide protection from chemicals used in a boiler room.*

What first aid procedure should be followed for a small steam or hot water burn?

The burn should be cooled immediately by placing it under cool water and keeping it there as long as practical. Immersion in cool water helps to ease the pain and quenches the burn by bringing the local temperature down. First-degree burns are accompanied by redness of the skin and minor pain. First aid ointments containing aloe are often soothing for first-degree burns. Second-degree burns are those resulting in more severe redness of the skin and blistering. Third-degree burns are those resulting in charring or penetration of skin tissue. Any second- or third-degree burns should be attended to by a medical professional immediately.

Why should accidents and/or injuries be reported immediately?

Failure to report an injury, no matter how small, may complicate insurance claims and result in additional suffering and inconvenience for the injured employee. Unreported injuries that later surface may also result in penalties to the employer from regulatory agencies due to failure to comply with recordkeeping regulations. A timely investigation of the accident or injury may also reveal important information that can be used to prevent a recurrence. Virtually all plants have some type of accident reporting procedure and form. These forms and procedures should be used so that a formal and timely record of the accident is made and future accidents may be prevented.

What are the three items that are necessary for combustion to occur?

Fuel, oxygen, and heat are traditionally known as the fire triangle. **See Figure 9-7.** If any one of these three elements is not present, combustion cannot occur. Combustion is the most common method of producing heat. In the combustion process, fuel is combined with oxygen from the air at a temperature high enough to start a chemical chain reaction. As heat is generated, more of the fuel is vaporized, the vapors ignite and generate more heat, and so on. Spontaneous combustion is the process where a material can self-generate heat until the ignition point is reached. This may occur in accumulations of oily waste rags or absorbent materials. Special receptacles are available for storing such materials to prevent spontaneous combustion from resulting in a fire.

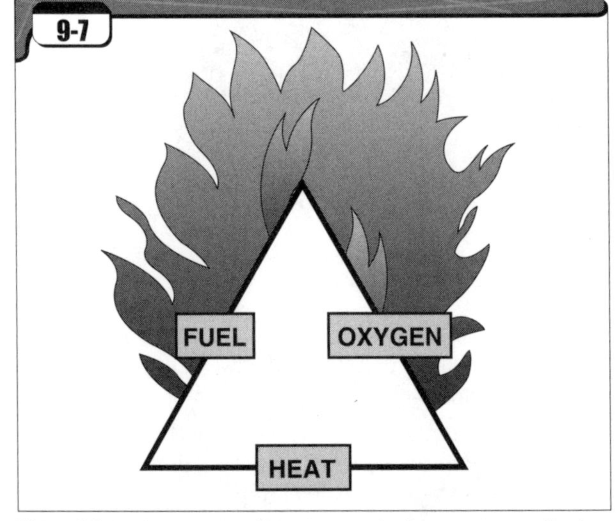

Figure 9-7. *Fuel, oxygen, and heat are required to support combustion. The fire will go out when any one of these is removed.*

13 What are the different classes of fires?

The class of a fire is determined by the type of combustible material involved. Fires are classified as Class A, Class B, Class C, Class D, or Class K. **See Figure 9-8.**

Class A fires involve common combustibles such as wood, paper, trash, textiles, and many plastics. Class B fires involve flammable liquids such as oils, gasoline, kerosene, paints, and greases. Class C fires involve energized electrical equipment such as transformers, motors, breakers, and conductors. Class D fires are rare specialized fires involving combustible metals such as titanium, zirconium, magnesium, sodium, and potassium. Class K fires involve commercial cooking grease.

Fire extinguishers are rated according to the class of fire they will put out. A fire extinguisher rated A:B:C can only be used on Class A, B, and C fires. Fire extinguishers rated A:B:C are charged with a multipurpose dry chemical. Fire extinguishers rated B:C can only be used on Class B and C fires. A B:C extinguisher is charged with dry chemical or CO_2. CO_2 leaves no messy powder to clean and does not obscure vision. A special dry powder is used to extinguish Class D fires and may be applied to the fire with a scoop or shovel.

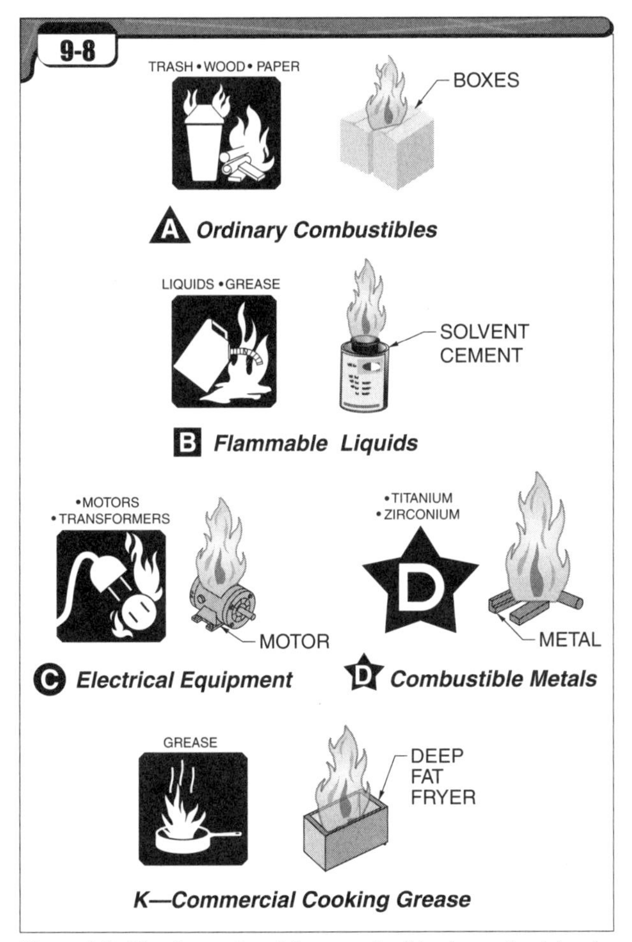

Figure 9-8. *The fire extinguisher required is determined by the combustible material to be extinguished.*

14 What are important safety rules that should be followed in a steam plant?

Do not wear loose-fitting clothing or jewelry around rotating equipment such as pumps or fans. Loose-fitting clothing or jewelry can become entangled in a rotating shaft, causing serious injury. Keep all coupling and shaft guards in place on rotating equipment at all times unless the equipment is properly locked out.

Do not climb on equipment, conduits, piping, or building steel. Use proper ladders or aerial lifts.

Assume that all piping is hot. Wear protective gloves, temporary aluminized sleeves, or other protective garments as appropriate to prevent burns. Place temporary insulation on any bare hot piping to prevent burns until the piping can be insulated properly. Before removing insulation to make repairs, confirm that the insulation does not contain asbestos. Specific abatement procedures must be executed by specially trained personnel if asbestos-containing insulation must be disturbed.

Double-check equipment for hazards such as loose flange bolts, uncovered shafts, or open drain valves before starting operation. Use two closed valves for protection rather than one when opening a pipeline wherever it is possible to do so. Vent between the valves where it is possible to do so. Use appropriately rated pipe, valves, and fittings in making any piping revisions or repairs. A minimum rule is to "replace in kind." If in doubt, contact the plant engineering department.

Do not place yourself in the line of fire. When working on a pipe or other equipment that could release unexpected energy, work in a position from which you have an escape route. Always drain and relieve pressure on piping before breaking joints and then open the joint so that any fluid that may blow out is directed away from you. For example, when opening a flanged joint, remove the bolts farthest from your body first. Remember that high-pressure steam leaks may be invisible for several inches or even feet. Use extreme caution near any steam leak. Be aware that piping may spring when taken apart. For example, if two pipe sections were pulled together so that flange bolts could be installed, the piping may suddenly spring apart when the bolts are removed. Similarly, piping that was aligned when cold may spring apart if a pipe joint is opened while the line is still hot.

Open manholes with extreme caution. Do not attempt to open a manhole on a boiler drum, deaerator, or other pressure vessel until it has been determined that no vacuum or hot water exists inside. Do not use pipe wrenches on valve handles to achieve greater torque. This may create sharp metal burrs on the valve handle, which could result in a hand injury. Use an approved valve wrench of minimal length if necessary. Always use a backing wrench when loosening or tightening pipe fittings to take the torque off the rest of the assembly. Otherwise, the piping may break. **See Figure 9-9.**

Always open steam valves very slowly. This practice is observed so that thermal strains on the downstream piping are minimized. This also allows the steam traps serving the downstream piping to drain condensate as quickly as it forms so that dangerous water hammer can be avoided. Large steam valves may take a half-hour or longer to open. Use equalizing valves where provided to warm up the downstream piping.

Use a "supervised warm-up" procedure for steam lines whenever feasible. Open a drain at the far end of the steam line before slowly admitting steam to the line. Allow the condensate to drain freely during piping warm-up. When live steam issues from the drain, the drain may be closed and the steam traps allowed to drain the condensate. This allows the large volume of condensate that occurs during line heat-up to drain.

Use handheld shields containing protective tinted glass or other proper eye protection when looking into furnace ports for any extended period to protect the eyes from infrared and ultraviolet rays. Use low-voltage droplights when working inside boilers. The entire boiler is a conductor. Low-voltage droplights minimize the risk of electrical shock.

Clean all spills immediately. Store all oily rags and other waste in approved containers to prevent fires due to spontaneous combustion. Empty these containers regularly. Survey the plant regularly for hazards. Establish and follow a procedure for regular inspection of safety equipment. Know the locations of all emergency showers and emergency eyewash fountains. Certain chemicals can cause serious skin and eye burns. Emergency showers and eyewash fountains should be installed and kept in good condition wherever these chemicals are used.

Figure 9-9. *A backing wrench should be used to prevent excessive torque on the adjacent piping assembly.*

15 What standards exist for color-coding and identification of plant equipment and piping systems?

OSHA requires that all industries color-code safety equipment locations, physical hazards, and protective equipment. Safety color codes were established by ANSI and adopted by OSHA for use in areas in which hazards exist or may exist. They are also consistent with those specified by the Compressed Gas Association for identification of piping containing compressed gases. Labels should be included to identify the contents of piping. The labels should also include arrows showing direction of flow as well as the normal pipeline pressures. ANSI A13.1-1981, *Scheme for the Identification of Piping Systems,* should be consulted for additional details regarding piping identification systems.

Red (white lettering on a red background) denotes the following:
- locations of fire protection equipment and apparatus
- piping for fire-quenching materials, including water, foam, and CO_2
- denotation of danger, such as portable containers of flammable liquids
- stop applications such as emergency stop buttons or bars on powered equipment

Orange denotes the following:
- dangerous parts of machines or energized equipment that may cause injury
- an intermediate level of hazard
- biological hazards (fluorescent orange or orange-red)

Yellow (solid yellow, yellow with black stripes, yellow with black checkers, yellow with a suitable contrasting background, black lettering on a yellow background) denotes the following:
- caution applications representing physical hazards including striking against, falling, tripping, or "caught between"
- piping for hazardous materials, including piping containing materials that are flammable, explosive, chemically active, toxic, or at extreme pressure or temperature

Green (black lettering on a green background) denotes the following:
- locations of safety equipment other than firefighting equipment including emergency respirators, stretchers, and protective clothing
- piping of liquids of low hazard

Blue (white lettering on a blue background) denotes the following:
- caution applications to warn against starting, use of, or movement of equipment being repaired
- piping of gases of low hazard

Purple (yellow lettering on a purple background) denotes the following:
- Radiation hazards including gamma rays, X rays, or several other types of radiation hazards

Black, white, or combination black/white (black and white stripes, black and white checkers, or other combination) denotes the following:
- traffic and housekeeping
- stairways and risers
- directional indicators

Normal Operating Duties

What are the boiler operator's first duties upon relieving a shift?

A boiler operator should review the information from the previous shift and discuss any important events or conditions from the previous shift with the boiler operator being relieved. In addition, the operator's logbook should be reviewed for important information. A boiler operator should check the water level of all on-line boilers and make sure that all boilers have enough water in them by blowing down the water column and gauge glass and watching the water level return in the gauge glass. **See Figure 9-10.** A boiler operator needs to tour the steam plant for the first time as early in the shift as possible and check the operating condition of all equipment.

Figure 9-10. *The water should return to the gauge glass immediately when the water column or gauge glass blowdown valve is closed.*

Why should all boiler operators maintain a boiler room logbook?

Boiler plants are almost always rotating-shift operations, and maintaining good communications in a rotating-shift environment is always a challenge. A logbook is used to ensure that important information is passed on to the relieving shift and to subsequent shifts. This may include information regarding equipment that has been removed from service for maintenance work, equipment testing that is in progress, or any other critical information.

In addition, a well-maintained logbook serves to provide very helpful historical and troubleshooting documentation. For example, the logbook can reveal how long a particular operating problem has existed, what the reasoning was for certain actions that were taken, and how effective a modification has proven to be. Note that if the log is maintained and saved in a computer file, it will provide the additional convenience of searching for key words when researching the history of an operational problem.

How should a boiler operator blow down the water column and gauge glass?

The water column should be blown down first to prevent mud or debris in the column or in the piping to the boiler from being carried into the small piping passages to the gauge glass. Mud or debris can block the free flow of water into the gauge glass and give false water level readings. If the water column also houses a low water fuel cutoff switch, a test switch, or "dead man switch," is usually provided. **See Figure 9-11.** This spring-loaded pushbutton switch must be held in while blowing down the water column and gauge glass in order to keep the low water fuel cutoff switch from sensing a low water condition in the water column and shutting the boiler down.

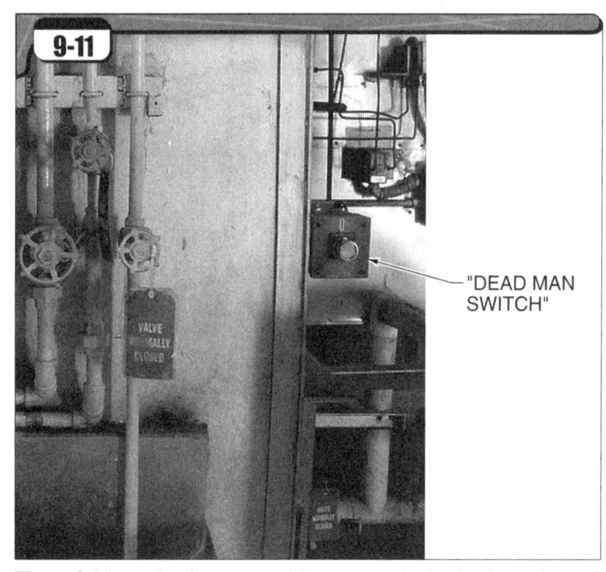

Figure 9-11. *A "dead man switch" prevents the boiler from shutting down while the water column is blown down.*

How is a low water fuel cutoff switch tested?

If possible, a low water fuel cutoff switch should be tested when the boiler burner is on low fire in order to prevent thermal strains on the boiler. A *quick-drain test* is a test that empties the float chamber or electric probe chamber while the burner is firing in order to test the low water fuel cutoff switch. The operator needs to confirm that the fuel shutoff valves close and the burner shuts down when the water level drops in the float chamber. The low water fuel cutoff switch needs to be checked at least daily. In a continuous 24-hour operation, the low water fuel cutoff switch should be tested on each shift. If the boiler has redundant low water fuel cutoff devices, the primary device should be disabled temporarily after testing so that the secondary device may be tested.

The low water fuel cutoff switch should be tested monthly with the evaporation test, or slow-drain test. An *evaporation test*, or *slow-drain test*, is a test that allows a low water condition to occur in the boiler in order to test the low water fuel cutoff switch. This can be done by closing the feedwater valve and allowing the water level to drop as the water is evaporated. If a makeup water feeder is present, it must also be secured.

The operator needs to confirm that the fuel shutoff valves close and the burner shuts down when the water level in the gauge glass falls to the cutoff point. The cutoff point should be set at approximately the level of the lower try cock. The burner must shut off while the water level is still visible in the gauge glass. If the burner does not shut off as the water level approaches the lowest visible point in the gauge glass, the low water fuel cutoff switch is defective. In this case, reopen the feedwater valve to prevent the water level from falling below the visible range of the gauge glass. The boiler should then be removed from service and the switch repaired. If it is not possible to remove the boiler from service immediately, a person should be assigned the sole duty of monitoring the water level until such time that the boiler can be shut down.

The evaporation test is generally considered the more reliable of the two tests because float switches and other moving devices are more likely to stick when the water level is reduced gradually. This is the same condition that would be encountered if the source of boiler feedwater were lost. In jurisdictions where licensed boiler operators are required, the evaporation test is commonly used to determine how long the boiler operator may be away from the boiler room. The test begins with the water level at the NOWL. The length of time before a low water cutoff condition is encountered is the maximum time that the boiler operator may be absent. As a worst-case scenario, some jurisdictional inspectors may require that this test be performed with the boiler at a high firing rate.

20 How is a flame-detection system tested?

A flame-detection system is tested by removing the flame scanner and covering it while the burner is firing. The burner management system (BMS), or flame safeguard control, should cause the fuel valves to close and shut down the burner when the flame scanner can no longer detect the flame. The detection of flame loss and valve closure should happen within 4 sec. If this does not occur, the boiler should be removed from service immediately and the cause of the device failure remedied before restarting. It is recommended that the flame scanners be tested at least weekly. After testing, the flame scanner should be cleaned and reinstalled. Pressing the reset button then resets the BMS and the boiler may again be fired. **See Figure 9-12.**

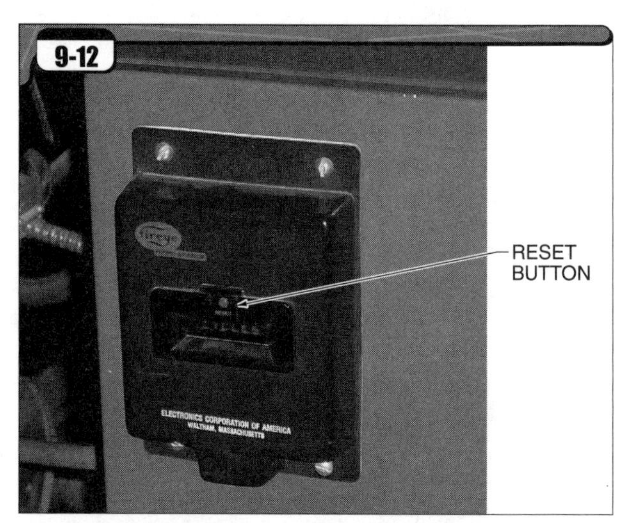

Figure 9-12. *The burner management system must be reset after testing the flame scanner.*

21 How is a safety valve tested?

On low- and medium-pressure boilers, a safety valve is tested by pulling the manual lever on the valve to ensure that the valve opens properly. The valve is opened with the test lever for only a second or so and then released. **See Figure 9-13.** The safety valve should close immediately. Manual testing should be scheduled on a regular basis, such as monthly, in facilities where it is feasible to do so.

The safety valve may also be tested by actually raising the pressure in the boiler to the point where the valves should pop. With gaseous, liquid, and pulverized coal fuels, automatic pressure controls will have to be bypassed or adjusted beyond normal limits to perform this test. Extreme care must be taken when doing this, and specific written procedures should be developed. In facilities where this test is performed, it is normally an annual test.

On many very high-pressure industrial and utility boilers, the safety valves are only tested during boiler outages. In these cases, testing is either performed by sending the valves out to an authorized safety valve service shop for testing and recalibration or by having authorized shop personnel hydraulically test the valves in place on the boiler. In the latter case, a hydraulic test apparatus is installed on the valve and the exact hydraulic pressure required to lift the valve is measured.

Trade Tip

A boiler room logbook, if properly used, becomes a valuable source of troubleshooting information.

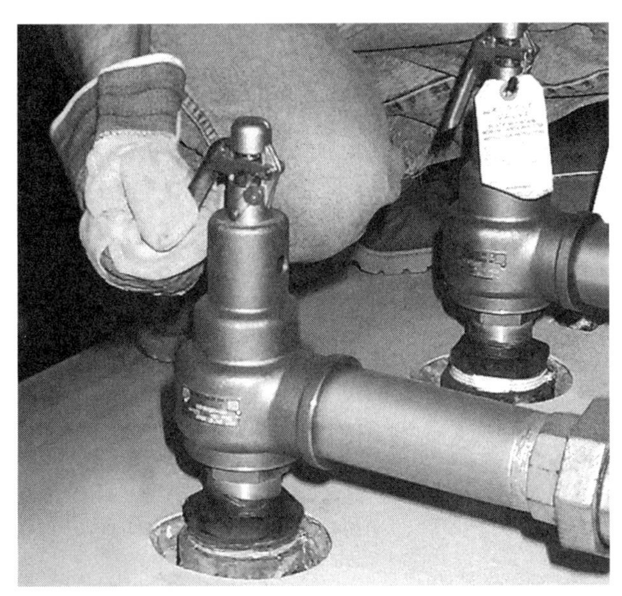

Figure 9-13. *A safety valve is tested by lifting the manual lever and allowing steam to flow through the valve.*

List additional items that should be watched or checked while the boiler plant is in normal operation.

The operator should observe the feedwater regulating system on each boiler to confirm that it is maintaining the NOWL. The percentage of condensate return should be calculated daily. The operator should observe the condition of the fires in the furnaces periodically for proper atomization, normal flame color and shape, soot accumulation, or flame impingement. The operator also needs to lubricate equipment according to the plant preventive maintenance program or as needed to ensure protection of moving parts.

Draft pressures and flows should be maintained at the values required for proper combustion and adequate removal of the gases of combustion without losing excess heat up the stack. Steam header pressure should be kept at a normal, constant value through proper regulation of the fuel firing rate. The flue gas outlet temperatures should be monitored regularly. Oil burner tips should be cleaned of debris on a regular basis to prevent carbon and ash buildup from causing poor atomization and a resulting loss of combustion efficiency.

The fuel supply should be checked for availability, proper pressures and flows, any evidence of leaks, etc. For coal-fired boilers, the coal conveying system should be inspected. Coal feeding systems should be monitored for plugging, especially when the coal is wet or fine. On heavy oil systems, the boiler operator should also confirm that the steam tracing system is maintaining the minimum oil temperature needed to ensure flow through the piping and that any strainers are clean.

The stack analysis of the flue gas and stack appearance should be checked several times during each shift to ensure that pollution control requirements are met. The operating condition of flue gas cleaning equipment should be evaluated and adjustments should be made as needed to obtain the highest feasible efficiency. Electrostatic precipitator rappers should be run through a test cycle daily to confirm that all are in working order. Filter baghouse cleaning systems should be run as frequently as needed to keep the differential pressure across the bags from exceeding the manufacturer's specified limits. Soot blowers or other tube cleaning equipment should be operated as necessary and observed for proper operation, leaks, or mechanical problems.

All rotating equipment, such as pumps and fans, should be checked for proper bearing temperatures, correct belt tension, correct seal flush or packing flush flows, abnormal noise, and excessive vibration. Makeup water, condensate, fuel, and steam flowmeter readings should be taken at specific times as determined by the plant supervision. These readings are used to monitor plant efficiency and to reveal developing problems. **See Figure 9-14.**

Figure 9-14. *The makeup water usage is metered and compared with the steam production over the same 24-hour period. This information is used in determining the percentage of condensate return.*

What water analyses should the boiler operator perform during the shift?

The water quality specifications and the water tests that are appropriate vary substantially from plant to plant because of differences in the steam applications. Although each plant differs, there are certain water tests that are appropriate for almost every boiler plant. These include raw water, softened makeup water, feedwater, condensate, and boiler water tests. **See Figure 9-15.**

The softened makeup water should be tested frequently. A common test is hardness. A boiler operator may also periodically test the conductivity before and after the softener to ensure that the brine is thoroughly rinsed out. In addition, a boiler operator should test for the quantity of iron in the softened water. The iron content should be 1 ppm or less.

Boiler operators also need to perform tests on the feedwater going to the boiler. Typically the feedwater should contain about 3 ppm to 6 ppm of sodium sulfite. Feedwater is a blend of condensate and makeup water. Therefore the quantity of hardness in the feedwater should be less than in the makeup water. A typical hardness reading in the feedwater should be about 0.5 ppm or less. A typical range for feedwater conductivity in medium-pressure boilers would be about 100 µmho to 500 µmho. Often the M alkalinity of the feedwater is tested as well.

The condensate should be tested before it is added back to the boiler. The pH of the condensate should be approximately 7.5 to 8.5. The condensate conductivity should be very low at about 5 µmho to 50 µmho. The quantity of iron should also be periodically tested. Boiler operators should periodically test the water in the boiler. The amount of sodium sulfite in the boiler water is tested to confirm that a residual of about 30 ppm to 60 ppm of sodium sulfite is maintained in the boiler water. A boiler operator should also run P alkalinity or OH alkalinity tests to ensure that a residual of about 300 ppm to 500 ppm of caustic soda is maintained. An operator should test the conductivity or TDS to ensure a typical range of about 2500 µmho to 3500 µmho. Finally, the boiler operator may test for the specific scale control agent being used in the boiler.

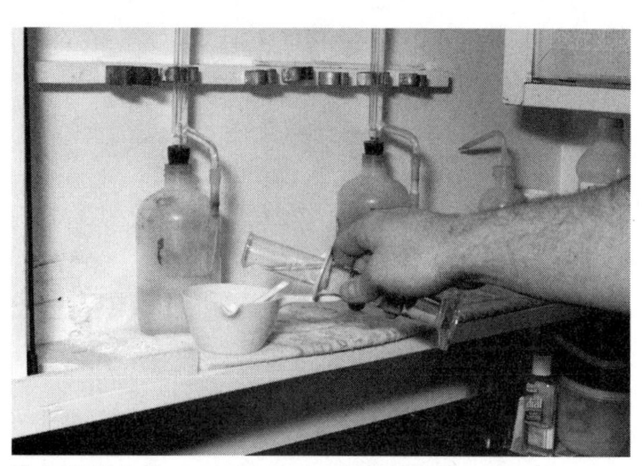

Figure 9-15. *Boiler water is tested to determine the required frequency of blowdowns and the chemicals needed.*

24 What procedure should be followed for performing a bottom blowdown?

The boiler operator should check the boiler water level before performing a bottom blowdown to ensure that adequate water volume exists in the boiler. In addition, the boiler operator should have a direct line of sight or a mirror to watch the gauge glass during the blowdown or have an assistant watch it.

If the boiler is equipped with two blowdown valves, the valve closest to the boiler should be opened first and closed last. The valves should be opened at a slow and deliberate rate until fully opened. Allow 3 sec to 5 sec with the valves fully open and then close the valves. This allows the outer valve to take the wear. Closing the valve closest to the boiler last also ensures that the piping between the valves is full of oxygen-free, chemically treated water at all times. If the blowdown valve manufacturer calls for a different procedure, follow the manufacturer's instructions.

25 What precautions should be taken when pulling bottom ash?

Pulling bottom ash is the process of performing the activities required to remove the ashes that accumulate in the bottom of a pulverized coal-fired furnace, ashes discharged from stoker equipment, or ashes from another solid fuel-fired unit. **See Figure 9-16.** The design of many boilers necessitates that these ashes be removed manually from the ash hoppers. Generally the boiler operator uses a large metal hoe to pull the ashes into a pneumatic ash removal system, a grate-covered auger system, or an ash sluice. A *sluice* is a trench through which water flows rapidly to carry away solid materials.

The primary concerns when pulling bottom ash are to prevent burn injuries and to avoid upsetting the furnace draft. Injury is possible any time the boiler operator is exposed to the furnace area. Control upsets, fan motor overloads, or any of a number of other problems could cause the furnace pressure to become positive without warning. For this reason, the boiler operator should wear appropriate personal protective equipment (PPE). This should include fire-retardant clothing, Nomex hood, face shield, and suitable protective gloves. It is also important to remain to the side of the furnace opening.

To avoid upsetting the furnace draft, the ash hopper door or other furnace opening should be opened very slowly. This gives the automatic control system time to respond by adjusting the forced draft and induced draft fan dampers. The boiler operator at the controls should also be notified prior to opening any furnace access door so that control adjustments may be made. Furnace access doors should never be opened while soot blowers are in operation.

Figure 9-16. *Boiler operators must use great care when pulling bottom ash from a solid fuel-fired boiler.*

Is it accepted industry practice to blow down the boiler from the waterwall headers?

Blowing down the boiler from the waterwall headers should not be done. This can disrupt the circulation in the waterwalls, and hot spots can result. The waterwall tubes are designed for a certain amount of water flow from natural circulation. If this flow is slowed or stopped, the movement of the water across the heating surfaces may be insufficient to carry away the heat and the tubes could overheat. A few older boilers have large waterwall tubes that provide adequate water circulation to allow this. The boiler manufacturer should be consulted for specific approval of this practice, however.

27 What precautions should be taken when blowing down a boiler that is in battery with other boilers?

Boilers in battery often use common blowdown headers. If one or more of these boilers is open for inspection and its blowdown valves were not locked out, it is possible that flash steam and hot blowdown water from the boiler being blown down could enter the boiler that is open. The blowdown valves on the boiler being inspected should be securely closed, locked, and tagged, in accordance with the OSHA Permit-Required Confined Spaces standard.

For extra safety, the blowdown valve closest to the idle boiler (upstream blowdown valve) may be removed and the blowdown line capped off with a blind flange so that it is completely disconnected from the boiler. **See Figure 9-17.** Thus, no flash steam can enter the idle unit through the blowdown. These lockout precautions must be taken on the bottom blowdown and continuous blowdown lines.

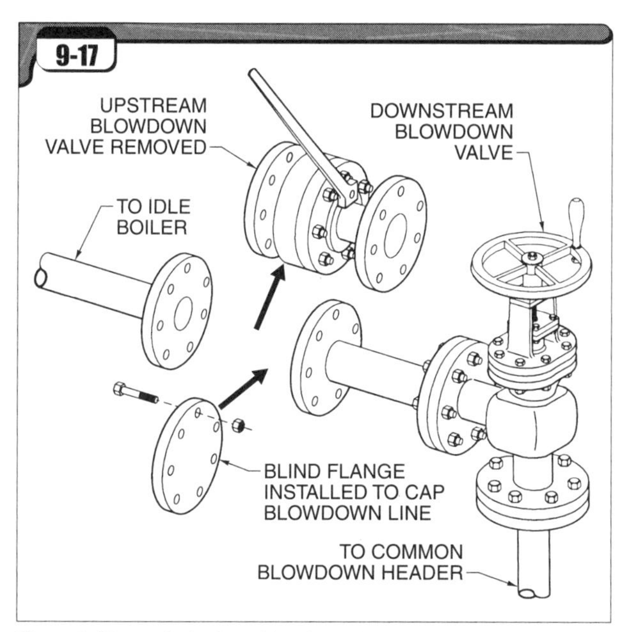

Figure 9-17. *An idle boiler's blowdown line is capped off with a blind flange for extra safety when a common blowdown header is used.*

After opening a gate or globe valve fully, why is it good practice to close it 1 to 1½ turns?

Back-seating is the situation where a slow-opening valve such as a gate or globe valve is fully opened until it stops. If the valve is left back-seated, thermal expansion may cause the valve to bind and make it very difficult to dislodge from the back-seated position and close again. By rotating the valve slightly off the back-seated position, the boiler operator ensures that the valve will not bind.

What is the boiler operator's duty upon leaving the shift?

The boiler operator being relieved is responsible for passing on important information to the next boiler operator concerning the boilers and auxiliaries. This information should be recorded in a boiler logbook and explained verbally. It is important that the relieving operator be informed about disturbances in the operation during the previous shifts, boilers open for inspection, control problems, and other current issues.

Boiler Plant Maintenance

How should a tubular gauge glass be cleaned?

A bottlebrush with a wooden or plastic core is typically used to clean a tubular gauge glass. Alternatively, a piece of rag can be pushed through the glass with a wooden dowel or plastic rod. A metallic object such as a bottlebrush with a metal core or a rag pushed by a piece of metal rod should not be used. The metal can easily scratch and score the glass and the steam and hot water may cause the glass to crack at this score mark. New gauge glass washers should be used whenever possible. Used washers will seldom seal properly and the torque required on the gland nuts to produce a seal will place undue stress on the glass.

What procedure should be followed when replacing a broken tubular gauge glass?

The gauge glass is isolated by closing the top and bottom valves. **See Figure 9-18.** The blowdown valve is opened to drain the remaining water in the gauge glass assembly, and the protective shield, if present, is removed. Any pieces of glass left in the packing glands can be blown out by cracking open the steam and water connections. The old gauge glass washers should be completely removed as well. The boiler water level can be monitored with the try cocks while the gauge glass is being replaced.

A new gauge glass can be cut with a gauge glass cutter so that it is about ¼″ shorter than the distance from the stop inside one valve to the stop inside the other valve. This will allow room for inserting the glass and space for the glass to expand as it heats up. Before installing a new gauge glass, some boiler operators recommend using a propane torch to heat-polish sharp burrs from the newly cut ends. This helps prevent the rough edges from cracking and causing premature failure. The packing gland nuts and new washers are placed on the gauge glass and the gauge glass is installed. The packing gland nuts should not be tightened more than hand tight plus a quarter turn with a wrench.

The blowdown valve from the gauge glass needs to be fully opened and the steam valve cracked open. The flow of steam through the gauge glass and down the blowdown line will heat the gauge glass evenly and prepare it for the thermal stress from the hot water. After warming the gauge glass for a few moments, the water valve to the gauge glass is opened and the drain is closed. If the packing glands leak, the isolation valves can be closed and the packing gland nuts tightened slightly. A few adjustments may need to be made until the leaking stops. The packing gland nuts must not be overtightened. If the gauge glass has a protective shield, it needs to be replaced.

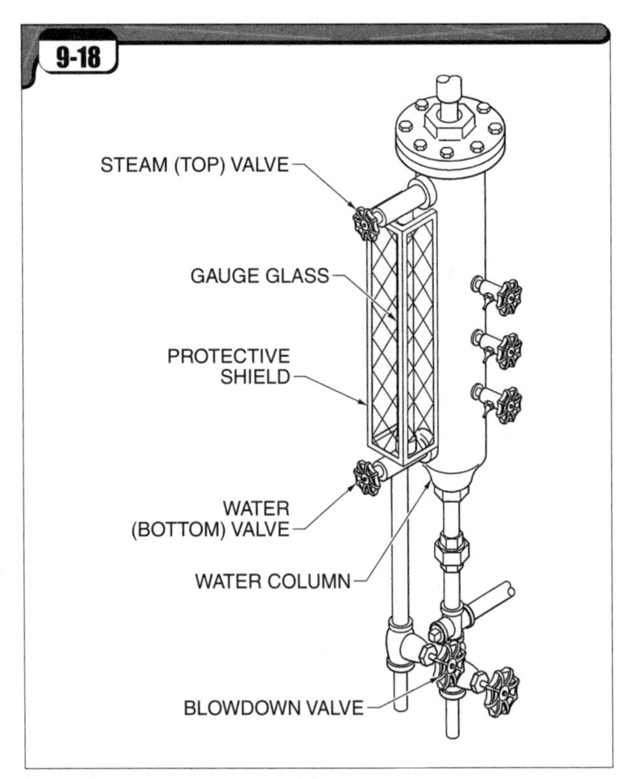

Figure 9-18. *Gauge glass protective shields minimize the danger of spray from hot water or steam if the gauge glass breaks.*

32 How should a leaking gauge glass washer be repaired?

The gauge glass needs to be isolated by closing the valves. The bottom valve should be closed first because the condensation of the steam in the glass will cause additional water to be pulled in from the boiler. The leaking packing gland nut can be tightened a small amount to stop the leak and the gauge glass can be put back into service to see if the leak has stopped. If this procedure does not stop the leak after a couple of tries, the washers need to be changed.

The packing gland nuts are hexagonal (six-sided) nuts. A flat is one of the sides on a nut or bolt head. When adjusting the packing gland nut, it should be tightened about ½ to 1 flat at a time. Excessive tightening will stress the gauge glass, and it may break after being returned to service.

33 Why is it important to have a preventive maintenance program in place in a boiler plant?

Preventive maintenance is the practice of performing maintenance activities on a piece of equipment to prevent breakdowns from normal or predictable causes. This includes activities such as replacing the lubricant in bearings, readjusting couplings and shaft alignments, cleaning filters, etc.

An enhanced form of preventive maintenance is predictive maintenance. *Predictive maintenance* is a study of the history of the plant components and determination of the expected service life of critical components. For example, the bearings or another part of a plant component may be replaced even before significant problems appear because history has shown that the end of that component's service life is near. This is especially appropriate when failure of that component may result in damage to other components, as is the case when failure of a bearing may also damage an expensive shaft.

The opposite of preventive or predictive maintenance is breakdown maintenance. Breakdown maintenance is simply responding to maintenance problems as they arise, with little or no effort made to avoid or plan for them. It is a poor approach to maintenance management.

34 How should equipment documentation, such as manufacturer's literature, be stored?

Manufacturer's literature and vendor literature in binders should be arranged on a bookshelf in alphabetical order by the manufacturer or vendor name. Loose materials such as small manuals and historical documentation should be arranged in hanging file folders in a filing cabinet sorted by manufacturer or vendor name. It is not good practice to sort the documentation by the equipment name, such as feedwater pumps and the deaerator, because certain equipment may not be called by the same name by everyone in the department.

Equipment information may also be filed by equipment number. Some facilities assign an equipment number to every piece of equipment in the facility and permanent tags are placed on each piece of equipment. However, some equipment may be provided as a package by a certain manufacturer, such as a chemical day tank, agitator, and metering pump, and it may not be feasible to separate the literature. Cross-referencing by equipment number, manufacturer name, and equipment name is helpful through the use of available computer software designed for tracking inventory.

Blueprints and sketches should be stored in hanging blueprint file racks or flat blueprint file drawers. They should be sorted first by type of blueprint (electrical, mechanical, or HVAC) and then by plant area and vendor.

Troubleshooting Techniques

35 Describe techniques that will help a new boiler operator develop good troubleshooting skills.

Though it may seem overly simplistic, the first question that should be asked when troubleshooting any mechanical device is whether it ever worked correctly. For example, if the component or system was working fine for the last six months, there is almost no chance that there is a problem with the design. There is also little or no chance that the problem lies in the equipment installation, assuming that the installation has not been changed.

If the device previously worked well, the next step is to consider what has changed since then. For this reason, good troubleshooting skills are based on normality. If the boiler operator doesn't know what conditions are normal, it will be nearly impossible to recognize conditions that are abnormal. Thus, troubleshooting can usually follow a predictable sequence as follows:

- identifying the equipment or system which is not operating normally
- identifying the nature of the abnormality
- determining the likely causes of the abnormal condition
- taking appropriate steps to eliminate each cause

In order to understand what equipment conditions are normal, the boiler operator should make an effort to become familiar with and list these conditions. In some cases, this is addressed through routine log readings. It is often a good idea to mark or label gauges, linkages, and other equipment to indicate their normal position or range of position. Some boiler operators mark damper and valve linkages with paint after they are set so that it will be apparent if the linkages have slipped or have been changed. Before making operational adjustments in a trouble condition, the indications and available data should be confirmed. For example, a pressure gauge or gauge glass indication is useless if the gauge or gauge glass is isolated from the system by a closed valve.

Before working on the equipment, it should also be determined whether the problem condition is a cause or an effect. For example, if a feedwater pump is cavitating, it could be that the pump's suction valve is not fully open. In addition, the water level in the deaerator could be low, the water temperature may be higher than normal, or the recirculation line may be shut off.

Factoid

Pride and ego may make some boiler operators believe that written procedures and checklists are unnecessary. This is a dangerous delusion.

36 Why should the boiler plant use written procedures and checklists?

Written procedures help ensure that important activities in the plant operation are performed correctly, consistently, and safely, and that critical steps are not missed. **See Figure 9-19.** Checklists also provide a record that may be helpful in later troubleshooting scenarios.

Some boiler operators discount the value of written procedures and checklists because they believe that their plant experience makes the procedures and checklists unnecessary. However, memory lapse and mistakes are always possible, and fatigue experienced on rotating shifts and overtime makes them more likely. The written procedures should be posted near the equipment when it is appropriate to do so.

Figure 9-19. *Severe boiler damage may result from failure to establish and follow written procedures.*

37 Why are new boiler system startups often very difficult to manage?

A great many small design mistakes and shortcomings, omissions, and improper installations always manifest themselves in a new facility startup. These include items such as check valves installed backwards, piping being undersized, and access to elevated components being unsafe. For these reasons, a new facility startup should follow a planned sequence of commissioning, shakeout, and normal operation.

The commissioning step is the most important. *Commissioning* is the process of inspecting, preparing, and testing each major component of a system prior to operation. For example, piping debris such as gravel, pipe cutting oils, and shavings should be flushed from new piping before the piping is placed into service. Even then, temporary strainers may be installed because such debris will damage critical components like control valves. Control devices such as float switches are often shipped to the site with packing materials inside to keep the floats from being damaged during transit. If these materials are not removed prior to service, the devices will not work properly. Gaskets and packing should be inspected for leakage. Wiring terminations need to be checked to confirm that they have been tightened. Damper and valve controls should be calibrated and stroked to confirm that the control devices function as planned. Interlocks should be tested to confirm that they work properly to ensure the safety of the equipment and the plant personnel. The commissioning step deserves careful planning, and it should be assumed that everything is improperly installed until it is carefully scrutinized and tested.

A *shakeout* is the process of operating a new system as needed to expose and correct the major impediments to reliable operation. During a shakeout, the new process is usually operated under low load, and it must be understood by the rest of the facility that the new process may be unreliable for a period of time. As the significant problems are overcome, the new process is operated under greater loads and for longer periods of time until normal operation is realistic. Small problems are addressed on a priority basis.

Trade Tip

Boiler plant management should ensure that all changes made to plant systems during the commissioning and shakeout are reflected by approved changes to the plant prints, especially the P&IDs. Although this is a significant investment of time, it will pay big dividends for many years in reduced troubleshooting time. These updated drawings are called as-built drawings.

Minor Upsets

What is swell in a boiler?

Swell is the rise in the boiler water level that occurs when the steam load on the boiler is increased or when the steam pressure drops. When the steam load is increased, the pressure in the boiler drops slightly. The drop in pressure causes a portion of the water in the boiler to flash into steam bubbles, and these bubbles displace the water around them. This results in an overall rise in the water level as these bubbles occupy more space than the water. When the combustion equipment's firing rate is increased in response to the pressure drop, the increased heat transfer causes additional steam bubbles to form in the water. These bubbles displace more water and cause the water level to rise further. Swell also occurs in a slower way when starting up a boiler. In this case, the water level rises as the comparatively cool water is heated and expands in volume slightly. When starting a cold boiler, the boiler should be filled to a level that is about 1″ to 2″ below the NOWL. The exact water level is dependent on the specific boiler. The water will swell when it is heated and as steam bubbles begin to form. If too much water is put into the boiler initially, part of it may have to be drained later.

How can a high water condition develop while a boiler is offline?

A boiler may become flooded if the main feedwater regulating valve or the bypass around it leaks through when the boiler is idle. This condition should be preventable by closing a manual valve in the feedwater line after the regulating valve, but any automatic or manual valves that leak through should be repaired as quickly as feasible. It is also good practice for a facility to use a warning tag system to identify valves or controls that are not in their normal positions. A vacuum condition inside the boiler will contribute to flooding of the boiler. For example, the boiler may develop a vacuum inside if the boiler vent is not opened as the pressure dissipates after shutdown.

What happens to the water level if the fires go out?

The sudden removal of the source of heat causes the water to settle down. As the heat transfer rate drops, fewer steam bubbles form in the water. The water settles down when the bubbles are no longer generated. This settling can cause the water level to drop to the low water alarm point or even below the low water fuel cutoff point.

Emergencies and Dangerous Situations

If a boiler has the proper water level, how might priming occur when the boiler comes online?

If there is a relatively large difference in pressure between the boiler and the steam header when the nonreturn valve opens, the sudden surge of steam from the boiler can entrain slugs of water. **See Figure 9-20.** This can lead to water hammer. This surge may occur if the nonreturn valve sticks. The nonreturn valve should be properly maintained so that it opens smoothly and puts the boiler online as gently as possible.

How may priming occur during boiler operation?

Priming commonly occurs as a result of boiler water contamination. Priming may also occur if a large steam demand is suddenly placed on an operating boiler. The boiler should be allowed to increase steam production slowly so that the water level and the steam header pressure may stabilize. This is a significant problem with smaller package boilers. These units often have no internal steam separation equipment. In addition, the water level is often close to the steam outlet, and the steam outlet nozzles are often marginally sized. These conditions result in very high velocity of the steam leaving the steam outlet nozzle, and that can entrain boiler water.

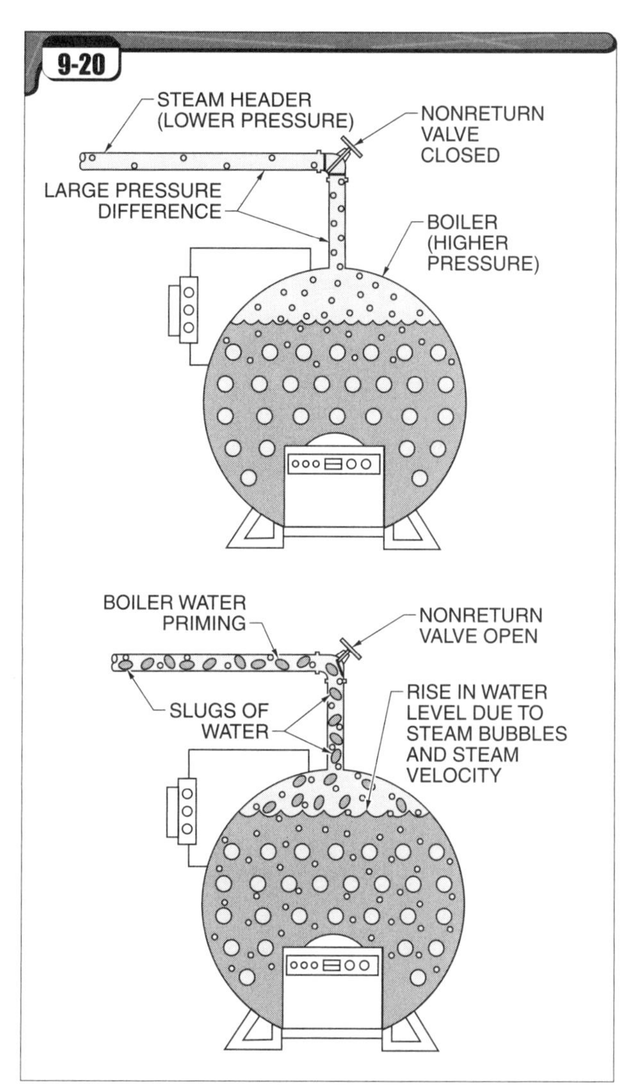

Figure 9-20. *A surge of steam flow from the boiler can cause priming.*

43 What action should be taken in the event of high water?

In the event of high water, the firing rate should be reduced quickly by cutting back on the fuel and combustion air. The reduction in heat transfer will allow the water to settle down somewhat as the volume of steam bubbles in the water is reduced. If the feedwater regulating system has allowed excess water into the boiler, the feedwater regulator may be isolated with manual valves so that the water level can be controlled manually through the bypass line. The boiler can be blown down as required to bring the water level down to normal.

The steam from the boiler may need to be shut off to protect personnel near steam lines. The decision to close the nonreturn valve or main steam stop valve to prevent water hammer and damage to the equipment downstream rests with the boiler operator. It may be appropriate to close one of these valves to prevent priming and subsequent water hammer. **See Figure 9-21.**

Figure 9-21. *A boiler operator may decide to close the nonreturn valve or main steam stop valve to stop water hammer.*

44 What action should be taken in the event of a low water condition?

The combustion should be stopped immediately. While procedures for stopping combustion vary depending on the type of fuel and type of equipment, the operator should shut down the fuel and air supplies. On solid fuel-fired boilers with fire doors on the front of the furnace, the operator should open the fire doors to let atmospheric air begin cooling the furnace. On boilers firing gas or liquid fuels, the operator should shut off the burner.

The boiler operator should not make sudden changes to the unit that may cause a change in the stresses in the boiler. Pulling the lever on the safety valve may cause the overheated, softened metal to be jolted. An operator should never put water into a hot boiler if the water level cannot be determined. The resulting thermal shock may cause an explosion. **See Figure 9-22.**

The operator should close a manual valve in the feedwater line. If the water level in the boiler has become dangerously low due to a control problem and the control suddenly begins working again, a large volume of water could be fed to the boiler. This could create enough thermal shock to result in a boiler explosion. For example, if a level-detecting float is stuck in the up position and then breaks free and falls, this could signal an automatic feedwater valve to open or a feedwater pump to start.

The building should be cleared of all personnel and the boiler allowed to cool very slowly. The boiler must be inspected for damage before the boiler is restarted because the strength of the boiler may be seriously impaired if it has been overheated.

45 How is water admitted to a boiler in the event of a feedwater regulating system failure?

Water may be admitted to a boiler by opening the bypass valve around the feedwater regulating valve. **See Figure 9-23.** The feedwater regulating system may fail due to loss of pneumatic control air pressure to the regulating valve actuator, loss of the signal to an electronic controller, sticking of the valve, or other reasons. On a low-pressure boiler, the makeup water feeder should admit city makeup water in the event that the primary feedwater regulating system fails. However, this is normally untreated water, so the boiler operator should be alert to this condition. The boiler operator should preplan the procedures to be followed in such emergencies.

Figure 9-22. *The addition of water to a severely overheated boiler may result in an explosion due to thermal shock.*

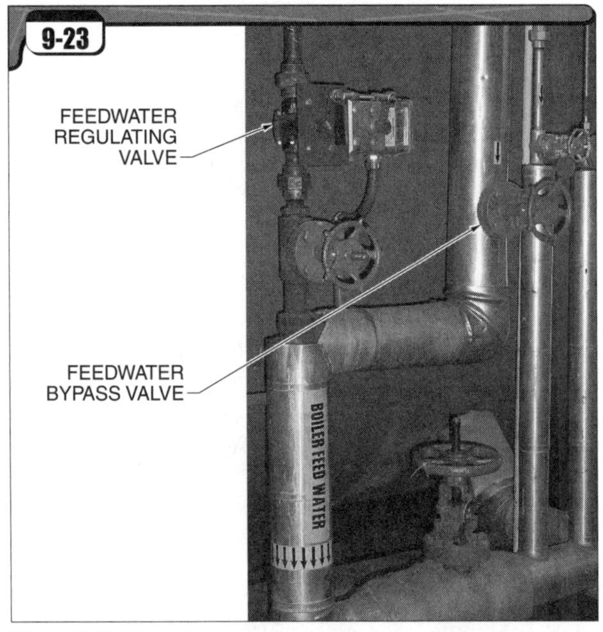

Figure 9-23. *Water is manually admitted to a boiler through the feedwater bypass.*

46 What action should be taken to remedy a cavitating centrifugal pump?

A number of possible actions may help reduce or stop cavitation. The boiler operator should verify that the level in the supply tank is normal. If the level has fallen below the normal level, the water level should be increased. For example, it may be necessary to temporarily open a bypass around a level control valve if the control valve has failed. The liquid being supplied to the pump can be cooled. For example, if a steam trap has failed open, the condensate in a condensate receiver may become hotter than normal due to the admission of live steam into the receiver. Until the trap is identified and repaired, cooler makeup water can be introduced into the receiver to cool the condensate. **See Figure 9-24.**

The boiler operator should confirm that the pump recirculation line is open. If the pump is deadheaded, friction from the impeller churning in the liquid will build up enough heat over time to make the pump cavitate. The pump discharge valve can be throttled slowly to reduce the output of the pump. This will reduce the tendency of the liquid to flash to vapor bubbles at the suction side of the impeller.

If there is a strainer installed on the suction side of the pump, it should be cleaned to allow free flow through the strainer. If the pump is equipped with a variable frequency drive or a steam turbine drive, the drive can be slowed down in order to reduce the liquid velocity entering the suction side of the pump. This may reduce pressure drop at the suction side and thus reduce flashing and cavitation.

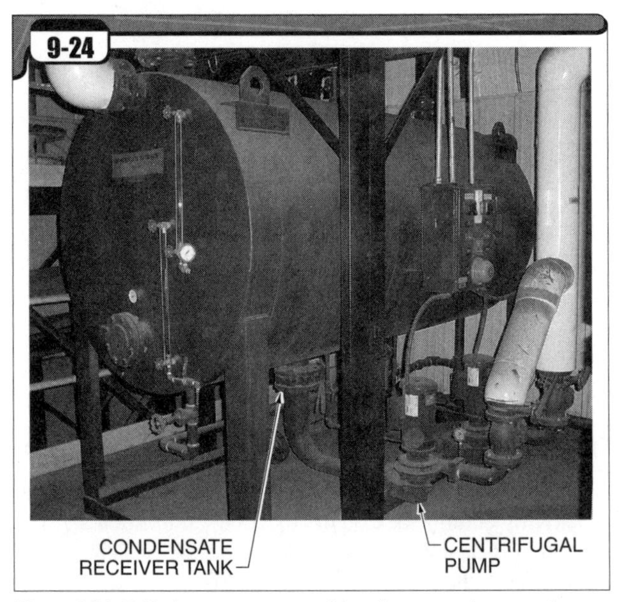

Figure 9-24. *Failed steam traps may blow live steam into condensate receivers, causing cavitation in the condensate pumps.*

47 If the boiler feedwater is lost and the boiler is needed for an additional period of time, what can be done to continue operation?

The boiler operator should determine whether the source of feedwater has been lost or whether the feedwater pump has failed. If the source of feedwater has been lost, the boiler must be shut down before the water level leaves the gauge glass. Modern feedwater pumps are generally installed in pairs. If feedwater is available and the primary pump fails, the backup pump should be started.

If the water column and gauge glass are installed properly, the lowest visible point in the gauge glass will be above the highest heating surface. Once the water drops below the gauge glass, it is impossible to measure the water level. A stoker-fired boiler must be shut down before the water leaves the gauge glass. An automatically fired boiler low water fuel cutoff control should shut off the combustion equipment when the water level reaches approximately the lower try cock, or about 2″ to 6″ below the NOWL. **See Figure 9-25.**

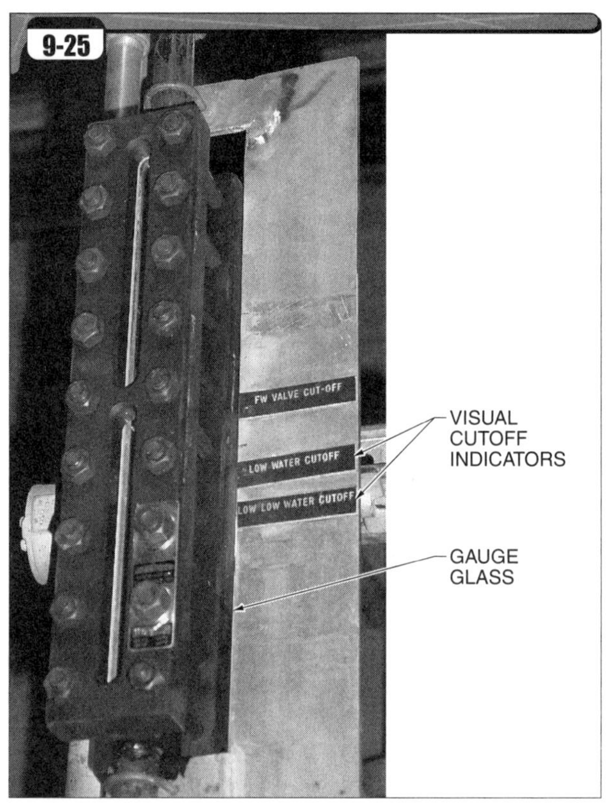

Figure 9-25. *The low water fuel cutoff control shuts down the combustion equipment when the water level drops to a preset level below the NOWL.*

48 What action should be taken if a tube bursts in a boiler in battery with other boilers?

The combustion should be stopped immediately. The operator should continue to supply feedwater to the boiler if possible. If all boilers on the battery receive water from a common feedwater header, the damaged boiler may draw too much water and other boilers may start to starve for feedwater. The undamaged boilers must be supplied with water first. If the supply of feedwater is inadequate and a backup feedwater pump is available, the second pump may be started to supply more feedwater. However, the boiler operator must be careful not to draw feedwater from the deaerator or other supply tank faster than it can be made available. This would cause cavitation and damage in the feedwater pumps. The steam load should be transferred from the damaged boiler to any other available boilers.

49 What should be done if a boiler that supplies steam to a turbine is foaming badly?

The firing rate should be reduced or shut down to get the water to settle down. The steam load should be transferred to other available boilers if possible. If the carryover is serious or if priming is beginning to occur, the boiler must be shut down. The steam flow from the boiler should be stopped as soon as feasible by closing the main steam stop valve or nonreturn valve if this can be done safely. The boiler should be blown down and fresh feedwater added to dilute the contaminants as much as possible before putting the unit back on line. The source of the contamination that led to the foaming must be found and repaired before starting the boiler again.

50 What two rules of thumb help the boiler operator avoid the dangers of water hammer?

"No water, no hammer." This is the rule of thumb for the prevention of water hammer. Both steam and water (condensate) are necessary in order for the most common forms of water hammer to occur. If the steam distribution system and heat exchange equipment are kept drained of condensate, then no high-velocity slugs of water can be created.

"No steam, no hammer." This is the rule of thumb for response to water hammer. If the source of steam is shut off, the water hammer will stop. Once isolated, the steam line or other equipment should be manually drained of condensate before attempting to readmit steam to the system. The reason why the piping or equipment was flooded should be determined and remedied.

51 What should be done if the boiler pressure rises above the safety valve popping point but the safety valve does not lift?

The fuel feed should be stopped so that the pressure drops as steam leaves the boiler. The safety valve lever should be lifted manually to pop the safety valve when the pressure drops to approximately 75% of the set pressure of the safety valve. After the safety valve reseats, the firing rate should be brought up and the boiler pressure raised to the popping point of the safety valve. On a solid fuel-fired boiler, this may be accomplished by carefully throttling the steam flow from the boiler. On automatically fired boilers, pressure controls may have to be bypassed to perform this test. If the safety valve pops as it should, normal operation may be resumed. If the safety valve does not pop as it should, the boiler should be taken out of service and the safety valve should be replaced. **See Figure 9-26.**

Figure 9-26. *If a safety valve does not operate correctly, the valve should be removed from the boiler and sent to an authorized valve repair shop for service.*

What should be done in the event of a fire in the pulverizer or in the pulverized coal ducts that lead to the furnace?

The air feed should be decreased and the coal feed increased to the pulverizer so that the mixture inside becomes too rich to support combustion. The air preheater should be bypassed (some types may be shut down) in order to supply the pulverizer with cold air. The outlet air-coal mixture temperature should be observed for several minutes to see that it drops. If the temperature does not drop, then the air supply should be closed off and the pulverizer shut down. The cleanout ports and access ports should be opened one at a time and the interior of the pulverizer drenched with water, fire fighting foam, or steam to put out the fire.

What would cause the fires to pulsate in a pulverizer installation?

If the air-fuel mixture is not balanced properly, the combustion of the pulverized coal may not be stable. For example, if the mixture is too lean (not enough fuel), pockets of coal dust may flash as the proper mixture is achieved in these areas. As the fuel flashes, the rapid expansion of the gases of combustion can upset the furnace draft. A similar situation can occur if the mixture is too rich (too much fuel). As pockets of fuel in the furnace finally get enough oxygen to achieve the proper mixture, they ignite. Both of these situations are dangerous. The fire can go out, coal dust can accumulate in the hot furnace, and the coal dust can cause a furnace explosion if ignition is suddenly reestablished.

What should be done if a large puddle of fuel oil is discovered in the furnace of an operating package boiler?

A puddle of fuel oil most likely indicates that the atomizing tip has come off the end of the burner assembly or the fuel oil safety shutoff valve assembly is leaking. Both of these conditions are dangerous. The power to the boiler should immediately be shut off by opening the main disconnect. This should stop the combustion air fan as well as the pilot solenoid valves and the ignition transformer. It is not enough to simply turn off the burner switch. This will cause the burner to go into postpurge and thus allow the large volume of fuel vapors in the furnace to become lean enough to explode. The main disconnect and control power to the boiler should be locked out. The main manual fuel oil supply valve should be closed to prevent any more oil from leaking into the furnace. The gas supply to the pilot should be shut off as well. All personnel should be cleared from the area until the boiler furnace has cooled enough that self-ignition from the hot refractory materials is no longer likely.

What does black smoke from the stack of a gas-fired boiler indicate?

Black smoke from a gas-fired boiler indicates a poor air-fuel mixture. This condition is very dangerous because it means the fuel mixture in the furnace is far too rich. If the burner start/stop switch is turned off when this condition is discovered, the burner control will start a postpurge cycle to move remaining combustibles out of the furnace. This could cause the large volume of gas in the furnace to be suddenly supplied with adequate oxygen, which could cause the remaining fuel to violently ignite.

The area should be cleared at once and the manual gas valve closed very slowly. **See Figure 9-27.** This will allow the mixture to lean down gradually so that the boiler can be shut down in a controlled manner. Once the boiler is shut down, it should be locked out so that it may not be operated until the air-fuel controls are readjusted. The manual gas valve should also be locked closed.

If the fire is lost, is it acceptable to bypass the purge cycle in order to resume the load quickly?

The furnace must always be purged before lighting a gas, fuel oil, or pulverized coal fire. Residual fuel in the furnace could ignite violently or explode if an ignition source (spark or pilot) is reintroduced prior to purging.

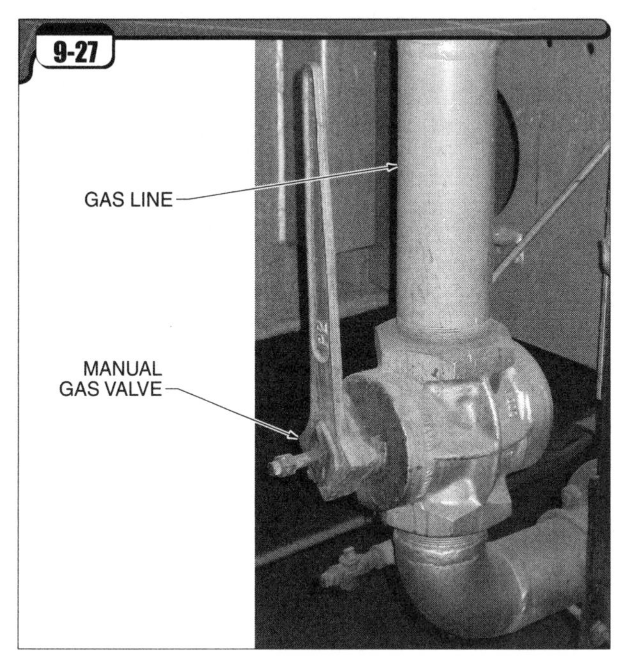

Figure 9-27. *A boiler operator should know the location of manual gas valves in the event of an emergency.*

Boiler and Vessel Inspection

57 How should a boiler be prepared for inspection?

Prior to shutting down, the boiler should be blown down frequently to reduce water side concentrations. The water treatment consultant should be contacted for other specific recommendations. If it is a large boiler with soot blowers, operate the soot blowers at twice their normal frequency to clean the fire side as much as feasible.

The manufacturer's instructions should be followed for proper shutdown and cooling procedures. After the boiler is shut down, it should be cooled slowly to allow even contraction of all metal components. Prior to draining, the boiler should be cool enough that the boiler operator can touch the manhole hatch or bare area of the shell without having to pull away. This ensures that any sludge in the boiler will not become baked on the metal when the unit is drained.

The air cock should be open when the boiler is drained to allow air to flow into the boiler and break the vacuum. If it is a watertube boiler with waterwalls, the waterwall drains need to be opened to empty the individual waterwall tube banks. All appropriate OSHA lockout/tagout procedures should be followed. All pipelines that could allow a fluid to flow into the boiler must be closed and locked out. The control power to the burner, blower motor, soot blowers, all fans, electrostatic precipitators, and attemperator should be locked out.

After confirming that the boiler contains no hot water or vacuum, the manhole hatch and all handhole covers can be removed. The fire doors and/or furnace access doors on the boiler can be opened. If there are scale or sludge deposits in the boiler water side, representative samples should be collected for analysis by the water treatment company. Be careful not to contaminate the samples with dirty tools or foreign matter. The interior of the boiler should be washed thoroughly with high-pressure water sprays to remove any sludge before it dries and hardens. Any waterwall header flanges should be removed and the pipe plugs in the water column piping and low water fuel cutoff piping should be removed so that the piping may be inspected.

58 What precautions should be taken when removing a manhole hatch?

The hazards associated with opening a manhole hatch include the possibilities that a vacuum exists in the boiler and that hot water is above the manhole level. A boiler vent valve is used to prevent a vacuum from forming when draining the boiler. Even so, the vent valve stem could be broken such that the handwheel turns but the valve disc stays in the closed position. The bolts and braces that hold the manhole hatch in place should not be removed completely at first to be sure that there is no vacuum inside or hot water above the hatch level. **See Figure 9-28.** The bolts should be loosened and the hatch unseated to see whether air draws into the drum or water runs out. It may be necessary to rap the hatch with a mallet to break the gasket seal. The hatch may be opened fully if there is no vacuum present and no water runs out.

Figure 9-28. *The braces on a manhole hatch should be left in place until it is determined that no vacuum or scalding water remains inside the vessel.*

Factoid

If a manhole gasket leaks as pressure is created in the boiler, the leak may often be stopped by tapping on the manhole hatch with a hammer or mallet. This slight bump helps the gasket mate to the sealing surfaces.

What items should be checked when inspecting the water side of a boiler?

The drum internals, such as cyclone separators, need to be removed as required to facilitate inspection of hidden areas and tube ends. The cyclone separators should be numbered as each is fitted to a specific location. Notes should be taken for comparison from year to year concerning the location and magnitude of scale or sludge deposits, corrosion or pitting locations, evidence of overheating, and tube leak locations. It is also a good idea to take photographs of the interior each time the boiler is inspected.

Representative samples of any sludge or corrosion products should be taken to be analyzed by the water treatment company. A *tubercle* is a bump on a steel boiler surface made up of corrosion products. If there are tubercles on the steel surfaces, representative ones should be scraped off for analysis. Black or gray material underneath indicates active oxygen pitting. Reddish material underneath indicates that corrosion was occurring at one time but is not currently occurring. Any scale accumulation greater than the thickness of an eggshell is cause for concern. It indicates that the water softening equipment is not functioning well, that the blowdown rate has been inadequate, or the scale control chemical treatment is not performing properly.

All the internal and connected piping, such as the safety valve passages, feedwater lines, and chemical injection lines, should be inspected for encrustation and debris. The water column piping, low water fuel cutoff piping, connections to instruments, and all float chambers should be inspected to determine that the piping is clear. Any separation equipment should be checked for deposits or plugged holes. The manhole and handhole sealing surfaces should be inspected for evidence of water infiltration, chemical concentration, or other deterioration. Welds should be inspected for loss of metal or heavy corrosion. The water level line inside the boiler should be well defined. A poorly defined water line may indicate that foaming has been taking place or that the water level controls are not holding a steady water level.

In firetube boilers, the flue furnace-to-tube sheet joints and tube ends should be carefully inspected for metal loss and stress corrosion cracking. **See Figure 9-29.** These are points of particular concern in firetube boilers. The outside of the pressure vessel should be thoroughly inspected as well. For example, it is important to investigate any evidence of leakage or water infiltrating into the area between the exterior insulation jacket and the boiler. This creates a hot, moist environment that can severely corrode the exterior of the vessel.

WARNING

Do not watch a burner light-off through an inspection peephole. If a malfunction occurs and a fuel explosion results, this location is very dangerous. Stand away from any inspection doors, burners, relief panels, or viewports.

Figure 9-29. *The flue furnace-to-tube sheet joints in a firetube boiler should be examined for metal loss due to stress corrosion.*

What is nondestructive testing (NDT)?

Nondestructive testing (NDT) is a method of determining the condition of components without causing damage and impairing their future usefulness. Typical methods for such evaluation include X-ray testing, dye penetrant testing, and ultrasonic thickness testing.

X-ray technology is used in evaluating the quality and condition of critical welds on boilers, on pressure piping, and on other metal components. When the X-ray shot is made, the internal condition of the component is captured on the film. X-ray photographs are used to identify cracks, inclusion of impurities in weld material, welding undercut, and many other integrity problems. X-ray photographs are taken during new construction of boilers and are also available at a reasonable cost for evaluating field repairs or other areas of concern.

Dye penetrant testing is used to expose cracks or other surface-breaking flaws in metal that are invisible or barely visible to the naked eye. A special dye penetrates any cracks or fissures in the metal. The excess dye is wiped away and a material called developer, usually a dry powder, is applied over the same area. If cracks are present, the dye that resides in the crack will migrate through the developer and reveal the cracks.

Ultrasonic thickness testing is used to determine the thickness of metal tank walls, pipe walls, and other components when access is available only to one side of the metal. A contact probe that is similar in appearance to that of a stethoscope is connected to the test instrument. **See Figure 9-30.** The test instrument generates an extremely high frequency tone and measures the time required for the tone to return as an echo reflected by the opposite surface of the material. The instrument converts this time delay into a readout of thickness of the metal. This is very helpful where it is suspected that significant corrosion has occurred but visual access to the other side of the metal is impossible.

Figure 9-30. *Ultrasonic thickness testing is used to determine the usable thickness of metal components such as boiler shells, tanks, and piping.*

61 What does redness inside a boiler indicate?

Redness in a boiler may indicate that oxygen is present in the steam and condensate system while the boiler is operating. This can happen because of poor deaeration or inadequate use of oxygen scavengers in the feedwater. If the oxygen problem is in the boiler and feedwater piping, it will generally be accompanied by pitting. The deaerator should be inspected internally and the oxygen scavenger program should be evaluated. Redness may also come from a high concentration of iron in the makeup water supply that is not removed prior to becoming boiler feedwater. The redness in the boiler may not be a boiler problem but may be from corrosion products (dissolved pipe) in the condensate system being washed back into the boiler with the condensate.

Oxygen corrosion may also occur in the steam system even when the deaerator and oxygen scavenger are working very well. This can occur when heat exchange equipment under modulating steam control is shut down and oxygen is drawn in as the steam collapses and forms a vacuum. This may be largely remedied through the use of thermostatic air vent fittings on affected heat exchange equipment.

62 What inspection should be made of the fire side of the boiler?

A fire side inspection should include looking for spalling and missing or damaged refractory brick. The furnace-sealing areas should be inspected for evidence of flue gas leakage. Any such areas should be repaired with new gasket materials or refractory. The interior of the breeching ductwork, economizer, air preheater, induced draft fan, and stack should be inspected for acid dewpoint corrosion. The operator should check for bags, blisters, and evidence of overheating on the boiler heating surfaces.

In solid fuel-fired boilers, ash and slag accumulations should be cleaned from the tubes, waterwalls, baffles, and brick so that any flaws will be revealed during the inspection. In boilers that fire oil or gas, burner alignment and insertion length should be set according to factory specifications. All burner linkages and dampers should be exercised. The condition of all baffles should be evaluated.

In watertube boilers, the tubes should be checked for warping and any evidence of leakage. Particular attention should be paid to areas in the immediate vicinity of the soot blowers to make sure the soot blowers are not directly impinging on the tubes. Close attention should be given to the tube ends at the top of waterwall headers and the mud drum for acidic corrosion. In firetube boilers, the tube ends should be checked for evidence of overheating or leakage. Stains on the tube sheet below a tube indicate that water has been leaking from that tube. These tubes should be rerolled or retubed if needed. **See Figure 9-31.**

The outside of the boiler casing should be inspected as well. Areas in which the paint has scorched, or warped areas in the casing indicate that excessive heat is migrating through the refractory materials. Locations where sections of the casing are joined by gasketed flanges should be examined, especially in packaged boilers using pressurized furnaces. Streaks or color changes on the metal often indicate leakage of hot gases through the gasket materials. On smaller boilers using atmospheric burners, the areas around burners should be checked for evidence of flame rollout. This is an indication of delayed ignition, probably due to pilot-related problems.

Figure 9-31. *Tube leaks in a firetube boiler are easy to identify by the boiler water solids that deposit below.*

63 What inspection and maintenance should natural gas valve trains receive?

Pressure switches should be visually inspected to determine that the setpoints are set for the proper value. A handheld pump can be used to create the trip conditions needed to cause the switch to respond. This test can reveal whether the device is faulty or responds sluggishly. Gaskets may be inspected for leaks using a soap bubble solution or electronic meters that detect hydrocarbon vapors.

SSOVs in the main gas line and the pilot line should be bubble tested for leak-through. This test is performed by blocking the downstream side of the SSOV using a downstream manual valve. A piece of flexible tubing is connected to a test port at the downstream side of the SSOV and immersed just below the surface of a container of water. **See Figure 9-32.** If bubbles appear in the water, the SSOV is leaking. All fluid power gas valve actuators should be inspected for hydraulic fluid leakage. The hydraulic fluid normally cannot be replenished. Visible hydraulic fluid beneath the actuator on an SSOV is an indication of imminent valve failure.

The upstream SSOV in a double block-and-bleed installation may also be tested by determining whether natural gas escapes up the vent while the burner is shut down. This can be tested by covering the end of the vent with a large balloon or a plastic bag and observing for any inflation. An odor of natural gas is not an indicator of leakage at the vent because natural gas is routinely vented through the vent line when the burner shuts off. Insects, spiders, birds, and even rodents may enter natural gas and propane vent lines and cause blockages as well.

Most natural gas valves used on boilers are plug valves. These valves generally contain a special sealant that both lubricates and seals the valve. This sealant can break down over time and should be replenished as necessary to maintain the valve seal. Valves in the natural gas line to combustion equipment should also be exercised periodically to confirm that they will work if needed.

A burner-tuning technician should adjust the natural gas burner so that it operates at maximum efficiency while maintaining safe conditions. The technician should perform a pilot turndown test. This test confirms that the smallest pilot flame that is capable of being sensed by the flame scanner is still strong enough to actually light the main burner. If the pilot flame is too weak, a substantial amount of air-fuel mixture can accumulate before being ignited by the weak pilot. This can result in an explosion even though the flame scanner proved the presence of the pilot flame.

The boiler operator should also perform a spark pickup test. There is ultraviolet light in the spark from the burner's igniter. A flame scanner that detects ultraviolet light could detect the spark and indicate that the pilot has lit when in fact only the spark is present. For this reason, ultraviolet-type flame scanners should never be installed in an orientation where they may "see" the ignition spark.

64 How are bags and blisters formed?

A *bag* is a protruding bubble or bulge in the steel plate of a boiler. A bag is commonly about 2″ to 4″ in diameter. **See Figure 9-33.** Bags are caused by scale deposits on the water side of the steel plate that inhibit proper heat transfer from the steel to the water. The bag area of the steel plate overheats and becomes soft. Internal pressure blows a bubble in the steel plate at that point. Bags can appear on tubes, but they are more commonly found on larger or flatter steel plates such as the flue furnace in a scotch marine boiler or on the water leg plate in a firebox boiler.

A *blister* is a lamination of steel plate or tube surfaces or where the steel plate splits into layers. Blisters are often caused by impurities in the steel during manufacture. These impurities in the steel cause the steel to split into layers. Blistering is most likely when the steel is overheated.

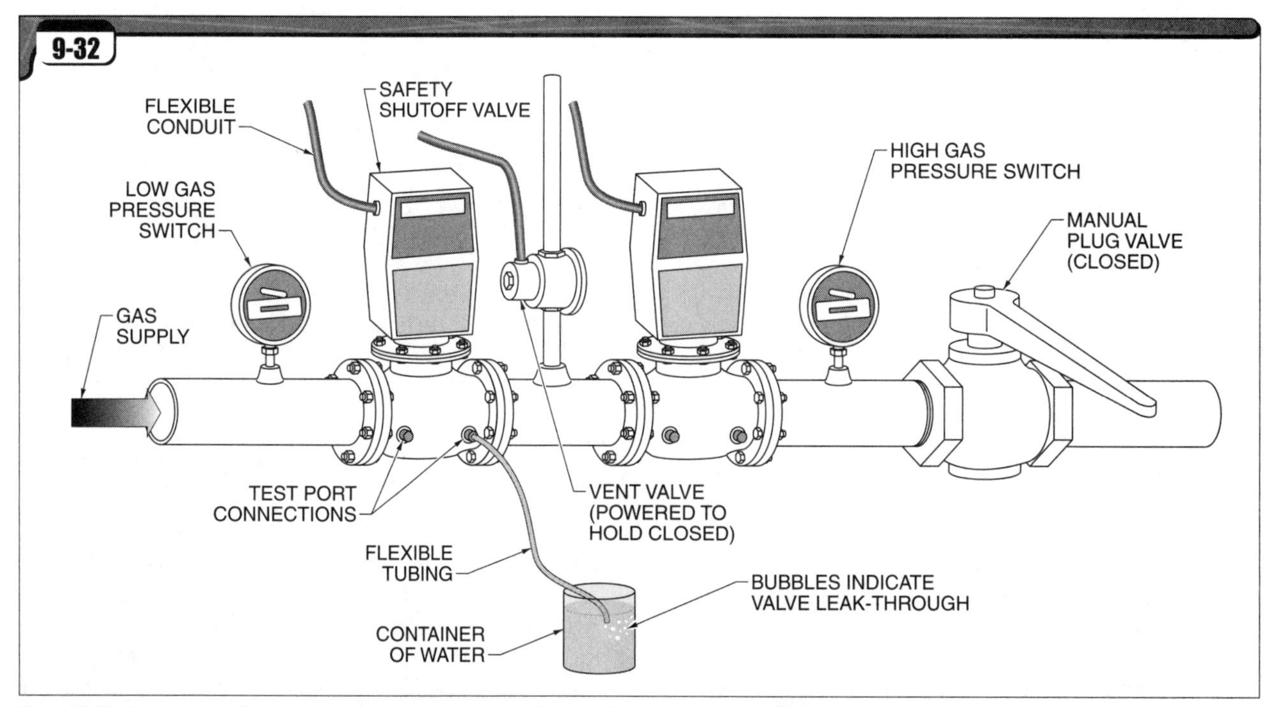

Figure 9-32. *Safety shutoff valves in gas valve trains should be tested for leakage on a scheduled basis.*

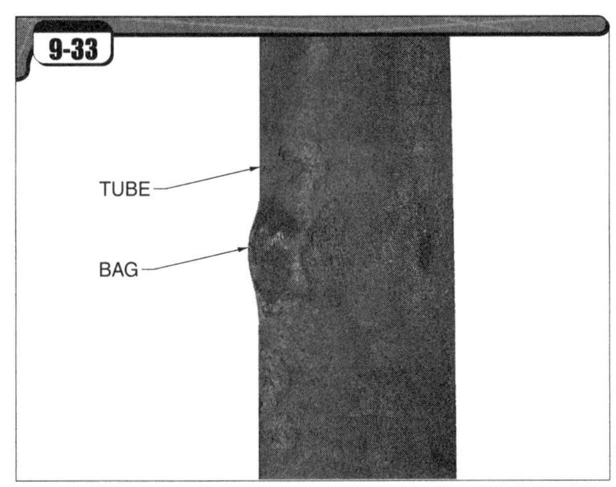

Figure 9-33. *A bag is a bubble caused by scale deposits that inhibit heat transfer.*

Cleaning, Maintenance, and Repairs

65 How can a bag on a heating surface be repaired?

If the bag is small, it can be heated and driven back into place. If the bag is larger or more severe, it can be cut out and a patch welded in. In either of these cases, qualified boiler repair personnel should perform the work.

66 How should a boiler be inspected after it has been shut down because of low water?

The insurance inspector should be notified. The fireside of the tubes should be thoroughly cleaned. **See Figure 9-34.** The tubes and shell should be inspected for unusual color changes, burned tube ends, warpage, bags, or blistering. Changes in the color of the steel, such as ashen gray areas, bluing, or iridescent sheens, may be a valuable indicator of how hot the steel has been. Such color gradations are often enhanced in photographs taken with a 35 mm film camera. Photographs should be taken of suspect areas and examined for discoloration undetectable by the naked eye.

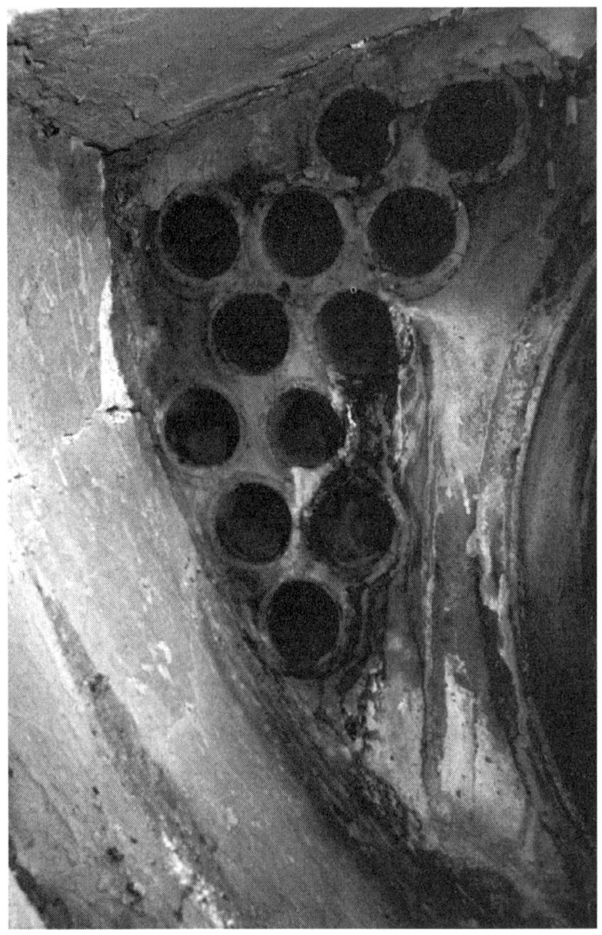

Figure 9-34. *Discolored, distorted, or leaking tube ends are evidence of boiler overheating.*

67 What precautions should be taken when gagging a safety valve?

During a hydrostatic test, the safety valves may be gagged to prevent them from opening during the test. **See Figure 9-35.** The gag only needs to be tightened enough to keep it in place. The gag does not force the safety valve closed. It only keeps it closed. The safety valve seat, disc, or stem may be damaged if the gag is overtightened. Gagging any safety valve for a hydrostatic test or removing it and blanking the opening is a matter that calls for meticulously followed procedures. It is critically important that the gag be removed or the safety valve replaced before the boiler is returned to service.

Figure 9-35. *Gags may be installed on safety valves to prevent them from lifting during a hydrostatic test.*

68 What is a turbine tube cleaner?

A *turbine tube cleaner* is a motorized mechanical cutter or knocker that removes scale from boiler tubes. **See Figure 9-36.** The motor is driven by compressed air or a flexible shaft. The turbine tube cleaner is fed through the tubes individually. For cleaning water tubes, a cutter end shaves the scale deposits from the interior of the tube. For cleaning fire tubes, a knocker end raps the inside of the tube, which cracks the scale from the outside of the tube. The turbine tube cleaner can easily damage the thin-walled tubes if left in one place for too long. It should be fed as smoothly as possible through the tube without stopping the motion. After cleaning, the boiler should be hydrostatically tested to ensure that the turbine tube cleaner did not damage any of the tubes. The accumulation of scale and debris that is knocked loose during cleaning must be removed from the low points to which it has fallen. If the debris is left in the boiler, it may redeposit during operation, or it may plug blowdown lines and instrument connections.

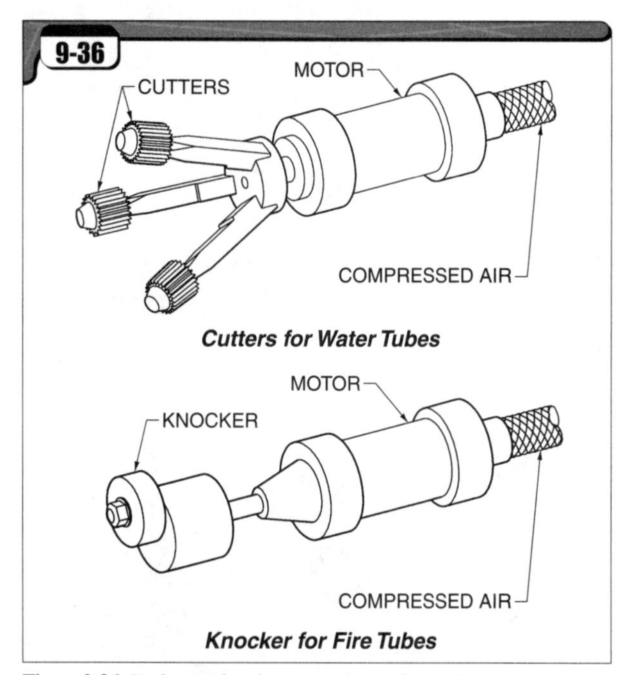

Figure 9-36. *Turbine tube cleaners remove the scale from the tubes.*

69 How is oil contamination removed from a boiler?

Oils or other organic contaminants that enter an existing boiler from a leak in a heat exchanger may cause problems. A *boil-out procedure* is the process of removing oily residues from a boiler by adding chemicals to the boiler water and boiling the mixture for a period of time.

A qualified water treatment specialist should be consulted for specific recommendations and procedures regarding the chemical solution to be used for cleaning the boiler. The chemicals to be used depend on the nature of the contaminants. After cleaning, the boiler is closed and filled to just above the normal water level to help clean residues that have adhered at the NOWL. A chemical solution consisting of trisodium phosphate or soda ash and sometimes dilute caustic soda is added to the water. This solution is boiled under low to moderate pressure in the boiler for a period of several hours to several days.

The chemical solution is designed to break down the oils into soluble soaps, which can be blown down. The boiler is blown down frequently to remove the contaminants. Fresh, hot feedwater is used to replenish the water lost through blowdown. After the prescribed time, the boiler is again shut down, drained, and inspected to see if the procedure was successful. If the boiler is not clean, the procedure is repeated.

70 How can scale be chemically cleaned?

A boiler can be cleaned with a mild acid solution that dissolves the scale deposits. The solution is generally about 5% to 10% hydrochloric acid (muriatic acid) that contains chemical inhibitors to prevent corrosion of the boiler. The procedure consists of filling the unit with the acid solution and circulating it in the boiler for a specified time. A competent water treatment specialist should be consulted for guidance when acid-cleaning. Specialty vendors are generally used to provide this service. The spent acid solution must be properly disposed of in compliance with EPA regulations.

71 What precautions should be taken before performing any work on natural gas piping?

The area of piping to be worked on should be securely locked out following appropriate lockout procedures. Any controls associated with this section of piping should be de-energized and the power source locked out as well. The piping should be purged with nitrogen to expel the captive natural gas before work commences. Tools made of beryllium or other nonferrous spark-resistant metals should be used until it is confirmed that no fuel remains in the piping. When the piping is reassembled, it should again be purged with nitrogen so that an explosive mixture of fuel and air cannot be developed inside the piping. This could allow flame to back up into the piping from the burner when it is lit off, causing an explosion. In addition, the piping should be pressure-checked for leaks. All work on natural gas piping should be performed by qualified personnel.

72 How should debris and rust be removed from new steam piping before connecting the piping?

Steam blowing is the process of cleaning impurities from new piping by blowing steam through the pipe. This is done before connecting the piping to a steam turbine or other critical piece of equipment. Temporary piping is installed to branch off as close as practical to the turbine inlet. **See Figure 9-37.** This piping is securely braced and run to the atmosphere in a safe area. A target, which is simply a piece of square carbon steel bar stock, is placed across the piping near the outlet. The target is held by a welded bracket.

Steam is blown through the new piping at approximately half the pressure at which the turbine will be operated. This steam blows out to the atmosphere, carrying the rust and debris with it. After blowing for about 2 min, the steam is shut off and the target is removed and inspected. Debris blowing through the lines will have made flecks and small dents in the target. The target is turned 90° to an unscarred side, and the process is repeated. This process continues, using new targets when necessary, until the target comes out unscathed. The temporary piping is then removed and the turbine is put into service.

73 What is the procedure for removing defective fire tubes?

Fire tube repair should be done by qualified personnel from a boiler repair company. A U- or V-shaped section that extends about 2″ to 3″ back from the tube sheet is cut out of the end of the fire tube with an acetylene torch. **See Figure 9-38.** This cut-out section provides space to collapse the remaining end portion of the fire tube. The repair person must not damage the tube sheet while cutting through the tube. A chisel is used to collapse the bead from the tube sheet so that it can pass through the tube sheet hole. The fire tube is driven toward the end of the boiler shell after the other bead is loosened from the tube sheet. An acetylene torch is used to cut a hole through the fire tube at the end where the fire tube is protruding through the tube sheet. This hole can be used as a point of attachment to pull the fire tube through the hole.

74 What is the procedure for repairing a defective water tube?

Water tube repair should be done by qualified personnel from a boiler repair company. In most cases the ends of the tube where it is rolled into the drums will still be in good condition. If this is the case, the middle portion of the tube or the damaged portion can be cut out with an acetylene torch and a new piece can be cut to fit into the space. **See Figure 9-39.** The new piece is then welded with a window weld. A *window weld* is a weld made through an opening, or window, in the tube. If the stubs at the drums must be removed, this is done in the same manner as with a fire tube. If the failed tube is surrounded by other tubes, it may be necessary to remove several others to get to the failed tube.

In some cases, it may be decided that the defective tube is simply too difficult to access for the repair. For example, it may not be desirable to cut through several good tubes to access the damaged tube. In such cases, the defective tube is plugged at both ends using a cone-shaped plug. The plugs are welded into place and this tube is simply abandoned. If needed, a hole is blown into the tube to prevent the buildup of pressure inside when the boiler is fired again. If the tube is in a hot enough area of the furnace, it will burn off and be destroyed.

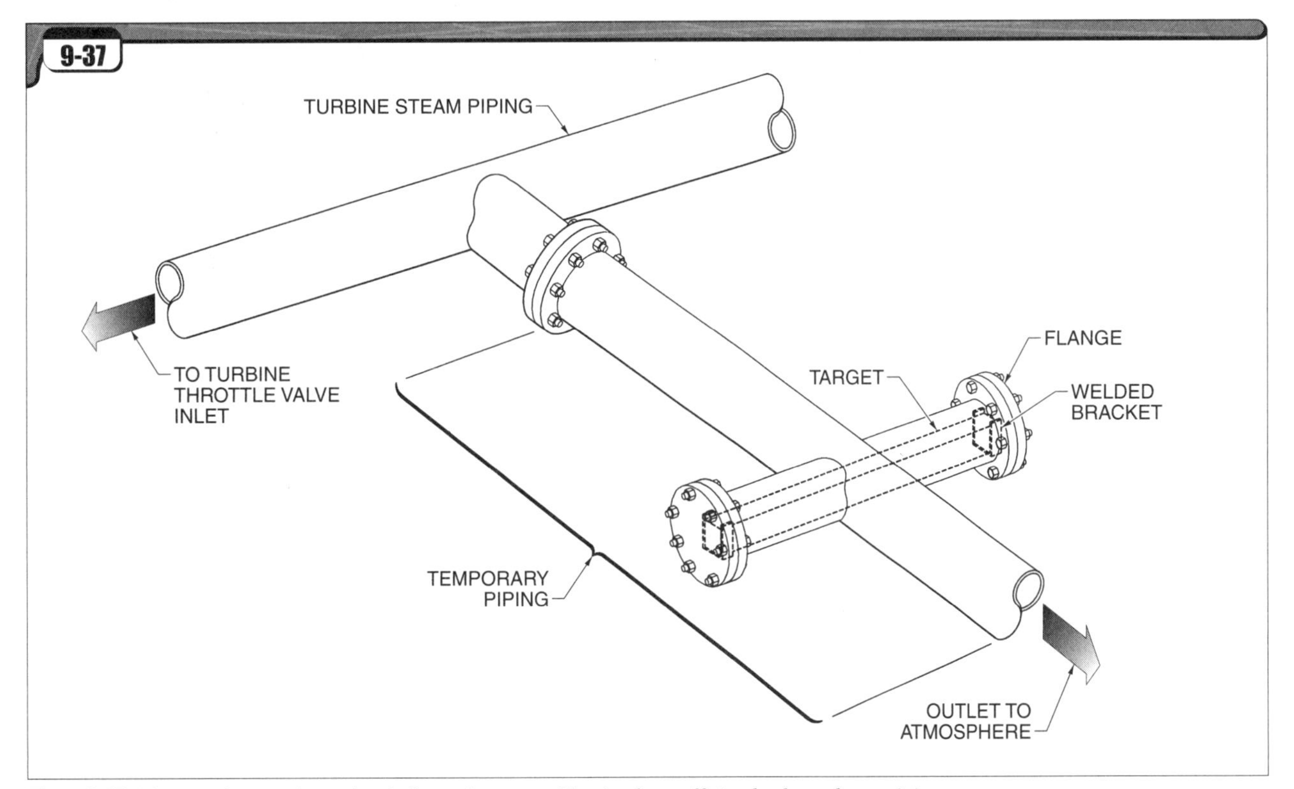

Figure 9-37. *A bar stock target is used to indicate that steam blowing has sufficiently cleaned new piping.*

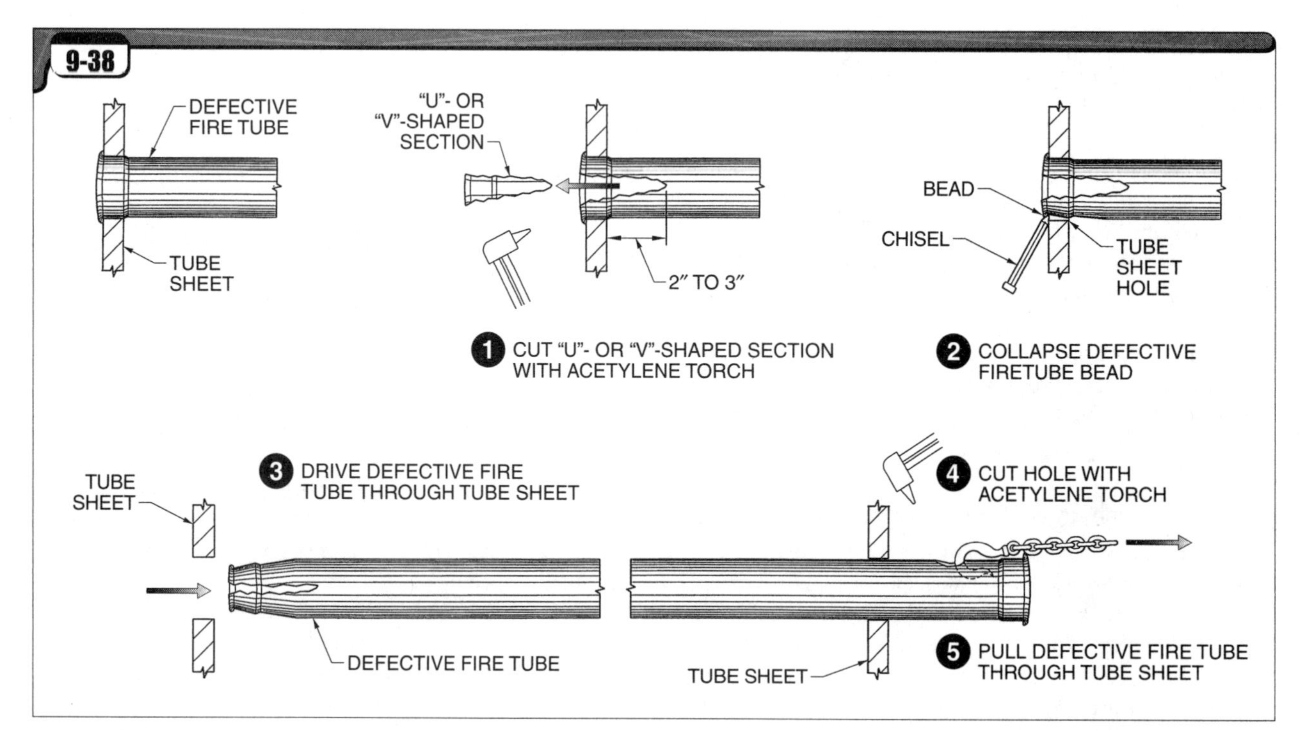

Figure 9-38. *Defective fire tubes are cut with an acetylene torch and pulled out.*

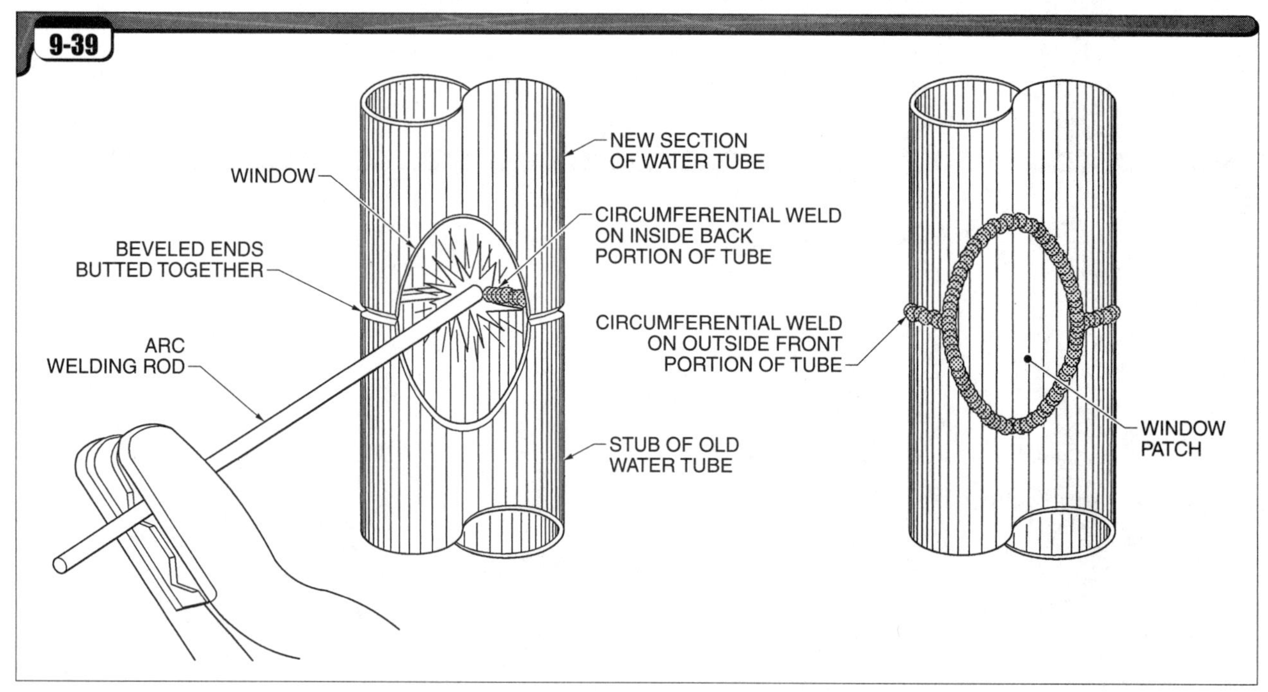

Figure 9-39. *Window welds may be used in repairing water tubes when access is limited to one side of the tube.*

Chapter 9

Boiler Operation and Maintenance

Trade Test

Name: ______________________________ Date: ______________________________

______________ **1.** The hazards associated with confined spaces include ___.
A. limited means for entry and exit
B. a hazardous atmosphere inside
C. unfavorable natural ventilation
D. all of the above

______________ **2.** Upon taking over a shift, a boiler operator should check the water level of all boilers by ___.
A. looking at a remote water level indicator on a panelboard or computer
B. blowing down the water columns and observing the water in the gauge glass
C. checking the feedwater pump
D. all of the above

______________ **3.** The low water cutoff switch should be tested ___ by blowing down the float chambers while the burner is firing.
A. hourly
B. daily
C. weekly
D. monthly

T F **4.** The safety valve may be checked by manually pulling the lever on a regular basis.

______________ **5.** The rotating mechanisms on natural gas valves should not be painted because ___.
A. the paint may have an adverse chemical reaction with any gas that may leak out
B. the paint may cause the valve to stick, making it difficult to operate in an emergency
C. only the local gas company is authorized to paint gas valves
D. none of the above

T F **6.** The steam header pressure should fluctuate frequently.

______________ **7.** Water, fuel, and steam flowmeter readings are used to monitor plant ___.

______________ **8.** If a fire occurs inside a coal pulverizer during operation, the boiler operator should ___.
A. increase the air flow to the pulverizer to make the air-fuel mixture too lean to burn
B. open the boiler safety valve to reduce steam pressure
C. decrease the air flow to the pulverizer to make the air-fuel mixture too rich to burn
D. shut down the feedwater pump

______________ **9.** A(n) ___ is one of the sides on a nut or bolt head.

______________ **10.** When a negative-pressure respirator must be worn on the job, the statement ___ is false.
A. employees must be medically examined and approved to use the respirator
B. the respirator may be worn by an employee who has a full beard
C. the specific respirator to be used must be fit-tested to the employee
D. the respirator must make an airtight seal to the employee's face

__________ **11.** ___ is the rise in the boiler water level when the load is increased.
A. Swell
B. Impingement
C. Pressure
D. none of the above

T F 12. Priming can occur if a large steam demand is suddenly placed on the boiler.

T F 13. As one boiler in a battery is removed from service, other boilers in a battery on automatic mode should pick up the additional steam load.

__________ **14.** A drop in pressure in the boiler during a hydrostatic test indicates that ___.
A. additional steam load has been placed on the boiler
B. the water in the boiler has increased in temperature
C. the pump being used to apply the pressure is too large in capacity
D. there is a leak of water from the boiler

__________ **15.** Blowing down a boiler from the waterwall headers could cause ___.
A. the main safety valve to pop
B. a rise in the water level of the gauge glass
C. the water flow across the tubes to decrease, which can overheat the tubes
D. the flue-gas outlet temperature to increase

16. Why is a sudden increase in steam demand on a boiler often accompanied by a rise in the water level?

__________ **17.** A boiler operator's primary duty when leaving the shift is to ___.
A. punch out on the time clock
B. leave the work area clean and orderly
C. vacate the premises in a timely manner
D. inform the relieving operator of any problems

T F 18. Corrosion and pitting are the main concerns during wet layup when a boiler is idle.

T F 19. If a dangerous low water level occurs on a solid fuel-fired boiler, the fire doors on the front of the furnace should be closed immediately.

__________ **20.** A boiler operator should set the continuous blowdown by ___.
A. comparing the actual boiler water conductivity or TDS with the target range
B. measuring the amount of sodium sulfite in the boiler water
C. observing the gauge glass for solid debris
D. setting it at the maximum possible rate

T F 21. A boiler should be shut down before the water level in the gauge glass drops below the lowest visible point in the gauge glass.

______________________ **22.** It is important to inspect the alignment of soot blowers during a maintenance shutdown to ___.
A. ensure that the soot blowers will not be damaged by excessive vibration
B. confirm that the soot blowers are blowing directly on the tube surfaces for maximum efficiency
C. verify that the soot blowers will not cause erosion of the tubes by blowing directly on the tube surfaces
D. ensure that the soot blowers will not be damaged by the furnace heat

T F **23.** Modern feedwater pumps are generally installed in pairs with one pump serving as a backup.

______________________ **24.** ___ can be caused by the introduction of relatively cold water into a very hot boiler.

T F **25.** Safety glasses provide very limited protection against splash hazards or blowing dusts.

T F **26.** If the fire is lost, the purge cycle should be bypassed in order to resume the load quickly.

______________________ **27.** If black smoke is produced when firing gas, the air-fuel mixture in the furnace is not getting enough ___.

______________________ **28.** When opening a flanged joint in a pipe, the first bolts that should be loosened are the bolts that are ___.
A. on the opposite side from the boiler operator's body
B. the easiest to access
C. on the top of the flange
D. on the bottom of the flange

T F **29.** Oxygen in the steam and condensate system could cause redness in a boiler.

T F **30.** Bags are caused by scale deposits on the water side of the steel plate.

______________________ **31.** Leaks of water or condensate into the insulation on the outside of a boiler should be stopped as quickly as possible because ___.
A. the constant wetness of the insulation reduces boiler efficiency
B. steam vapor from the wet insulation can obscure vision in the boiler room
C. the constant wetness of the insulation can cause severe corrosion on the outside of the boiler
D. the water running out the bottom of the insulation jacket creates an unsightly housekeeping problem

______________________ **32.** The inside of a boiler can be cleaned with a solution of 5% to 10% ___ acid.

______________________ **33.** ___ of a new operation is the process of inspecting, preparing, and testing each major component of the system prior to operation.
A. Startup
B. Shakeout
C. Commissioning
D. Pre-staging

______________________ **34.** A turbine tube cleaner ___ if it is allowed to remain in one place too long while running.
A. may damage the tube
B. may be damaged
C. will slow down its rotational speed
D. will accelerate its rotational speed to a dangerous degree

________________ **35.** The percentage of oxygen in the air inside a tank should be at least ___% before personnel enter the tank.

A. 9.5 B. 19.5
C. 29.5 D. none of the above

________________ **36.** Fuel, oxygen, and heat are known as the fire ___.

T F **37.** Waist belts are not acceptable as fall arrest devices because of the potential to cause serious spinal and organ damage if a fall occurs.

________________ **38.** ___ is the most common method of producing heat.

________________ **39.** Class ___ fires involve wood, paper, and trash.

________________ **40.** Class ___ fires involve combustible metals.

________________ **41.** Class ___ fires involve energized electrical equipment.

________________ **42.** Class ___ fires involve flammable liquids.

T F **43.** A boiler operator should never assume that a locked valve holds tightly and allows no fluid to leak through.

T F **44.** The strength of the boiler steel may be seriously impaired if it has been overheated due to a low water condition.

T F **45.** Loose-fitting clothing should not be worn around rotating equipment.

________________ **46.** ___ droplights should be used when working inside a boiler.

A. Mercury vapor B. Fluorescent
C. High-voltage D. Low-voltage

T F **47.** When a piece of equipment is locked out for maintenance work, personal safety locks must be installed by every person who will be performing work on the equipment.

T F **48.** If a valve is left back-seated, thermal expansion may cause the valve to bind, making it difficult to dislodge from the back-seated position.

________________ **49.** A(n) ___ regulating system should maintain the NOWL of a boiler during normal operation.

________________ **50.** In a solid fuel-fired boiler, a gradual increase in the flue gas exit temperature over the course of a shift indicates to the boiler operator the need to ___.

A. increase the blowdown B. inspect the air preheater
C. operate the soot blowers D. bypass the economizer

________________ **51.** The correct feedwater temperature helps to ensure ___.

A. oxygen removal B. maximum thermal strains
C. correct flue gas temperature D. all of the above

________________ **52.** An unusually aggressive plume of vapor blowing from a condensate receiver's vent pipe indicates that ___.

A. the vent pipe is oversized
B. the condensate level in the receiver is too low
C. the condensate pump on the receiver is cavitating
D. steam traps are failed open and blowing live steam into the receiver

______________ **53.** A(n) ___ fire is started on the stoker grates to start the coal fire.

______________ **54.** When a high water condition is discovered in an operating boiler, why should the fuel firing equipment be shut down?

T F **55.** Overheating may result from heavy scale or low water.

______________ **56.** The usable thickness of a pitted boiler shell may be determined by drilling a hole in the shell and measuring the thickness or by using a(n) ___ tester.

______________ **57.** Describe at least three actions that may help reduce or stop cavitation of a centrifugal pump.

______________ **58.** The blue box of the NFPA hazard signal system represents ___.

A. the health hazard
B. the reactivity hazard
C. the flammability hazard
D. any special hazards

T F **59.** Tubular gauge glasses should be cleaned with metal brushes to ensure thorough removal of mineral buildup.

______________ **60.** If a boiler is found to be overheating from a dangerously low water level, why should the feedwater line be shut off by closing a manual valve?

______________ **61.** The OSHA Fall Protection standard stipulates that the points to which fall arrest devices are attached be capable of supporting a weight of ___.

A. 500 lb
B. 5000 lb
C. at least 10% greater than the weight of the person using the attachment point
D. at least twice the weight of the person using the attachment point

T F **62.** The low water alarm may sound if the fires were to go out suddenly.

______________ **63.** Describe the "supervised warm-up" process used to bring a major steam line up to operating temperature and pressure.

______________ **64.** When starting a package boiler and placing it on-line with other boilers already in service, the boiler should remain on low fire until the flue gas outlet temperature reaches at least ___°F.

A. 100
B. 200
C. 300
D. none of the above

______________ **65.** Filter baghouses are normally bypassed during the startup of a coal-fired boiler in order to ___.

A. conserve energy until the filter baghouse is needed
B. avoid damage to the bags due to excessive temperature
C. avoid unnecessary restriction to the flue gas flow during startup
D. keep the oily nature of the coal smoke during startup from blinding the bags

______________ **66.** When electrical work is being performed on boilers, the ___ means should be padlocked in the OFF (open) position.

______________ **67.** If a large valve is difficult to open or to close completely, a(n) ___ should be used on the handwheel to provide more torque.

A. approved valve wrench of minimal size
B. aluminum pipe wrench
C. lever fashioned from a long piece of ¾″ pipe
D. adjustable wrench

T F **68.** A boiler being prepared for inspection should be cooled rapidly to allow even contraction of all metal components.

______________ **69.** A(n) ___ is a lamination of steel plate or tube surfaces.

T F **70.** Whenever possible, the boiler operator should observe the light-off of burners through a furnace inspection port.

______________ **71.** The process of removing the oily residues from a newly installed boiler or one that has been contaminated with organic oils is called a(n) ___.

A. acid cleaning procedure
B. solvent wash
C. boil-out procedure
D. shakeout

T F **72.** A feedwater pump failure will result in a loss of feedwater pressure considerably before the water level begins to run low in the boiler.

______________ **73.** ___ is the process of operating a new system as needed at low loads to expose and correct the major impediments to reliable operation.

A. Startup
B. Shakeout
C. Commissioning
D. Commencement

T F **74.** Turbine tube cleaners remove scale from boiler tubes.

_______________ **75.** Why are cotton and leather gloves unsuitable hand protection when handling liquid chemicals?

_______________ **76.** The proper first aid procedure for a small steam or hot water burn is to ___.
A. wrap it to prevent exposure to air
B. immerse it in cool water
C. wait for the burning to stop
D. none of the above

T F **77.** Oil burner tips should be cleaned of debris as often as needed to prevent loss of atomization effectiveness.

_______________ **78.** Scorched paint on the outside of the boiler furnace casing usually indicates ___.
A. damaged refractory materials inside the casing
B. standard trade paint was used instead of the correct high temperature paint
C. acid dewpoint corrosion is occurring on the other side of the casing metal
D. the furnace refractory has been subjected to thermal cycling

_______________ **79.** ___ maintenance involves a study of the history of the plant components and determination of critical components' expected service life.
A. Preventive
B. Breakdown
C. Planned
D. Predictive

_______________ **80.** If a boiler has two separate bottom blowdown valves, a boiler operator can ensure that the piping between the valves is subjected to as little corrosion as possible by ___.
A. closing the valve farthest from the boiler last
B. opening the valve farthest from the boiler first
C. closing the valve closest to the boiler last
D. closing the valve closest to the boiler first

_______________ **81.** Steam ___ is the process of cleaning impurities from new piping.

_______________ **82.** The evaporation test is considered more reliable than the quick-drain test when testing a float-type low water fuel cutoff control because the ___.
A. slow drop in the water level causes the float to drop more reliably
B. slow drop in the water level makes the float more likely to stick in the up position
C. rush of water and steam down the drain line during the quick-drain test makes the float more likely to stick in the up position
D. float is less likely to be damaged by water hammer

Matching

______	**83.** Bag	**A.** Mechanical cutter or knocker
______	**84.** Blister	**B.** Protruding bubble in steel plate
______	**85.** Lamination	**C.** Layer
______	**86.** Spalling	**D.** Separation of steel plate or tube surface
______	**87.** Steam blowing	**E.** Cleaning impurities from new piping
______	**88.** Turbine tube cleaner	**F.** Breaking off chips

Chapter 10

Boiler System Optimization

High energy costs require that industry place much more emphasis on steam system efficiency than in years past. Energy may be wasted in all parts of a steam system. Therefore, it is critical that the steam distribution system, steam applications, steam traps, and condensate return system all be examined for losses. Steam system surveys can provide valuable guidance toward achieving efficient operation. Surveys allow the facility to rank potential improvements in terms of capital costs and potential returns. Knowledgeable boiler operators can make valuable contributions toward energy efficiency simply by implementing improved operating techniques. Sophisticated control devices and add-ons often produce disappointing results if the fundamentals of efficient operation are not followed.

Facility management personnel are responsible for the energy efficiency of the plant but often do not have all the first-hand knowledge needed to understand efficiency losses and their remedies. As a means of encouraging production departments to conserve energy, it is common to meter steam usage and make a charge against department budgets, and allow credit for hot condensate returned. Boiler operators should learn to perform energy-related calculations associated with flash steam use, condensate return, and similar operations in order to justify improvements and encourage management support.

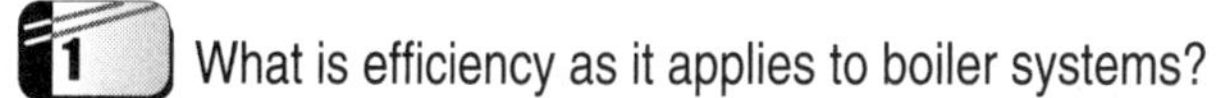

Efficiency Measurement

1 What is efficiency as it applies to boiler systems?

Efficiency is the ratio of energy output to energy input in a piece of equipment or in a system. The term "efficiency" is used in a number of contexts with regard to boiler systems. Energy can take multiple forms and the expression of efficiency often reflects those forms. *Combustion efficiency* is the percentage of the Btu content of fuel that is liberated as heat by the boiler fuel-burning equipment. *Boiler thermal efficiency* is the percentage of the heat liberated that is transferred into the boiler water. It is possible to have high combustion efficiency but low thermal efficiency and vice versa.

When combustion efficiency and thermal efficiency are combined, the result is fuel-to-steam efficiency. *Fuel-to-steam efficiency* is the percentage of the heat content of the fuel that is transferred into the boiler water. However, some boiler manufacturers use this term to refer to the performance of a particular boiler at its most efficient fixed firing rate.

The *steam rate* of a boiler is the combination of combustion efficiency and thermal efficiency at the full range of loads and conditions that the boiler encounters over a typical period of time. The steam rate of a boiler is normally expressed as the average number of pounds of steam produced by that boiler per unit of fuel. One boiler manufacturer uses the term "in-service efficiency" in much the same way as steam rate. However, in-service efficiency is expressed as a percentage.

2 What measurement is used in expressing the cost of steam production?

The industry standard used in expressing the cost of steam production is dollars per thousand pounds of steam generated. This basic measurement of the cost of steam generally takes into account the cost of the fuel, the water, and the chemicals used to treat the water. However, in specific discussions at the plant level, it is necessary to clarify the measurement being used. For example, some facilities that burn coal also consider the cost of coal storage and handling, flue gas cleaning, ash disposal, and other associated costs. Others use accounting formulas that consider the cost of boiler room labor, equipment depreciation, and other factors.

3 What methods are used to gauge the efficiency of a boiler?

The two standard methods of measuring boiler efficiency are the input-output method and the heat loss method. Both are spelled out in ASME *Power Test Code 4.1*. The input-output method records all the energy that is provided to and exits from the boiler from all sources. The input-output method is a direct method of testing boiler efficiency, but often is not feasible in industrial and commercial boiler plants due to the lack of accurate metering equipment. The input-output method requires the accurate measurement of many flows and conditions including the following:

- fuel flow
- fuel temperature
- steam pressure and temperature
- steam flow
- feedwater temperature
- stack temperature
- combustion air temperature

The heat loss method identifies the efficiency losses from the boiler and subtracts them from 100% to obtain the overall percent efficiency. **See Figure 10-1.** This method is recognized as the standard method of determining efficiency where measuring instrumentation is otherwise inadequate. In addition to identifying the losses, this method helps quantify the losses. This allows for assessment of potential improvements. The losses that are typically measured include the following:

- heat loss due to combustible fuel not burned
- heat loss in blowdowns
- heat loss due to radiation
- heat loss due to moisture in the fuel
- heat loss due to the formation of water vapor when hydrogen is burned
- heat loss due to dry stack gases

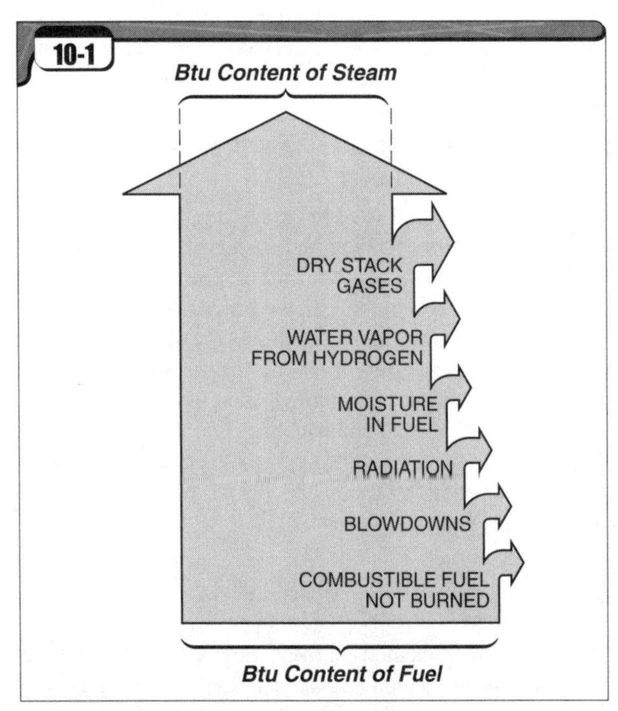

Figure 10-1. *The available Btu in fuel may be lost in a number of ways instead of contributing to steam generation in the boiler.*

4 Why is efficiency of a steam system important even when low-cost fuel is available?

The operation of the equipment involved in the generation of steam results in costs to the facility beyond the cost of the fuel. Unneeded operation of the equipment increases the need for maintenance services and parts and shortens the life of the equipment. In addition, combustion of the fuel creates stack emissions. This creates unnecessary pollutants and increases costs associated with environmental compliance. If the capacity of the steam system is marginal, inefficient use of steam can result in periods when the steam system cannot meet the steam demand. By operating the steam system more efficiently, boiler operators help the plant avoid or delay the large expense of installing additional boilers.

5 What is steam system efficiency?

Steam system efficiency is a measurement of steam usage that takes into account both the equipment supplying the steam and the equipment demanding the steam. It is important to remember that steam system efficiency is not the same thing as the fuel-to-steam efficiency of the boilers. All the equipment on either the supply side or the demand side should be evaluated for potential efficiency improvements. The supply side of the steam system consists of the boilers and the equipment that supports the operation of the boilers. The demand side of the steam consists of all equipment that uses steam.

Factoid

The steam system consists of both the supply side and the demand side. Both should be monitored for efficient operation.

6 What metering equipment is needed to determine supply-side efficiency?

The boilers and the combustion equipment represent the majority of the energy-related costs on the supply side of the steam system. It is necessary to measure the main energy inputs and outputs to clearly reveal how well the boiler/combustion equipment combination is performing.

The primary energy input is the fuel. A reliable fuel meter should be provided, such as a natural gas meter, a fuel oil meter, or a coal consumption scale. The fuel meter should be a totalizing flowmeter as well as an instantaneous flowmeter. A totalizing flowmeter is a meter that measures the total amount of material that has passed through the meter. This may consist of a digital counter, an automatic summed calculation in an electronic meter, or rotary dials in a natural gas meter. An *instantaneous flowmeter* is a meter that displays the current rate of flow. Many modern flowmeters display both instantaneous and totalized flow readings.

Many plants only have a single fuel meter that indicates the total fuel usage of the plant. This makes it difficult or impossible to determine the fuel usage of the boilers from the single meter. In these cases, metering should be installed for the boilers themselves. If this is not feasible, clamp-on mass flowmeters may be rented for short-term efficiency evaluations.

A steam flowmeter should also be provided for each boiler. Like the fuel meter, this meter should provide both instantaneous and totalized flow readings. A boiler operator can determine the steam rate by dividing the totalized steam flow reading by the totalized fuel flow reading. In facilities where this metering is available, the boiler operator should perform this calculation every day or every shift to confirm that the equipment is performing efficiently and to reveal developing problems.

Water softening equipment should be equipped with totalizing flowmeters. If all the condensate returns to a common main condensate receiver, having a condensate return meter is also recommended. Feedwater is made up of a mixture of condensate and the required amount of makeup water. This is the total water entering the boilers. The steam flow out of the boiler plus the blowdown flow represents the total water flow leaving the boiler. Therefore, the sum of condensate return plus makeup water equals the sum of steam flow plus blowdown. All these meters must be dependable and get regular calibration.

What metering equipment is needed to determine demand-side efficiency?

The management of each plant must make a decision about the demand-side metering. In the best case, totalizing steam flowmeters are installed to measure the steam consumed by each of the major steam-using departments in the plant, and each department budget is charged for the steam energy that department consumes. If it is not feasible to install metering, then it is usually appropriate to use a competent engineering firm to perform a plant steam survey.

How may the losses in a steam system be classified or categorized?

A good way to begin a search for lost or wasted energy in the steam system is to consider the system as a number of steps that occur in a particular order. **See Figure 10-2.** Thus, the losses may be categorized as follows:

- combustion
- steam generation
- steam distribution
- steam application
- condensate drainage
- condensate return

Combustion losses occur for several reasons. There may be extensive losses from a poor air-fuel ratio. For example, if there is inadequate combustion air to burn all the fuel, part of the fuel will be wasted. Excess combustion air causes part of the heat to be carried up the stack before it can be transferred into the boiler water. Additional fuel can be wasted due to poor atomization of fuel oil.

Steam generation losses can occur if soot, scale, or fly ash buildup interferes with heat transfer through the boiler tube surfaces. The heat exchange efficiency of the boiler (thermal efficiency) is also a function of the boiler design. Another cause of steam generation losses is lowered steam quality due to carryover.

In the steam distribution system, energy is lost due to missing or inadequate insulation, steam leaks, inadequate pipe sizes, and numerous other factors.

Losses associated with steam applications include poor heat exchange within the process itself and heat losses from the process. Poor heat exchange within the process may result from numerous conditions such as excessive steam pressure, fouled process heat exchanger surfaces, poor heat exchanger design, or the insulating effects of an accumulation of air trapped in the equipment. Heat losses from the process may occur due to bypass valves leaking through, excessive venting from a deaerator, and poor insulation of the equipment.

Condensate drainage losses include such items as losses due to steam traps that are piped to discharge onto the ground or to drain. Other losses can occur due to missing or inadequate insulation.

Losses related to the condensate return system may include heat loss due to inadequate insulation, flash steam that is allowed to vent to the atmosphere, and loss of live steam from condensate receiver vents due to failed-open steam traps.

Trade Tip

Purging a boiler's furnace is absolutely necessary for safety reasons, but is undesirable for efficiency reasons.

Combustion

How does the hydrogen content of the fuel contribute to efficiency losses?

Boiler fuels are composed primarily of the elements hydrogen and carbon, with trace amounts of other elements like sulfur. The burning of the carbon and hydrogen creates the vast majority of the heat. Combustion is the process of oxidizing, or burning, a fuel where the elements in the fuel mix with oxygen in a chemical reaction that gives off heat. The reaction also yields water as the product of oxidized hydrogen and carbon dioxide as the product of the oxidized carbon. **See Figure 10-3.** The water evaporates to steam and becomes superheated. This requires 970 Btu per pound of latent heat, as well as more energy for superheat. The latent heat and superheat to evaporate the water are obtained from the heat released during the combustion process. This heat remains in the flue gases instead of heating the water.

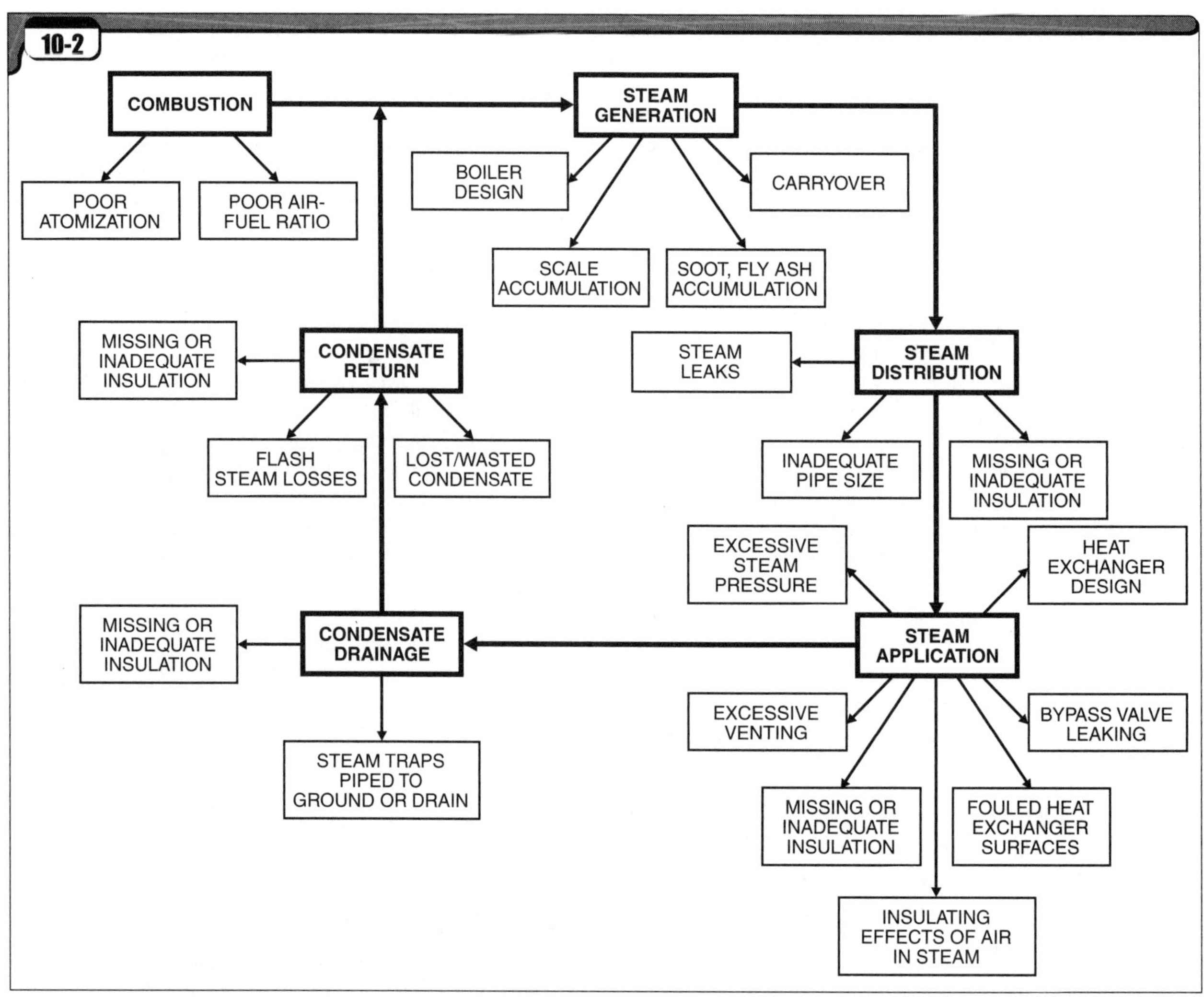

Figure 10-2. *Energy losses may occur during any part of the steam and condensate return cycle.*

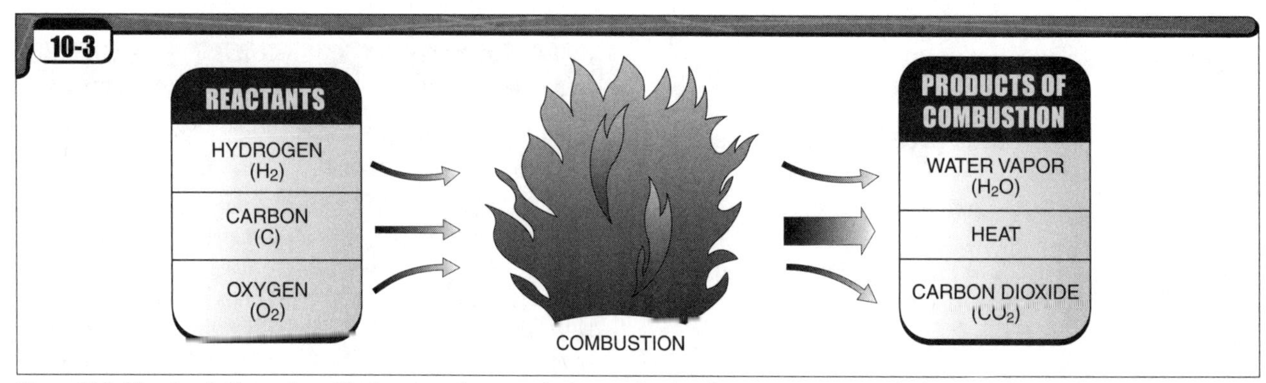

Figure 10-3. *The chemical reaction of hydrogen and oxygen during combustion forms water vapor, which absorbs some of the heat.*

10 How does fuel moisture contribute to efficiency losses?

The moisture present in a fuel can be removed before combustion or it can be evaporated during the combustion process itself. For example, coal may be wet from rain. In a power plant, energy must be used in the form of preheated air in a pulverizer to drive off the moisture so that the pulverized coal will transport properly. In other coal-firing equipment such as stokers, moisture on or in the coal is evaporated and superheated during combustion. Most of this heat is lost up the stack.

Why should excessive purging be identified and avoided?

For safety reasons, steam boilers that use burners must be purged. When purging the furnace, heat is removed from the boiler water and passed into the cooler air flowing through the furnace and boiler passages. The warmed air then exits the stack, carrying expensive heat with it.

The primary way that attentive boiler operators prevent excessive purging is by placing in service boilers of appropriate capacity to meet the plant steam demand without excessive ON/OFF cycling. For example, if an automatically fired boiler is too large for the plant demand, it will tend to purge, run for a short time, purge, and shut off. If this type of cycling is occurring, the boiler operator should place a smaller-capacity boiler in service and remove the larger-capacity boiler from service. If multiple identical boilers are in service, another approach is to remove one of the boilers from service so that the other boilers will run more continuously at nearer their design capacity.

How may incomplete combustion cause efficiency losses?

When burning natural gas, incomplete combustion most often occurs because of an air-fuel mixture that is too rich. The majority of natural gas- and fuel oil-fired boilers in industry use jackshafts and linkages that position both the fuel valve and the forced draft fan damper from a single actuator. It is important that the linkages be adjusted to provide the correct air-fuel mixture over the full arc of movement of the actuator. **See Figure 10-4.**

Inadequate or poor atomization of fuel oils can result in delayed combustion and in poor mixing of the fuel oil vapors with oxygen. These problems can allow fuel oil vapors to exit the furnace before the combustion process has finished. The boiler operator should inspect and clean the atomizing tips on fuel oil burners frequently enough to ensure good atomization.

When firing coal on grates, unburned coal is often discharged from the furnace due to an inadequate supply of combustion air and resulting clinkering. When the flow of air from beneath the grates (forced draft) is excessive, coal fines and fly ash containing unburned carbon can be carried from the furnace and into the flue gas passages of the boiler.

How may the air supply to the boiler room cause efficiency losses?

If the combustion air for the boilers enters the boiler room through louvers or screened vents, the openings must be large enough for an adequate volume of air to enter. The same is true if the combustion air is delivered directly to the boiler through ductwork. Undersized ducts will restrict the volume of combustion air available. **See Figure 10-5.** If either the boiler room air inlet vents or the screens over the ductwork inlets become blocked by debris, this may also starve the combustion equipment for air. A warning sign of inadequate air flow is a substantial inrush of air into the boiler plant when an exterior door is opened.

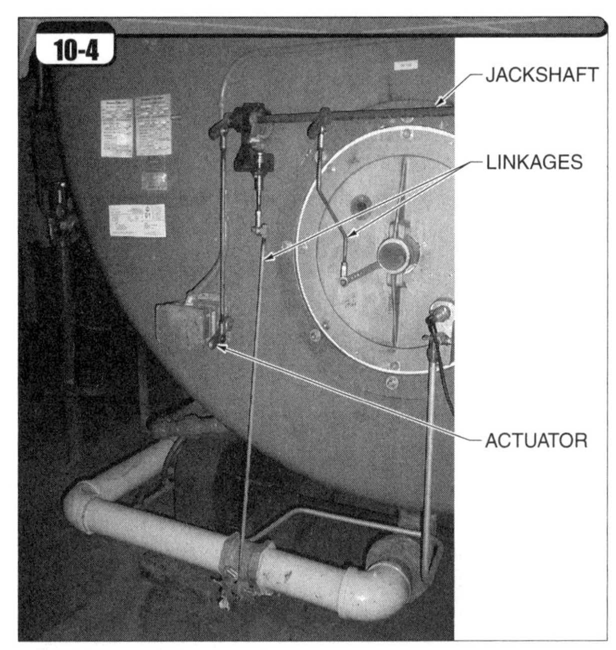

Figure 10-4. *A competent burner-tuning specialist should adjust the burner linkages.*

Figure 10-5. *Combustion air supply ducts must be adequately sized to prevent starving the combustion equipment for air.*

How can the boiler draft conditions cause efficiency losses?

There are a number of draft-related problems that will affect the performance of a boiler. Most are fairly simple to correct and vigilant boiler operators should minimize them.

If the furnace is kept at an excessively negative pressure, cool ambient air is drawn into the furnace and flue gas passages. This cooler air lowers the flue gas temperature and cools and dilutes the air-fuel mixture. The furnace pressure should be maintained as close to atmospheric as feasible. The pressure should be only negative enough to prevent "puffing" of hot flue gases and noxious vapors into the operating area during firing rate changes or minor upsets. These leakage paths should be eliminated to the greatest extent possible using gaskets, high-temperature sealants, or other means as appropriate. Cool air entering through these paths also can cause localized condensation of the moisture vapor in the gases, which results in corrosion. This reduces the differential temperature between the flue gases and the boiler water and reduces the heat transfer rate. It also reduces the time allowed for heat transfer before the combustion gases are released to the stack.

On larger boilers where test ports are available, the oxygen or carbon dioxide content of the flue gases may be checked at successive points along the flue gas path. If the oxygen content of the flue gases increases as the flue gases flow through the ductwork, or if the carbon dioxide content decreases, leakage is occurring. Smoke-generating wicks and ultrasonic listening devices are useful in finding the leakage points.

If steel or refractory baffles in the flue gas passages are cracked or otherwise damaged, hot flue gases are allowed to bypass part of the boiler's heating surfaces. The baffles should be inspected closely during boiler outages and kept in good repair.

Damaged damper vanes or damper linkages that are improperly adjusted or poorly characterized will result in poor air-fuel mixtures at certain firing rates while the mixture may be acceptable at other firing rates. For this reason, burner performance should be tested at multiple firing rates over the whole modulating range of the burner.

Why should the amount of excess air used be optimized?

The fuel and air in a furnace do not mix perfectly. As the combustion process continues, the oxygen and hydrocarbon molecules become more and more sparse in the mixture and it becomes more difficult for hydrocarbon molecules to come into contact with oxygen molecules. For this reason, more air than the theoretical amount needed must be provided in order to make sure all the fuel is burned. However, any excess air is simply heated and thrown away.

> **Factoid**
>
> ***Optimizing excess air is one of the most important steps in optimizing a boiler's efficiency.***

Why must the amount of excess air used be higher at lower firing rates?

At low firing rates, it is more difficult to create the turbulence needed to thoroughly mix the fuel with the air due to the lower air velocities. For this reason, a somewhat greater ratio of air to fuel is needed at low firing rates to ensure that all the fuel is burned. This is one reason why boilers tend to be less efficient at low firing rates, and prolonged firing at low rates should be avoided.

How do oxygen trim (O_2 trim) systems function to optimize combustion efficiency?

Oxygen trim systems automatically make adjustments to the combustion air flow independently of the fuel flow in order to optimize the combustion conditions for maximum efficiency.

Why should the flue gases be analyzed regularly for carbon monoxide and oxygen?

The flue gases should be analyzed for relative percentages of carbon monoxide and oxygen at several load points on a regular basis, such as monthly or quarterly. Periodic testing will reveal when the efficiency has dropped and signal the need for tuning the combustion equipment. **See Figure 10-6.**

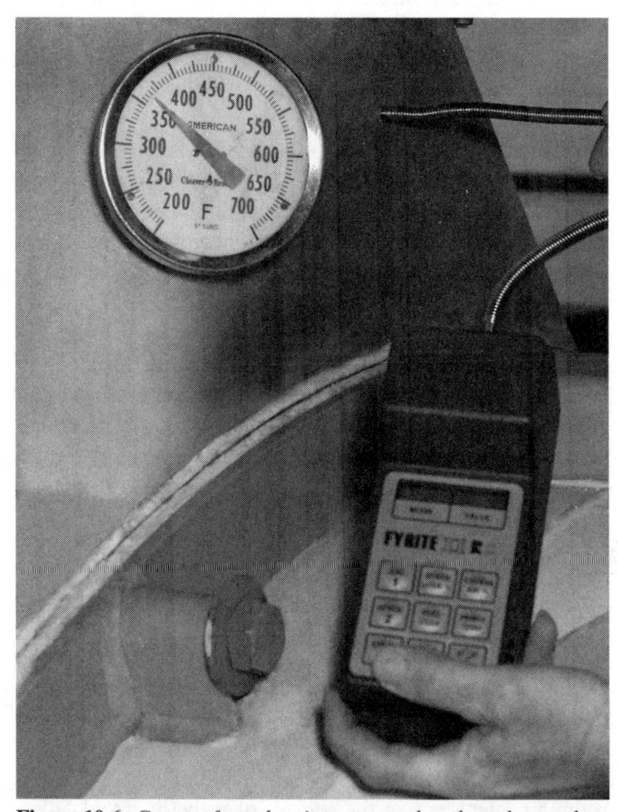

Figure 10-6. *Gases of combustion are analyzed at the stack to determine combustion efficiency.*

19 Why is compressed air usually more desirable than steam as an atomizing medium?

Compressed air is a less expensive atomizing medium in facilities in which the fuel must be purchased. Compressed air has the added benefit of serving as primary air. In facilities that generate steam from waste fuels such as refinery gas or blast furnace gas, steam will likely be the less expensive of the two atomizing mediums.

Steam Generation

20 How does soot cause efficiency losses?

Soot is the black residue formed when unburned carbon in the combustion gases sticks to the boiler's tube surfaces. Soot is often slightly oily, which causes it to stick to surfaces. Because of its composition and its texture, soot is a poor conductor of heat. As a blanket of soot forms, it tends to keep the flue gases from coming into contact with the tube surfaces, so the heat does not transfer well into the boiler water.

21 How is the accumulation of soot detected?

Soot is difficult to detect visually since the furnace and flue gas passages need to be opened to inspect the heating surfaces. Therefore, it is necessary to monitor for the conditions that lead to the formation of soot. Unburned carbon is released when there is inadequate oxygen or the oxygen and fuel are not thoroughly mixed. Boiler operators can detect these conditions by monitoring the oxygen and carbon monoxide concentrations in the flue gases. The appearance of smoke at the stack is also an indication of conditions conducive to soot formation. It is desirable to operate with the minimum feasible oxygen concentration in the flue gases without starving the fire.

Once soot has begun to accumulate, less of the heat from the fuel will go into the boiler water and more will go up the stack. The key to recognizing these conditions is normality. The boiler operator should trend and log the stack temperature at various burner firing rates over time when the boiler is clean, such as after a maintenance outage. Comparison of these charted values to the actual conditions on a day-to-day basis will then allow the operators to recognize deteriorating conditions. Boiler operators should also trend the quantity of fuel consumption to be expected at various steam loads; more fuel consumption than normal may indicate the development of soot.

22 How does scale cause efficiency losses?

The accumulation of scale inhibits heat transfer from the flue gases into the boiler water. However, it is not feasible to visually inspect the boiler for scale accumulation while the boiler is in operation. When scale buildup occurs, the temperature of the flue gases exiting the boiler will be higher than when the boiler is free of scale. Boiler operators should trend the flue gas temperature against the boiler load while the boiler is known to be clean so as to establish best-case conditions. When these conditions deteriorate, such as when the flue gas outlet temperature rises and steam production drops at various firing rates, this is an indication of fouled heating surfaces.

23 How may the best pressure at which to run a steam boiler system be determined?

The system should be operated at the lowest pressure at which steam meets the required operating conditions. Typical required conditions are that an adequate steam temperature be provided, acceptable flow velocities and pressure drops within the piping network be maintained, and proper operation of the steam traps and condensate return system be maintained.

Because higher-pressure steam is smaller in volume (in cu ft/lb), raising the steam pressure on an existing system can result in additional steam-carrying capacity of the piping network. This is especially important in large operations where steam must be transported long distances. **See Figure 10-7.** Before doing this, however, the entire steam distribution system should be examined to confirm that all components are able to withstand the additional pressure.

Figure 10-7. *Steam piping may need to be designed to deliver steam over considerable distances.*

24 How may the radiant heat losses from the boiler itself be minimized or recovered?

The boiler and associated piping and attachments should be adequately insulated to minimize heat losses. A good method of reclaiming much of the heat lost from the boilers is to have the combustion air fans pull the combustion air supply from a point high in the boiler room. Drawing the combustion air from this elevation helps preheat the combustion air and thus less energy is used. However, this warmer air will be less dense and may require slightly larger ductwork and somewhat more fan horsepower. As a rule of thumb, a 40°F increase in the combustion air temperature will result in a 1% improvement in overall boiler efficiency.

25 Why is it important to properly load a boiler?

As a general rule, boilers are the most efficient in their absorption of heat from the fuel when they are operating at around 75% of their capacity. **See Figure 10-8.** It is inefficient to operate a large boiler at a low firing rate for extended periods of time. If a package boiler is greatly oversized for the steam load, it will frequently shut down and restart. Each time the boiler starts it must purge the furnace. This results in a substantial amount of heat energy being lost up the stack.

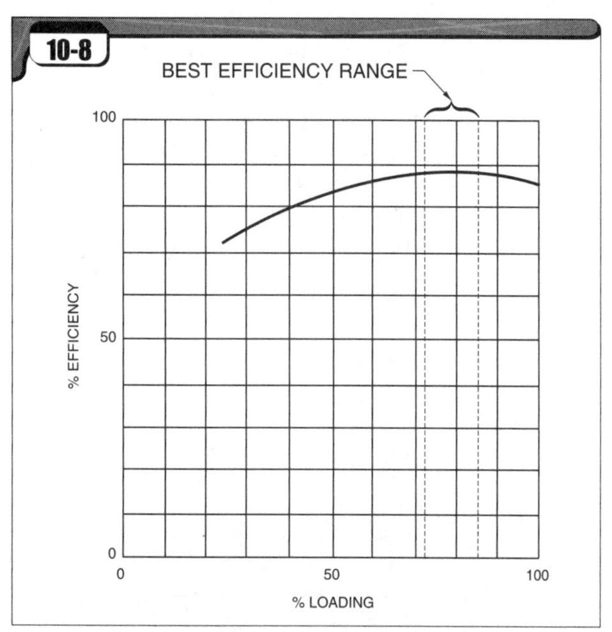

Figure 10-8. *Boilers typically achieve their best efficiency when operating at about 75% of their rated capacity.*

26 Why should the use of boiler blowdown be limited to only that which is necessary?

When the boiler is blown down, impurities are removed from the boiler. However, expensive heat energy, water treatment chemicals, and water also go down the drain. For this reason, the amount of blowdown used should be set to the minimum necessary to remove the impurities.

27 How should the continuous blowdown be operated to minimize losses?

The continuous blowdown system is normally used to control the conductivity of the boiler water. Boiler operators should be attentive and maintain the boiler water conductivity near the top of the prescribed range. This is much more realistic when an automatic continuous blowdown system is used. When a manual continuous blowdown approach is used, the blowdown may be too low or too high for some period of time before the boiler operator makes another adjustment. Automatic systems adjust the blowdown flow as soon as it is appropriate.

28 How should the bottom blowdown be operated to minimize losses?

The bottom blowdown should normally be used only to remove sludge that may settle to the bottom of the boiler. A good test to confirm that the amount of bottom blowdown used is appropriate is to capture the bottom blowdown water in a tank or containment basin and visually inspect this water for sludge. If only negligible sludge exists, the amount and/or frequency of bottom blowdown should be reduced.

The amount of bottom blowdown needed may also be reduced by ensuring that the sludge in the boiler is removed very quickly when the bottom blowdown valves are opened. This may be made possible by using a blowdown collector. **See Figure 10-9.** A blowdown collector consists of a length of angle iron that is capped off on the ends and has small notches cut into it along its length. This angle iron is configured to cover the bottom blowdown outlet and should run most or all of the length of the boiler drum or shell. By using this approach, sludge is removed very quickly from the entire length of the boiler when the bottom blowdown valves are opened. This allows the blowdown period to be minimized.

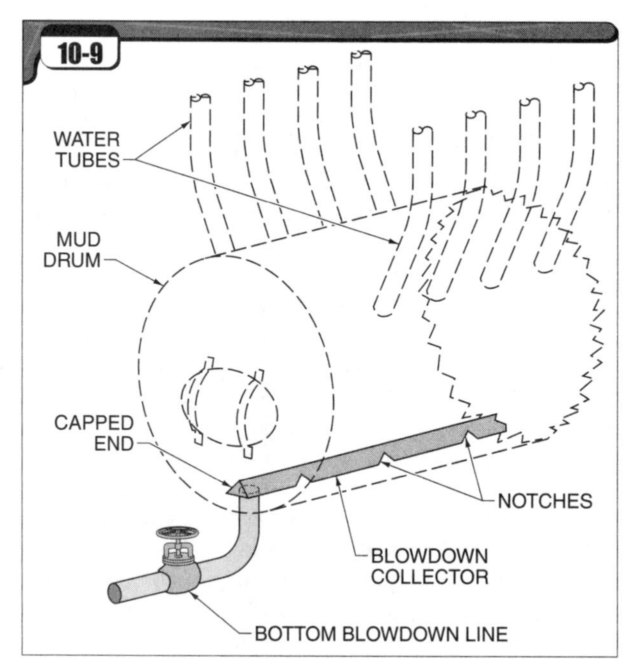

Figure 10-9. *A blowdown collector helps remove sludge from the boiler quickly and reduces the amount of blowdown needed.*

Trade Tip

The amount of bottom blowdown needed in modern boilers is often much less than in the past due to advances in water treatment technology. For this reason, the amount of bottom blowdown used is often excessive.

29 How is the amount of recoverable flash steam calculated?

Flash steam may be recovered from any source of high-temperature water. The amount of the continuous blowdown stream that will flash into steam may be calculated as follows:

$$FS\% = \frac{S_H - S_L}{L_L} \times 100$$

where

$FS\%$ = amount of water that flashes into steam (in %)

S_H = sensible heat at the higher pressure (in Btu/lb)

S_L = sensible heat at the lower pressure (in Btu/lb)

L_L = latent heat at the lower pressure (in Btu/lb)

The values of sensible and latent heat may be obtained from the table, "Properties of Saturated Steam" in the Appendix. These tables are commonly called the Steam Tables.

For example, the continuous blowdown from a boiler is directed into a flash tank. The pressure in the boiler is 100 psig (114.7 psia), and the pressure in the flash tank is maintained at 10 psig (24.7 psia). From the Steam Tables, the sensible heat at 115 psia is 309 Btu/lb. At 24.7 psia, the sensible heat is 208 Btu/lb and the latent heat is 952.9 Btu/lb. The percentage of the continuous blowdown stream that will flash into steam is calculated as follows:

$$FS\% = \frac{S_H - S_L}{L_L} \times 100$$

$$FS\% = \frac{309 - 208}{952.9} \times 100$$

$$FS\% = \frac{101}{952.9} \times 100$$

$$FS\% = 0.10 \times 100$$

$$FS\% = \mathbf{10\%}$$

Under these conditions, 10% of the blowdown water is recovered as usable steam at atmospheric pressure. This steam can be used for other purposes. The results of this type of calculation can be summarized in tables and graphs. **See Figure 10-10.**

30 How may the savings be calculated for the recovery of flash steam?

With good metering, the blowdown flow rate may be measured. Therefore, the quantity of flash steam may be quantified. For example, a boiler is online and the continuous blowdown is in service for 24 hr/day, 350 days/year. Metering indicates that the steam flow from the boiler averages 75,000 lb/hr and the feedwater flow averages 82,400 lb/hr. There is no condensate return. The blowdown flow is the difference between the feedwater flow and the steam flow, or 7400 lb/hr (82,400 – 75,000 = 7400).

Based on the boiler pressure and flash tank pressure, it is determined that 10% of the blowdown water will flash into steam. If the cost of steam at this plant is $6.90/Klb and the bottom blowdown is considered to be negligible, the value of the flash steam produced from the continuous blowdown is calculated as follows:

$7400 \times 10\% = 740$ lb/hr flash steam

$740 \div 1000 = 0.740$ Klb/hr flash steam

$0.740 \times \$6.90 = \$5.10$/hr

$\$5.10 \times 24 \times 350 =$ **$42,840/yr**

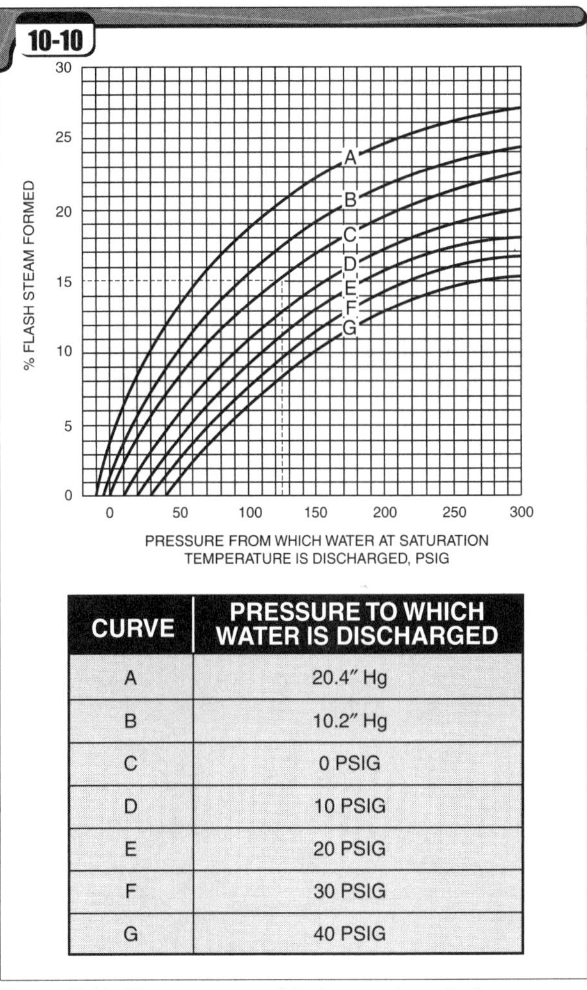

CURVE	PRESSURE TO WHICH WATER IS DISCHARGED
A	20.4″ Hg
B	10.2″ Hg
C	0 PSIG
D	10 PSIG
E	20 PSIG
F	30 PSIG
G	40 PSIG

Figure 10-10. *The percentage of flash steam formed when water at saturation temperature is discharged to a lower pressure is predictable.*

31 How may heat be recovered from blowdown water after flash steam has been removed?

The heat that remains in the blowdown water after the steam has flashed off in the flash tank may be recovered by passing the remaining hot water stream through a heat exchanger. For example, the remaining blowdown water may be used to preheat the makeup water being pumped to the deaerator. The cooled blowdown water then flows down the drain.

How may a boiler be kept hot and ready to place into service quickly?

In plants where the capacity of an additional boiler may be needed on short notice, it may be necessary to keep one boiler in standby condition. That is, the boiler is kept warm or hot so that it may be brought up to operating conditions more quickly than if started from cold conditions. The problem with this mode of operation is that it takes an appreciable amount of fuel to maintain the idle boiler in standby condition.

If the boilers are equipped with continuous blowdown, the blowdown from an operating boiler may be used to keep the standby boiler warm or hot. **See Figure 10-11.** This requires the addition of some piping and valves that are affected by ASME Code, but the end result is that the standby boiler may be kept warm or hot at no cost.

The continuous blowdown line from the operating boiler runs into the bottom blowdown line from the standby boiler. The air cock (drum vent or boiler vent) connection on the standby boiler is used as the outlet. Thus, the hot blowdown water from the operating boiler passes through the standby boiler on its way to the blowdown tank and drain. By using this approach, the standby boiler is also maintained in a wet storage condition because it is kept filled to the top with this treated water. When the standby boiler is needed, the boiler operator diverts the continuous blowdown flow from the operating boiler to other heat recovery equipment or to a blowdown tank. Then the water level in the standby boiler is lowered to the NOWL and the standby boiler is fired.

How may package boilers be kept warm during periods when there is little or no steam demand?

If it is not feasible to use waste heat to keep a boiler warm during short-term standby periods, the boiler may be maintained in warm condition using an automatic temperature switch installation. In these cases, the operating limit pressure control is disabled and control of the boiler is switched to temperature control. A temperature-sensing probe is installed to sense the temperature of the boiler water. This probe causes a temperature switch to open or close, which in turn causes the boiler's burner to start, run briefly, and shut down. This keeps the boiler water at about 200°F to 250°F so that the boiler may be started quickly when it is needed.

Factoid

Flash steam may be recovered from high-pressure condensate, boiler blowdown, or other sources of high temperature water. The economic payback period for boiler blowdown heat recovery equipment is usually very short.

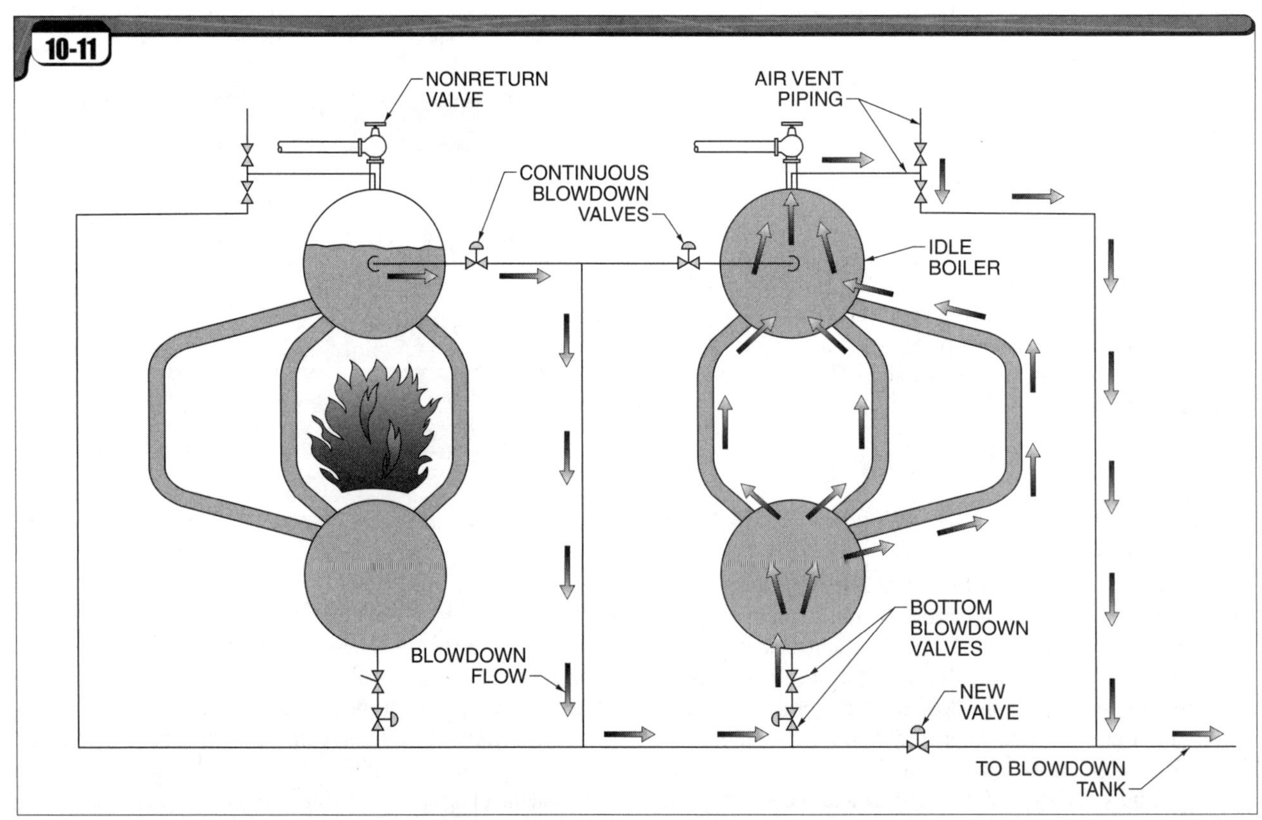

Figure 10-11. *The continuous blowdown from an operating boiler may be used to keep a standby boiler hot.*

Steam Distribution

34 How may the benefits of adding insulation to piping be quantified?

In order to quantify the benefits of adding insulation, the cost of losses from uninsulated piping must be calculated. If a 250′ section of uninsulated piping were losing $17,782/yr, adding insulation could prevent most of this loss. For example, if the piping were properly insulated, a conservative rule of thumb used in industry is that the loss would be reduced by 75%. The energy savings may be calculated as follows:

$$savings = \$17,782 \times 0.75$$

$$savings = \mathbf{\$13,336}$$

35 How may radiant heat losses from valves or other components of irregular shape be minimized?

Large steam system components with irregular shape, such as valves and large steam traps, are often left uninsulated. In these cases, removable and reusable insulation jackets should be used to prevent energy losses. **See Figure 10-12.** These insulation jackets may be custom fitted for covering large or irregularly shaped components. This allows the insulation to be reused many times after making repairs or adjustments.

Figure 10-12. *Removable and reusable insulation jackets make it easy and convenient to insulate irregularly shaped objects.*

> **Factoid**
>
> ***Boiler operators should avoid the tendency to run "favorite" boilers and instead use those capable of burning the least expensive fuel and meeting the current steam demand while operating in their most efficient load range.***

> **Factoid**
>
> ***Low firing rates require more excess air to ensure adequate turbulence for complete combustion. This is a less efficient mode of operation; therefore, firing at low firing rates for extended periods of time should be avoided.***

36 How may the heat losses from uninsulated piping be quantified?

Industry tables, charts, and software are available that quantify the heat losses from both insulated and uninsulated piping. **See Figure 10-13.** These charts are available from organizations such as the American Society of Heating, Refrigeration, and Air-Conditioning Engineers (ASHRAE), the National Insulation Association (NIA), and manufacturers of equipment such as steam traps and heat transfer equipment.

The number of hours that the piping is in service multiplied by the condensation rate in pounds per hour (lb/hr) equals the total steam loss. Since the cost of the steam is normally given in dollars per thousand pounds of steam, the loss per year is then divided by 1000 and multiplied by the cost of steam. This determines the cost of the heat loss and may be used to justify the cost of insulating the piping.

For example, a plant has identified 250′ of uninsulated 4″ steam piping that is used for 180 psig steam. This section of piping is in service 24 hr/day, 365 days/yr. The piping is inside a building and is surrounded by still air at 70°F. The amount of condensate can be determined from the table. The number 1.25 is found at the intersection of the row for 4″ piping and the column for 180 psig steam. This means that 1.25 lb of steam will condense per hour per foot of uninsulated pipe due to the loss of radiant heat from the piping. Since the uninsulated section of piping is 250′ in length, the steam loss will be 312.5 lb/hr (1.25 × 250 = 312.5).

This may be multiplied by the cost of steam at the particular facility to determine the value of the lost steam. For example, if the cost of steam is $6.50/1000 lb, the cost of lost steam can be calculated as follows:

$$cost = \frac{W_w}{1000} \times steam\ cost$$

where

$cost$ = cost of lost heat (in $)

W_w = weight of steam lost (in lb/hr)

$steam\ cost$ = cost of steam (in $/lb)

$$cost = \frac{312.5}{1000} \times 6.50$$

$$cost = 0.3125 \times 6.50$$

$$cost = \$2.03/\text{hr}$$

$2.03/hr × 24 hr/day × 365 day/yr = **$17,782/yr**

Note: The tables in Figure 10-13 assume that the piping is surrounded by a still atmosphere at 70°F. If the piping is located outdoors in cold conditions, the losses may be several times higher than shown.

CONDENSATION IN UNINSULATED PIPES*										
PSIG		15	30	60	125	180	250	450	600	900
Pipe Size	Area†	Pounds of Steam Condensed per Hour Per Lineal Foot								
1	0.344	0.19	0.22	0.26	0.33	0.37	0.41	0.51	0.57	0.69
2	0.622	0.32	0.40	0.47	0.59	0.66	0.74	0.91	1.04	1.25
3	0.916	0.47	0.58	0.70	0.87	0.97	1.09	1.35	1.52	1.84
4	1.178	0.56	0.75	0.90	1.11	1.25	1.40	1.73	1.96	2.36
6	1.735	0.82	1.10	1.32	1.64	1.85	2.06	2.55	2.89	3.48
8	2.260	1.04	1.44	1.72	2.14	2.40	2.69	3.32	3.76	4.53
10	2.810	1.18	1.79	2.14	2.66	2.99	3.34	4.13	4.68	5.64
12	3.340	1.51	2.13	2.55	3.16	3.55	3.97	4.91	5.56	6.70
14	3.670	1.64	2.34	2.80	3.47	3.90	4.36	5.39	6.11	7.36
20	5.250	2.34	3.34	4.00	4.96	5.59	6.24	7.71	8.74	10.53
24	6.280	2.80	4.00	4.79	5.94	6.68	7.46	9.22	10.45	12.60

CONDENSATION IN INSULATED PIPES*										
PSIG		15	30	60	125	180	250	450	600	900
Pipe Size	Area†	Pounds of Steam Condensed per Hour Per Lineal Foot								
1	0.344	0.05	0.06	0.07	0.10	0.12	0.14	0.186	0.221	0.289
2	0.622	0.08	0.10	0.13	0.17	0.20	0.23	0.320	0.379	0.498
3	0.916	0.12	0.14	0.18	0.24	0.28	0.33	0.460	0.546	0.714
4	1.178	0.15	0.18	0.22	0.30	0.36	0.43	0.578	0.686	0.897
6	1.735	0.20	0.25	0.32	0.44	0.51	0.59	0.809	0.959	1.253
8	2.260	0.27	0.32	0.41	0.55	0.66	0.76	1.051	1.244	1.628
10	2.810	0.32	0.39	0.51	0.68	0.80	0.94	1.301	1.542	2.019
12	3.340	0.38	0.46	0.58	0.80	0.92	1.11	1.539	1.821	2.393
14	3.670	0.42	0.51	0.65	0.87	1.03	1.21	1.688	1.999	2.624
20	5.250	0.58	0.71	0.91	1.23	1.45	1.70	2.387	2.830	3.725
24	6.280	0.68	0.84	1.09	1.45	1.71	2.03	2.833	3.364	4.434

* Carrying saturated steam in still air at 70°F
† sq ft/lineal ft

Figure 10-13. *The heat loss from insulated and uninsulated steam piping may be quantified using industry standard charts.*

37 How can the losses be calculated for unused steam mains or branch piping?

When steam-using equipment is decommissioned or removed, the steam branch piping and/or steam mains feeding that equipment are often left in place. This piping is subsequently kept filled with steam, but serves no useful purpose. These unused steam pipes are referred to as "dead piping." Even though a steam main or branch line may be properly insulated, it still radiates some heat that could be stopped by shutting off or removing the dead piping.

For example, a section of 6″ diameter steam line 140′ in length is no longer used, but is left pressurized with 250 psig steam. This piping is properly insulated, is inside a building, and is surrounded by still air at 70°F. This dead piping is pressurized with steam 24 hr/day, 360 days/yr, and the cost of steam is $6.75 per 1000 lb. From the table of condensate in insulated pipes, the heat loss through the insulated piping will allow 0.59 lb of steam to condense per hour, per lineal foot of pipe. The cost of the wasted steam is calculated as follows:

$$cost = \frac{W_w}{1000} \times steam\ cost$$

$$cost = \frac{0.59}{1000} \times \$6.75$$

$$cost = 0.00059 \times \$6.75$$

$$cost = \$0.0040\text{/hr per ft}$$

$$cost = \$0.0040\text{/hr per ft} \times 140' \times 24\text{ hr/day} \times 360\text{ day/yr}$$

$$cost = \mathbf{\$4838/yr}$$

38 What is the justification for performing a steam trap survey?

A *steam trap survey* is the process of identifying, testing, and documenting the condition of all steam traps in a facility. This process serves many purposes. One purpose is to identify all the steam traps and their duty. During the steam trap survey process, all traps are tagged with a permanent numbered identification tag. **See Figure 10-14.** This reduces the time and cost required for future trap survey work.

A steam trap survey also serves to reveal the condition of all steam traps. Failed steam traps are identified as failed closed, failed open, or leaking. A steam trap survey helps to quantify the amount of steam being wasted by failed-open traps. In addition, the survey helps to identify poor steam trap applications. For example, steam traps that are undersized, oversized, or improperly installed are identified.

A steam trap survey also helps to prioritize repairs and testing of steam traps. The service vendor performing the steam trap survey should provide a report that documents the application and condition of all the steam traps. Most importantly, the report provides a list of the failed traps, prioritized by the cost of wasted steam and their impact on the plant processes. Even in a facility that performs steam trap surveys annually, a normal failure rate is 5% to 7%.

Steam trap surveys also help the plant reduce spare parts costs. For example, a steam trap surveyor is often able to help the plant standardize on a small number of steam trap brands and types that serve the great majority of the plant's needs. This allows the plant to inventory fewer models and sizes of steam traps and repair parts.

Trade Tip

The cost of a steam trap survey is almost always paid back many times over because the unnecessary waste of steam through failed-open traps is identified and can be stopped. During the course of a day, a steam trap may open and close hundreds of times. If the trap fails open, it blows steam continuously, costing thousands of dollars annually. If the trap fails closed, condensate will flood the system and cause extensive damage.

39 How does wasted steam have an impact on air emissions and environmental permit compliance?

Steam that is wasted through leaks, poor or nonexistent insulation, failed-open steam traps, flash steam losses, etc., must be replaced by the boiler. This involves burning more fuel and creates unnecessary stack emissions. **See Figure 10-15.** In addition, flue gas cleaning equipment must work harder and must capture more waste material, such as sulfur products and ash, that must be disposed of. In addition, more boiler blowdown and cooling water are required and this increases the wastewater treatment costs for the plant.

Figure 10-14. *All steam traps are identified with a permanent numbered tag as part of a steam trap survey.*

40 How may the steam loss through a steam trap or other orifice or leakage path be quantified?

The quantity of steam that will pass through an orifice can readily be calculated. This information is commonly presented in tabular form. **See Appendix.** The quantity of steam flow through a trap or orifice is based on the differential pressure across the orifice and can be estimated using Napier's formula as follows:

$$W_w = \frac{A \times dP}{70} \times 3600$$

where

W_w = weight of wasted steam (in lb/hr)
A = area of orifice (in sq in.)
dP = differential pressure across orifice (in psi)

For example, what is the steam loss through a ¼″ diameter orifice if the steam pressure is 100 psig and the steam is escaping to the atmosphere? The differential pressure is the difference between the pressure entering the orifice and the pressure leaving the orifice. The differential pressure is 100 psi (100 – 0 = 100). The orifice is circular and the area is calculated as follows:

$$A = \pi \times r^2$$

where

A = area (in sq in.)
r = radius (in in.)

$$A = \pi \times r^2$$
$$A = 3.14 \times (0.25 \div 2)^2$$
$$A = 3.14 \times (0.125)^2$$
$$A = 3.14 \times 0.0156$$
$$A = 0.049 \text{ sq in.}$$

$$W_w = \frac{A \times dP}{70} \times 3600$$

$$W_w = \frac{0.049 \times 100}{70} \times 3600$$

$$W_w = \frac{4.9}{70} \times 3600$$

$W_w = 0.070 \times 3600$
$W_w =$ **252 lb/hr**

Factoid

Even though condensate is steam that has given up its latent heat, it still contains sensible heat that came from the fuel. Condensate piping should therefore be insulated to allow this sensible heat to be reused.

41 What is the steam loss through a failed-open steam trap that has a ⅛″ orifice if the steam pressure is 150 psig and there is 15 psig of backpressure on the steam trap?

The differential pressure is 135 psi (150 – 15 = 135). The area is calculated as follows:

$A = \pi \times r^2$
$A = 3.14 \times (0.125 \div 2)^2$
$A = 3.14 \times (0.0625)^2$
$A = 3.14 \times 0.003906$
$A = 0.0123$ sq in.

$$W_w = \frac{A \times dP}{70} \times 3600$$

$$W_w = \frac{0.0123 \times 135}{70} \times 3600$$

$$W_w = \frac{1.66}{70} \times 3600$$

$W_w = 0.0237 \times 3600$
$W_w =$ **85.3 lb/hr**

Condensate Return

42 How may the costs be quantified for wasted condensate?

The costs that are easiest to quantify are the costs of the water, the heat energy in the water, and the chemical treatment. Of these three, the heat energy generally makes up about 80% to 90% of the costs. By determining the value of 1% of the condensate, the savings to be realized through various improvement projects may be calculated. The cost of wasted condensate is calculated as follows:

$$value = \frac{(T_C - T_M) \times L \times Hr \times C_f}{1{,}000{,}000 \times FSE \times V_C} \times 0.01$$

where
T_C = temperature of condensate (in °F)
T_M = temperature of makeup water (in °F)
L = average steam load (in lb/hr)
Hr = time of operation of the steam plant (in hr)
C_f = fuel cost (in \$/million Btu)
FSE = fuel-to-steam efficiency fraction for boiler
V_C = fraction of condensate value attributable to fuel

For example, a boiler plant has an average steam load of 50 Klb/hr. The returning condensate averages 200°F and the makeup water averages 55°F. The steam plant runs 24 hr/day, 360 days/yr. The boilers use natural gas at a cost $8.50 per million Btu. The fuel-to-steam efficiency of the boilers averages 70% and a typical value of the fraction of condensate value attributable to the value of the fuel is 85%. What is the value of 1% of the condensate return on a daily basis?

$$value = \frac{(T_C - T_M) \times L \times Hr \times C_f}{1{,}000{,}000 \times FSE \times V_C} \times 0.01$$

$$value = \frac{(200 - 55) \times 50{,}000 \times 24 \times 8.50}{1{,}000{,}000 \times 0.70 \times 0.85} \times 0.01$$

$$value = \frac{1{,}479{,}000{,}000}{595{,}000} \times 0.01$$

$value = 2486 \times 0.01$
$value =$ **\$24.86/day**

ENVIRONMENTAL POLLUTANTS—BLOW-THROUGH FAILURE OF ONE STEAM TRAP*							
Fuel Type	Firing Method	Quantity	Particulate Matter†	Sulfur Oxides†	Nitrogen Oxides†	Carbon Monoxide†	Total Carbon‡
Bituminous coal	spreader stoker	11.9 tons	715	465	167	60	9.7
#2 fuel oil	industrial commercial boiler	2531 gal.	5.1	359	51	12.7	7.9
Natural gas	industrial bolier	314,000 cu ft	0.94	0.19	44	11	5.3

* Live steam loss of 27 lb/hr, operating 8400 hrs/yr, drip trap with 5⁄64″ orifice, 150 psig inlet, 0 psig to 65 psig outlet
† lb/yr
‡ ton/yr

Figure 10-15. *Wasted steam results in increased emissions of environmental pollutants.*

How can the costs be justified for improvements to the condensate return system?

From the previous example, the boiler plant could save \$24.86/day if 1% of the condensate could be recovered and reused. This is \$8950/yr (24.86 × 360 = 8950). If the percentage of condensate return could be increased through improvement projects, the savings would total as follows:

Condensate Recovery	Annual Savings
1%	1 × 8950 = \$8950
10%	10 × 8950 = \$89,500
25%	25 × 8950 = \$223,750
50%	50 × 8950 = \$447,500

Why should condensate return piping be insulated?

The heat in the condensate comes from the fuel that is burned in the plant's boilers. Therefore, even though condensate is no longer useful in the plant processes, the residual heat in the condensate should be returned to the boilers. It takes less heat energy to make steam from water that is already hot than from cold makeup water. Insulating the condensate piping helps the plant reuse this expensive heat energy.

Why is condensate sometimes intentionally diverted to the plant sewer system rather than being returned to the boiler plant?

A leak in heat exchange equipment could result in the process material contaminating the condensate. For example, a leak in a refinery heat transfer process could result in petroleum oil products being returned to the boilers with the condensate. This could result in foaming, carryover and resulting water hammer, steam system outages to clean the residue from the boilers, associated production plant downtime, and many other problems.

Sometimes the condensate may have been diverted to drains or to the ground in years past simply to reduce the maintenance workload and costs. For example, steam traps that discharge directly onto the ground do not require maintenance of condensate return piping or repair of frozen piping.

In many cases, the difficulties associated with returning the condensate are at least partially a result of lack of knowledge of alternatives. For example, misapplied or improperly piped steam traps may not keep the system properly drained. This results in freezing, accelerated corrosion, and water hammer. A decision may be made to simply waste the condensate because of a lack of knowledge as to how to remedy the problems.

How may the vent losses from a deaerator be minimized?

The volume of air/steam mixture that may be vented is often limited through the use of a small, appropriately sized orifice in the vent line. In cases where a manually adjusted vent valve is used, boiler operators often leave the vent open too much. This allows an excessive amount of steam to be lost while venting the non-condensable gases. When a manual vent valve is used, the vent opening should be restricted a little at a time under steady state, high load conditions until the point is reached where the dissolved oxygen content of the feedwater leaving the deaerator begins to increase. Then the vent is reopened slightly.

If the deaerator vent must be left open excessively to control pressure in the deaerator, this indicates maintenance or operational issues that should be addressed. For example, the pressure reducing valve or control valve supplying steam to the deaerator may not be controlling the steam pressure adequately. Steam traps may be failed open, allowing excessive steam to return to the deaerator with the condensate or with flash steam from a flash tank. In some plants, steam turbines are used to drive equipment such as pumps or fans, and the exhaust steam from the steam turbines is returned to the deaerator. If this exhaust steam results in excessive pressure in the deaerator, one or more of the steam turbines should be taken off line and an electric motor drive used instead.

To further minimize steam losses, a vent condenser may be used. **See Figure 10-16.** A vent condenser uses the steam that would otherwise escape from the vent pipe to preheat the incoming feedwater. Air and other non-condensable gases will pass through the vent condenser and discharge from the vent pipe, while much of the steam is condensed.

Figure 10-16. *Vent condensers help minimize the amount of steam lost through the deaerator vent.*

Chapter **10** — **Trade Test**

Boiler System Optimization

Name: ______________________ Date: ______________________

T F **1.** Efficiency is the comparison of a piece of equipment's energy output to its energy input.

______________ **2.** Thermal efficiency of a boiler is the percentage of the ___.
A. heat liberated in the furnace that is transferred into the boiler water
B. Btu content of the fuel that is liberated as heat in the furnace
C. Btu content of the fuel that is converted to Btu content of the steam
D. heat transferred into the boiler water over a typical range of loads

T F **3.** It is impossible to have high combustion efficiency and low thermal efficiency at the same time.

______________ **4.** The steam rate of a boiler is normally expressed as the average ___.
A. fuel consumption divided by the steam production
B. steam production divided by the rated capacity
C. number of pounds of steam produced per unit of fuel
D. percentage of steam generated that is used for its intended purpose

______________ **5.** The industry standard used in expressing the cost of steam production is ___.
A. pounds of steam generated per dollar of fuel
B. dollars per thousand pounds of steam generated
C. dollars worth of steam consumed per manufactured unit
D. the sum of labor costs, chemical costs, and steam system equipment depreciation

______________ **6.** The waste of steam results in short-term or long-term costs due to ___.
A. increased equipment maintenance costs
B. reduced life of the steam system equipment
C. increased costs associated with environmental emissions
D. all of the above

T F **7.** The overall efficiency of a facility's steam system must take into account both the supply side and the demand side.

T F **8.** It is difficult or impossible to accurately determine the amount of fuel consumed only by boilers when only one fuel meter is used to measure fuel consumed by all fuel users in a plant.

______________ **9.** The two standard ASME International methods used for calculating the efficiency of a boiler are the ___ method and the ___ method.
A. supply; demand
B. input-output; heat loss
C. thermal efficiency; combustion efficiency
D. steam rate; fuel-to-steam

10. List six categories of losses that occur in a typical steam system.

T F **11.** Hydrogen in hydrocarbon fuels is converted to water vapor during combustion.

______________ **12.** Frequent purging of a boiler furnace is undesirable because ___.

A. it causes excessive stresses on the boiler metal due to expansion and contraction
B. extra heat is lost up the stack during the purge period
C. it is an indication that the boiler capacity is too great for the load
D. all of the above

______________ **13.** A substantial inrush of air into a boiler room when an entry door is opened is a sign that the ___.

A. room is well insulated against heat loss
B. door should remain open at all times
C. boiler efficiency is reduced while the door is open
D. ability of combustion air to enter the boiler room is inadequate

______________ **14.** The furnace pressure in a boiler configured for balanced-draft or induced-draft should be ___.

A. below atmospheric only enough to prevent combustion vapors from escaping during firing rate changes or minor upsets
B. –1.0″ WC at all times
C. below atmospheric enough to entrain all possible fly ash into the flue gas stream so that it can be captured by the pollution-control equipment
D. –1.5″ WC except when operating sootblowers

______________ **15.** Leaks of atmospheric air into the boiler furnace cause ___.

A. reduced efficiency due to cooling of the flue gases in the furnace
B. potential for localized corrosion due to condensation of moisture by the cool air stream
C. increased load on the combustion air and flue gas fans
D. all of the above

T F **16.** When testing the combustion efficiency of a modulating burner, the flue gases should be analyzed at several points over the full range of the burner.

T F **17.** The amount of excess air used is usually higher at higher firing rates.

18. By what basic method does an oxygen trim system improve the combustion efficiency of a boiler?

19. How can a boiler operator detect the development of scale on the boiler?.

____________ **20.** As a rule of thumb, a ___°F increase in the combustion air temperature will result in a 1% overall boiler efficiency improvement.
A. 1
B. 15
C. 40
D. 100

____________ **21.** Boilers are generally the most efficient at absorbing heat from the fuel when they are operating at about ___% of their capacity.
A. 25
B. 50
C. 75
D. 100

____________ **22.** Operating a large boiler at a low firing rate for extended periods of time is ___.
A. efficient because there is very little wear on the boiler under this condition
B. inefficient because the heat absorption of the boiler is not optimum at low fire
C. efficient because the heat absorption of the boiler is best at low fire
D. not applicable because the firing rate has only a negligible effect on efficiency in a properly designed boiler

____________ **23.** The proper amount of bottom blowdown to be used is ___.
A. two minutes per shift
B. three times per shift for 10 sec at each use
C. only as needed to remove the actual amount of sludge that accumulates in the boiler
D. only as needed to remove impurities that lead to high surface tension

____________ **24.** The continuous blowdown flow should be increased or decreased as needed in order to ___.
A. maintain a blowdown flow rate as dictated by the boiler manufacturer
B. maintain the boiler water conductivity within a specified range
C. ensure that no sludge accumulates in the bottom of the boiler
D. prevent the hardness in the boiler water from exceeding 10 ppm

T F **25.** Flash steam that separates from blowdown water is not suitable for use in plant applications due to the high concentration of impurities in the blowdown water.

____________ **26.** A total of 640 lb/hr of flash steam may be recovered by sending condensate to a flash tank. The steam system operates 350 days/yr and 24 hr/day. If the cost of steam is $7.75/Klb of steam, the annual value of the flash steam is ___.

T F **27.** A boiler operator should always keep an extra boiler on-line to guard against the loss of steam pressure in case of a boiler failure.

_______________ **28.** A 160′ section of uninsulated 8″ piping contains steam at 250 psig. The piping is inside a building and surrounded by still air at about 70°F. The amount of steam wasted as a result of heat loss is ___ lb/hr.

A. 2.69
B. 21.52
C. 430.4
D. none of the above

29. List five things that are accomplished during a properly performed steam trap survey.

T F **30.** The waste of steam results in increased emissions of pollutants from the boiler stacks as additional fuel is burned to replace the wasted steam.

_______________ **31.** A steam trap with a ¼″ orifice is blowing through continually into a condensate return system. The steam pressure supplied to the steam trap is 125 psig and the backpressure at the trap is 10 psig. The failed steam trap wastes ___ lb/hr of steam.

A. 0.8
B. 290
C. 461
D. 1478

_______________ **32.** A steam trap with a ⅛″ orifice has failed open and discharges onto the ground. The trap is supplied with steam at 400 psig. If it discharges 24 hr/day, 365 days/yr, and the cost of steam is $7.80 per 1000 lb, the annual cost of the wasted steam is ___.

A. $2847
B. $5490
C. $17,240
D. $551,700

33. List several costs associated with the waste of condensate.

_______________ **34.** A boiler plant generates 145,000 lb/hr for 363 days/yr. The returning condensate averages 180°F and the makeup water averages 60°F. The fuel cost is $7.35 per million Btu. The boiler efficiency is 76%, and 85% of the cost of the condensate is attributed to the heat from the fuel. The annual savings that will be realized if the plant increases its condensate return by 10% is ___.

A. $35,550
B. $146,602
C. $172,473
D. $320,166

______________________ **35.** The continuous blowdown from a large boiler is directed into a flash tank, from which the flash steam is directed to the plant heating system. The pressure in the boiler is 150 psig and the pressure maintained in the flash tank is 10 psig. At 150 psig, the blowdown water contains 338.6 Btu/lb of sensible heat. At 10 psig, the water contains 207.9 Btu/lb of sensible heat and 952.9 Btu/lb of latent heat. The percentage of the blowdown water that will flash to steam is ___%.

A. 2.2 B. 13.7

C. 38.6 D. 63.5

Chapter 11

Licensing

The requirements related to the licensing of boiler operators vary. Some states in the U.S. require boiler operators to become licensed, while others do not. Some municipal and county governments require operator licensure even though the states in which they are located do not. In the locations requiring licensed boiler operators, both the exams and the requirements associated with the exams vary considerably. This includes variations in exam-related fees, prerequisite practical experience, and the formats of the exams themselves. Even in jurisdictions in which boiler operator licensure is not required, some employers use similar exams as selection or advancement criteria.

The likelihood of success in licensing exams may be enhanced through proven preparation methods. Some methods help in understanding and retaining needed information. Others help in successfully completing the exam itself. Taking several practice exams often helps reduce anxiety and improve the chances for success.

Exam Requirements

What jurisdictions are responsible for administering Boiler Operator licensing exams?

About one third of the states in the United States maintain a state boiler operator licensing program. There are also a number of individual municipalities and counties that require licensure even though the states they are in do not. The levels of licensure vary from jurisdiction to jurisdiction. For example, some jurisdictions require licensure for high-pressure boilers but not low-pressure boilers.

Most of the state agencies in the U.S. and provincial agencies in Canada that are responsible for licensing of boiler operators are also responsible for boiler inspections. If it is unclear whether a particular jurisdiction maintains a boiler operator licensing program, the agency having jurisdiction over boiler inspections should be able to provide information. These agencies usually also have a representative to the National Board of Boiler and Pressure Vessel Inspectors, and that individual can provide information about boiler operator licensing in the jurisdiction.

How do boiler operator exams vary between jurisdictions?

There are considerable differences in the boiler operator licensing exams administered by various jurisdictions. These differences include the nature of the exams, the costs associated with taking the exams, and the various levels of license that may be obtained.

The formats of the exam questions vary between jurisdictions and may include multiple choice, fill-in-the-blank, true/false, and essay questions. In Arkansas, the exam may be either written or oral. The exams in Massachusetts also include some oral portions. Some jurisdictions have examiners who administer and grade the exams, while others use exam proctors. These exam proctors distribute and collect the exam papers and then use computerized scanners to grade the answer sheets. Private companies are sometimes contracted by the jurisdiction to administer the exams. It is always advisable to contact the jurisdiction administering the exam in advance to find out what to expect.

In general, the licensing exams are written or written/oral, and do not require the demonstration of any tasks or procedures. Some questions may require the identification of boiler components on a sketch. New Mexico allows textbooks to be used during the exam, but in most cases resource materials are not allowed.

The costs associated with taking a licensing exam are established by the jurisdiction. There is often a fee for the processing of the application to take the exam, as well as a fee for the exam itself. Once the exam is completed successfully, there may be an additional fee for issuance of the license and an annual renewal fee. The agencies giving the exam often require that these fees be paid by certified check or money order, rather than personal check or cash. Both the fees and the required method of payment should be confirmed in advance of the exam.

Most jurisdictions that maintain boiler operator licensing programs have separate levels of licensure, such as for operators of low-pressure boilers and of high-pressure boilers. There may also be additional licensing for those who operate equipment of a specified minimum horsepower. New Jersey, for example, uses color-coding for the various levels of licensing. Boiler operators may hold a black seal, blue seal, red seal, or gold seal license. These various levels of licensure have specific requirements with regard to experience time, training, and horsepower of the equipment with which the boiler operator has accumulated experience.

Is there any form of national license for boiler operators?

There is currently no uniform boiler operator license that is recognized throughout the United States. However, part of the Clean Air Act required the U.S. Environmental Protection Agency (EPA) to develop and promote a model program for the qualification and certification of operators in high-capacity fossil fuel-fired boiler plants. This model program applies to boilers with a heat input of 10 million Btu/hr or greater. In response, the EPA requested that ASME International develop and manage a nationwide certification program for boiler operators. This program is known as the High Capacity Fossil Fuel Fired Plant Operator program, or QFO. It became available in October 1994.

The program is not mandatory, and it is not a licensing program but a certification program. The intent is to allow boiler operators to move between jurisdictions while maintaining certification of their knowledge. There is no affiliation with state or other governmental jurisdictions.

Two nongovernmental organizations that promote the idea of national licensing are ASOPE and NIULPE. ASOPE is the American Society of Power Engineers. NIULPE is the National Institute for the Uniform Licensing of Power Engineers. The objective of these organizations is to establish uniform national standards and a formal exam process for operators of boilers and related equipment.

What resources or study aids may be provided by the jurisdiction that administers the exams?

The various agencies administering the licensing exams often provide information as to the texts and materials from which they have taken exam questions. They may also make available a listing of the subjects that could be included on the exams. In some locations, these resources are available on web sites maintained by the jurisdiction.

What qualifications or prerequisites must a person satisfy in order to sit for a jurisdictional licensing exam?

Each jurisdiction has specific requirements covering job experience and other prerequisites. Typically, the applicant must be at least 18 years of age. A number of jurisdictions require that the applicant be able to document hands-on experience working with steam boilers and auxiliary equipment under the guidance of experienced personnel.

This experience may be obtained in a number of various ways. Most applicants learn the trade by working with experienced boiler operators at their places of work. For example, large plants such as refineries, steel mills, and chemical manufacturers have sophisticated boiler systems. Those new to the operation spend considerable time in secondary capacities under the guidance of lead operators or supervisors before being allowed to operate on their own. It is not unusual for a large facility to hire unlicensed trainees with the understanding that the trainee must study and accumulate appropriate experience so as to be ready to sit for the licensing exam within a prescribed period of time.

In Ohio, steam boilers must have 360 sq ft of heating surface or more to require attendance by a licensed boiler operator. To qualify to take the various Ohio license exam levels, applicants must be able to document a minimum amount of practical experience. However, experience gained operating most boilers with less than 360 sq ft of heating surface is acceptable toward the licensing exam qualifications. Therefore, many boiler operators work in facilities that use these smaller boilers but do not require licensed boiler operators. Once they have accumulated sufficient experience, they are then qualified to take the licensing exams.

The majority of jurisdictions that administer licensing programs require that an exam candidate's application to take the exam be endorsed or confirmed by one or more licensed boiler operators, the plant's chief engineer, or a representative of the plant management. Often this documentation must be notarized because it is a legal document.

When an applicant has obtained boiler operating experience in the U.S. Navy as a Boiler Technician or Machinist's Mate, the DD214 form (the form documenting a person's rank and classification upon discharge) provides adequate proof of experience. Other specific requirements may be made by the particular jurisdiction giving the exam. For example, the exam candidate may be required to show photo identification on the day of the exam.

If the exam is not successfully passed, there may be a waiting period before the exam may be taken again. In Minnesota, for example, the waiting period is 10 days. In Ohio it is 25 days.

Exam Preparation

Why is it helpful to use a variety of texts and study materials when preparing for a licensing exam?

Each textbook author, like most industry professionals, has a limited career exposure to the many types and styles of boilers and auxiliary equipment in the industry. In producing a trade textbook, an author often calls upon this personal experience and perspective. Authors and publishers also use different formats in presenting the material. In addition, the various texts provide a wide array of practice quizzes and other evaluation materials.

As a result, reviewing a variety of texts and study materials helps provide a broader understanding of each subject, a wider scope of terminology, and a more complete range of appropriate subject matter. No single text is likely to provide a comprehensive understanding of all potential exam material.

Some aspiring boiler operators prepare for the exam primarily by memorizing lists of questions and answers, making comparatively little effort to clearly understand either. Though this may be productive in the effort to pass the exam, it is poor practice in the long run. The exam is only a one-day gateway to a career. Working with boilers and associated equipment involves significant responsibility and potential hazards. Boiler operators should try to clearly understand all aspects of safety, efficiency, and longevity of the equipment.

What proven methods and techniques will best help a person prepare for a boiler operator licensing exam?

Preparatory courses are available in many locations and can be of great help in preparing for the exams. These courses are offered through vocational schools and community colleges, or by proprietary, privately run training schools. **See Figure 11-1.** In general, these programs greatly improve the chances of success on the exams. This is because the instructors of these courses are usually well informed and current as to the expected subject matter and the logistical issues associated with taking the exams and they structure the courses to provide good coverage of the relevant topics. These programs also generally provide the benefit of having many visual aids such as photographs, sectional models or parts, videos, and sometimes field trips to view operating equipment. Possibly the largest benefit of such a program is in its order and discipline. The class structure encourages a learning mindset and diligence in working toward the goal of passing the exam.

However, not all such courses have a good track record and it is advisable to do some research to find a program with a good reputation before enrolling. Ask for referrals from others who have taken such courses and inquire at the agency administering the exam as to which programs prepare the students well.

For individual studying, it is wise to develop a schedule with a block of time set aside each day. This block of time should be long enough to get organized and become focused on productive studying. In general, it is better to avoid marathon sessions. Once the mind is fatigued, forced study is of little benefit.

A productive study tool is to try to study during the period of the day when the brain is at its best. For example, some people are mentally sharp during the early morning hours; others find that they tend to get a "second wind" in the early evening, and so on. By capitalizing on these periods when the brain is most receptive, the benefit of studying may be enhanced.

Trade Tip

States and other governmental jurisdictions often provide helpful information about licensing exams on their web sites.

11-1

Gateway Community College

- TEST PREPARATION SHOULD BE PACED; DEVELOP A STUDY SCHEDULE FOR AN EXTENDED PERIOD OF TIME
- VERIFY THAT TEST IS CORRECT TEST FOR THE CERTIFICATION EXPECTED
- LEARN AS MUCH AS POSSIBLE ABOUT TEST; TALK TO PEOPLE WHO HAVE PASSED TEST
- REVIEW STUDY MATERIALS OVER SEVERAL DAYS; LIMIT ANY REVIEW OF MATERIAL THE NIGHT BEFORE
- IF TEST LOCATION IS UNFAMILIAR, DRIVE TO TEST SITE IN ADVANCE OF TEST DATE
- ALLOW AMPLE TIME FOR TRAVELING TO TEST SITE THE DAY OF TEST
- SCHEDULE A NORMAL AMOUNT OF SLEEP THE NIGHT BEFORE TEST

Figure 11-1. *Licensing exams are commonly given at local community colleges.*

List several study methods that may help in the retention of new material.

Set realistic reading goals. Most working adults tire after reading 8 to 12 pages of technical material. Take frequent short breaks to refresh the mind. There is little to gain from studying material that is already understood. Review this information only briefly, a few days before the exam. Have several different texts available while studying. A point that is unclear in one text may be well explained in another. Also, having a technical point explained in multiple ways often makes it become clear. Master the quizzes and tests found in the text materials. These are usually items to which the author assigned some importance.

Use a highlighter while reading new material. Highlight important items as they are spotted. This will make these items easier to find and reinforce later. Many people repeatedly try to read through an entire textbook in an unsuccessful attempt to complete it. This results in wasted time. Read a section of the material, calling attention to items that are not understood. Then research and resolve those items. Don't get frustrated; some points may take time to understand or require asking questions of a number of people before the answer is found. This effort often pays off in a solid retention of that material.

Explain things out loud. For example, while in the car, pretend to be explaining technical points to someone. This method often identifies aspects of the subject that are not clearly understood. Practice explaining the material in ways that are consistent with the method used on the licensing exam itself.

For example, if the exam is predominantly essay questions, practice writing out answers. This is especially important for those to whom writing does not come easily.

When possible, make a point of applying knowledge gained to actual practice. Perform related hands-on tasks under the guidance of an experienced mentor. For example, disassemble a component on a workbench and determine how it works. Discuss with the mentor how to tell if it wasn't working properly and what would need to be done to troubleshoot and remedy the problem.

Reinforce by repetition. For example, if a certain math formula is troublesome, have someone who is proficient with it provide a list of 10 to 20 exercises that require the correct use of that formula and check the results.

9 What tools are typically useful in memorizing technical materials?

Flash cards are often helpful in reinforcing materials by repetition. Sketching flow charts helps clarify sequences of events and interrelationships between pieces of equipment. Labeling components in sketches and photographs helps in learning nomenclature.

Pie charts are particularly helpful in memorizing conversions and math calculations having three components. **See Figure 11-2.** For example, the answer to a conversion and the two factors used in the conversion are written in a triangular style with the answer on top. The variable to be determined is covered (it is the unknown). The configuration of the other two shows how to calculate the unknown. If the top variable is covered, the bottom two are multiplied together. If one of the bottom two is covered, the top variable is divided by the visible bottom variable. For example, total force (TF) = pressure (P) × area (A). TF is written on the top of the pie chart and the P and A are written on the bottom. When TF is unknown, it is covered and $P \times A$ is shown. If P is the unknown, it is covered and $TF \div A$ is shown.

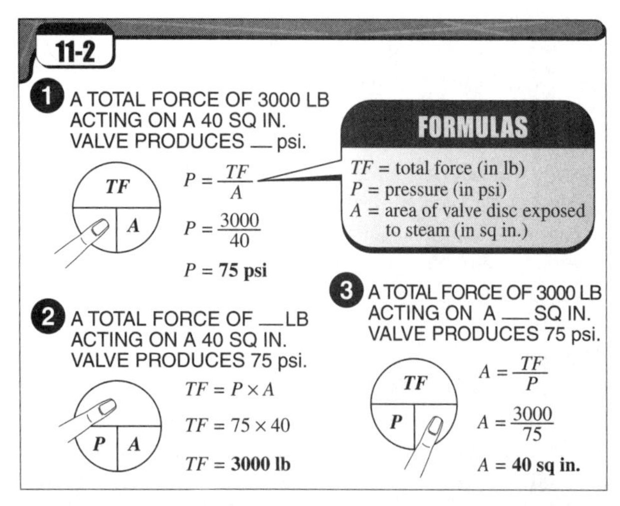

Figure 11-2. *Pie charts are useful in solving equations with three variables.*

10 What logistical aspects of the licensing exam should be prepared in advance?

Learn what to expect. Talk with individuals who have successfully completed the exam about the types of questions asked and the format of the exam. Confirm in advance any materials that are required. This may include documentation, personal identification, and checks for fees. The licensing jurisdictions usually require that necessary paperwork be submitted a minimum period before the exam in order to allow time for processing. Be sure to submit the paperwork in plenty of time.

Drive to the exam site prior to the exam day. This will help ensure timely arrival on the day of the exam and avoid the tension that would result from becoming lost on the way there. Get plenty of rest the night before the exam. Eat a sensible breakfast before the exam and avoid excessive caffeine. This will help ensure clarity of thought.

If calculators are permitted, take a solar-type calculator rather than a calculator with a battery to avoid the potential of the battery failing during the exam. Take extra pencils and erasers.

Taking the Exam

11 What proven methods will help ensure the best possible result on the exam itself?

Read each question carefully. Nervousness has a tendency to make the person taking the exam rush through the questions, and important points are often missed. Read the question word for word. If the test has multiple portions with a specified time allowed for each, pace your work according to the time allotted for each portion.

If the exam is graded by an electronic scanner and an erasure is necessary, erase thoroughly. Electronic scanners used for this purpose are sensitive and will detect any significant pencil residue on the paper. The software used is programmed to count as incorrect any question on which more than one answer is detected.

Do not become uncomfortable if others complete the exam first. Take enough time to be confident that the best possible answers were given. Do not leave any answers blank. Even a guess is better than no answer. If time permits, it is a good idea to read the entire exam before answering any of the questions. This is because a later question may reveal an answer to an earlier question. For this reason, you may wish to leave an unknown answer blank and come back to it. Be sure to return to this answer later and complete it. If the exam is of the essay type, leave room for writing the answer. Some jurisdictions use a pamphlet for the exam questions and a separate answer sheet. If a question is skipped, it is important to also skip the answer on the answer sheet. Otherwise the answers will not be in the correct order on the answer sheet.

If the answer to a multiple-choice question is not immediately known, two of the supplied answers can usually be eliminated quickly. Then make an educated, reasoned deduction as to the correct answer. The answer that comes to mind initially is more likely to be correct than an afterthought. Therefore, once an answer has been provided, resist the temptation to change it unless you are absolutely sure. Nervousness can make you second-guess yourself.

If a question calls for a math calculation, work the calculation several times. Confirm that the same answer is reached each time. This will eliminate mistakes due to a stray calculator button being pressed or a one-time mental error. In questions requiring math calculations, be aware that more information may be provided than is needed to perform the calculation. It is up to the person taking the exam to determine which factors are needed to arrive at the correct answer.

If a math calculation involves a common conversion such as converting Fahrenheit to Celsius, and you are not positive about the formula, substitute factors for which the answer is known. For example, 100°C equals 212°F. If these factors are entered into the formula and the answer is correct, then the formula must be correct. By using that formula with the variables provided, a correct answer will be reached.

When finished, review the answer sheet for any missing information or answers, misplaced answers, or stray marks that might be detected by a scanner.

Trade Tip

Those who prepare for licensing exams by simply memorizing questions and answers are likely to be poorly prepared for the challenges faced in actually operating boiler system equipment.

Sample Test 1

Name: ______________________ Date: ______________________

______ **1.** ___ is the hardness of coal.
A. Grade
B. Rank
C. Quality
D. Clinker

______ **2.** A temperature of 92°F is equal to ___°C.
A. 33.3
B. 61.8
C. 108
D. none of the above

______ **3.** Flues are ___.
A. smaller than 4″ I.D.
B. smaller than 4″ O.D.
C. greater than 4″ I.D.
D. greater than 4″ O.D.

______ **4.** A force of 1600 lb exerted on an area of 1.2 sq in. equals ___ psi.
A. 1333
B. 1599
C. 1920
D. 2667

______ **5.** On a balanced draft boiler installation equipped with interlocks, if the induced draft fan suddenly shuts down, the ___ should also shut down.
A. forced draft fan and fuel feed
B. feedwater control valve and pump
C. deaerator
D. blowdown

______ **6.** The minimum pressure at which a low-pressure boiler safety valve should be tested is ___ psig.
A. 3
B. 5
C. 15
D. 16.5

______ **7.** If priming occurs, a boiler operator should ___.
A. bypass the feedwater regulator and add water to the boiler manually
B. reduce the firing rate and blow the boiler down to gain control of the water level
C. increase the feed rate of boiler chemicals
D. increase the firing rate and minimize blowdown

______ **8.** A low pH reading in boiler water would indicate that ___ is likely to occur.
A. corrosion
B. caustic embrittlement
C. scale
D. carryover

______ **9.** The purpose of an attemperator is to control the temperature of ___.
A. saturated steam
B. liquid in a process that is heated by steam
C. superheated steam
D. the water in a hot water boiler

__________ **10.** A chain-grate stoker that is 13′-0″ wide is burning bituminous coal. The grate speed is 3″ per min and the coal being fed onto the grates is 4″ deep. The weight of the coal is 48 lb/cu ft. The coal is being burned at ___ ton/hr.

A. 1.56 B. 124.8
C. 3120 D. 7488

__________ **11.** The maximum pressure that may be developed in a low-pressure boiler during an accumulation test is ___ psig.

A. 15 B. 16.5
C. 20 D. none of the above

__________ **12.** The area of a 32′ × 80′ warehouse is ___ sq ft.

A. 17.8 B. 112
C. 2560 D. none of the above

__________ **13.** ___ is prevented by turning calcium- and magnesium-containing compounds into a nonadhering sludge.

A. Carryover B. Scale
C. Pitting D. Acidic corrosion

__________ **14.** A(n) ___ boiler is not a type of firetube boiler.

A. Stirling B. scotch marine
C. firebox D. HRT

__________ **15.** Steam separators work by ___.

A. passing the steam through a fine screen
B. throwing the steam through a swirl or abrupt change of direction
C. causing the steam to slow down, allowing water to drop out
D. causing an increase in pressure, which re-evaporates water droplets

__________ **16.** One purpose of the feedwater storage area below the deaerator is to supply about ___ min to ___ min of feedwater to the boiler if the water supply is lost.

A. 5; 10 B. 20; 30
C. 45; 60 D. none of the above

__________ **17.** If a boiler operates at 215 psi, this is equal to a static head of ___ ft.

A. 93.1 B. 497
C. 1791 D. 3161

__________ **18.** Disregarding friction and other losses, ___ HP are required to move 500 lb/min of water against a discharge head of 297′.

A. 0.54 B. 4.5
C. 16.2 D. 37.5

__________ **19.** If the water level is at the lowest visible point in the gauge glass of a firetube boiler, the water level in the boiler is at least ___.

A. 2″ above the lowest permissible water level
B. 3″ above the NOWL
C. 3″ above the highest point of the tubes, flues, or crown sheets
D. 3″ below the boiler's highest heating surface

_______________ **20.** Particulates may be removed from the flue gases with a(n) ___.
A. electrostatic precipitator
B. combustor
C. overflow air header
D. all of the above

_______________ **21.** Two bottom blowdown valves in series must be installed when the boiler pressure is ___ psig or more.
A. 15
B. 100
C. 125
D. 250

_______________ **22.** An electric spark from electrodes powered by an ignition transformer can be used to ignite #___ fuel oil.
A. 2
B. 4
C. 5
D. 6

_______________ **23.** Soot blowers are normally used in a ___ boiler.
A. firetube
B. watertube
C. cast iron
D. electric

_______________ **24.** Preheated combustion air is usually limited to about ___°F in boiler furnaces using stokers.
A. 200
B. 350
C. 500
D. 825

_______________ **25.** Silica gel is commonly used to prevent ___.
A. carbonic acid attack of condensate return piping
B. silica carryover from steam boilers to steam turbines
C. foaming of the boiler water
D. moisture in the boiler from causing corrosion during dry layup

_______________ **26.** A ___ is an instrument that measures draft by comparing pressure at two locations.
A. manometer
B. spectrophotometer
C. pyrometer
D. rotameter

_______________ **27.** Bicolor gauge glass view ports containing water will appear ___.
A. black
B. red
C. green
D. none of the above

_______________ **28.** A differential pressure of 29.0″ Hg equals ___ psi.
A. 0.45
B. 14.2
C. 28.9
D. 59.0

_______________ **29.** ___ is the exertion of equal forces from opposite sides of an object that push toward the middle.
A. Malleability
B. Tension
C. Ductility
D. Compression

_______________ **30.** A ___ is an open heat exchanger that removes oxygen from the feedwater going to a boiler.
A. steam header
B. flash tank
C. vacuum breaker
D. deaerator

_______________ **31.** The three main types of boilers are firetube, watertube, and ___.

_______________ **32.** ___ boilers are generally used for pressures over 250 psi.

______________ **33.** ___ is controlling the amount of flow that passes through a valve by partially closing the valve.

______________ **34.** ___ water is water that is added to the system to replace water that is lost or drained.

______________ **35.** A(n) ___ control is a control device that is installed a considerable distance from the equipment on which it is being used.

______________ **36.** Most modern water softeners work by the process of ___ exchange.

______________ **37.** ___ is formed when semimolten flyash cools after it sticks to furnace walls and boiler tubes.

______________ **38.** A(n) ___ of natural gas contains 100,000 Btu.

______________ **39.** A(n) ___ is another name for a low gas pressure switch.

______________ **40.** A(n) ___ is the desired point to which an automatic controller maintains a variable condition within a process.

T **F** **41.** A fluid is any material that can flow from one point to another.

T **F** **42.** Steel can withstand thermal stresses from expansion and contraction better than cast iron.

T **F** **43.** The boiler vent valve is connected to the highest point on the boiler shell.

T **F** **44.** The flow of steam through a superheater should never be less than 25% of the capacity of the boiler.

T **F** **45.** The low water fuel cutoff should be set to operate at a level below the lowest visible level in the gauge glass.

T **F** **46.** Bottom blowdown is more effective when it is used during heavy-load periods.

T **F** **47.** Gravity condensate return piping is commonly pitched about ¼″ to 1″ per 10′-0″.

T **F** **48.** Vent condensers minimize steam loss to the atmosphere.

T **F** **49.** Temporary hardness of water can be reduced by heating the water.

T **F** **50.** Foaming is the entrainment of small water droplets with the steam leaving the boiler.

51. Describe the procedure for handling a boiler with a dangerously low water level.

52. Describe the procedure for blowing down a water column and gauge glass.

53. What is the purpose of a sample cooler?

54. What is the purpose of a flame scanner? How is a flame scanner tested?

55. At what level in the gauge glass should the water be when starting to fire a cold boiler? Why?

56. What does a blowing whistle on top of the water column indicate?

57. What is flame impingement?

58. What is the return condensate percentage in the feedwater if the makeup water conductivity is 1000 micromhos, the feedwater conductivity is 360 micromhos, and the condensate conductivity is 19 micromhos?

59. What is an Orsat analyzer?

60. What problems could cause the quantity of water delivered by a centrifugal pump to decrease?

Sample Test 2

Name: ______________________ Date: ______________________

______ **1.** A container measuring 24″ × 30″ × 96″ is ___ cu ft in volume.
A. 40 B. 480
C. 5760 D. 69,120

______ **2.** A 55 gal. drum holds ___ lb of water.
A. 99 B. 458
C. 809 D. 880

______ **3.** The pressure at the bottom of a column of water 175′-0″ high is ___ psi.
A. 75.8 B. 85.9
C. 175 D. 404

______ **4.** The total force on a 16 sq in. surface with 120 psi acting upon it is ___ lb.
A. 7.5 B. 136
C. 1920 D. none of the above

______ **5.** Some firetube boilers use ___ to support the furnace area to keep it from collapsing.
A. buckstays B. horizontal through stays
C. staybolts D. telltales

______ **6.** ___ valves are recommended for fluid flow regulation.
A. Ball B. Gate
C. Globe D. Nonreturn

______ **7.** A(n) ___ boiler is a watertube boiler with a top steam and water drum and two bottom mud drums.
A. "A" style B. Firebox
C. Stirling D. "D" style

______ **8.** Tubular gauge glasses are made stronger by ___.
A. increasing the O.D. B. increasing the I.D.
C. increasing the wall thickness D. decreasing the O.D.

______ **9.** FM Global recommends that for boilers up to 400 psig, including low-pressure boilers, the safety valves should be manually tested by pulling the testing levers once per ___.
A. day B. week
C. month D. year

______ **10.** The radius of a circle 32′-6″ in diameter is ___.
A. 10.4′ B. 16′-3″
C. 65′-0″ D. none of the above

_______________ **11.** When water changes to steam, its ___ increases.
A. viscosity
B. temperature
C. sensible heat
D. volume

_______________ **12.** A low water cutoff switch should be tested ___ by blowing down the float chamber while the burner is firing.
A. daily
B. weekly
C. monthly
D. quarterly

_______________ **13.** ___ is the measure of the ability of an electric circuit to oppose current flow.
A. Conductance
B. Capacitance
C. Resistance
D. Voltage

_______________ **14.** ___ of a new operation is the process of inspecting, preparing, and testing each major component of the system prior to operation.
A. Shakeout
B. Commissioning
C. Overhaul
D. Trial run

_______________ **15.** The cycles of concentration is ___ when the boiler water chloride content is 200 ppm and the feedwater chloride content is 25 ppm.
A. 2
B. 4
C. 6
D. none of the above

_______________ **16.** ___ and ___ are the desirable elements in fossil fuels that release heat when burned with oxygen.
A. Carbon dioxide; carbon
B. Hydrogen; carbon monoxide
C. Hydrogen; carbon
D. Carbon dioxide; carbon monoxide

_______________ **17.** A(n) ___ is a flue gas cleaning device that removes particulates from the flue gas stream by attracting them to grounded collector plates.
A. dry scrubber
B. filter baghouse
C. mechanical dust collector
D. electrostatic precipitator

_______________ **18.** An HRT firetube boiler is 7′-6″ in diameter and 18′ in length. It contains 210 2″ tubes. Half of the shell is exposed to the furnace area. Disregarding the tube sheets, the heating surface of the boiler is ___ sq ft.
A. 1979
B. 2194
C. 2406
D. 6519

_______________ **19.** Incomplete combustion is a cause of ___.
A. particulate emissions
B. soot accumulation
C. scale accumulation
D. acid vapor emissions

_______________ **20.** To avoid water hammer, a(n) ___ procedure involves manually draining the condensate from steam piping while the piping is being placed into service.
A. automatic warmup
B. accelerated startup
C. supervised warmup
D. throttled velocity

_______________ **21.** A ___ is a protruding bubble or bulge on a boiler surface.
A. bag
B. blister
C. rupture
D. fracture

______________ **22.** If the blowdown valve on the water column is closed and water fills up quickly to the center of the gauge glass, this is an indication that the ___.
A. top gauge glass valve or line is plugged
B. water column piping is free of debris
C. bottom gauge glass valve or line is plugged
D. bottom blowdown line from the water column is restricted

______________ **23.** A(n) ___ feedwater regulating system is one that continually adjusts the position of a feedwater regulating valve as needed to maintain a constant boiler water level.
A. ON/OFF
B. feedforward
C. cycling
D. modulating

______________ **24.** A ___ pump uses the force of a secondary fluid to pump the primary fluid.
A. motive fluid
B. centrifugal
C. vane
D. metering

______________ **25.** As a general rule of thumb for natural gas-fired boilers, the flue gas outlet temperature before any heat recovery equipment should be between ___°F and ___°F above the saturated steam temperature.
A. 10; 20
B. 10; 100
C. 50; 150
D. 100; 300

______________ **26.** A boiler operator should adjust the amount of ___ in order to keep the boiler water within the specified conductivity range.
A. blowdown
B. feedwater
C. phosphate
D. sodium sulfite

______________ **27.** ___ is the plasticity exhibited by a material under tension loading.
A. Compressive strength
B. Tensile strength
C. Malleability
D. Ductility

______________ **28.** ___ is the entrainment of small water droplets with steam leaving the boiler.
A. Carryover
B. Foaming
C. Turbulence
D. Water hammer

______________ **29.** The top try cock of a water column should blow ___.
A. steam
B. mixed water and steam
C. water
D. air

______________ **30.** ___ is the ability of a liquid to resist flow.
A. Pressure
B. Malleability
C. Force
D. Viscosity

______________ **31.** One gallon of water weighs ___ lb.

______________ **32.** When applying a hydrostatic test to a boiler, the boiler is filled with water and then pressurized to ___ times its MAWP.

______________ **33.** ___ circles have the same centerpoint.

______________ **34.** The letters WOG on the side of a valve stand for ___, ___, and ___.

______________ **35.** The diameter of the liquid piston in a 6 × 3 × 10 reciprocating pump is ___″.

______________ **36.** ___ is a solution of salt and water.

______________ **37.** ___ is the process where a material can self-generate heat until the ignition point is reached.

______________ **38.** A quick-acting valve requires only a(n) ___° turn of the lever to move from fully closed to fully open.

______________ **39.** A 400 BHP package boiler is designed to produce approximately ___ lb/hr of steam.

______________ **40.** A high-pressure boiler has an MAWP higher than ___ psi.

T F 41. A horizontal return tubular boiler is a firetube boiler.

T F 42. Tube ends are normally beaded in a watertube boiler and flared in a firetube boiler.

T F 43. Shutoff valves may be installed between the boiler and the safety valve(s).

T F 44. A fusible plug is a pressure-sensitive device that causes an audible alarm when exposed to high temperature.

T F 45. The temperature of the water leaving a deaerator should be within 3°F to 5°F of the temperature of the steam entering the deaerator.

T F 46. Hot process water softeners are designed for pressure, not vacuum.

T F 47. The high velocity of secondary air creates turbulence in the burning combustion gases, which helps to ensure thorough mixing of the oxygen and fuel.

T F 48. Volatiles are flyash particles that are carried along with the flue gases.

T F 49. A flat gauge glass is normally used with boilers operating at 125 psi or less.

T F 50. An ion is an atom or molecule with a positive or negative charge.

51. What could cause a low water fuel cutoff to fail to work when required?

52. Why should sudden and substantial steam demands on a boiler be avoided?

53. Why is a steam valve that is under pressure opened slowly?

54. How is a low water fuel cutoff tested?

55. List three indicators of incomplete combustion.

56. Why is superheated steam used in plants that utilize large steam turbines?

57. What causes scale in a boiler? How can scale in a boiler be prevented?

58. What is cavitation? Under what conditions would cavitation be likely to occur?

59. What can cause the temperature of the gases of combustion at the boiler outlet to gradually increase? Explain.

60. What is an economizer? Where is it located? What are some problems encountered in its use?

Sample Test 3

Name: ____________________ Date: ____________________

1. The ___ is the seam running the length of a boiler shell or drum.
   A. longitudinal joint B. circumferential joint
   C. girth joint D. none of the above

2. When 200 lb of water is heated from 62°F to 75°F, ___ Btu is absorbed.
   A. 165 B. 2600
   C. 12,400 D. none of the above

3. A ___ is used to reduce sulfur dioxide emissions.
   A. scrubber B. electrostatic precipitator
   C. mechanical dust collector D. cyclone separator

4. A ___ protects a Bourdon tube inside a pressure gauge from warping due to the high temperature of the steam.
   A. capillary tube B. thermowell
   C. siphon loop D. displacer

5. A column of water 150′-0″ high exerts a pressure of ___ psi at the bottom of the column.
   A. 65 B. 150
   C. 346 D. none of the above

6. The short stubs of piping that are connected to the boiler during construction to provide attachment points for piping and appliances are called ___.
   A. stub ends B. nozzles
   C. butt ends D. protrusions

7. A dryback scotch marine firetube boiler is 8′-6″ in diameter and has a flue furnace that is 30″ in diameter. The boiler is 17′-3″ long, and has 122 tubes that are 2″ in diameter. If the tube sheets are disregarded, the boiler has ___ sq ft of heating surface.
   A. 188 B. 1063
   C. 1239 D. 1699

8. ___ is the erosion that occurs as steam or another high-velocity fluid streaks through a small opening like a throttled valve.
   A. Channeling B. Grooving
   C. Wire drawing D. Scoring

9. Boiler tubes can fail due to ___.
   A. encrustation with scale B. corrosion and pitting
   C. low water D. all of the above

______ **10.** The circumference of a boiler shell that is 6′-3″ in diameter is ___.
A. 12.5′
B. 15.5′
C. 19.625′
D. none of the above

______ **11.** When a boiler is equipped with a superheater, the safety valve that should lift at the lowest pressure is ___.
A. the superheater safety valve
B. either of the boiler drum safety valves
C. the boiler drum safety valve with the greatest capacity in lb/hr
D. it makes no difference as long as the safety valves prevent the boiler pressure from exceeding the MAWP

______ **12.** A(n) ___ line provides a minimum flow through a centrifugal pump to prevent overheating and cavitation when the pump must operate against a closed discharge valve for a period of time.
A. return
B. equalizing
C. impulse
D. recirculation

______ **13.** An area of 32 sq ft contains ___ sq in.
A. 382
B. 416
C. 4608
D. none of the above

______ **14.** The acronym MATT stands for ___.
A. mixture, air, temperature, and turbulence
B. methyl orange alkalinity titration test
C. modulation, atomization, turbulence, and temperature
D. mixture, atomization, time, and temperature

______ **15.** Grades of fuel oil are designated by number, from number ___ through number ___.
A. 1; 6
B. 2; 6
C. 1; 10
D. 2; 10

______ **16.** The order in which a sodium zeolite water softener regenerates is ___.
A. rinse, brine, and rinse
B. rinse, backwash, and brine
C. backwash, brine, and rinse
D. backwash, brine, and backwash

______ **17.** A ⅛″ layer of soot on tubes and heating surfaces can increase fuel consumption by over ___%.
A. 3
B. 6
C. 9
D. 12

______ **18.** Large amounts of combustible material in the furnace ash are most likely to be caused by ___.
A. not enough forced draft
B. low furnace temperatures
C. high furnace temperatures
D. defective grates

______ **19.** Class ___ fires involve flammable liquids like oils and gasoline.
A. A
B. B
C. C
D. D

______ **20.** ___ heat is Btu content of a substance that represents the heat absorbed or given up as it changes temperature.
A. Latent
B. Sensible
C. Total
D. Low-pressure

_______________ **21.** Water with a pH of ___ is regarded as neutral.
A. 0 B. 7
C. 8.3 D. 14

_______________ **22.** According to the ASME Code, boilers with more than ___ sq ft of heating surface require at least two safety valves.
A. 100 B. 500
C. 1000 D. 5000

_______________ **23.** One square foot contains ___ sq in.
A. 12 B. 24
C. 144 D. 1728

_______________ **24.** A low-pressure boiler has an MAWP of ___ psi or less.
A. 5 B. 15
C. 75 D. 150

_______________ **25.** ___ is the collection of mineral deposits formed on the heating surfaces of a boiler.
A. Soot B. Temporary hardness
C. Permanent hardness D. Scale

_______________ **26.** A ___ is used to vent air from the boiler while the boiler is filling.
A. boiler vent B. try cock
C. continuous blowdown valve D. bottom blowdown valve

_______________ **27.** ___ is the hydraulic shock that can result from water buildup in steam piping.
A. Water hammer B. Carryover
C. Foaming D. Priming

_______________ **28.** ___ are chemicals that cause hardness crystals in boiler water to settle out as a heavy sludge.
A. Phosphates B. Zeolites
C. Chelants D. Polymer dispersants

_______________ **29.** ___ air is the initial volume of air that enters the furnace with the fuel for most of the combustion process.
A. Primary B. Secondary
C. Overfire D. Pilot

_______________ **30.** ___ is the transfer of heat by actual physical contact.
A. Conduction B. Natural convection
C. Radiation D. Forced convection

_______________ **31.** ___ is the ability of a material to deform permanently under compression without rupture.

_______________ **32.** ___ pressure gauges are pressure gauges that may be used to read pressure or vacuum.

_______________ **33.** The main pressure gauge on the boiler must be able to read at least ___ times the MAWP of the boiler.

_______________ **34.** ___ is pressure lower than atmospheric pressure.

_______________ **35.** ___ is the development of froth on the surface of the boiler water.

_______________ **36.** A(n) ___ regulator automatically maintains a constant safe water level in the boiler.

_______________ **37.** ___ is the term commonly used for expansion of water when it is heated.

_______________ **38.** ___ is the measure of the ability of an electric circuit to allow current flow.

_______________ **39.** Field devices such as fuel pressure switches must be installed that inform the flame safeguard control of the status of various measured conditions. This is called ___ the condition.

_______________ **40.** One horsepower equals ___ ft-lb of work in 1 min.

T F 41. Cast iron boilers are used for closed, high-pressure heating systems only.

T F 42. An alloy is a blend of two or more metals.

T F 43. Tensile strength is the amount of force required to pull an object apart by stretching it.

T F 44. In the event of a dangerous low water condition, the boiler should be isolated by closing a manual valve in the feedwater line.

T F 45. A check valve is a one-way flow valve for fluids.

T F 46. A closed centrifugal pump impeller has a shroud on one side of the vanes.

T F 47. The three types of draft are natural, induced, and gravity.

T F 48. Baffles direct the flow of gases of combustion along a desired path.

T F 49. A pyrometer is an instrument that measures temperatures above the temperature range of a mercury thermometer.

T F 50. Priming can cause water hammer.

51. Two package boilers used for heating are frequently cycling on and off. What effect does this have on boiler efficiency and maintenance? How can this be improved?

52. List three conditions that could cause tube rupture in a watertube boiler.

53. List and explain two causes for poor fuel oil atomization.

54. What chemical reactions can occur in a boiler if proper storage procedures have not been followed?

55. When is the best time to blow down the waterwalls of a watertube boiler? Why?

56. Why should excessive amounts of air be kept from entering the boiler furnace?

57. What problems are caused by oxygen and carbon dioxide in the boiler water? How are these problems prevented?

58. Why may a bypass damper be used on an air preheater?

59. What can cause a furnace explosion?

60. What is the heating surface of a boiler?

Appendix

PROPERTIES OF SATURATED STEAM . . .									
Absolute Pressure*	Temperature†	Specific Volume‡		Enthalpy‖			Entropy		
p	t	Saturated Liquid v_f	Saturated Vapor v_g	Saturated Liquid h_f	Evap. h_{fg}	Saturated Vapor h_g	Saturated Liquid s_f	Evap. s_{fg}	Saturated Vapor s_g
1	101.74	0.01614	333.6	69.70	1036.3	1106.0	0.1326	1.8456	1.9782
2	126.08	0.01623	173.73	93.99	1022.2	1116.2	0.1749	1.7451	1.9200
3	141.48	0.01630	118.71	109.37	1013.2	1122.6	0.2008	1.6855	1.8863
4	152.97	0.01636	90.63	120.86	1006.4	1127.3	0.2198	1.6427	1.8625
5	164.24	0.01640	73.52	130.13	1001.0	1131.1	0.2347	1.6094	1.8441
6	170.06	0.01645	61.98	137.96	996.2	1134.2	0.2472	1.5820	1.8292
7	176.85	0.01649	53.64	144.76	992.1	1136.9	0.2581	1.5586	1.8167
8	182.86	0.01653	47.34	150.79	989.5	1139.3	0.2674	1.5383	1.8057
9	188.28	0.01656	42.40	156.22	985.2	1141.4	0.2759	1.5203	1.7962
10	193.21	0.01659	38.42	161.17	982.1	1143.3	0.2835	1.5041	1.7876
14.696	212.00	0.01672	26.80	180.07	970.3	1150.4	0.3120	1.4446	1.7566
15	213.03	0.01672	26.29	181.11	969.7	1150.8	0.3135	1.4415	1.7549
20	227.96	0.01683	20.089	196.16	960.1	1156.3	0.3356	1.3962	1.7319
25	240.07	0.01692	16.303	208.42	952.1	1160.6	0.3533	1.3606	1.7139
30	250.33	0.01701	13.746	218.82	945.3	1164.1	0.3680	1.3313	1.6993
35	259.28	0.01708	11.898	227.91	939.2	1167.1	0.3807	1.3063	1.6870
40	267.25	0.01715	10.498	236.03	933.7	1169.7	0.3919	1.2844	1.6763
45	274.44	0.01721	9.401	243.36	928.6	1172.0	0.4019	1.2650	1.6669
50	281.01	0.01727	8.515	250.09	924.0	1174.1	0.4110	1.2474	1.6585
55	287.07	0.01732	7.787	256.30	919.6	1175.9	0.4193	1.2316	1.6509
60	292.71	0.01738	7.175	262.09	915.5	1177.6	0.4270	1.2168	1.6438
65	297.97	0.01743	6.655	267.50	911.6	1179.1	0.4342	1.2032	1.6374
70	302.92	0.01748	6.206	272.61	907.9	1180.6	0.4409	1.1906	1.6315
75	307.60	0.01753	5.816	277.43	904.5	1181.9	0.4472	1.1787	1.6259
80	312.03	0.01757	5.472	282.02	901.1	1183.1	0.4531	1.1676	1.6207
85	316.25	0.01761	5.168	286.39	897.8	1184.2	0.4587	1.1571	1.6158
90	320.27	0.01766	4.896	290.56	894.7	1185.3	0.4641	1.1471	1.6112
95	324.12	0.01770	4.652	294.56	891.7	1186.2	0.4692	1.1376	1.6068
100	327.81	0.01774	4.432	298.40	888.8	1187.2	0.4740	1.1286	1.6026
110	334.77	0.01782	4.049	305.66	883.2	1188.9	0.4832	1.1117	1.5948
120	341.25	0.01789	3.728	312.44	877.9	1190.4	0.4916	1.0962	1.5878
130	347.32	0.01796	3.455	318.81	872.9	1191.7	0.4995	1.0817	1.5812
140	353.02	0.01802	3.220	324.82	868.2	1193.0	0.5069	1.0682	1.5751
150	358.42	0.01809	3.015	330.51	863.6	1194.1	0.5138	1.0556	1.5694
160	363.53	0.01815	2.834	335.93	859.2	1195.1	0.5204	1.0436	1.5640
170	368.41	0.01822	2.675	341.09	854.9	1196.0	0.5266	1.0324	1.5590
180	373.06	0.01827	2.532	346.03	850.8	1196.9	0.5325	1.0217	1.5542
190	377.51	0.01833	2.404	350.79	846.8	1197.6	0.5381	1.0116	1.5497
200	381.79	0.01839	2.288	355.36	843.0	1198.4	0.5435	1.0018	1.5453
250	400.95	0.01865	1.8438	376.00	825.1	1201.1	0.5675	.9588	1.5263
300	417.33	0.01890	1.5433	393.84	809.0	1202.8	0.5879	.9225	1.5104

... PROPERTIES OF SATURATED STEAM									
Absolute Pressure* p	Temperature† t	Specific Volume‡		Enthalpy‖			Entropy		
		Saturated Liquid v_f	Saturated Vapor v_g	Saturated Liquid h_f	Evap. h_{fg}	Saturated Vapor h_g	Saturated Liquid s_f	Evap. s_{fg}	Saturated Vapor s_g
350	431.72	0.01913	1.3260	409.69	794.2	1203.9	0.6056	.8910	1.4966
400	444.59	0.0193	1.1613	424.0	780.5	1204.5	0.6214	.8630	1.4844
450	456.28	0.0195	1.0320	437.2	767.4	1204.6	0.6356	.8378	1.4734
500	467.01	0.0197	0.9278	449.4	755.0	1204.4	0.6487	.8147	1.4634
550	476.94	0.0199	0.8424	460.8	743.1	1203.9	0.6608	.7934	1.4542
600	486.21	0.0201	0.7698	471.6	731.6	1203.2	0.6720	.7734	1.4454
650	494.90	0.0203	0.7083	481.8	720.5	1202.3	0.6826	.7548	1.4374
700	503.10	0.0205	0.6554	491.5	709.7	1201.2	0.6925	.7371	1.4296
750	510.86	0.0207	0.6092	500.8	699.2	1200.0	0.7019	.7204	1.4223
800	518.23	0.0209	0.5687	509.7	688.9	1198.6	0.7108	.7045	1.4153
850	525.26	0.0210	0.5327	518.3	678.8	1197.7	0.7194	.6891	1.4085
900	531.98	0.0212	0.5006	526.6	668.8	1195.4	0.7275	.6744	1.4020
950	538.43	0.0214	0.4717	534.6	659.1	1193.7	0.7355	.6602	1.3957
1000	544.61	0.0216	0.4456	542.4	649.4	1191.8	0.7430	.6467	1.3897
1100	556.31	0.0220	0.4001	557.4	630.4	1187.8	0.7575	.6205	1.3780
1200	567.22	0.0223	0.3619	571.7	611.7	1183.4	0.7711	.5956	1.3667
1300	577.46	0.0227	0.3293	585.4	593.2	1178.6	0.7840	.5719	1.3559
1400	587.10	0.0231	0.3012	598.7	574.7	1173.4	0.7963	.5491	1.3454
1500	596.23	0.0235	0.2765	611.6	556.3	1167.9	0.8082	.5269	1.3351
2000	635.82	0.0257	0.1878	671.7	463.4	1135.1	0.8619	.4230	1.2849
2500	668.13	0.0287	0.1307	730.6	360.5	1091.1	0.9126	.3197	1.2322
3000	695.36	0.0346	0.0858	802.5	217.8	1020.3	0.9731	.1885	1.1615
3206.2	705.40	0.0503	0.0503	902.7	0	902.7	1.0580	0	1.0580

* in psi
† in °F
‡ in cu ft/lb
‖ in Btu/lb

INSTRUMENT TAG IDENTIFICATION					
	First Letter		**Second Letter**		
	Measured or Initiating Variable	**Modifier**	**Readout or Passive Function**	**Output Function**	**Modifier**
A	Analysis		Alarm		
B	Burner Flame		User's Choice	User's Choice	User's Choice
C	Conductivity (Electrical)			Control	
D	Density (Mass) or Specific Gravity	Differential			
E	Voltage (EMF)		Primary Element		
F	Flow Rate	Ratio (Fraction)			
G	Gaging (Dimensional)		Glass		
H	Hand (Manually Initiated)				High
I	Current (Electrical)		Indicate		
J	Power	Scan			
K	Time or Time Schedule			Control Station	
L	Level		Light (Pilot)		Low
M	Moisture or Humidity				Middle or Intermediate
N	User's Choice		User's Choice	User's Choice	User's Choice
O	User's Choice		Orifice (Restriction)		
P	Pressure or Vacuum		Point (Test Connection)		
Q	Quantity or Event	Integrate or Totalize			
R	Radioactivity, radiation		Record or Print		
S	Speed or Frequency	Safety		Switch	
T	Temperature			Transmit	
U	Multivariable		Multifunction	Multifunction	Multifunction
V	Viscosity, Vibration			Valve, Damper, or Louver	
W	Weight or Force		Well		
X	Unclassified		Unclassified	Unclassified	Unclassified
Y	Event or State			Relay or Compute	
Z	Position			Drive, Actuate, or Unclassified Final Control Element	

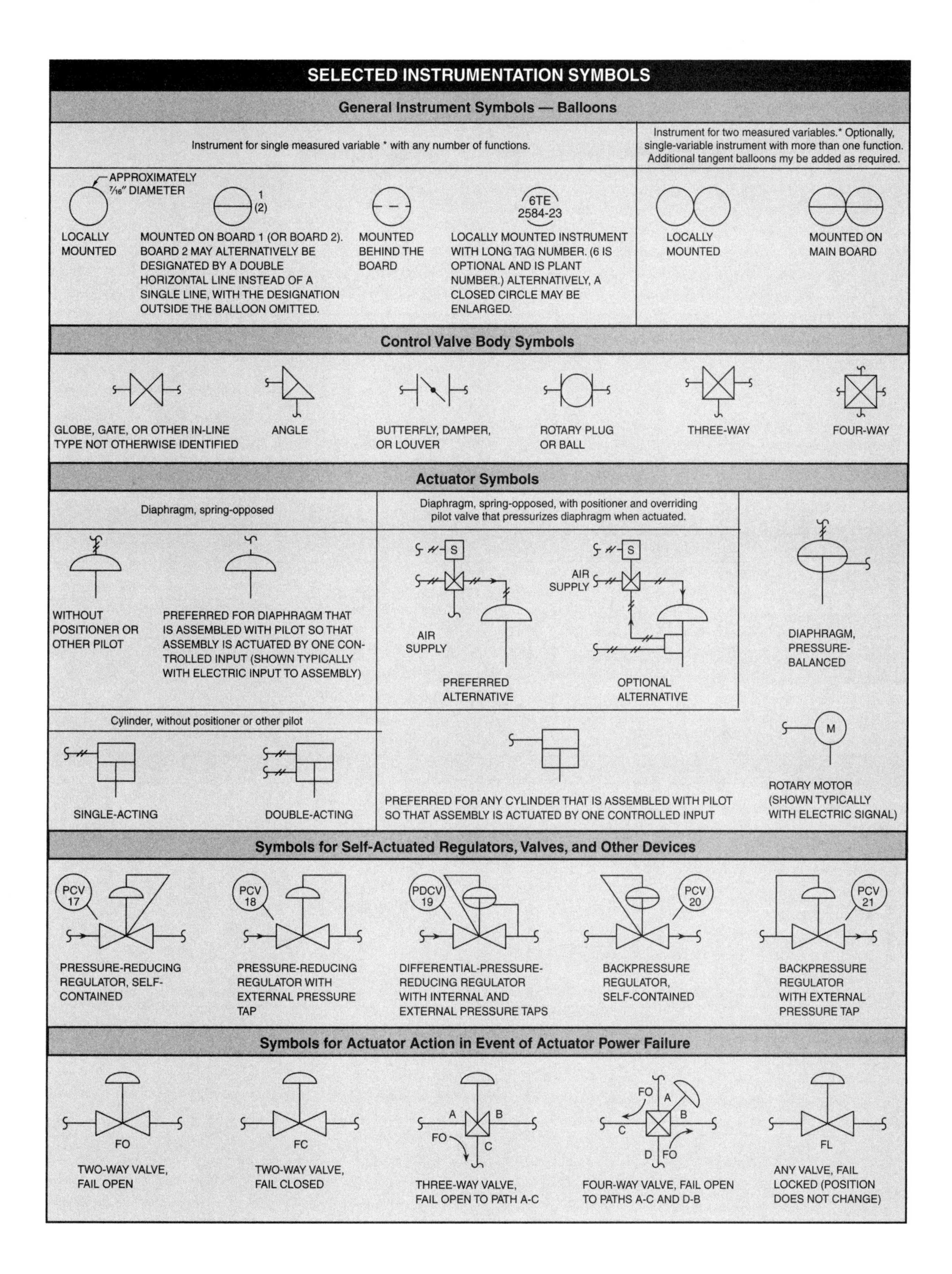

SELECTED INSTRUMENTATION SYMBOLS
General Instrument Symbols — Balloons
Instrument for single measured variable * with any number of functions.
Instrument for two measured variables.* Optionally, single-variable instrument with more than one function. Additional tangent balloons my be added as required.
APPROXIMATELY 7/16" DIAMETER
LOCALLY MOUNTED
1 (2)
MOUNTED ON BOARD 1 (OR BOARD 2). BOARD 2 MAY ALTERNATIVELY BE DESIGNATED BY A DOUBLE HORIZONTAL LINE INSTEAD OF A SINGLE LINE, WITH THE DESIGNATION OUTSIDE THE BALLOON OMITTED.
MOUNTED BEHIND THE BOARD
6TE 2584-23
LOCALLY MOUNTED INSTRUMENT WITH LONG TAG NUMBER. (6 IS OPTIONAL AND IS PLANT NUMBER.) ALTERNATIVELY, A CLOSED CIRCLE MAY BE ENLARGED.
LOCALLY MOUNTED
MOUNTED ON MAIN BOARD
Control Valve Body Symbols
GLOBE, GATE, OR OTHER IN-LINE TYPE NOT OTHERWISE IDENTIFIED
ANGLE
BUTTERFLY, DAMPER, OR LOUVER
ROTARY PLUG OR BALL
THREE-WAY
FOUR-WAY
Actuator Symbols
Diaphragm, spring-opposed
WITHOUT POSITIONER OR OTHER PILOT
PREFERRED FOR DIAPHRAGM THAT IS ASSEMBLED WITH PILOT SO THAT ASSEMBLY IS ACTUATED BY ONE CONTROLLED INPUT (SHOWN TYPICALLY WITH ELECTRIC INPUT TO ASSEMBLY)
Diaphragm, spring-opposed, with positioner and overriding pilot valve that pressurizes diaphragm when actuated.
S
AIR SUPPLY
PREFERRED ALTERNATIVE
OPTIONAL ALTERNATIVE
DIAPHRAGM, PRESSURE-BALANCED
Cylinder, without positioner or other pilot
SINGLE-ACTING
DOUBLE-ACTING
PREFERRED FOR ANY CYLINDER THAT IS ASSEMBLED WITH PILOT SO THAT ASSEMBLY IS ACTUATED BY ONE CONTROLLED INPUT
M
ROTARY MOTOR (SHOWN TYPICALLY WITH ELECTRIC SIGNAL)
Symbols for Self-Actuated Regulators, Valves, and Other Devices
PCV 17
PRESSURE-REDUCING REGULATOR, SELF-CONTAINED
PCV 18
PRESSURE-REDUCING REGULATOR WITH EXTERNAL PRESSURE TAP
PDCV 19
DIFFERENTIAL-PRESSURE-REDUCING REGULATOR WITH INTERNAL AND EXTERNAL PRESSURE TAPS
PCV 20
BACKPRESSURE REGULATOR, SELF-CONTAINED
PCV 21
BACKPRESSURE REGULATOR WITH EXTERNAL PRESSURE TAP
Symbols for Actuator Action in Event of Actuator Power Failure
FO
TWO-WAY VALVE, FAIL OPEN
FC
TWO-WAY VALVE, FAIL CLOSED
A B FO C
THREE-WAY VALVE, FAIL OPEN TO PATH A-C
FO A B C D FO
FOUR-WAY VALVE, FAIL OPEN TO PATHS A-C AND D-B
FL
ANY VALVE, FAIL LOCKED (POSITION DOES NOT CHANGE)

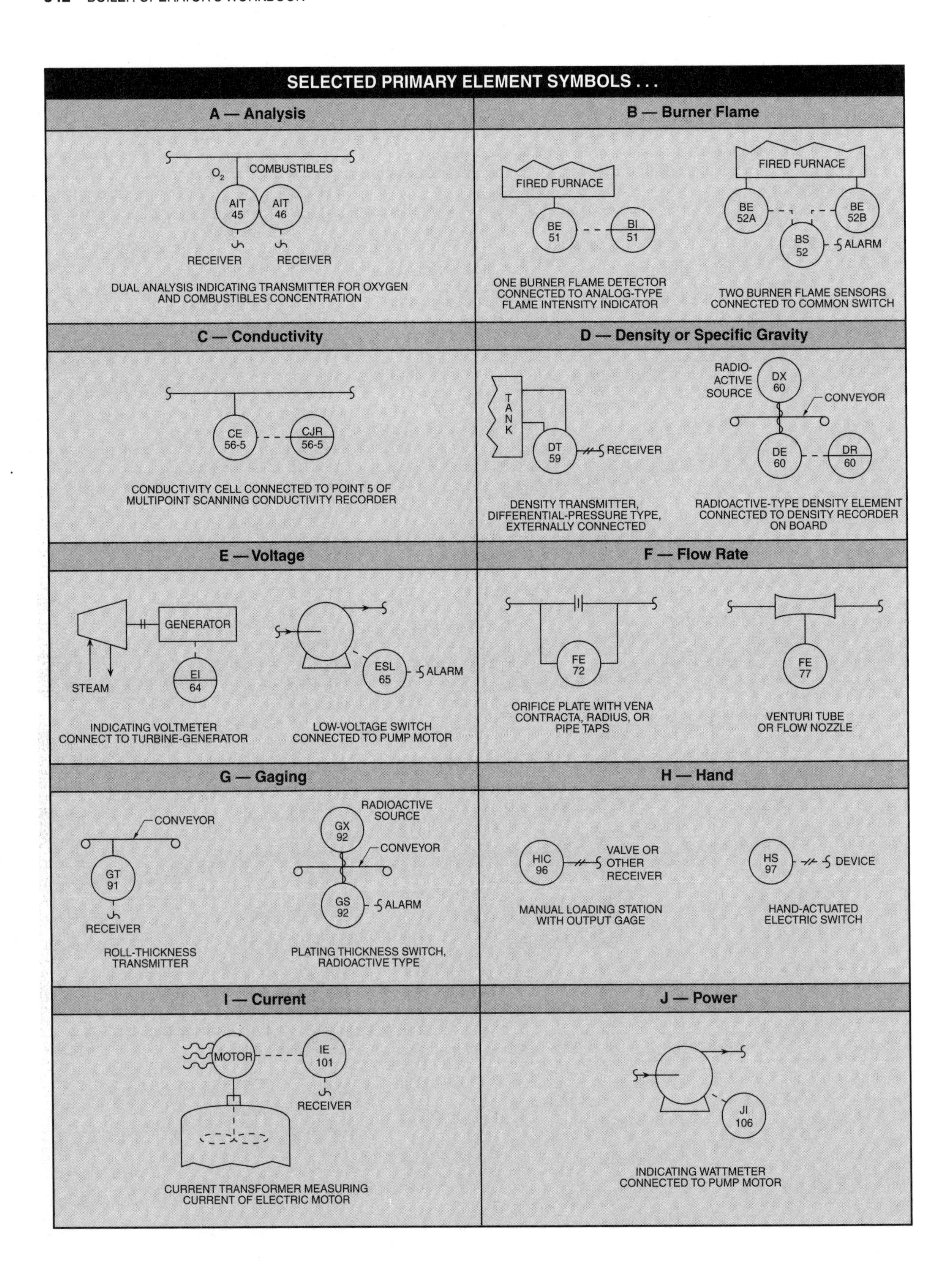

SELECTED PRIMARY ELEMENT SYMBOLS . . .
A — Analysis
O2
COMBUSTIBLES
AIT 45
AIT 46
RECEIVER
RECEIVER
DUAL ANALYSIS INDICATING TRANSMITTER FOR OXYGEN AND COMBUSTIBLES CONCENTRATION
B — Burner Flame
FIRED FURNACE
BE 51
BI 51
ONE BURNER FLAME DETECTOR CONNECTED TO ANALOG-TYPE FLAME INTENSITY INDICATOR
FIRED FURNACE
BE 52A
BE 52B
BS 52
ALARM
TWO BURNER FLAME SENSORS CONNECTED TO COMMON SWITCH
C — Conductivity
CE 56-5
CJR 56-5
CONDUCTIVITY CELL CONNECTED TO POINT 5 OF MULTIPOINT SCANNING CONDUCTIVITY RECORDER
D — Density or Specific Gravity
TANK
DT 59
RECEIVER
DENSITY TRANSMITTER, DIFFERENTIAL-PRESSURE TYPE, EXTERNALLY CONNECTED
RADIO-ACTIVE SOURCE
DX 60
CONVEYOR
DE 60
DR 60
RADIOACTIVE-TYPE DENSITY ELEMENT CONNECTED TO DENSITY RECORDER ON BOARD
E — Voltage
GENERATOR
STEAM
EI 64
INDICATING VOLTMETER CONNECT TO TURBINE-GENERATOR
ESL 65
ALARM
LOW-VOLTAGE SWITCH CONNECTED TO PUMP MOTOR
F — Flow Rate
FE 72
ORIFICE PLATE WITH VENA CONTRACTA, RADIUS, OR PIPE TAPS
FE 77
VENTURI TUBE OR FLOW NOZZLE
G — Gaging
CONVEYOR
GT 91
RECEIVER
ROLL-THICKNESS TRANSMITTER
RADIOACTIVE SOURCE
GX 92
CONVEYOR
GS 92
ALARM
PLATING THICKNESS SWITCH, RADIOACTIVE TYPE
H — Hand
HIC 96
VALVE OR OTHER RECEIVER
MANUAL LOADING STATION WITH OUTPUT GAGE
HS 97
DEVICE
HAND-ACTUATED ELECTRIC SWITCH
I — Current
MOTOR
IE 101
RECEIVER
CURRENT TRANSFORMER MEASURING CURRENT OF ELECTRIC MOTOR
J — Power
JI 106
INDICATING WATTMETER CONNECTED TO PUMP MOTOR

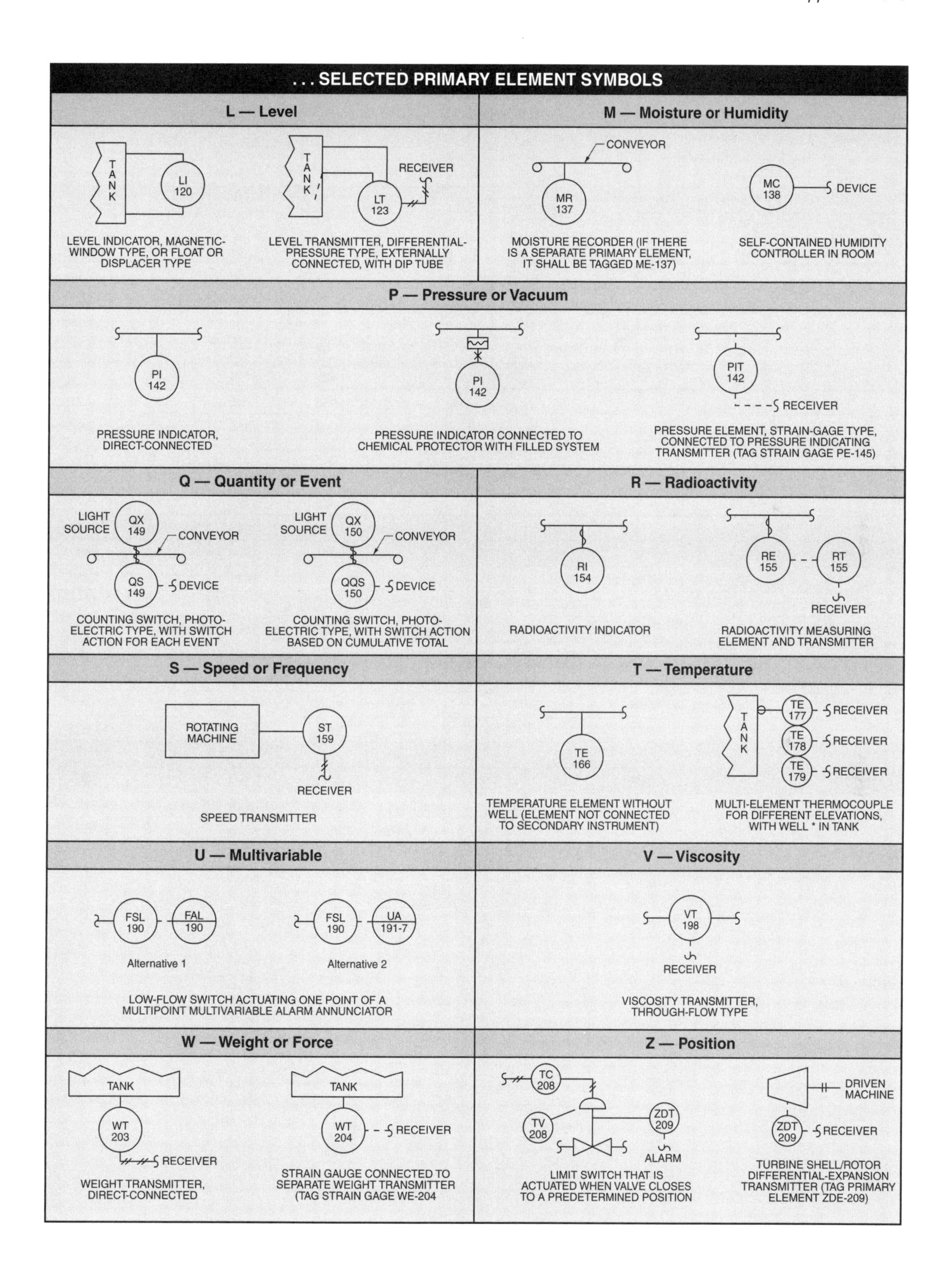

. . . SELECTED PRIMARY ELEMENT SYMBOLS
L — Level
TANK
LI 120
LEVEL INDICATOR, MAGNETIC-WINDOW TYPE, OR FLOAT OR DISPLACER TYPE
RECEIVER
LT 123
LEVEL TRANSMITTER, DIFFERENTIAL-PRESSURE TYPE, EXTERNALLY CONNECTED, WITH DIP TUBE
M — Moisture or Humidity
CONVEYOR
MR 137
MOISTURE RECORDER (IF THERE IS A SEPARATE PRIMARY ELEMENT, IT SHALL BE TAGGED ME-137)
MC 138
DEVICE
SELF-CONTAINED HUMIDITY CONTROLLER IN ROOM
P — Pressure or Vacuum
PI 142
PRESSURE INDICATOR, DIRECT-CONNECTED
PRESSURE INDICATOR CONNECTED TO CHEMICAL PROTECTOR WITH FILLED SYSTEM
PIT 142
RECEIVER
PRESSURE ELEMENT, STRAIN-GAGE TYPE, CONNECTED TO PRESSURE INDICATING TRANSMITTER (TAG STRAIN GAGE PE-145)
Q — Quantity or Event
LIGHT SOURCE
QX 149
CONVEYOR
QS 149
DEVICE
COUNTING SWITCH, PHOTO-ELECTRIC TYPE, WITH SWITCH ACTION FOR EACH EVENT
QX 150
QQS 150
COUNTING SWITCH, PHOTO-ELECTRIC TYPE, WITH SWITCH ACTION BASED ON CUMULATIVE TOTAL
R — Radioactivity
RI 154
RADIOACTIVITY INDICATOR
RE 155
RT 155
RECEIVER
RADIOACTIVITY MEASURING ELEMENT AND TRANSMITTER
S — Speed or Frequency
ROTATING MACHINE
ST 159
RECEIVER
SPEED TRANSMITTER
T — Temperature
TE 166
TEMPERATURE ELEMENT WITHOUT WELL (ELEMENT NOT CONNECTED TO SECONDARY INSTRUMENT)
TE 177
TE 178
TE 179
RECEIVER
MULTI-ELEMENT THERMOCOUPLE FOR DIFFERENT ELEVATIONS, WITH WELL * IN TANK
U — Multivariable
FSL 190
FAL 190
Alternative 1
UA 191-7
Alternative 2
LOW-FLOW SWITCH ACTUATING ONE POINT OF A MULTIPOINT MULTIVARIABLE ALARM ANNUNCIATOR
V — Viscosity
VT 198
RECEIVER
VISCOSITY TRANSMITTER, THROUGH-FLOW TYPE
W — Weight or Force
TANK
WT 203
RECEIVER
WEIGHT TRANSMITTER, DIRECT-CONNECTED
WT 204
STRAIN GAUGE CONNECTED TO SEPARATE WEIGHT TRANSMITTER (TAG STRAIN GAGE WE-204
Z — Position
TC 208
TV 208
ZDT 209
ALARM
LIMIT SWITCH THAT IS ACTUATED WHEN VALVE CLOSES TO A PREDETERMINED POSITION
DRIVEN MACHINE
ZDT 209
RECEIVER
TURBINE SHELL/ROTOR DIFFERENTIAL-EXPANSION TRANSMITTER (TAG PRIMARY ELEMENT ZDE-209)

STEAM LOSS THROUGH AN ORIFICE*									
Orifice Size	**Orifice Size**	**Differential Pressure Across Orifice, PSIG**							
		15	**30**	**60**	**100**	**150**	**250**	**400**	**600**
#60	0.04	1	2	4	6	10	16	26	39
3/64″	0.0469	1	3	5	9	13	22	36	53
1/16″	0.0625	2	5	9	16	24	39	63	95
5/64″	0.0781	4	7	15	25	37	62	99	148
3/32″	0.0938	5	11	21	36	53	89	142	213
#38	0.1015	6	12	25	42	62	104	166	250
7/64″	0.1094	7	15	29	48	73	121	193	290
1/8″	0.125	9	19	38	63	95	158	252	379
9/64″	0.1406	12	24	48	80	120	200	319	479
5/32″	0.1562	15	30	59	99	148	246	394	591
3/16″	0.1875	21	43	85	142	213	355	568	852
7/32″	0.2188	29	58	116	193	290	483	773	1160
1/4″	0.25	38	76	151	252	379	631	1010	1515
9/32″	0.2812	48	96	192	319	479	798	1278	1916
5/16″	0.3125	59	118	237	394	592	986	1578	2367
11/32″	0.3438	72	143	286	477	716	1194	1910	2865
3/8″	0.375	85	170	341	568	852	1420	2272	3408
7/16″	0.4375	116	232	464	773	1160	1933	3093	4639
1/2″	0.5	151	303	606	1010	1515	2525	4039	6059
9/16″	0.5625	192	383	767	1278	1917	3195	5112	7668
5/8″	0.625	237	473	947	1578	2367	3945	6311	9467
11/16″	0.6875	286	573	1145	1909	2864	4773	7637	11,455
3/4″	0.75	341	682	1363	2272	3408	5680	9088	13,632
7/8″	0.875	464	928	1856	3098	4639	7731	12,370	18,555
1- 1/16″	1.0625	684	1368	2736	4560	6840	11,400	18,240	27,359
1- 1/8″	1.25	947	1893	3787	6311	9467	15,778	25,245	37,868
1- 5/8″	1.625	1600	3200	6400	10,666	15,999	26,665	42,664	63,996

* in lb/hr

COMMON ASME INTERNATIONAL BOILER CLASSIFICATIONS	
Name	**Description**
Automatic boiler	Equipped with certain controls and limit devices per ASME code
Boiler	Closed vessel used for heating water or liquid, or for generating steam or vapor by direct application of heat
Boiler plant	One or more boilers, connecting piping, and vessels within the same premises
Hot water supply boiler	Low pressure hot water heating boiler having a volume exceeding 120 gal. or heat input exceeding 200,000 Btu/hr, or an operating temperature exceeding 200°F that provides hot water to be used externally to itself
Low pressure hot water heating boiler	Boiler in which water is heated for the purpose of supplying heat at pressures not exceeding 160 psi or temperatures not exceeding 250°F
Low pressure steam heating boiler	Boiler operated at pressures not exceeding 15 psi for steam
Power hot water boiler	Boiler used for heating water or liquid to pressure exceeding 160 psi or a temperature exceeding 250°F
Power steam boiler	Boiler in which steam or vapor is generated at pressures exceeding 15 psi
Small power boiler	Boiler with pressures exceeding 15 psi but not exceeding 100 psi and having less than 440,000 Btu/hr input

HEATING VALUES AND CHEMICAL COMPOSITION OF STANDARD GRADES OF COAL							
Rank	**Heating Value***	**Chemical Composition****					
		Oxygen	**Hydrogen**	**Carbon**	**Nitrogen**	**Sulfur**	**Ash**
Anthracite	12,910	5.0	2.9	80.0	0.9	0.7	10.5
Semi-anthracite	13,770	5.0	3.9	80.4	1.1	1.1	8.5
Low-volatile Bituminous	14,340	5.0	4.7	81.7	1.4	1.2	6.0
Medium-volatile Bituminous	13,840	5.0	5.0	79.0	1.4	1.5	8.1
High-volatile Bituminous A	13,090	9.2	5.3	73.2	1.5	2.0	8.8
High-volatile Bituminous B	12,130	13.8	5.5	68.0	1.4	2.1	9.2
High-volatile Bituminous C	10,750	21.0	5.8	60.6	1.1	2.1	9.4
Subbituminous B	9150	29.5	6.2	52.5	1.0	1.0	9.8
Subbituminous C	8940	35.8	6.5	46.7	0.8	0.6	9.6
Lignite	6900	44.0	6.9	40.1	0.7	1.0	7.3

* in Btu/lb
** in %

BOILER FORMULAS...

Problem	Formula	Solution
Boiler Horsepower		
What is the boiler horsepower of a boiler generating 21,500 lb of steam per hour at 155 psi? The factor of evaporation is 1.08.	$BHP = \frac{lb/hr \times FE}{34.5}$ where BHP = boiler horsepower lb/hr = pounds per hour FE = factor of evaporation 34.5 = constant	$BHP = \frac{lb/hr \times FE}{34.5}$ $BHP = \frac{21,500 \times 1.08}{34.5}$ $BHP = \frac{23,220}{34.5}$ $BHP =$ **673 BHP**
Cycles of Concentration of Boiler Water		
What is the cycles of concentration if the chloride content of boiler water is 186 ppm and the feedwater chloride content is 38 ppm?	$CYC = \frac{BCl}{FCl}$ where CYC = cycles of concentration BCl = boiler water chlorides (in ppm) FCl = feedwater chlorides (in ppm)	$CYC = \frac{BCl}{FCl}$ $CYC = \frac{186}{38}$ $CYC =$ **4.89**
Degrees API Conversion		
Convert a specific gravity of 0.88 to °API.	$°API = \frac{141.5}{SG @ 60 °F} - 131.5$ where $°API$ = degrees American Petroleum Institute 14.5 = constant SG @ 60°F = specific gravity at 60°F 131.5 = constant	$°API = \frac{141.5}{SG @ 60 °F} - 131.5$ $°API = \frac{141.5 - 131.5}{0.88}$ $°API = 160.80 - 131.5$ $°API =$ **29.3 °API**
Differential Setting		
What is the differential setting of an automatic pressure control that turns the burner ON at 80 psi and OFF at 105 psi?	$\Delta S = P_1 - P_2$ where ΔS = differential setting P_1 = cut-out pressure P_2 = cut-in pressure	$\Delta S = P_1 - P_2$ $\Delta S = 105 - 80$ $\Delta S =$ **25 psi**
Differential Temperatures		
The temperature of the flue gas entering a boiler is 2060°F. The temperature of the same gas as it leaves the boiler is 812°F. What is the differential temperature?	$T = T_1 - T_2$ where ΔT = differential setting T_1 = higher pressure T_2 = lower pressure	$\Delta T = T_1 - T_2$ $\Delta T = 2060 - 812$ $\Delta T =$ **1248°F**
Force		
What is the force of 260 lb pressure exerted on 8 sq in.?	$F = \frac{P}{A}$ where F = force (in psi) P = pressure A = area	$F = \frac{P}{A}$ $F = \frac{260}{8}$ $F =$ **32.5 psi**

...BOILER FORMULAS...

Horsepower

What is the horsepower of a pump that moves 450 lb of water against a discharge head of 220′ in 1 minute? Disregard friction and other losses.

$$HP = \frac{d \times f}{t \times 33{,}000}$$

where
HP = horsepower
d = distance (in ft)
f = force (in lb)
t = time (in minutes)
33,000 = constant

$$HP = \frac{d \times f}{t \times 33{,}000}$$
$$HP = \frac{220 \times 450}{1 \times 33{,}000}$$
$$HP = \frac{99{,}000}{33{,}000}$$
$$HP = \mathbf{3\ HP}$$

Inches of Mercury

How many inches of mercury are there at an atmospheric pressure of 14.5 psi?

$$in.\ Hg = \frac{P}{0.491}$$

where
$in.\ Hg$ = inches of mercury
P = pressure (in psi)
0.491 = constant (psi @ 1″ Hg)

$$in.\ Hg = \frac{P}{0.491}$$
$$in.\ Hg = \frac{14.5}{0.491}$$
$$in.\ Hg = \mathbf{29.53\ in.\ Hg}$$

Percent of Blowdown

What is the percent of blowdown for a safety valve set to pop at 300 psi and reseat at 275 psi?

$$\%BD = \frac{PP - RP}{PP}$$

where
$\%BD$ = percent of blowdown
PP = popping pressure
RP = reseat pressure

$$\%BD = \frac{PP - RP}{PP}$$
$$\%BD = \frac{300 - 275}{300}$$
$$\%BD = \frac{25}{300}$$
$$\%BD = \mathbf{0.083\ or\ 8.3\%}$$

Rate of Combustion for Gaseous or Liquid Fuels

A scotch marine boiler has a furnace volume of 45.5 cu ft. If 3825.2 cu ft of natural gas is burned per hour and each cubic foot of natural gas contains 1100 Btu, what is the rate of combustion?

$$RC = \frac{H}{V_f \times t}$$

where
RC = rate of combustion (in Btu/hr)
H = heat released (in Btu)
V_f = volume of furnace (in cu ft)
t = time (in hours)

$$RC = \frac{H}{V_f \times t}$$
$$RC = \frac{3825.2 \times 1100}{45.5 \times 1}$$
$$RC = \mathbf{92{,}477.36\ Btu/hr}$$

Rate of Combustion for Solid Fuels

What is the rate of combustion if 3.1 tons of coal is burned per hour in a furnace with a 12′ × 16′ grate surface?

$$RC = \frac{F}{A_g \times t}$$

where
RC = rate of combustion (in Btu/hr)
F = fuel burned (in lb)
A_g = area of grate (in sq ft)
t = time (in hours)

$$RC = \frac{F}{A_g \times t}$$
$$RC = \frac{3.1 \times 2000}{(12 \times 16) \times 1}$$
$$RC = \frac{6200}{192}$$
$$RC = \mathbf{32.29\ lb/hr}$$

...BOILER FORMULAS...

Return Condensate Percentage in Feedwater

What is the return condensate percentage in feedwater if the makeup conductivity is 834 micromhos, the feedwater conductivity is 185 micromhos, and the condensate conductivity is 65 micromhos?

$$\%RC = \frac{MC - FC}{MC - CC}$$

where
$\%RC$ = return condensate percentage
MC = makeup conductivity (in micromhos)
FC = feedwater conductivity (in micromhos)
CC = condensate conductivity (in micromhos)

$$\%RC = \frac{MC - FC}{MC - CC}$$
$$\%RC = \frac{834 - 185}{834 - 65}$$
$$\%RC = \frac{649}{769}$$
$$\%RC = \mathbf{0.84 \text{ or } 84\%}$$

Specific Gravity of Fuel Oil

What is the specific gravity of a brine solution that weighs 10.2 lb/gal. at 60°F?

$$SG = \frac{Lb/V\ liquid}{Lb/V\ water}$$

where
SG = specific gravity
$Lb/V\ liquid$ = weight of volume of liquid at 60°F
$Lb/V\ water$ = weight of equal volume of water at 60°F

$$SG = \frac{Lb/V\ liquid}{Lb/V\ water}$$
$$SG = \frac{10.2}{8.33}$$
$$SG = \mathbf{1.22}$$

Equivalent Static Head

What is the equivalent static head of a boiler operating at 275 psi?

$SH = Boiler\ pressure \times 2.31$

where
SH = static head (in ft)
$Boiler\ pressure$ = boiler pressure (in psi)
2.31 = multiplier

$SH = Boiler\ pressure \times 2.31$
$SH = 275 \times 2.31$
$SH = \mathbf{635.25'}$

Steam Generation

How much steam will a 150 HP boiler make in 2.5 hours?

$S = HP \times 34.5 \times t$

where
S = steam (in lb)
HP = horsepower
34.5 = constant (lb produced per hour)
t = time (in hours)

$S = HP \times 34.5 \times t$
$S = 150 \times 34.5 \times 2.5$
$S = \mathbf{12{,}937.5\ lb}$

Temperature Conversion

Fahrenheit to Celsius

$$°C = \frac{(F° - 32)}{1.8}$$

Convert 92°F to Celsius

$$°C = \frac{(F° - 32)}{1.8}$$
$$°C = \frac{(92 - 32)}{1.8}$$
$$°C = \frac{60}{1.8}$$
$$°C = \mathbf{33.3°C}$$

Celsius to Fahrenheit

$°F = (1.8 \times °C) + 32$

Convert 30°C to Fahrenheit

$°F = (1.8 \times °C) + 32$
$°F = (1.8 \times 30) + 32$
$°F = 54 + 32$
$°F = \mathbf{86°F}$

...BOILER FORMULAS		
Total Force		
What is the total force of 120 psi acting on 4 sq in.?	$TF = P \times A$ where TF = total force (in lb) P = pressure (in psi) A = area of valve disc exposed to steam (in sq in.)	$TF = P \times A$ $TF = 120 \times 4$ $TF =$ **480 lb**
Water Column		
How high is a colomn of water that exerts 42.43 psi at the bottom of the column?	$WC = \frac{P}{0.433}$ where WC = water column P = pressure 0.433 = constant (force per 1′ of water depth)	$WC = \frac{P}{0.433}$ $WC = \frac{42.43}{0.433}$ $WC =$ **97.99′ or 98′**

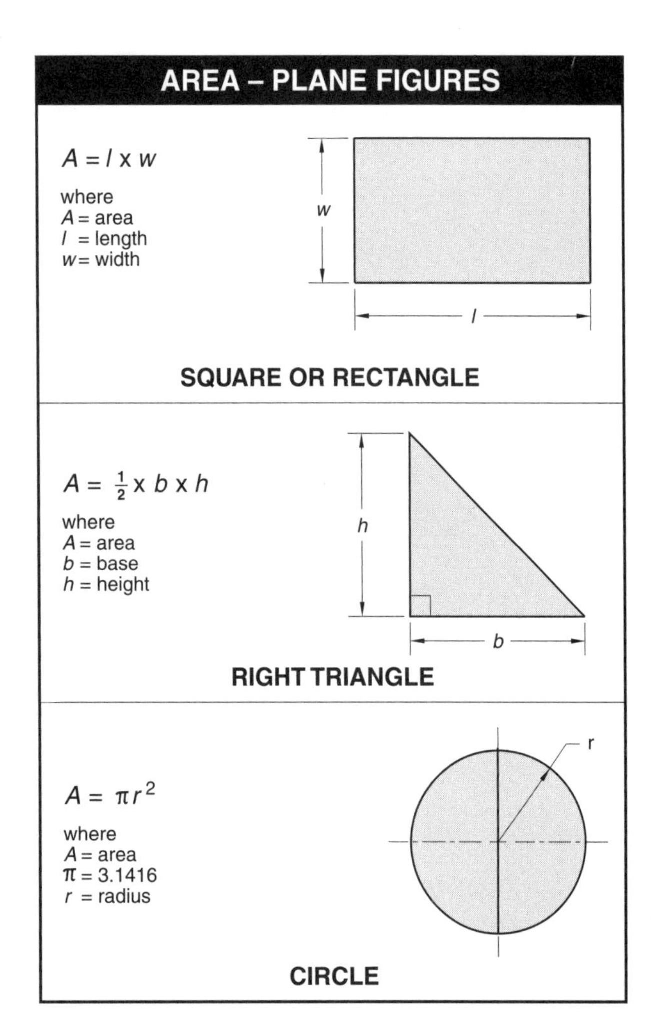

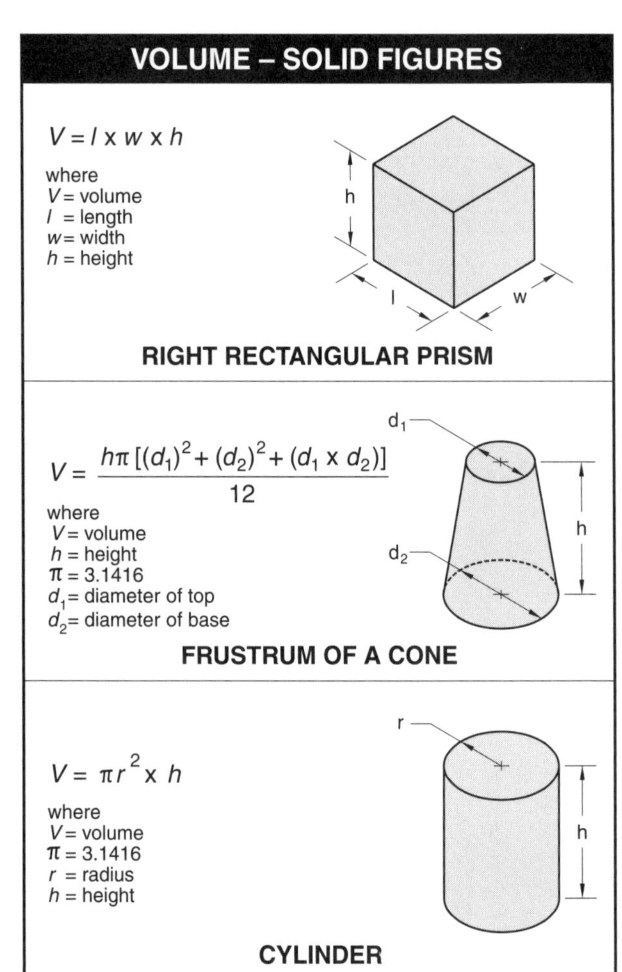

DECIMAL EQUIVALENTS OF AN INCH							
Fraction	Decimal	Fraction	Decimal	Fraction	Decimal	Fraction	Decimal
1/64	0.015625	17/64	0.265625	33/64	0.515625	49/64	0.765625
1/32	0.03125	9/32	0.28125	17/32	0.53125	25/32	0.78125
3/64	0.046875	19/64	0.296875	35/64	0.546875	51/64	0.796875
1/16	0.0625	5/16	0.3125	9/16	0.5625	13/16	0.8125
5/64	0.078125	21/64	0.328125	37/64	0.578125	53/64	0.828125
3/32	0.09375	11/32	0.34375	19/32	0.59375	27/32	0.84375
7/64	0.109375	23/64	0.359375	39/64	0.609375	55/64	0.859375
1/8	0.125	3/8	0.375	5/8	0.625	7/8	0.875
9/64	0.140625	25/64	0.390625	41/64	0.640625	57/64	0.890625
5/32	0.15625	13/32	0.40625	21/32	0.65625	29/32	0.90625
11/64	0.171875	27/64	0.421875	43/64	0.671875	59/64	0.921875
3/16	0.1875	7/16	0.4375	11/16	0.6875	15/16	0.9375
13/64	0.203125	29/64	0.453125	45/64	0.703125	61/64	0.953125
7/32	0.21875	15/32	0.46875	23/32	0.71875	31/32	0.96875
15/64	0.234375	31/64	0.484375	47/64	0.734375	63/64	0.984375
1/4	0.250	1/2	0.500	3/4	0.750	1	1.000

Glossary

A

absolute pressure: The pressure above a perfect vacuum. It is expressed as psia (pounds per square inch absolute).

accumulation test: A pressure test used to ensure that the safety valve has sufficient relieving capacity to vent all of the excess steam that the boiler can produce.

acid dewpoint: The temperature at which acidic vapors begin to condense out of the flue gases.

actuator: A device that receives a control signal from a controller and converts the signal into a proportional movement of the valve.

after-treatment: The water treatment processes that occur during or after steam generation.

air cock: See *boiler vent.*

air pollutant: A substance that produces an adverse effect on humans, animals, vegetation, or materials.

air-to-fuel ratio: Amount of air and fuel supplied to the burner over high and low fire.

alkalinity: Determined by boiler water analysis. Boiler water with a pH over 7 is considered alkaline.

amine: A chemical that prevents corrosion in condensate and steam piping.

analog signal: A continuous signal used in the ongoing control of a continuous process.

anion: An ion that has a negative electrical charge.

annunciator: An audible alarm that is created electrically or electronically.

anthracite coal: A geologically older coal that contains a high percentage of fixed carbon and a low percentage of volatiles. Anthracite is also known as hard coal.

ash fusion temperature: The temperature at which ash begins to become molten.

ash hopper: Large receptacle used to store ashes until they can be disposed of.

ASME Boiler and Pressure Vessel Code* (ASME Code)*: Code written by ASME International (formerly the American Society of Mechanical Engineers) that governs and controls the types of material, methods of construction, and procedures used in the installation of boilers.

"A" style watertube boiler: A watertube boiler design with a top steam and water drum and two smaller bottom mud drums.

atmospheric pressure: The force exerted by the weight of the atmosphere bearing on the Earth's surface.

atomization: The process of breaking a liquid fuel stream into a mist of tiny droplets.

B

back-seating: The situation where a slow-opening valve such as a gate or globe valve is fully opened until it stops.

badge plate (nameplate): A data plate attached to a boiler.

baffle: A metal or refractory-covered panel that directs the flow of gases of combustion for maximum boiler heating surface contact.

bag: A protruding bubble or bulge in the steel plate of a boiler.

ball check valve: An automatic self-closing gauge glass valve.

ball valve: A quick-acting, two-position shutoff valve.

banking a fire: The process of greatly slowing the burning of coal or some other solid fuel.

barometric damper: A free-swinging adjustable balanced damper used on smaller boilers to automatically limit the amount of air pulled through the combustion chamber.

battery: A group of boilers that feed steam into the same steam header.

bellows: A flexible device that expands and contracts with changes in pressure.

bent-tube watertube boiler: A boiler design in which the tubes are bent (curved) to some degree.

bituminous coal: A geologically younger coal that contains a high percentage of volatiles and a low percentage of fixed carbon. Bituminous coal is also known as soft coal.

blended fuel oil: A mixture of distillate oils and residual oils that may contain some crude oil.

blister: A lamination of steel plate or tube surfaces or where the steel plate splits into layers.

blowback: The drop in pressure in the boiler that occurs after the safety valve has opened.

blowdown: 1. The amount of pressure in a pressure vessel that must be reduced before a safety valve reseats. **2.** Removal of impurities from the boiler water by draining some of the water.

boiler heating surface: Any part of the boiler metal that has hot gases of combustion on one side and water on the other.

boiler horsepower (BHP): The energy required for the evaporation of 34.5 lb of water at 212°F into steam at atmospheric pressure and at 212°F in 1 hr.

boiler load: The amount of steam being produced by a boiler.

boiler master: A controller that determines the appropriate quantities of fuel and air that need to be provided to a boiler's combustion equipment in order to produce the required quantity of steam.

boilers in battery: Two or more boilers connected to a common steam header.

boiler thermal efficiency: The percentage of the heat liberated that is transferred into the boiler water.

boiler vent: A section of steel pipe about ½″ to 1″ ID coming off the top of the vessel with one or two valves in it.

boil-out procedure: The process of removing oily residues from a boiler by adding chemicals to the boiler water and boiling the mixture for a period of time.

bottom blowdown: The process of periodically draining part of the boiler water to remove heavy sludge that settles to the bottom of the boiler.

Bourdon tube: The tube inside a mechanical pressure gauge. It is a bronze or stainless steel tube bent into a question mark shape and flattened into an elliptical shape.

branch line: The piping that takes steam from the steam mains to individual pieces of steam-using equipment.

breeching: The ductwork that carries cooled flue gases from the exit of the boiler to the stack.

bridge wall: A firebrick wall built across a boiler furnace.

brine: A solution of salt and water.

British thermal unit (Btu): The amount of heat necessary to raise the temperature of 1 lb of water by 1°F.

bubbler control: A set of components that use a small flow of compressed air or another gas to detect the level of liquid in a vessel such as a storage tank.

buckstay: A metal brace used to attach a wall to a steel framework that supports the wall.

bunker oil: One of the heavy oils formed as crude oil is stabilized after the lighter components have been distilled off.

butterfly valve: A valve that consists of a circular disc that is rotated by the valve stem so that the disc is parallel to the flow through the valve, perpendicular to the flow, or somewhere in between.

bypass damper: Controls the air temperature in air heaters to prevent corrosion.

bypass line: A pipeline that passes around a control, heater, or steam trap. Used so that a plant can operate while equipment is serviced or repaired.

calibrate: Adjusting a gauge to conform to a test gauge.

calorie: The amount of heat necessary to raise the temperature of 1 g of water by 1°C.

calorimeter: A laboratory instrument used to measure the heat content of a substance, such as a sample of a fuel to be used for a boiler.

capacity: 1. The amount of steam in pounds per hour that the safety valve is capable of venting at the rated pressure of the valve. **2.** The volume of fluid that can be delivered by a pump over a given unit of time.

capillary tube: A long, small-diameter tube, usually of stainless steel or a copper alloy.

carryover: The entrainment of small water droplets with steam leaving the boiler.

cascading control: A control scheme in which one process variable is measured and used to set the setpoint of another controller.

cascading flash steam system: A piping and tank arrangement where condensate is allowed to flash several times in progressively lower pressure flash tanks.

cation: An ion that has a positive electrical charge.

caustic embrittlement: A problem in which boiler metal becomes brittle and weak because of cracks in the crystalline structure of the metal (crystalline cracking).

cavitation: The condition caused when a portion of the water or other liquid entering the eye of a pump impeller flashes into steam bubbles.

centrifugal pump: A pump in which a rotating impeller throws liquid from its vanes through centrifugal force.

chain-grate stoker: A stoker in which the grates are composed of thousands of small staggered segments that are interlaced by support bars or rods, forming a heavy chain conveyor.

characterization: The relationship between the amount a valve is open and the amount of flow through the valve.

check valve: A one-way flow valve for fluids.

chelant: A chemical that helps keep the hardness in the water dissolved so that it does not crystallize on heating surfaces.

chemical concentration: The amount of a specific chemical in the boiler water.

chemical energy: Energy in the fuel that converts to heat energy during the combustion process.

chimney: Used to create draft. Also an outlet to the atmosphere for the gases of combustion.

classifier: A spinning set of vanes located at the coal and air outlet from the pulverizer that separates very fine coal dust from larger coal particles.

class of fire: The five classes of fires are Class A, started from wood, paper, or other combustible materials containing carbon; Class B, started from oil, grease, or flammable liquids; Class C, started from electrical devices; Class D, started from combustible metals; and Class K, started from grease in commercial cooking equipment.

clinker: A mass of coal and ash that has fused together during burning.

clinker grinder: A large set of steel rollers with heavy teeth that grind ash and clinkers to reduce their size before they enter the ash hoppers.

closed feedwater heater: Feedwater heater in which steam and feedwater do not come into direct contact. Steam is in the shell of the heater while water passes through tubes.

closed heat exchanger: A heating unit in which the heating medium and the fluid being heated do not mix but are separated by tube walls or other heating surfaces.

closed impeller: An impeller that has shrouds on both sides of the vanes.

closed system: A steam system in which the condensate is recovered and returned to the boiler.

coal bunker: An overhead bin where large quantities of coal are stored.

coal feeder: Controls the flow of coal entering the pulverizer.

coal gate: Used to control the depth of coal entering the boiler furnace on chain grate stokers.

coal ram: Distributes coal evenly into the center retort on underfeed stokers and forces the coal up to the top where it is burned.

coal scale: Measures and records the amount of coal fed to stoker-fired or pulverized coal-fired boilers.

code: A regulation or law.

coefficient of expansion: The property of a given material that expresses how much a standard unit of length of the material expands or contracts under a specific change in temperature.

coil watertube boiler: A boiler design in which the tubes are formed into a continuous coil, with the combustion gases passing through the interior of the coil.

coking: The separation of heavy carbon-based fractions from the oil, resulting in precipitation of solids that may plug the piping.

color wheel comparator test: A relatively simple test used to determine the quantity of iron in the condensate or makeup water.

column of water: Water of some specified depth or height.

combined-cycle boiler system: An electric power generating system that uses both a gas turbine-driven generator and a steam turbine-driven generator.

combustion air preheater: A piece of equipment provided to preheat the combustion air to some degree before the combustion air enters the furnace.

combustion chamber: The area of a boiler where the burning of fuel occurs.

combustion efficiency: The percentage of the Btu content of fuel that is liberated as heat by the boiler fuel-burning equipment.

commissioning: The process of inspecting, preparing, and testing each major component of a system prior to operation.

complete combustion: A fire where the fuel is burned with a slight excess of oxygen so that the fuel is completely consumed without forming any smoke, and only a minimal amount of oxygen is left over.

compression: The exertion of equal forces from opposite sides of an object that push toward the middle.

condensate: The water formed when steam condenses to water.

condensate polisher: An ion-exchange water softener similar to a sodium zeolite water softener but with a resin that can withstand the high temperatures encountered with condensate.

condensate pump: Used to return condensed steam to the open feedwater heater.

condensate receiver: A tank or other collection point where condensate is accumulated and saved for reuse.

condensate return line: A pipe that carries the condensate and air discharged by the steam traps.

condensate tank: Where condensed steam (water) is stored before it is delivered back to the open feedwater heater by the condensate pump.

condensation: The process of steam being changed back into water.

conductance: The quantitative measure of the ability of an electric circuit to allow current flow.

conduction: A method of heat transfer in which heat moves from molecule to molecule.

conductivity: The ability of a conductor to allow current flow.

confined space: A space that has limited or restricted means for entry and exit, has unfavorable natural ventilation such that a dangerous atmosphere could exist inside that does not naturally vent out, and is not intended for continuous employee occupancy.

continuous blowdown: The process of continuously draining water from a boiler to control the quantity of impurities in the remaining water.

controller: A device that takes an input signal, compares the signal to a setpoint, performs a computation, and sends an output signal to a final control element.

control loop: The collection of control devices and other components necessary for automatic control of a process or subprocess.

control valve: A valve used to modulate the flow of fluid.

convection: A method of heat transfer that occurs as heat moves through a fluid.

corresponding pressure: The pressure at which both the water and steam are at the same temperature.

corrosion: The pitting and channeling of metal caused by gases in water.

counterflow: Principle used in heat exchangers where the medium being heated flows in one direction and the medium supplying the heat flows in the opposite direction.

critical pressure: The pressure at which the density of the water and the density of the steam are the same.

cross-limiting: The configuration of boiler combustion controls such that, with an increase or decrease in the firing rate, the combustion conditions never become oxygen-deficient.

cross "T": Used on connections on a water column for inspection of steam and water lines to ensure they are clean and clear.

cycles of concentration: Measure the concentration of the solids in the boiler water compared to the concentration of the solids in the feedwater.

cyclone separator: Separates water droplets from steam by centrifugal force and by changing direction.

damper: A device that partially or totally inhibits the flow of air or combustion gases through ductwork.

data plate: A plate that must be attached to a safety valve containing data required by the ASME Code.

deaerating feedwater heater: Type of open feed-water heater equipped with a vent condenser.

deaerator: An open heat exchanger that removes dissolved gases from the feedwater going to a boiler.

deaerator surge tank: A tank or pressure vessel used to even out flow into a deaerator and provide a continuous, modulated flow.

dealkalizer: An ion-exchange unit that works in a manner similar to a sodium zeolite water softener, but exchanges anions rather than cations.

decatherm: 10 therms, or 1,000,000 Btu.

demineralizer: A highly efficient ion-exchange process generally used for high-pressure boilers.

desiccant: A drying agent.

desuperheating: Removing heat from superheated steam to make it suitable for process.

diagonal stays: Braces that are installed in a firetube boiler to keep the upper portions of the tube sheets above the tubes from bulging outward due to the internal pressure.

diaphragm valve: A valve that uses a flexible diaphragm as the movable sealing surface.

differential pressure: The difference between two pressures at different points.

differential temperature: The difference in temperature between two different points.

diffusion ring: A stationary vane in the pump casing.

digital signal: An instantaneous signal used to implement a one-time action.

dirt pocket: The pipe nipple installed on the bottom of the drip leg that catches rust and weld slag.

discharge piping: Piping attached to the outletside of a safety valve that conveys steam to the atmosphere.

disc steam trap: A comparatively small steam trap that uses a flat, round disc as the means of opening and closing the outlet orifice.

displacement: The volume of fluid forced out of a full container when another body is forced into the container.

displacer: A long, fairly heavy cylindrical float of small diameter used to measure liquid level in a tank or vessel.

dissolved gases: Gases that have gone into solution in water.

dissolved solids: Solid impurities that have gone into solution in water.

distillate fuel oil: A fuel produced by distilling crude oil.

double-acting pump: A reciprocating pump that moves fluid in both directions of stroke.

double block and bleed: A valve configuration that consists of two automatic shutoff valves arranged in series, with a vent or bleed valve between them that vents outdoors.

double-seated valve: A control valve that has two discs on one stem and two seats in the body.

double strap and butt joint: A riveted joint made by rolling the steel plate into a cylinder to make the shell of the boiler (or drum) and butting the two edges of the plate together to form the seam of the cylinder.

double-suction pump: A pump with a casing and impeller designed to allow a liquid to flow into both sides (eyes) of the impeller at once.

downcomer tubes: Tubes that contain the cooler, descending water.

draft: The movement of air and/or gases of combustion from a point of higher pressure to a point of lower pressure.

draft loss: The loss of available draft due to friction and other pressure losses as the flue gases flow through the combustion gas passageways.

draft system: Consists of the equipment, controls, and ductwork that deliver air to the boiler's furnace area for combustion of the fuel, and then conduct the spent combustion gases to the atmosphere.

drainable superheater: A superheater configured so that condensate in the superheater tubes migrates to a low point from which it can be drained.

drip leg: A downward extension from a steam distribution line or piece of heat exchange equipment where condensate is allowed to drain.

drum desuperheater: An attemperator that diverts part of the superheated steam through a heat exchanger in the boiler mud drum.

drum pressure control valve: A control valve that is configured so as to maintain a constant pressure on the steam drum of a watertube boiler or the shell of a firetube boiler at all times.

drum vent: See *boiler vent.*

dryback scotch marine boiler: A firetube boiler with a refractory-lined chamber at the rear of the boiler that is used to direct the combustion gases from the flue furnace to the first pass of tubes.

dry pipe: An upside-down T-shaped pipe connected to the main steam outlet from either a firetube or watertube boiler.

dry sheet: The metal that is an extension of the cylindrical shell of a firetube boiler.

"D" style watertube boiler: is a watertube boiler similar to the "O" style, except that the steam-generating tubes on one side are extended to leave an open area close to the center.

D-slide valve: A valve that controls the movement of steam into and out of the steam cylinder in a duplex pump.

ductility: The plasticity exhibited by a material under tension loading.

duplex pump: A steam-driven, reciprocating, positive-displacement, double-acting pump with two steam cylinders and two liquid cylinders.

duplex strainers: Remove solid particles from the fuel oil in fuel oil systems.

economizer: A series, or bank, of boiler tubes used to recover heat from the boiler flue gas.

efficiency: The comparison of energy output to energy input in a piece of equipment or in a system.

electric boiler: Boiler that has heat produced by electric resistance coils or electrodes.

electronic flue gas analyzer: Device used to analyze flue gas for temperature, gases, draft, and smoke.

electrostatic precipitator: A device used to separate flyash particles from the flue gas stream before it goes out the stack.

emergency plan: A document that details procedures, exit routes, and assembly areas for facility personnel in the event of an emergency.

emission factor: An expression of the rate of pollutant production per unit of fuel input.

enthalpy: Total heat in the steam.

entrainment: The process where solid or liquid particles are carried along with steam flow.

equalizing line: Line used to warm up the main steam line and equalize the pressure around the main steam stop valve.

error: The amount of deviation of a measurement from the setpoint.

evaporation test: A test that allows a low water condition to occur in the boiler in order to test the low water fuel cutoff switch.

evaporator: A set of heat exchangers that produces water suitable for boiler use from water that contains large quantities of impurities.

excess air: The amount of air added to the combustion process over and above that which is theoretically necessary.

expansion bends: Installed on boiler main steam lines to allow for expansion and contraction of the lines.

exposed-tube (dry-top) vertical boiler: A firetube boiler with fire tubes extending several inches through the steam space at the top before ending at the tube sheet.

external header cast iron sectional boiler: Contains cast iron sections individually connected to external manifolds (headers) with screwed (threaded) nipples.

externally fired firetube boiler: A boiler with a separate furnace area that is usually built of refractory brick.

external treatment: Boiler water treated before it enters the boiler to remove scale-forming salts, oxygen, and noncondensable gases.

extractive CEMS: A monitoring system that withdraws a sample of the flue gas stream, conditions the sample, analyzes the conditioned sample, and then provides a readout of the flue gas condition.

factor of evaporation (FE): The heat added to the water in an actual boiler in Btu/lb and divided by 970.3.

feathering: The point when a safety valve is about to lift.

feedback transmitter: A transmitter used to provide confirmation that a valve or damper actuator has made a change as commanded.

feedwater: A mixture of condensate and makeup water that is provided to the boiler to make steam.

feedwater line: The piping that carries the feedwater from the feedwater pump to the boilers.

feedwater pump: A pump that sends the returned condensate and any makeup water into the boiler.

feedwater system: Consists of all the equipment, controls, and piping that prepare and treat the water for use in the boiler, put the water into the boiler, and maintain a normal, safe amount of water in the boiler.

field-assembled boiler: A boiler of large size that cannot be shipped as a completed unit by the manufacturer to the site where it will be placed in service.

field-erected boiler: Boiler that must be erected in the field because of its size and complexity.

filming amine: A chemical that prevents corrosion of piping by providing a protective barrier.

final element: The device that actually causes the change in the process.

firebox: The part of the boiler where combustion of fuel takes place.

firebox boiler: A firetube boiler in which an arch-shaped furnace is surrounded on the sides by a water leg area.

firecrack: A crack in a riveted joint that runs from the rivet to the edge of the plate or from rivet to rivet.

fired vessel: A pressure vessel that includes a burner or combustion equipment of some kind.

fire extinguisher: Portable unit used to put out small fires or contain larger fires until the company fire brigade or the fire department arrives.

fire point: The lowest temperature at which the vapor given off by a substance will ignite and burn for at least 5 sec when exposed to an open flame.

firetube boiler: Has heat and gases of combustion passing through tubes that are surrounded by water.

firing rate: Amount of fuel the burner is capable of burning in a given unit of time.

first law of thermodynamics: Energy cannot be created or destroyed, but can be changed from one form to another.

fittings: Trim found on the boiler that is used for safety, and/or efficiency.

fit-up: The process of fitting and rolling the tubes into the drums of a new watertube boiler.

fixed carbon: The burnable remainder of coal left when coal is heated and the volatile matter is driven off.

flame failure: When the flame in a furnace goes out.

flame impingement: A condition where flame from burning fuel continually strikes the boiler surfaces or refractory brick.

flame propagation rate: The rate at which the flame can ignite the incoming air-fuel mixture, in feet per second.

flame safeguard system: The collection of automatic control devices that ensure safe operation of the combustion equipment.

flame scanner: A device that proves that the pilot and main burner flames have been established and remain in service.

flame sensor: Sensing device in a flame scanner used to sense the pilot and the main flame in the burner.

flareback: Flames discharging from the boiler through access doors or ports. Caused by delayed ignition or furnace pressure buildup.

flashback: The condition where a flame travels upwind and into the burner assembly.

flash economizer: A heat recovery system used to reclaim the heat from the boiler blowdown water and used in conjunction with the continuous blowdown system.

flash point: The lowest temperature at which the vapor given off by a substance will make a flash of flame, but not continue to burn, when an open flame is passed over it.

flash steam: Steam that is instantly produced when very hot water is released to a lower pressure and, thus, a lower boiling temperature.

flash tank: A pressure vessel in which condensate or other very hot water under a high temperature and pressure is allowed to partially flash into steam.

flat gauge glass: Type of gauge glass used for pressures over 250 psi.

flexible joint: Used to allow for expansion and contraction of steam or water lines.

flexible-tube watertube boiler: A boiler design in which the tubes exposed to the combustion gases are sharply bent to provide the maximum possible flexibility.

float and thermostatic steam trap: Contains a thermostatic bellows or other thermostatic element and also contains a steel ball float connected to a discharge valve by a linkage.

fluid: Any material that can flow from one point to another. Fluids can be liquids or gases.

fluidized bed boiler: A boiler in which fuels are burned in a bed of inert materials such as limestone pellets or sand.

fly ash: Ash from the combustion of coal or other solid fuels that is carried along with the draft through a boiler furnace and ductwork.

foaming: The development of froth on the surface of the boiler water.

foot-pound: A unit of work equal to the movement of a 1-lb object over a distance of 1 ft.

foot valve: A check valve installed at the bottom of the suction line on a negative-suction pump that keeps the suction line primed when the pump shuts down.

forced circulation: A variation in watertube boiler design in which boiler water circulation through the tubes is enhanced by a pump.

forced draft: The discharge of combustion air from a fan into a furnace to combine with the fuel for combustion.

friction head: The pressure loss associated with friction in the pump piping and fittings, converted into equivalent feet of static head.

fuel system: Consists of all the equipment, controls, and piping that deliver the fuel to the boiler's combustion equipment and control the combustion process.

fuel-to-steam efficiency: The percentage of the heat content of the fuel that is transferred into the boiler water.

furnace volume: Amount of space available in a furnace to complete combustion.

fusible plug: A temperature sensitive device that causes an audible alarm when exposed to excessive temperature.

Fyrite® analyzer: Instrument used to measure percentage of carbon dioxide in the gases of combustion.

gagging: Application of a clamp on a safety valve spindle to keep the valve in full closed position during a hydrostatic test.

galvanometer: Used to measure small electric currents.

gas analyzer: Used to analyze the gases of combustion to determine combustion efficiency.

gas calorimeter: Used to determine the Btu content of natural gas.

gas cock: A manual quick-closing shutoff valve.

gases of combustion: Gases produced by the combustion process.

gas leak detector: Device used to locate gas leaks in a boiler room.

gas mixing chamber: Where air and gas mix before they enter the furnace in low pressure gas burners.

gas pressure regulator: Used to supply gas to the burner at the pressure needed for combustion ofthe gas.

gas tungsten arc welding (GTAW): A welding process in which a virtually nonconsumable tungsten electrode is used to provide the arc for welding.

gate valve: A valve used to stop or start flow. It has a wedge-like disc that is lowered into or raised out of the path of the fluid that flows through the valve.

gauge glass: A tubular or flat glass connected to a water column that allows an operator to see the water level in the water column, and thus in the boiler, at a glance.

gauge glass blowdown valve: Valve used to remove any sludge and sediment from gauge glass lines.

gauge pressure: The pressure above atmospheric pressure. It is expressed as psig (pounds per square inch gauge).

gear pump: A rotary positive-displacement pump in which the liquid being pumped fills the open spaces between the teeth of rotating cylindrical gears and the pump housing.

globe valve: A valve that has a tapered, rounded, or flat disc held horizontally on the stem.

grade: The size of the pieces of coal.

grates: Where the combustion process starts in a coal-fired furnace.

greensand: A dark, coarse, sandy material.

gunning materials: Plastic refractory materials that are gunned, or sprayed, under pressure onto a surface.

hammer test: A test performed on the stays inside the boiler to check the integrity of each, relative to the others.

handhole: A small access hole used for looking and reaching into the boiler shell during inspections.

hardness: The measurement, in ppm, of calcium and magnesium in water.

hazardous material: A substance that could cause injury to personnel or damage to the environment.

head: A vertical column of liquid that, due to its weight, exerts pressure on the bottom and sides of its container.

heat energy: Kinetic energy caused by molecular motion within a substance.

heat exchanger: Any piece of equipment that transfers the heat of the steam into some other material.

heating surface: That part of the boiler that has heat and gases of combustion on one side and water on the other.

heating value: Expressed in Btu per gallon or per pound. Heating value varies with the type of fuel used.

heat transfer: Movement of heat from one substance to another that can be accomplished by radiation conduction or convection.

higher heating value (HHV): The total heat obtained from the combustion of a specified amount of the fuel under perfect (stoichiometric) combustion conditions. Also known as gross heating value.

high pressure steam boiler: Boiler that operates at a steam pressure over 15 psi and over 6 BHP.

high temperature water boiler: A boiler in which the maximum operating temperature of the water may reach temperatures in excess of 250°F and the operating pressure may exceed 160 psig.

horizontal return tubular (HRT) boiler: A firetube boiler consisting of a horizontal shell set above a refractory brick-lined furnace.

horizontal split-case pump: A pump that has a horizontally split pump casing where the top half of the pump casing can be lifted off for inspection and maintenance without disturbing the shaft, impeller, or bearings.

horizontal through stays: Braces that are installed in a firetube boiler to keep the upper portions of the tube sheets above the tubes from bulging outward due to the internal pressure.

horsepower (HP): A unit of power equal to 33,000 foot-pounds (ft-lb) of work done in 1 minute.

hot process water softener: A pressure vessel that uses steam to heat makeup water by direct contact and uses chemical injection to precipitate the hardness out of the water.

hot water boiler: A boiler that generates hot water, not steam. The heated water produced by a hot water boiler is usually between approximately 170°F and 190°F.

hot well: A reservoir located at the bottom of a condenser where condensate collects.

huddling chamber: Part on a safety valve that increases the area of the safety valve disc, thus increasing the total upward force, causing the valve to pop open.

huddling ring: An adjustment in a safety valve that controls the degree to which the escaping steam is directed against the safety valve disc.

hydrostatic test: A test in which the boiler is filled with approximately 70°F water and then pressurized to 1½ times its MAWP. Any leaks are exposed by observing water drips.

ignition: The lightoff point of a combustible material.

ignition arch: The curved refractory brick arch directly above the location where green coal enters the furnace in a chain-grate stoker-fired boiler.

impeller: The rotating element found in a centrifugal pump that converts centrifugal force into pressure.

impingement: 1. Fuel oil striking brickwork or the boiler heating surface that results in the formation of carbon deposits and smoke. **2.** Steam that strikes the boiler heating surface, causing erosion of boiler metal.

implosion: An inward collapse from external pressure.

incomplete combustion: A fire where the fuel is burned without the proper amount of oxygen, without enough mixing of fuel and oxygen, or at a temperature too low to allow satisfactory reaction of the fuel and oxygen.

induced draft: The use of a fan to simulate the effect of a stack by drawing the combustion gases from the furnace and through the flue gas passages.

injector: A motive fluid pump that uses the velocity of steam to draw water and pump it into a boiler.

inleakage air: Air that leaks into a furnace.

in-line steam separator: A cylindrically shaped vessel that is installed in a steam pipe to remove moisture droplets after the steam has left the boiler.

input signal: The flow of control information provided to a control device.

in situ CEMS: A monitoring system that directly measures the concentration of a specific constituent in the stack, without conditioning, and provides a readout of that concentration for the boiler operator.

instrument (boiler): Device that measures, indicates, records, and controls boiler room systems.

insulation: Material used to cover steam, water, and fuel oil lines to cut down on radiant heat losses.

integrator: A calculating device that combines multiplication and addition.

interlock: Used with burner controls to ensure proper operating sequence.

intermittent pilot: A pilot that is lit at the appropriate time to light the main burner and then stays on during the entire period that the burner is on.

internally fired firetube boiler: A boiler with a furnace area surrounded by the pressure vessel.

internal treatment: The addition of chemicals directly into the boiler water to control pitting, scale, and caustic embrittlement.

interrupted pilot: A pilot that is lit at the appropriate time to light the main burner and then extinguished as soon as the main burner is lit.

inverse solubility: The tendency of certain impurities in water to crystallize and precipitate as the temperature of the boiler water increases.

inverted bucket steam trap: Contains an upside-down steel cup, called a bucket, that is attached to a linkage that opens and closes a discharge valve as the cup rises and falls inside the trap.

ion: An atom or molecule with an electrical charge.

ion exchange: An exchange of one ion for another ion in a solution.

iron filter: A pressure vessel through which raw water flows on its way to the water softeners.

lantern ring: A spacer installed between two of the rings of packing in a pump.

lap joint: A riveted joint with two overlapping plates that are drilled through and riveted together at the edges.

latent heat: The Btu content of a substance that represents the heat absorbed or given up as it changes state between solid, liquid, and gas.

laying up: The procedure used to take a boiler out of service for a longer than normal period of time.

lift: The condition where the level of the liquid to be pumped is below the elevation of the pump.

ligament: The portion of the drum wall between the tube holes.

lighting off: The initial ignition of the fuel.

lignite: Very young and very soft coal that has a high moisture content and low heat value.

lime-soda process: A process that uses lime and soda ash to soften water.

line desuperheater: A device that automatically removes superheat from superheated steam so that the steam becomes saturated steam.

live steam: Steam in its pure, invisible form.

lobe pump: A rotary positive-displacement pump in which the liquid being pumped fills the open spaces between the lobes of matched rotors and the pump housing.

local control: A control device that is installed directly on or very near the equipment on which it is being used.

lockout: The use of locks, chains, and other physical restraints to prevent the operation of specific equipment.

lower heating value (LHV): The quantity of heat remaining after subtracting the latent heat used in evaporating the water formed in the combustion of the hydrogen. Also known as net heating value.

low pressure steam boiler: Boilers that operate at a steam pressure of no more than 15 psi.

low water fuel cutoff: A device located slightly below the NOWL that turns off the boiler burner in the event of low water.

lug and roller method: A support method where steel lugs (support plates projecting horizontally from the boiler shell) are welded or riveted to the front and rear of the boiler shell.

main steam outlet lines: Consist of the piping and valves that direct the steam from the boilers to the steam header.

main steam stop valve: A gate valve in the main steam line between the boiler and the steam header used for isolating the steam side of a boiler that is to be out of service.

main trial for ignition or **MTFI:** A period of about 5 sec to 10 sec for the flame scanner to sense the presence of flame from the main flame.

makeup water: Water used to replace condensate that is not returned to the boiler.

makeup water feeder: An automatic float-operated valve that feeds makeup water to a low-pressure heating boiler to replace condensate that has been lost from the system or water that has been lost in the form of steam leaks.

makeup water line: A city water pipe or well water pipe through which makeup water is added to a boiler.

malleability: The ability of a material to deform permanently under compression without rupture.

manhole: Opening found on the steam and water side of a boiler that is used for cleaning and inspection of the boiler.

manometer: An instrument that measures draft by comparing pressure at two locations.

manual reset valve: Used to secure the gas in the event of a low water condition or a pilot flame failure on a low pressure gas system.

master control: Unit that receives the primary signal and relays signals to individual control units.

material safety data sheet (MSDS): Printed material used to relay chemical hazard information from the manufacturer, importer, or distributor to the employer.

MAWP (maximum allowable working pressure): The highest pressure at which a boiler or pressure vessel may be safely operated.

maximum capacity: The maximum rating in pounds of steam that a boiler is designed to produce in 1 hr at a given pressure and temperature.

mechanical draft: Draft created by fans and blowers.

mechanical seal: An assembly installed around a pump shaft that prevents leakage of the pumped liquid along the shaft.

membrane watertube boiler: A boiler design in which rows of tubes are formed into solid panels through the use of welded steel strips that fill the spaces between the tubes.

membrane waterwalls: Waterwalls that are formed into a solid, airtight panel by welding a strip of steel between each of the waterwall tubes.

mercury switch: A switch that uses the movement of mercury in a glass tube to start or stop electrical current flow in a circuit.

metering control: An approach to control where the flows of fuel and air are precisely measured by flow-measuring devices and then adjusted by the control system so as to always be in the correct proportions.

metering feedwater regulating system: A control that continually measures the boiler conditions and adjusts the feedwater control valve.

metering pump: A small-capacity pump used to pump a closely measured amount of a liquid.

miniature boiler: A small boiler that meets several criteria for dimensions and capacity.

modulating feedwater regulating system: Continually adjusts the position of a feedwater regulating valve as needed in an effort to maintain a constant boiler water level by matching the position of the valve to the change in the boiler water level.

modulating motor: A small electric motor and reduction gear assembly enclosed in a metal box.

modulating pressure control: A control device that regulates the burner for a higher or lower fuel-burning rate, depending on the steam pressure in the boiler.

motive fluid pump: A pump that uses the force of a secondary fluid to pump the primary fluid.

multiple-pass boiler: Boilers that are equipped with baffles to direct the flow of the gases of combustion so that the gases make more than one pass over the heating surfaces.

natural draft: Draft that occurs without mechanical aid.

needle valve: A valve that is very similar to a globe valve, except that the opening/closing mechanism on the end of the valve stem is usually a sharp tapered cone that seats in a matching cone-shaped seat.

negative-suction pump installation: Any installation where the pump must draw (lift) liquid up from a source below the pump.

neutralizing amine: A chemical that neutralizes the pH of the condensate.

noncondensable gas: A gas that does not change into a liquid when its temperature is reduced to room temperature.

nondestructive testing (NDT): A method of determining the condition of components without causing damage and impairing their future usefulness.

nondrainable superheater: A superheater that does not have condensate drain connections.

non-permit confined space: A confined space that does not contain or, with respect to atmospheric hazards, have the potential to contain, any hazards capable of causing death or serious physical harm.

nonreturn valve: A combination shutoff valve and check valve that allows steam to pass out of the boiler.

nonrising stem valve: A valve that has a disc in the valve that threads up onto the stem as the stem is turned, and the stem does not back out of the valve.

normal operating water level (NOWL): Water level carried in the boiler gauge glass during normal operation (approximately one-third to one-half glass).

nozzle: One of the short stubs of piping that are connected to the boiler during construction.

OFF/low/high control: An approach to control where a burner is either OFF or operating with a low flame or with a high flame.

ON/OFF control: An approach to control in which a burner is either ON or OFF.

ON/OFF feedwater regulating system: A level control system that uses a water level-detecting device to turn the feedwater pump ON when the boiler water level drops to a preset point, and turn the pump OFF when the water level has risen to an upper setting.

ON/OFF with modulation control: An approach to control in which the amount of flame is changed to a degree that is proportional to the need.

opacimeter: An automatic indicator that measures the amount of light blocked by the smoke and ash going up the stack.

open heat exchanger: A heating unit in which steam or another heating medium and the fluid being heated come into direct contact.

open impeller: An impeller that has vanes that are not enclosed or supported by a shroud (side wall) on either side.

operating range: Range that must be set when using an ON/OFF combustion control in order to prevent extremes in firing rate.

orifice plate: Plate with a fixed opening that is installed in a pipeline to give a certain pressure drop across the opening where liquid or steam is flowing.

Orsat analyzer: A flue gas analyzer that measures the percentages of carbon dioxide, carbon monoxide, and oxygen in flue gases.

OSHA: The Occupational Safety and Health Administration.

"O" style watertube boiler: A watertube boiler design with a top steam and water drum and a bottom mud drum that are interconnected by banks of symmetrical tubes in an "O" shape.

output signal: The flow of control information leaving a control device and traveling to another device.

outside stem and yoke valve (os&y): Shows by the position of the stem whether it is open or closed. Used as boiler main steam stop valves.

overfire air: Air introduced over the fire to aid in complete combustion. Used mostly when burning soft coal that has a high volatile content with its own feedwater pumps, fuel system, and draft fans.

overfiring: Forcing a boiler beyond its designed steam-producing capacity.

overhead suspension method: A support method where the boiler is suspended from an overhead steel beam structure by sling members.

oxyacetylene welding (OAW): A welding process that uses acetylene (C_2H_2), which is combined with oxygen (O_2) to produce a flame with a temperature over 6000°F.

oxygen (O_2) trim system: An automatic control system that makes fine adjustments in the amount of combustion air used in order to minimize excess air.

oxygen scavenger: A chemical that reacts with any oxygen remaining in boiler feedwater and changes it into a form that does not cause corrosion.

package boiler: A boiler that is supplied from the manufacturer complete with controls, burner(s), and appliances attached.

packing gland: Holds packing or seals in place on valves and pumps to minimize leakage.

particulate: Fine ash particles from a burner that ultimately settle back to earth.

part per billion (ppb): The concentration of a solution equal to one part of a chemical in one billion parts of the solution ($^1/_{1,000,000,000}$).

part per million (ppm): The concentration of a solution equal to one part of a chemical in one million parts of the solution ($^1/_{1,000,000}$).

passes: The number of times the gases of combustion flow the length of the boiler as they transfer heat to the water.

perfect combustion: A fire where the fuel is burned with precisely the right quantity of oxygen so that no fuel or oxygen remains and the maximum possible heat results.

permanent hardness: A type of hardness that can be reduced only by the use of chemicals or distillation.

permissive: A process condition that must be met before a certain action may be taken.

permit-required confined space: A confined space that has specific health and safety hazards associated with it.

personal protective equipment: Any device worn by a boiler operator to prevent injury.

phosphate: A chemical that causes hardness in boiler water to precipitate and settle out as a heavy sludge.

phosphonate: An organic phosphate that provides multiple functions in water treatment.

pi (π): The ratio of the circumference of a circle to its diameter. The circumference of a circle is always equal to the diameter multiplied by pi.

pilot: Used to ignite fuel at the proper time in a firing cycle.

pilot trial for ignition or **PTFI:** A period of about 5 sec to 10 sec for the flame scanner to sense the presence of flame from the pilot.

piston valve: A valve that contains a finely machined piston that moves up or down in the interior of a cylindrical steel cage.

plant master: The master controller that calculates and distributes the steam production requirements to two or more boilers when they are used to maintain the pressure in a common steam header, such as when the boilers are installed in battery.

plug valve: A valve that is similar to a ball valve, except that a plug valve contains a semi-conical plug through which the flow passes.

pneumercator: An air-actuated liquid level measuring device.

polymer dispersant: A synthetic compound that prevents scale deposits by dispersing the scale before it deposits on boiler surfaces.

popping pressure: Predetermined pressure at which a safety valve opens and remains open until the pressure drops.

pop-type safety valve: Valve with a predetermined popping pressure. Commonly found on steam boilers.

positioning control: An approach to control in which a master device, or controller, senses the pressure in the steam header and uses compressed air to modulate power units, or actuators, which in turn position control linkages.

positive-displacement pump: A pump that moves the same amount of liquid with every stroke or rotation.

positive draft: The condition wherein the pressure inside the boiler furnace becomes greater than the pressure outside the furnace.

positive-suction pump installation: Any installation where the pump receives liquid on the suction side from a source above the pump—that is, under head.

pour point: The lowest temperature at which a liquid will flow from one container to another.

power: The rate at which work is done.

predictive maintenance: A study of the history of the plant components and determination of the expected service life of critical components.

pressure reducing valve (PRV): A valve that is designed to reduce the pressure of a fluid flowing through a pipe to a desired lower pressure, and constantly maintain this desired pressure downstream of the valve.

pretreatment: The water treatment processes that occur before the water enters the boiler.

preventive maintenance: The practice of performing maintenance activities on a piece of equipment to prevent breakdowns from normal or predictable causes.

primary air: The initial volume of air that enters the furnace with the fuel for most of the combustion process.

primary control: A flame safeguard control that consists of the relays and electronics required to safely start, run, and stop the burner under orders from an external control device such as a pressure switch.

primary element or **sensor:** A device that measures the process variable and produces a usable output in the form of a mechanical movement, electrical output, or instrument air pressure output.

priming: A severe form of carryover in which large slugs of water leave the boiler with the steam.

priming pump: A vacuum pump that ejects air from the suction line of a larger negative-suction pump installation.

private organization: An organization that develops standards from an accumulation of knowledge and experience with materials, methods, and practices.

process: The collection of equipment and actions required to accomplish a desired objective.

process variable: The condition that is being controlled. It is the condition that may vary within the process.

programmable logic controller (PLC): A small computer that may be configured, or programmed, to control a wide variety of processes.

programmer: See *programming control.*

programming control: A flame safeguard control that consists of all the components needed to safely perform a desired sequence of operations for a larger commercial or industrial burner.

proof of closure (POC) switch: A sensor that detects the position of a valve to ensure that the valve closes properly.

proximate analysis: The percentages of moisture, volatiles, fixed carbon, and ash.

pulverized coal: Coal that has been pulverized to the consistency of talcum powder and that is highly explosive.

pulverizer: A mill that grinds coal to a very fine powder.

pump controller: Starts and stops a feedwater pump, depending on the water level in the boiler.

purge period: A short period of time (typically 30 sec to about 2 min) when air is blown through the furnace area to ensure that no volatile fuel vapors are present.

push-nipple cast iron sectional boiler: Contains hollow cast iron sections joined with tapered nipples and pulled tightly together with tie rods or bolts.

pyrite: A common mineral consisting of iron disulfide (FeS_2).

pyrites trap: A compartment or box in the pulverizer that catches nuggets of pyrite as they are separated from coal.

pyrometer: An instrument that measures temperatures above the temperature range of mercury thermometers.

quality of steam: The dryness of the steam.

quick-closing valve: A valve that requires a one- quarter turn to be fully open or closed.

quick-drain test: A test that empties the float chamber or electric probe chamber while the burner is firing in order to test the low water fuel cutoff switch.

quicklime: Limestone that has been thoroughly dried.

quick-opening valve: A valve that requires only a 90° change in the position of a lever arm to move the valve from fully closed to fully open.

radiant superheater: A nest of tubes that the saturated steam passes through to acquire heat.

ramming materials: Plastic refractory materials that are rammed into place using heavy bars and other tools.

rank: The hardness of coal.

rate of combustion: The amount of fuel that is being burned in the furnace per unit of time.

reagent: A chemical used in a water treatment test to show the presence of a specific substance, such as hardness.

rear header: Found on straight-tube watertube boilers. Connected to front header by water tubes.

reciprocating pump: A displacement pump that uses a reciprocating piston or diaphragm to repeatedly displace fluid from a cylinder or chamber.

recirculation line: A line that provides a minimum flow through a pump to prevent overheating.

recorder: An instrument that records data such as pressures and temperatures over a period of time.

rectifier: An electrical device that converts alternating current (AC) into direct current (DC).

refractory: Brickwork used in boiler furnaces and for boiler baffles.

regenerative air preheater: Consists of a rotating segmented wheel that is 8′ to 20′ in diameter and about 3′ to 6′ high.

reinsurance: Insurance purchased by insurance companies for large-loss contingencies.

relative humidity: The ratio of the actual amount of water vapor in the air to the greatest amount possible at the same temperature.

relief valve: A valve that opens in proportion to the excess pressure, rather than popping open fully.

remote control: A control device that is installed a considerable distance from the equipment on which it is being used.

representative sample: A sample that is exactly the same as the item being tested.

reset: Switch that must be reset manually after it is tripped.

residual fuel oil: Fuel oil that remains after the lighter, more volatile hydrocarbons have been distilled off.

resiliency: The ability of a material to return to its original shape after being deformed.

resistance: The measure of the ability of an electric circuit to oppose current flow.

retarder: See *turbulator.*

retort: Space below the grates of an underfeed stoker.

retort chamber: A V-shaped trough, usually about 5′ to 12′ long, with a back plate enclosing the rear end of the retort opposite the feed ram.

retort stoker: A stoker in which both the fuel and the combustion air are fed from below the combustion zone (the surface of the grates).

reverse osmosis (RO): A water-purification process in which the water to be treated is pressurized and applied against the surface of a semipermeable membrane.

Ringelmann chart: A comparison chart used to measure opacity.

riser tubes: Tubes exposed to the highest temperatures in the furnace area and contain rising water.

rising stem valve: A valve that has a handwheel and stem that move outward from the body of the valve as the valve is opened.

rivet pitch: The distance from the center of one rivet to the center of the next rivet in the same row.

rotameter: A variable-area instrument used for measuring rate of flow.

rotary cup burner: A burner that has a cone-shaped cup, usually made of brass or stainless steel, that mixes fuel oil with air.

rotary pump: A pump with a rotating shaft. A rotary pump may be either a dynamic pump or a displacement pump.

RTD: An acronym for resistance temperature device.

safety relief valve: A valve specially designed to serve as either a safety valve or relief valve, depending on the application for which it is used.

safety valve: A valve that opens fully and instantly and causes a definite, measured drop in pressure before closing.

sample cooler: A small, closed heat exchanger that cools condensate or other hot water to a temperature below about 130°F to 140°F before the water emerges from the cooler and into a sample container.

sample coupon: A small, flat strip of steel that is inserted into an elbow or tee fitting in a piping network.

saturated: Means that a substance has absorbed as much of another substance as it can absorb.

saturated steam: Steam that contains no liquid water and is at the temperature of the boiling water that formed the steam.

scale: Deposits caused by improper boiler water treatment.

scanner: Device that monitors the pilot and main flame of the furnace. The scanner is used to prove the pilot and main flame.

scotch marine boiler: A firetube boiler that has a flue furnace and horizontal shell.

scrubber: Device that removes undesirable gaseous elements from flue gas.

secondary air: Air mixed with the fuel to ensure that enough oxygen is available to complete the burning.

second law of thermodynamics: Heat always flows from a body having a higher temperature toward a body having a lower temperature.

sediment: Particles of foreign matter present in boiler water.

semiclosed impeller: An impeller that has a shroud on one side of the vanes.

sensible heat: The Btu content of a substance that represents the heat absorbed or given up as it changes temperature.

setpoint: 1. The desired point at which an automatic controller maintains a variable condition within a process. **2.** The desired point at which an automatic controller maintains a variable condition within a process.

shakeout: The process of operating a new system as needed to expose and correct the major impediments to reliable operation.

shaft sleeve: A replaceable sacrificial part covering a pump shaft.

shear: The exertion of equal forces in opposite directions in the same plane.

shear pin: A link used in a mechanical drive that is designed to break under a specific amount of shear stress.

shielded metal arc welding (SMAW): A welding process that uses an electric arc to heat the metal in the weld area; metal from the electrode is added to the weld pool.

signal: The language that the control devices use to communicate with each other.

simplex pump: A steam-driven, reciprocating, positive-displacement, double-acting pump with one steam cylinder and one liquid cylinder.

single-acting pump: A reciprocating pump that moves fluid in only one direction of the stroke.

single-loop control: The use of a controller to control a single process variable without any influence from any other controller or process variable.

siphon: Protective device used between the steam and Bourdon tube in a steam pressure gauge.

slag: The solid deposits that accumulate on furnace walls and boiler tubes.

slag screen: A loosely spaced bank, or several rows, of water tubes placed between the superheater and the combustion area of the furnace.

slaker: A conveyor in which lime is mixed with water to make a soluble paste.

slinger ring: A metallic ring that hangs on the rotating shaft and is considerably larger in diameter than the shaft.

slip: The difference between calculated and actual displacement of a pump.

slow-drain test: See *evaporation test.*

slow-opening valve: A valve that requires five or more full turns of a handwheel to move the valve from fully closed to fully open.

sludge: Accumulated residue produced from impurities in water.

sluice: A trench through which water flows rapidly to carry away solid materials.

smoke box: The area at the end of a firetube boiler where the flue gases are allowed to reverse direction for a subsequent pass.

smoke indicator: An indicating or recording device that shows the density of the smoke leaving the chimney.

sodium zeolite water softener: An ion-exchange water softener that uses resin beads and a brine solution to soften water.

solenoid: An electric actuator that consists of an iron plunger surrounded by an encased coil of wire.

solenoid valve: A valve that is snapped open or closed by an electric actuator.

soot: Carbon deposits caused by incomplete combustion.

soot blowers: Used to remove soot from around tubes to increase boiler efficiency. Mostly found on watertube boilers.

spalling: Hairline cracks in boiler brickwork (refractory) due to changes in furnace temperature.

specific gravity: The weight of a given volume of a material divided by the weight of an equal volume of water when both are measured at the same temperature.

specific volume: The space occupied by a fluid or gas of a particular unit of weight under specified conditions of pressure and temperature.

spectrophotometer: An instrument that measures the ability of different frequencies of light to pass through a sample of liquid.

spontaneous combustion: The process where a material can self-generate heat until the ignition point is reached.

staged combustion: The process of mixing fuel and air in a way that reduces the flame temperature and NO_x generation in a burner.

staging: The placement of more than one impeller on the same shaft in a centrifugal pump.

standard: An accepted reference, practice, or method.

standing pilot: A gas pilot that is always lit.

static discharge head: The vertical distance from the centerline of the pump up to the surface of the liquid in the tank or vessel into which the piping discharges.

static suction head: The vertical distance from the centerline of the pump up to the level of the liquid in the supply tank.

static suction lift: The vertical distance from the centerline of the pump down to the level of liquid in the supply source below.

staybolt: A short bolt brace that passes through the water leg of a boiler.

steam: The gaseous form of water.

steam and water drum: The pressure vessel in a steam boiler that contains both steam and water.

steam blanketing: A condition that occurs when steam bubbles are generated so quickly from a boiler heating surface that a layer of steam is formed between the water and the heating surface.

steam blowing: The process of cleaning impurities from new piping by blowing steam through the pipe.

steam boiler: A closed vessel in which water is transformed into steam under pressure through the application of heat.

steambound: Condition that occurs when the temperature in the open feedwater heater gets too high and the feedwater pump cannot deliver water to the boiler.

steam header: A manifold that receives steam from two or more boilers and provides a single location from which steam may be routed through the steam mains and branch lines to various points of use.

steam main: The piping that carries steam to a section of a building or plant.

steam rate: The combination of combustion efficiency and thermal efficiency at the full range of loads and conditions that the boiler encounters over a typical period of time.

steam separator: Device used to increase the quality of steam. Found in the steam and water drum.

steam space: The space above the water line in the steam and water drum.

steam strainer: Used before steam traps and turbine throttle valves to remove solid impurities.

steam system: Consists of the equipment, controls, and piping that carry the steam generated by the boiler to its points of use.

steam system efficiency: A measurement of steam usage that takes into account both the equipment supplying the steam and the equipment demanding the steam.

steam tracing: A small copper or steel tube which is supplied with steam and is usually run alongside a process pipe to keep the fluid within the pipe warm.

steam trap: A mechanical device used for removing condensate and/or air from steam piping and heat exchange equipment.

steam trap survey: The process of identifying, testing, and documenting the condition of all steam traps in a facility.

steam turbine: Used to drive boiler auxiliaries or generators in large plants.

steam working pressure (SWP): The maximum steam working pressure of the valve.

Stellite®: An alloy of chromium, cobalt, and tungsten.

Stirling boiler: A watertube boiler design with three steam and water drums on the top and a mud drum beneath, interconnected by a large number of water tubes.

stoker: A device that automatically feeds green, unburned coal or other solid fuel to a furnace.

stoichiometric combustion: The process of burning a fuel with precisely the amount of air required so that no unburned fuel or unused oxygen remains.

stopcock: A quick-opening or closing valve usually found on gas lines.

straight-tube watertube boiler: A boiler design in which the steam-generating tubes are straight rather than bent or curved.

submerged arc welding (SAW): A welding process in which an electric arc is submerged or hidden beneath a granular material (flux).

submerged conveyor: A heavy steel pan conveyor or apron conveyor immersed in a water trough.

submerged-tube (wet-top) vertical boiler: A firetube boiler with fire tubes completely covered with water all the way to the upper tube sheet.

suction pressure: Pressure on the liquid at the suction side of a pump.

superheated steam: Steam that has been heated above the saturation temperature (the temperature that corresponds to the steam pressure).

superheater: A bank of tubes through which only steam passes, not water.

superheater drain: Valve found on the superheater header outlet. Used to maintain flow throughout the superheater during startup and shutdown.

superheater header: Main inlet and outlet line to and from the superheater tubes in the superheater.

surface blowdown: The process of intermittently removing water from a boiler to control the quantity of impurities in the remaining water or to remove a film of impurities on the water.

suspended solids: Solid impurities that are suspended in water.

suspension sling: Used to support the drum of an HRT boiler.

swell: The rise in the boiler water level that occurs when the steam load on the boiler is increased or when the steam pressure drops.

synchronize: To balance out combustion controls before switching to automatic.

tagout: The use of a danger tag at the source of the hazardous energy to indicate to other personnel that the device is not to be operated until personnel working on the equipment have removed their lockout devices and the equipment is safe to operate.

tandem valve: A blowdown valve configuration with two valves in series machined into a common valve body.

technical society: An organization made up of personnel having expertise in a particular subject and a common professional interest.

temporary hardness: A type of hardness that can be reduced by heating the water.

tensile strength: The amount of force required to pull an object apart.

tension: The exertion of equal forces pulling in opposite directions that can stretch an object.

tertiary air: Combustion air added to a burner in addition to the secondary air.

therm: A unit of measure indicating 100,000 Btu.

thermal efficiency: The ratio of the heat absorbed by the boiler to the heat available in the fuel per unit of time.

thermal liquid boiler: A fire tube or watertube boiler that uses a liquid chemical solution instead of water.

thermal shock: The stress imposed on boiler metal by a sudden and drastic change in temperature.

thermocouple: A device used to measure temperature consisting of two dissimilar metals joined together.

thermohydraulic feedwater regulator: A modulating control that controls feedwater flow in direct response to changes in the boiler water level.

thermometer: Instrument used to measure temperature (degree of heat). Calibrated in degrees Celsius or degrees Fahrenheit.

thermostatic expansion tube feedwater regulator: A modulating control that controls feedwater flow in direct response to changes in the boiler water level.

thermostatic steam trap: A steam trap that contains a temperature-operated device, such as a corrugated bellows, that controls a small discharge valve.

thermowell: A receptacle into which a temperature-sensing instrument is inserted.

three-element feedwater regulating system: A water level control system that measures the steam flow from the boiler and the feedwater flow into the boiler in addition to the water level.

throttling: Controlling the amount of flow that passes through a valve by partially closing it.

through stays: Found on firetube boilers (HRT and scotch marine) to keep front and rear tube sheets from bulging.

titration test: A test that determines the concentration of a specific substance dissolved in water.

total dissolved solids (TDS): A measurement of the concentration of impurities in boiler water.

total dynamic head: The total amount of head produced by the pump and available to perform useful work, after losses have been subtracted.

total force: The pressure (in psi) being exerted on a surface multiplied by the area of the surface.

total heat: The sum of sensible heat and latent heat.

totalizing flowmeter: A flow measuring device that not only measures on an instantaneous basis (flow rate), but also measures total flow over time.

trade association: An organization that represents the producers of specific products.

tramp air: See *inleakage.*

transducer: A device that converts one type of control signal into another type of control signal.

transmitter: **1.** An instrument used to send information about the condition of a process to a control device. **2.** A device that conditions a low-energy signal from the primary element and produces a suitable signal for transmission to other components and devices.

try cock: A valve used as a secondary water level indicator.

tubercle: A bump on a steel boiler surface made up of corrosion products.

tubular air preheater: Consists of tubes enclosed in a shell where flue gases heat up incoming combustion air.

turbine pump: A rotary positive-displacement pump that uses a flat impeller with small flat perpendicular fins machined into the impeller rim.

turbine stages: That part of the turbine where steam gives up its energy to the turbine blades. As the steam pressure drops, the stages (blades) become larger.

turbine tube cleaner: A motorized mechanical cutter or knocker that removes scale from boiler tubes.

turbulator: Device that swirls the hot gases of combustion as the gases pass through the center of the tube so that the gases come into more efficient contact with tube walls where heat transfer occurs.

turbulence: Movement of water in the steam and water drum.

turndown ratio: The ratio of the maximum firing rate of the burner to the minimum firing rate.

tuyere (tweer): A special air-admitting grate designed to start combustion of the entering fuel.

two-element feedwater regulating system: A water level control system that measures the steam flow from the boiler in addition to the water level.

ultimate analysis: The percentages of nitrogen, oxygen, carbon, ash, sulfur, and hydrogen (NOCASH) in the coal.

unfired vessel: A pressure vessel without combustion equipment, such as compressed air tanks, feedwater heating tanks, steam piping, steam-jacketed heat exchange equipment, and similar vessels.

utility watertube boiler: An extremely large watertube boiler that generates steam at a very high pressure and temperature.

U-tube manometer: When filled with mercury, used to measure vacuum. U-tube manometers are calibrated in inches.

vacuum: A pressure lower than atmospheric pressure.

vacuum breaker: A check valve that prevents the formation of a vacuum in a tank, pressure vessel, or piping system.

vacuum gauge: Pressure gauge used to measure pressure below atmospheric pressure that is calibrated in inches of mercury.

vacuum pump: A pump that withdraws gases or vapors from a closed container and creates a vacuum in the container.

vane pump: A rotary positive-displacement pump that uses a rotating drum located eccentrically inside a cylindrical pump casing.

vapor pressure: The equilibrium pressure where the number of molecules evaporating from a liquid surface equals the number of molecules condensing back to the liquid.

vaporstat: Control with a large diaphragm that makes it highly sensitive to low pressure.

variable-area flow meter: Measures the flow of a substance by how much resistance is created by a float or piston that changes the area (size) of the flow path.

vent condenser: An in-line heat exchanger installed in the vent from a deaerator to the atmosphere.

venturi: A nozzle with a slight hourglass-shaped taper.

vertical firetube boiler: One-pass boiler that has fire tubes in a vertical position. Vertical firetube boilers are classified as wet-top or dry-top.

vibrating-grate stoker: Inclined grates that vibrate, causing the fuel bed to move slowly toward the lower end.

viscosity: A measurement of a liquid's resistance to flow.

viscous liquid: A liquid that is thick and resists flow.

volatile matter: Gas given off when coal burns.

volute: A spiral-shaped form.

warping: Bending or distortion of boiler or superheater tubes, usually caused by overheating.

waste heat recovery boiler: A fire tube or watertube boiler in which heat that would otherwise be discarded is used to make steam.

water column: A metal vessel installed on the outside of the boiler shell or drum at the normal operating water level (NOWL) for the purpose of determining the location of the water level.

water column blowdown valve: Valve on the bottom of the water column used to remove sludge and sediment that might collect at the bottom of the water column.

water hammer: The hydraulic shock in piping caused by the presence of liquids in the steam flow.

water, oil, or gas (WOG): The maximum pressure under which the valve may be used with these fluids.

water softening: The removal of scale-forming salts from water.

watertube steam boiler: Boiler that has water in the tubes with heat and gases of combustion around the tubes.

waterwall: Many tubes placed side by side to create a large, flat surface against the furnace walls in a watertube boiler.

waterwall blowdown valve: Approved valve used to remove sludge and sediment from waterwalls and waterwall headers.

weight-type alarm whistle: Alarm whistle that signals high or low water by the gain or loss of buoyancy of weights in water within the water column.

wetback scotch marine boiler: A firetube boiler with a water-cooled reversing chamber used to direct the combustion gases from the flue furnace to the first pass of tubes.

windbox: The plenum to which the forced draft fan (primary air fan) supplies air in order to maintain enough pressure to provide proper air flow through the furnace.

window weld: A weld made through an opening, or window, in the tube.

wire drawing: The erosion that occurs as steam or another high-velocity fluid flows through a small opening like a throttled valve.

working pressure: A shortened term for maximum allowable working pressure, but it may also be used to mean the pressure at which the boiler is normally operated.

zeolite: A synthetic sodium aluminosilicate cation-exchange material based on naturally occurring clays called zeolites.

Index

C

G

H

USING THE *BOILER OPERATOR'S WORKBOOK* CD-ROM

Before removing the CD-ROM from the protective sleeve, please note that the book cannot be returned for refund or credit if the CD-ROM sleeve seal is broken.

System Requirements

The *Boiler Operator's Workbook* CD-ROM is designed to work best on a computer meeting the following hardware/software requirements:

- Intel® Pentium® processor
- Microsoft® Windows® 95, 98, 98 SE, Me, NT®, 2000, or XP® operating system
- 64 MB of free available system RAM (128 MB recommended)
- 90 MB of available disk space
- 800 × 600 16-bit (thousands of colors) color display or better
- Sound output capability and speakers
- CD-ROM drive
- Internet Explorer™ 3.0 or Netscape® 3.0 or later browser software

Opening Files

Insert the CD-ROM into the computer CD-ROM drive. Within a few seconds, the home screen will be displayed allowing access to all features of the CD-ROM. Information about the usage of the CD-ROM can be accessed by clicking on USING THIS CD-ROM. The Chapter Quick Quizzes™, Illustrated Glossary, Sample Licensing Exam, Master Math™, Media Clips, and Reference Material can be accessed by clicking on the appropriate button on the home screen. Clicking on the American Tech web site button (www.go2atp.com) accesses information on related educational products. Unauthorized reproduction of the material on this CD-ROM is strictly prohibited.

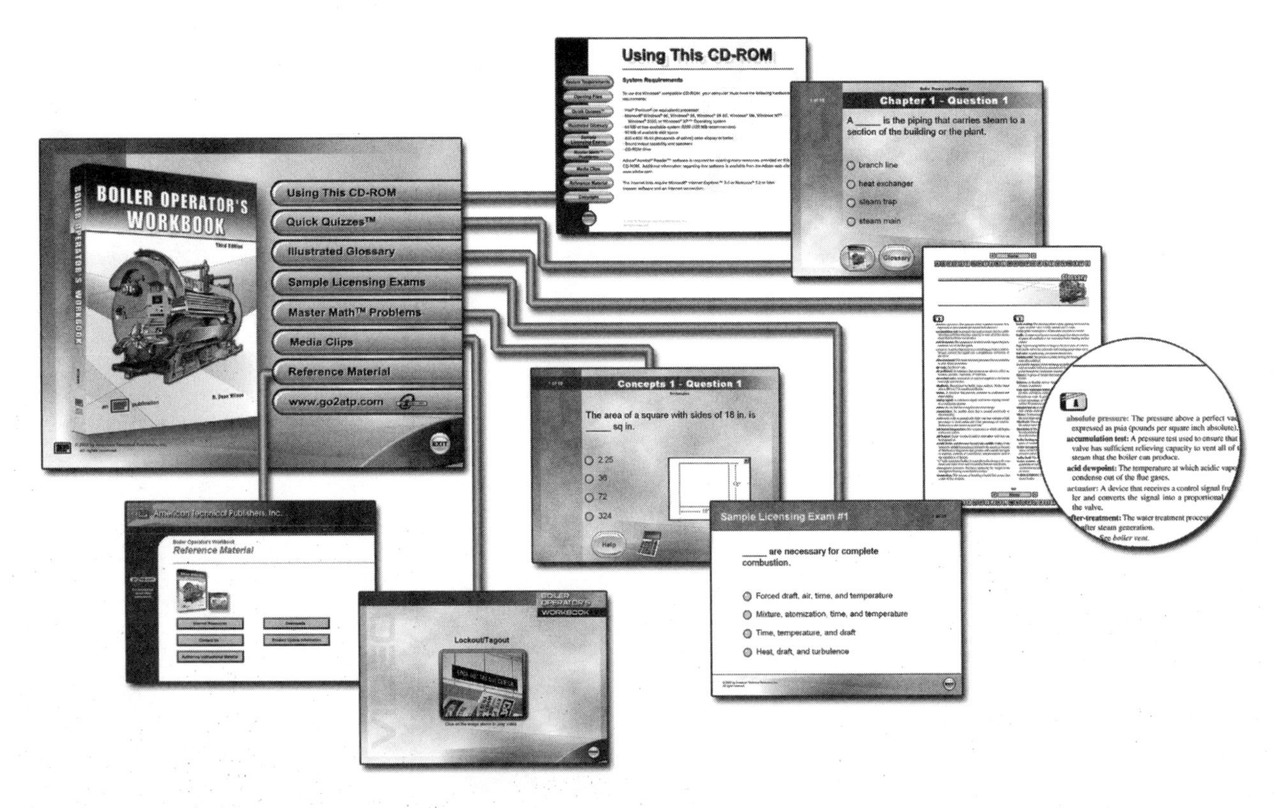